PRINTED CIRCUIT BOARDS
Design and Technology

Editor
Arvind V Shah

Contributors
S R Bhat
Walter C Bosshart
V K Hingne
H S Jamadagni
V Mitter
Arvind V Shah
J Vaid

McGraw-Hill Offices

New Delhi New York St Louis San Francisco Auckland Bogotá
Guatemala Hamburg Lisbon London Madrid Mexico Montreal Panama
Paris San Juan São Paulo Singapore Sydney Tokyo Toronto

PRINTED CIRCUIT BOARDS
Design and Technology

WALTER C BOSSHART
El. Ing. HTL, GRETAG Ltd.
8105 Regensdorf
Switzerland

Formerly Visiting Expert
Centre for Electronics Design and Technology (CEDT)
Indian Institute of Science
Bangalore

Tata McGraw-Hill Publishing Company Limited
NEW DELHI

Centre for Electronics Design and Technology
Indian Institute of Science
Bangalore

© 1983, TATA McGRAW-HILL PUBLISHING COMPANY LIMITED

Sixteenth reprint 1995
RCDCQDDSRZXQL

No part of this publication can be reproduced in any form or by any means
without the prior written permission of the publishers

This edition can be exported from India only by the publishers,
Tata McGraw-Hill Publishing Company Limited

ISBN 0-07-451549-7

Published by Tata McGraw-Hill Publishing Company Limited,
4/12 Asaf Ali Road, New Delhi 110 002, and printed by
Mohan Makhijani at Rekha Printers Pvt. Ltd., New Delhi 110 020

FOREWORD

The Centre for Electronics Design Technology (CEDT) was set up at the Indian Institute of Science in cooperation with the Swiss Federal Institute of Technology, to meet the need of engineers trained in the design and technology of electronic equipment. In about six years, the Centre has established laboratory and other facilities for teaching and training. The Centre also conducts long- and short-term courses on different aspects of electronic equipment design and technology. These courses have attracted the attention of engineers working in industries, R & D laboratory as well as educational institutions. They have been also well received by other students of the Institute.

The instructional material required for these courses is not readily available. Therefore, Indian and Swiss engineers on the staff of CEDT have brought out this material, mainly in cyclostyled/offset printed form. Over 25 publications covering these courses on different topics have been brought out so far and they are popular both inside and outside the Institute.

Recognizing the usefulness of these publications in engineering education, the Institute has seriously considered how best to make them available to the engineering public at a comparatively low price. As a first step, revised and updated versions of three of the most widely used books are being published in this series under the supervision of an Editorial Board. The grant for this purpose was made available by the Government of Switzerland, to whom our thanks are due.

I am certain that this series of books will be of use to students in the Institute and to all working engineers.

S RAMASESHAN
Director
Indian Institute of Science
Bangalore

EDITOR'S NOTE

Printed Circuit Boards (PCBs) are certainly the most important element in the fabrication of electronic equipment. It is the *design* of properly laid-out PCBs that determine many of the limiting properties with respect to noise immunity, as well as to fast-pulse, high frequency and low-level characteristics of the equipment. High-power PCBs in their turn require a special design strategy. The present volume gives specific guidelines for all these cases.

The *fabrication process* of the printed circuit board will determine to a large extent the price and reliability of the equipment. A common target aimed at is the fabrication of small series of highly-reliable, professional-quality PCBs with low investment cost. This target becomes especially important for custom-tailored equipment in the area of industrial electronics. It is also important for R&D laboratories in telecommunication and biomedical electronics. The present volume tries to give practical tips on how to set-up and operate such a PCB fabrication process.

The Centre for Electronics Design Technology (CEDT) at the Indian Institute of Science, Bangalore, recognized from the beginning the cardinal role of PCBs in the production of electronic equipment. CEDT engineers were trained to design competently PCB layouts and understand in depth the fabrication of PCBs. This can be considered to be a model in the education of engineering students.

Mr. Walter C. Bosshart, a Swiss engineer working at CEDT, within the framework of a Indo-Swiss Cooperation Agreement, along with many of his Indian colleagues at the CEDT and in associated industries played a key role in setting these standards. It is the crowning achievement of his job to bring out the present volume. I am personally quite sure that this volume will be of great use to practising engineers and advanced students in various countries.

ARVIND V SHAH

PREFACE

Electronics has become a growing influence in our daily lives. Considering all possibilities of applying electronics in a useful and responsible manner, we recognise that the so-called 'electronics age' is still in its infancy. Among the various discoveries and developments to bring electronics to the level it has reached until now, printed circuit boards (PCBs) have definitely contributed in a significant manner as a means to interconnect electronic components. Also, in the years to come, PCBs will undergo further development in order to satisfy the growing demand to interconnect electronic components of the latest technology.

The field of PCBs combines a wide range of disciplines, such as mechanics, chemistry, electronics design, production and process management, etc. Such widespread constraints may be the reason why only little literature on the subject has become available so far. This book is therefore an attempt to bridge a few of these gaps at the engineer's and engineering student's level. Special emphasis has been given to an industrial and technological environment as typically existing and further developing in a country like India: The basic design procedures and production processes are not only dealt with in sufficient depth and details but latest developments are also included.

The book is divided into two major parts: Part A contains information for the PCB designer starting from special hints for the electronic circuit designer to automated artwork generation. Part B deals with the aspects of the actual PCB production processes right from the film master manufacturing to the assembly and soldering of finished boards. At the end of the book is given a list of addresses which will be useful to anyone directly connected with practical PCB processes. This address list is neither complete nor does it represent an evaluation on the suppliers: it is just what was found available at that time in CEDT.

The views expressed in this book generally represent the personal views of the author of the particular chapter. Although every care has been taken while preparing the book, it is not possible to assume any liability for its contents. This equally applies for data and process details given. Any suggestions and comments to improve this book are much welcome by the different authors.

This book would not have come into existence without the active support given to me from various sides. My sincere thanks must particularly go to the following persons:

—Prof. Arvind V. Shah, University of Neuchatel, Switzerland: He stood behind as a promoter editor and sincere adviser right from the initial idea up to the completion of the manuscript; the valuable ideas and suggestions he gave after going through the draft have helped me significantly to bring the book to this form and content.

—Co-authors: Their fruitful cooperation was a continuous source of encouragement to me.

—Mr. M. Badrinath, I.I.Sc. Campus: He did all the draft and manuscript typing in a most efficient manner.

—Mr. A. Vijayendra, ECE Department, I.I.Sc.: For drawing and redrawing of the figures in a helpful and patient way.

—Colleagues of the Process Technology Group at CEDT: Thanks for all understanding shown for my preoccupation. This book reflects many results of their own efforts.

—Swiss Technical Cooperation, Berne, Switzerland: This book was written on their request and their sympathetic support formed the essential base not only for this book but also for my assignment at CEDT.

Last but not least, I am thankful to my wife Ursula and children Maja, Philipp and Susanne for all the goodwill and understanding they showed for my new 'hobby' of bookwriting. Their continuous moral support was most important for me to carry on this task.

WALTER C BOSSHART

LIST OF CONTRIBUTORS

S.R. Bhat, Senior Scientific Officer, CEDT, Indian Institute of Science, Bangalore
   Chapter 5 Design rules for analog circuit PCBs

Walter C. Bosshart, Sales Manager, GRETAG Limited, Regensdorf, Switzerland
   Chapter 1 Layout planning
           2 Layout, general rules and parameters
           6 Design rules for PCBs in power electronics applications
           8 Artwork
           9 Automation and computers in PCB design
          10 Film master production
          12 Board cleaning before pattern transfer
          13 Photoprinting
          14 Screen-printing
          16 Etching
          17 Mechanical machining operations
          18 PCB technology trends
          21 Component assembly techniques
          22 Guidelines for starting PCB facilities
          23 Appendix

V.K. Hingne, Technical Service Manager, Formica India Division, The Bombay Burmah Trading Corp. Ltd., Bombay
   Chapter 11 Properties of copper-clad laminates

H.S. Jamadagni, Senior Scientific Officer, CEDT, Indian Institute of Science, Bangalore
   Chapter 7 Design rules for PCBs in microwave applications

V. Mitter, Fellow I.C.T.; Director, Hegde & Golay Ltd., Kanakapura Road, Bangalore
   Chapter 15 Plating
           19 Multilayer boards

Arvind V. Shah, Professor, Institute of Microtechnics, University of Neuchatel, 2000 Neuchatel, Switzerland. Formerly Scientific Advicer, Centre for Electronic Design and Technology (CEDT), Indian Institute of Science, Bangalore

Chapter 3 Design rules for digital circuit PCBs
       4 Design rules for PCBs in high-frequency and fast-pulse applications

J. Vaid, Chief Chemist, Peico Electronics & Electricals Ltd., Loni-Kalbhor
Chapter 20 Soldering techniques

CEDT

CENTRE FOR ELECTRONICS DESIGN AND TECHNOLOGY

CEDT was established in the Indian Institute of Science, Bangalore, as an attempt to bridge the gap between academic institutes and industries and also to fulfil the needs of small-scale industries. This is achieved through giving highly practically-oriented postgraduate training programmes, offering short-term courses on topics of current interest for participants from industries and undertaking development work for industries under contract.

The activities of CEDT are mainly concentrated in the areas of microprocessor systems, power electronics, data communication systems, instrumentation and process control, electromechanics and product design. The Centre is supported by the Electronics Commission, University Grants Commission (UGC) and the Swiss Development Co-operation.

CONTENTS

Foreword — v
Editor's Note — vii
Preface — ix
List of Contributors — xi

Part A: Design of Printed Circuit Boards

1 LAYOUT PLANNING 4

1.1 Introduction — 4
1.2 General Considerations — 4
   1.2.1 Layout Scale *4*; 1.2.2 Grid System *5*; 1.2.3 Board Types *5*;
   1.2.4 PCB Production Facilities *7*; 1.2.5 Standards *7*
1.3 PCB Sizes — 7
   1.3.1 Mechanical Stress *7*; 1.3.2 Other Board Size Constraints *7*;
   1.3.3 Standard Sizes *9*
1.4 Layout Approaches — 12
   1.4.1 Materials and Aids *12*; 1.4.2 Procedures *13*
1.5 Documentation — 14
   1.5.1 Scope *14*; 1.5.2 Circuit Diagram *14*; 1.5.3 Component List *14*;
   1.5.4 Layout Sketch *15*; 1.5.5 Mechanical Drawing *15*; 1.5.6 Assembly
   Drawing *16*
Bibliography — 16

2 LAYOUT, GENERAL RULES AND PARAMETERS 17

2.1 Introduction — 17
2.2 Resistance — 17
   2.2.1 Resistance in General *17*; 2.2.2 Resistance and Temperature *18*;
   2.2.3 Practice *18*
2.3 Capacitance — 20
   2.3.1 Capacitance between Conductors on Opposite Sides of the PCB *20*;

2.3.2 Capacitance between Adjacent Conductors *22*
2.4 Inductance of PCB Conductors — 23
2.5 Conductor Spacing — 24
2.6 Realising Supply and Ground Conductors — 25
2.6.1 Width *25*; 2.6.2 Configuration *27*
2.7 Component Placing and Mounting — 29
2.7.1 Component Placing Approaches *29*; 2.7.2 Mounting Considerations *29*
2.8 Cooling Requirements and Package Density — 30
2.8.1 Heatsinks *30*; 2.8.2 Package Density *30*
2.9 Layout Check — 31
2.9.1 General Considerations *31*; 2.9.2 Mechanical Considerations *31*
2.9.3 Electrical Considerations *32*
Bibliography — 32

3 DESIGN RULES FOR DIGITAL CIRCUIT PCBs 33

3.1 Reflections — 33
3.1.1 TTL Integrated Circuits *33*; 3.1.2 C-MOS Integrated Circuits *37*; 3.1.3 ECL Integrated Circuits *40*; 3.1.4 Matching with Additional Circuit Elements or with Line-Driver/Receiver ICs *40*
3.2 Crosstalk — 44
3.3 Ground- and Supply-Line Noise — 48
3.4 Electromagnetic (E.M.) Interference from Pulse-Type E.M. Fields (Mains Noise) — 50
3.5 Recommendations and Summary — 55
Bibliography — 57

4 DESIGN RULES FOR PCBs IN HIGH-FREQUENCY AND FAST-PULSE APPLICATIONS 59

4.1 Introduction — 59
4.2 Matching of Conductors: Effect of Mismatch in different Cases — 59
4.2.1 Transmission Lines and Wave Impedances *59*; 4.2.2 Effect of Mismatch in the Fast-Pulse Case *61*; 4.2.3 Effect of Matching and Mismatch in the High-Frequency Case *61*
4.3 Pulse Circuits — 67
4.3.1 Inductive and Capacitive Approximations for Lines with Multiple Reflections *67*; 4.3.2 Effect of Mismatch on Rise-Time *72*; 4.3.3 Coupling Capacitance between Two Conductors *72*
4.4 Line Defects: The Inductive and the Capacitive Approximation — 73
4.5 Printed Capacitors and Printed Inductors — 75
4.5.1 Realisation *75*; 4.5.2 General Remarks *77*
4.6 Ground and Supply Lines — 79
4.7 Concept of Guard Line: Example of a Cascode Stage — 79
4.8 Rise-Time Limitations and High-Frequency Losses due to Skin-Effect, Dielectric Losses and Radiation Losses — 81
4.8.1 Skin-Effect Losses *83*; 4.8.2 Radiation Losses *86*; 4.8.3 Dielectric Losses *89*; 4.8.4 Concluding Remarks on Line Losses *91*

4.9	**Summary and Recommendations**	91
	4.9.1 Summary of Effects *91*; 4.9.2 Recommendations for Design *92*	
Acknowledgement		92
Bibliography		93

5 DESIGN RULES FOR ANALOG CIRCUIT PCBs 95

5.1	**Component Conductors**	95
5.2	**Signal Conductors**	96
	5.2.1 In General *96*; 5.2.2 High Frequency Amplifiers/Oscillators *96*; 5.2.3 Multistage Amplifiers especially with High-Power Output Stage *98*; 5.2.4 Feedback Amplifiers/Regulators with Remote Sensing Lines *99*; 5.2.5 High-Gain DC Amplifiers (Thermal Effects) *99*; 5.2.6 Amplifiers handling Low-Level Signals *100*; 5.2.7 Precision Differential Amplifiers *102*	
5.3	**Supply and Ground Conductors**	104
	5.3.1 Ground Lines *104*; 5.3.2 Supply Lines *104*	
Bibliography		104

6 DESIGN RULES FOR PCBs IN POWER ELECTRONICS APPLICATIONS 105

6.1	**Introduction**	105
6.2	**Dividing Circuit into High- and Low-Power Part**	105
6.3	**Copper-Clad Laminates**	106
6.4	**PCB Terminal Connections and their Assembly**	107
6.5	**Conductors**	109

7 DESIGN RULES FOR PCBs IN MICROWAVE APPLICATIONS 114

7.1	**Introduction**	114
7.2	**Some Definitions**	116
	7.2.1 Characteristic Impedance of a Transmission Line *116*; 7.2.2 Propagation Constant *117*; 7.2.3 Reflection Coefficient *117*; 7.2.4 Voltage Standing Wave Ratio (VSWR) *117*; 7.2.5 Modes of Propagation *117*	
7.3	**Strip Line and Microstrip Line**	119
	7.3.1 Strip Line *119*; 7.3.2 Microstrip Line *122*; 7.3.3 Attenuation in Strip- and Microstrip-Lines *123*	
7.4	**Applications of Strip- and Microstrip-Lines**	123
	7.4.1 Filters *125*; 7.4.2 Coupled Transmission Lines *125*	
7.5	**Materials for Microwave PCBs**	127
7.6	**Fabrication of PCBs for Microwave Use**	128
7.7	**Use of Strip and Microstrip Lines in Pulse Circuits**	128
Bibliography		128

8 ARTWORK 129

8.1	**Introduction**	129
	8.1.1 Personnel *129*; 8.1.2 Rooms *129*; 8.1.3 Artwork Scale *130*	

8.2 Basic Approaches — 130
8.2.1 Ink Drawing on White Cardboard Sheet *131*; 8.2.2 Black Taping on Transparent Base Foil *133*; 8.2.3 Red/Blue Taping on One Transparent Base Foil *136*; 8.2.4 Black Taping on Diazo Films *139*; 8.2.5 Cut-and-Strip Artwork *141*

8.3 Artwork Taping Guidelines — 142
8.3.1 Equipment *142*; 8.3.2 Practical Hints *143*

8.4 General Artwork Rules — 145
8.4.1 Conductor Orientation *145*; 8.4.2 Conductor Routing Practice *145*; 8.4.3 Spacing *145*; 8.4.4 Hole Diameter, Solder Pad Diameter and Solderability *148*; 8.4.5 Component Polarity Identification *148*

8.5 Artwork Check and Inspection — 148
8.5.1 Check *148*; 8.5.2 Inspection *150*

Bibliography — 150

9 AUTOMATION AND COMPUTERS IN PCB DESIGN — 151

9.1 Limitations of Manual Designing — 151
9.2 Automated Artwork Draughting — 152
9.2.1 Layout Sketch *152*; 9.2.2 Equipment for Automated Artwork Draughting *152*; 9.2.3 Scope *155*

9.3 Computer-Aided Design — 156
9.3.1 Input Data *156*; 9.3.2 Component Placement *158*; 9.3.3 Conductor Routing *159*; 9.3.4 Checking *160*; 9.3.5 Scope *160*

9.4 Design Automation — 161
9.4.1 Conductor Routing *163*; 9.4.2 Component Package Placement *163*; 9.4.3 Performance and Scope *163*

9.5 Summary and Limitations of Automation in PCB Design — 164
Bibliography — 165

Part B: Technology of Printed Circuit Boards

10 FILM MASTER PRODUCTION — 170

10.1 Introduction — 170
10.2 Emulsion Parameters — 171
10.3 Film Emulsions — 175
10.4 Dimensional Stability of Film Masters — 176
10.5 Reprographic Cameras — 181
10.6 Darkroom — 184
10.7 Film Processing — 187
10.8 Increasing and Decreasing Line Widths — 193
10.9 Film Registration — 195
Bibliography — 197

11 PROPERTIES OF COPPER-CLAD LAMINATES 199

- 11.1 **Introduction** — 199
- 11.2 **Manufacture of Copper-Clad Laminates** — 199
  - 11.2.1 Materials *199*; 11.2.2 Manufacturing Process *201*
- 11.3 **Properties of Laminates** — 203
  - 11.3.1 Electrical Properties *204*; 11.3.2 Copper-to-Base Laminate Bond Characteristics *208*; 11.3.3 Physical Characteristics *208*; 11.3.4 Copper Surface Standards *210*
- 11.4 **Types of Laminates** — 211
  - 11.4.1 Phenolic Laminates *211*; 11.4.2 Epoxy Laminates *213*; 11.4.3 Polyester Laminates *214*; 11.4.4 Diallyl Pthalate Laminates *214*; 11.4.5 PTFE Laminates *215*; 11.4.6 Silicone Laminates *215*; 11.4.7 Melamine Laminates *215*; 11.4.8 Polyimide Laminates *215*
- 11.5 **Specifications and Test Methods** — 216
  - 11.5.1 Test Method Details *218*; 11.5.2 Solder Float Test by NEMA *221*; 11.5.3 Underwriters Laboratories Tests for Flame Resistance *221*
- **Bibliography** — 224

12 BOARD CLEANING BEFORE PATTERN TRANSFER 225

- 12.1 **Categories of Soils** — 225
- 12.2 **Manual Cleaning Processes** — 226
- 12.3 **Machine Cleaning Processes** — 228
- 12.4 **Equipment Trends** — 231
- **Bibliography** — 231

13 PHOTOPRINTING 232

- 13.1 **Basic Processes for Double-Sided PCBs** — 232
- 13.2 **Photoresists in General** — 235
- 13.3 **Wet-Film Resists** — 236
- 13.4 **Coating Processes for Wet-Film Resists** — 239
- 13.5 **Exposure and further Processing of Wet-Film Resists** — 244
  - 13.5.1 Exposure *244*; 13.5.2 Developing *247*; 13.5.3 Dyeing *248*; 13.5.4 Touch-up *248*; 13.5.5 Postbaking *248*; 13.5.6 Stripping *248*
- 13.6 **Dry-Film Resists** — 250
  - 13.6.1 Features of Dry-Film Resists *250*; 13.6.2 Categories of Dry-Film Resists *251*; 13.6.3 Exposure and further Processing of Dry-Film Resists *253*
- **Bibliography** — 253

14 SCREEN-PRINTING 255

- 14.1 **Scope of Screen-Printing** — 255
- 14.2 **Screen Fabrics** — 256
- 14.3 **Screen and Frame Preparation** — 258
  - 14.3.1 Frames *258*; 14.3.2 Fixing the Screen onto the Frame *259*

14.4 Pattern Transfer onto the Screen 260
14.4.1 Direct Method *261*; 14.4.2 Indirect Method *263*
14.5 Reclamation of the Screen Fabrics 264
14.6 Printing 265
14.6.1 Inks (Resists) *265*; 14.6.2 Hand Screen-printing *266*; 14.6.3 Screen-Printing with Machines *267*; 14.6.4 Screen Cleaning *268*
14.7 Trouble-Shooting 268
Bibliography 269

15 PLATING 270

15.1 Introduction 270
15.2 Immersion Plating 271
15.2.1 Tin Immersion Plating *271*; 15.2.2 Immersion Plating of Gold *272*; 15.2.3 Rinsing *272*
15.3 Electroless Plating 272
15.3.1 Electroless Plating Processes *272*; 15.3.2 Electroless Copper Plating *272*; 15.3.3 Chemical Reduction *273*; 15.3.4 A Typical Electroless Copper Bath *273*
15.4 Electroplating 274
15.4.1 Principle *274*; 15.4.2 Faraday's Laws *274*; 15.4.3 Preplate Treatment *274*; 15.4.4 Purity of Water *275*; 15.4.5 Typical Preplate Treatment *275*; 15.4.6 Plating Installation Consideration *276*; 15.4.7 Copper Plating *276*; 15.4.8 Tin Plating *278*; 15.4.9 Nickel Plating *279*; 15.4.10 Gold Plating *281*; 15.4.11 Tin-Lead Plating *281*
15.5 Alternative Finishes 283
15.5.1 Limitation of Tin-Lead Plating *283*; 15.5.2 Roller Tinning *283*; 15.5.3 Centrifugal Tinning *283*; 15.5.4 Hot-Air-Levelling *286*
15.6 Relative Performance of different Coatings 286
15.7 Plating Quality Control 286
15.7.1 Microsection *286*; 15.7.2 Porosity Test *286*; 15.7.3 Solderability Test *288*; 15.7.4 Thickness Test *288*
Bibliography 290

16 ETCHING 292

16.1 Introduction 292
16.2 Etching Machines 293
16.3 Etchant Systems 296
16.3.1 Comparison of Etchants *296*; 16.3.2 Ferric Chloride *298*; 16.3.3 Cupric Chloride *300*; 16.3.4 Chromic Acid *303*; 16.3.5 Alkaline Ammonia *305*
16.4 Minimising Pollution 307
16.4.1 Optimising Etchant Economy *308*; 16.4.2 Waste-Water *311*
Bibliography 313

17 MECHANICAL MACHINING OPERATIONS 314

17.1 Introduction 314

17.2	Shearing	*315*
17.3	Sawing	*316*
17.4	Punching	*318*
17.5	Blanking	*320*
17.6	Milling	*320*
17.7	Routing	*320*
17.8	Drilling	*320*

17.8.1 Hole Diameter Tolerances *321*; 17.8.2 Drilling Machines *321*; 17.8.3 Drill Bits *328*; 17.8.4 Drilling Practice *329*

Bibliography *332*

18 PCB TECHNOLOGY TRENDS 333

18.1 Fine-Line Conductors with Ultra-Thin Copper Foil *333*

18.1.1 Ultra-Thin Copper Foils *333*; 18.1.2 Pattern-Plate Processing with Ultra-Thin Foils *334*; 18.1.3 Advantages and Disadvantages with Ultra-Thin Copper Foils *335*

18.2 Multilayer Boards *336*

18.3 Multiwire Boards *337*

18.3.1 Concept *337*; 18.3.2 Techniques in Multiwire Board Production *338*; 18.3.3 Where to Apply Multiwire Boards *341*; 18.3.4 High Conductivity Multiwiring *341*

18.4 Subtractive-Additive Process *342*

18.4.1 Principle of the Subtractive-Additive Process *342*; 18.4.2 Process Details *342*; 18.4.3 Electroless Copper Bath for Subtractive Additive Process *344*

18.5 Semi-Additive Processes *345*

18.5.1 Base Materials for Semi-Additive and Additive Processes *345*; 18.5.2 Semi-Additive Processing *347*; 18.5.3 Why Semi-Additive Processes *348*

18.6 Additive Processes *348*

18.6.1 Scope *348*; 18.6.2 Base Materials and their Preparation *350*; 18.6.3 Additive Processing *351*

18.7 Flexible Printed Circuit Boards *352*

18.7.1 Scope *352*; 18.7.2 Flexible Base Materials *353*; 18.7.3 Design Constraints *354*; 18.7.4 Processing *356*

18.8 Metal Core Circuit Boards *357*

18.8.1 Scope *357*; 18.8.2 Metal Core Board Production *359*; 18.8.3 Generation of the Circuit Pattern *361*

18.9 Mechanical Milling of PCBs *361*

18.9.1 Equipment *361*; 18.9.2 Circuit Patterns which can be Realised *362*; 18.9.3 Where to Employ PCB Production by Mechanical Milling *364*

Bibliography *365*

19 MULTILAYER BOARDS 367

19.1 Introduction *367*

19.1.1 Background *367*; 19.1.2 Where to Apply Multilayers *367*;

19.2 Design and Test Considerations — 368
19.2.1 Design 368; 19.2.2 Artwork Methods 369; 19.2.3 Tolerances 370; 19.2.4 Board Thickness 370; 19.2.5 Tests to be Incorporated 372

19.3 Multilayer Construction — 375
19.3.1 Specifications 375; 19.3.2 Interconnection Techniques 375

19.4 Equipment — 376

19.5 Laminating Process — 378
19.5.1 Preparations 378; 19.5.2 Preheating 378; 19.5.3 Post-Laminating Inspection 379

19.6 Further Processing — 379
19.6.1 Inspection of the Holes 379; 19.6.2 Etch Back 380; 19.6.3 Copper Plating 380

Bibliography — 381

20 SOLDERS AND SOLDERING TECHNIQUES 382

20.1 Introduction — 382
20.2 Principles of Solder Connections — 382
20.3 Solder Alloys — 386
20.3.1 Tin-Lead 386; 20.3.2 Other Solder Systems 387; 20.3.3 Influence of Impurities on Tin-Lead Solder 389; 20.3.4 Mechanical Properties of Solder Alloys 389

20.4 Soldering Fluxes — 390
20.4.1 Corrosive Fluxes 390; 20.4.2 Intermediate Fluxes 391; 20.4.3 Non-Corrosive Fluxes 392; 20.4.4 Testing of Fluxes 392

20.5 Soldering Techniques — 392
20.5.1 Iron Soldering 393; 20.5.2 Mass Soldering 394; 20.5.3 Flux Removal after Soldering 397

20.6 Solder Mask — 398
20.7 Reflow Soldering Practice — 398
20.8 Testing and Quality Control — 400
20.8.1 Solderability Test Methods 400; 20.8.2 Inspection 403

20.9 Safety, Health and Medical Aspects in Soldering Practice — 406
20.9.1 Safety 406; 20.9.2 Health and Medical Aspects 406

Bibliography — 407

21 COMPONENT ASSEMBLY TECHNIQUES 409

21.1 Preparation and Mounting of Components — 409
21.1.1 Component Lead Preparation 409; 21.1.2 Component Mounting 410

21.2 Organisation of Non-Automatic PCB Assembly — 415
21.3 Lead Cutting and Soldering — 416
21.4 PCB Cleaning after Soldering — 417
Bibliography — 418

22 GUIDELINES FOR STARTING PCB FACILITIES 419

22.1 Introduction — 419

22.2 Market Constraints	*419*
22.3 Equipment	*420*

22.3.1 Local versus Imported Make *420*;   22.3.2 Simple Professional PCB Facilities *421*;   22.3.3 Simple Consumer PCB Facilities *423*

22.4 Room and Building Requirements *424*

22.4.1 Darkroom *424*;   22.4.2 Room for Photoresist Work *424*;   22.4.3 Screen-Printing Room *425*;   22.4.4 Etching/Plating Room *425*;   22.4.5 Chemical Store Room *426*;   22.4.6 Mechanical Fabrication Room *426*

22.5 Control of Pollution and Health Hazards *428*

22.5.1 Etchant and Plating Solutions *428*;   22.5.2 Rinsing Water *431*;   22.5.3 Air *433*;   22.5.4 Health of Personnel *434*

Bibliography *438*

APPENDIX: LIST OF ADDRESSES *439*

A.1 Professional Organisations *439*
A.2 Standard Organisations *439*
A.3 PCB Equipment and Material Manufacturers *440*

A.3.1 Introduction *440*;   A.3.2 Alphabetical Key-Word Index *441*;   A.3.3 Addresses *442*

Index *459*

Part A

Part A

DESIGN OF PRINTED CIRCUIT BOARDS

The design of a Printed Circuit Board (PCB) can, as many will agree, be considered as the last step in electronic circuit design as well as a first major step in the production of PCBs. It forms a distinct factor in electronic circuit performance and reliability. The producibility of a PCB and its assembly and serviceability also depends on the design. All these factors finally get reflected in the price for the electronic equipment of which, PCBs take away approximately 20% of the cost. From this, it is evident that the task of engineers involved in PCB design is not very simple or always straightforward: Intimate knowledge of all the implications is very much required and can only be gradually acquired, mostly by experience.

However, there are certain standard practices in PCB designing which are followed by most designers. They can serve as a guideline. The author makes an attempt to include them in this book. Hereby, we shall confine ourselves to the design of single- and double-sided PCBs: They are by and large the most important categories.

The designing of a PCB consists of the designing of the layout followed by the generation or preparation of the artwork. The layout, therefore, should include all the relevant aspects and details of the PCB design, while the artwork preparation brings it to the form required for the production process. We can say that layout design is the stage where engineering capacity combined with creativity are the governing inputs, while practical skills paired with technical commonsense have to be predominant in artwork preparation.

1
LAYOUT PLANNING

1.1 INTRODUCTION

The layout of a PCB has to incorporate all the information on the board before one can go on to the artwork preparation. This means that a concept, that clearly defines all the details of the circuit and partly also of the final equipment, is a prerequisite before the actual layout can start. The detailed circuit diagram is very important for the layout designer but he must also be familiar with the design concept and with the philosophy behind the equipment. Only with this in mind will he be able to bring out results which do not repeatedly call for modifications. This problem does not arise so much in small companies where the same engineer might get involved at all steps right from circuit design to production. In larger companies with their inherent trend for specialisation, appropriate ways have consciously to be searched for, in order to keep the layout designer informed about the various constraints between equipment design concept and testing of the final equipment.

1.2 GENERAL CONSIDERATIONS

1.2.1 Layout Scale

Depending on the accuracy required, artwork should be produced at a 1:1 or 2:1 or even 4:1 scale. The layout is best prepared on the same scale as the artwork. This prevents all the problems which might be caused by redrawing of the layout to the artwork scale. It just needs some rethinking in order to imagine all the dimensions of components on an enlarged scale. Since most of the artworks made in a particular organisation are of the same scale, it is not very difficult to get accustomed to the scaling practice for layouts, and also the final layout checking will be easier.

The layout/artwork scale commonly applied is 2:1. It offers a reasonable compromise between accuracy gained and handling convenience. Remember that a 2:1 artwork has 4× the actual PCB area and a 4:1 artwork is of 16× the PCB area. The 4:1 scale is therefore applied only where very high precision is required. With a 1:1 scale, non-demanding single-sided boards

LAYOUT PLANNING

can be designed but sufficient care should be taken, particularly during the artwork preparation.

1.2.2 Grid System

The use of a grid system in layout/artwork design is a commonly accepted practice. Components are manufactured with their leads fitting into grid intersections and numerically controlled drilling equipment can easily be programmed accordingly.

For the layout designer, too, the use of a grid gives more convenience in placement of components and conductors. Less accuracy in drawing the hole locations is easily rectified at the artwork preparation stage where the pads are placed exactly in centre of the grid intersections.

The pattern in Fig. 1-1 shows a DIP (*D*ual-*I*nline-*P*ackage) in the artwork stage and how it is drawn in the layout. If the layout is realised on a 2:1 scale, one unit (grid equidistance) will be 0.2" or 5.08 mm. Where such grid sheets are not available, ordinary paper with a 5 mm grid can also serve the purpose for sketching of layouts.

Where the grid system based on 0.1" is found to be too coarse, a grid equidistance of 0.025" or even 0.1 mm is recommended [1].

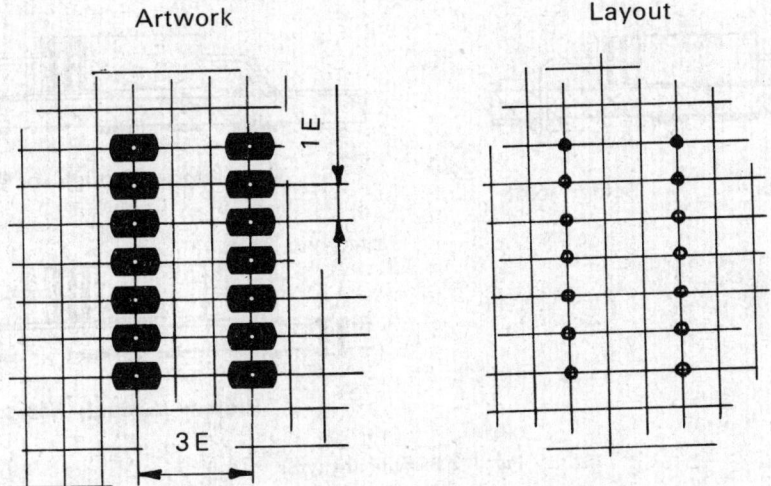

1 E = 1 Unit = 0.1" = 2.54 mm at actual PCB scale

Fig. 1-1 Component holes following a grid pattern in artwork and layout

1.2.3 Board Types

Only the two most popular PCB types are mentioned here:

Single-Sided Boards

The single-sided PCBs are mostly used in entertainment electronics where manufacturing costs have to be kept at a minimum. However, in industrial electronics also, cost factors cannot be neglected and single-sided boards should be used wherever a particular circuit can be accommodated on such boards. To jump over conductor tracks, components have to be utilised (Fig. 1-2A). If this is not feasible, jumper wires are used. The number of jumper wires on a board, however, is restricted by economic reasons. If their number is more than a few, the use of a double-sided PCB should be considered.

Double-Sided Boards

Double-sided PCBs can be made with or without plated-through holes (Fig. 1-2B). The production of boards with plated-through holes is fairly expensive. Therefore, plated-through hole boards are only chosen where the circuit complexity and density does not leave any other choice. Even on such boards, the total number of plated-through holes, in particular of via-holes (holes utilised only for through-contact and not for component mounting), should be kept to the minimum for reasons of economy and reliability.

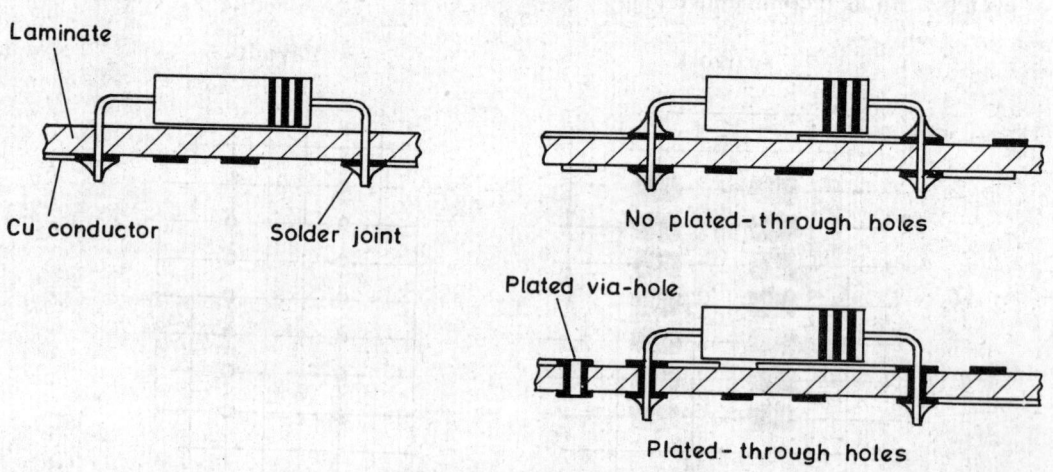

Fig. 1-2 Basic board types

The cost factor for double-sided PCBs without plated-through holes is considerably lower because plating can be avoided. Through-contacts are made by soldering the component leads on both the board sides where required. Jumper wires may still be added. However, hand soldering must be applied for soldering of the component side joints.

In the layout design of such boards, solder joints on the component sides have to be kept minimum in number because the replacing of such components is extremely difficult. A typical strategy is therefore to realise the conductors as much as possible on the non-component side and to put only the remaining ones on the component side. Such boards are therefore a compromise between serviceability and electrical design optimum on the one hand and the cost factor on the other.

1.2.4 PCB Production Facilities

Exact knowledge of all the limiting factors of the concerned PCB production set-up is essential. As an example: because of the limitations in the copyboard size, a layout/artwork scale of 4:1 might hardly be compatible with a small vertical-type of reprographic camera for film master production. Other limitations can be given in the artwork table size, minimum or maximum board processing size, drilling accuracy, fine-line etching performance etc. The author considers it to be very important for any PCB designer to have spent a certain amount of time in the actual PCB production process. With this background, a much better understanding of the practical design implications becomes possible than with daylong studies of internal rules and recommendations.

1.2.5 Standards

Standards are laid down to achieve repetitive results for satisfying specified requirements. Standardisations can be introduced at any level and for almost any purpose. For the layout design, the designer should keep two points in mind; firstly to have an exact knowledge of the standards that have to be applied and secondly to follow them strictly.

As an example of internal standardisation for PCB layout preparation, Fig. 1-3 shows a 0.2″ grid sheet with standardised PCB outlines, connector and test point location and other details. The corresponding PCB forms the standard for 19″ racks at the Centre for Electronics Design Technology (CEDT).

1.3 PCB SIZES

1.3.1 Mechanical Stress

Comparatively small PCBs, with a size of less than 100 × 150 mm and the standard thickness of 1.6 mm, will hardly pose any problems in mechanical strength if assembled with the usual electronic components. Care is needed with bigger board sizes or thinner laminates or if heavy parts like transformers have to be mounted on the board. In any case heavy parts should be mounted near a supporting device like cardguide, connector or stiffener.

Mechanical stress does not only occur during day-to-day use of the equipment: Probably more dangerous are the resonances caused during the transport or shipment of the equipment from the factory to the user. When designing large-size boards, to avoid board failures because of short circuiting with nearby conductive elements, cracked solder joints or component fatigue, the enormous deflections possible should be controlled. This can be done by providing stiffeners or supporting devices.

1.3.2 Other Board Size Constraints

Electrical Functioning

If more circuitry has to be realised on the same board, then, as a rule, the board size will become disproportionately larger. This is because more and more space has to be provided for

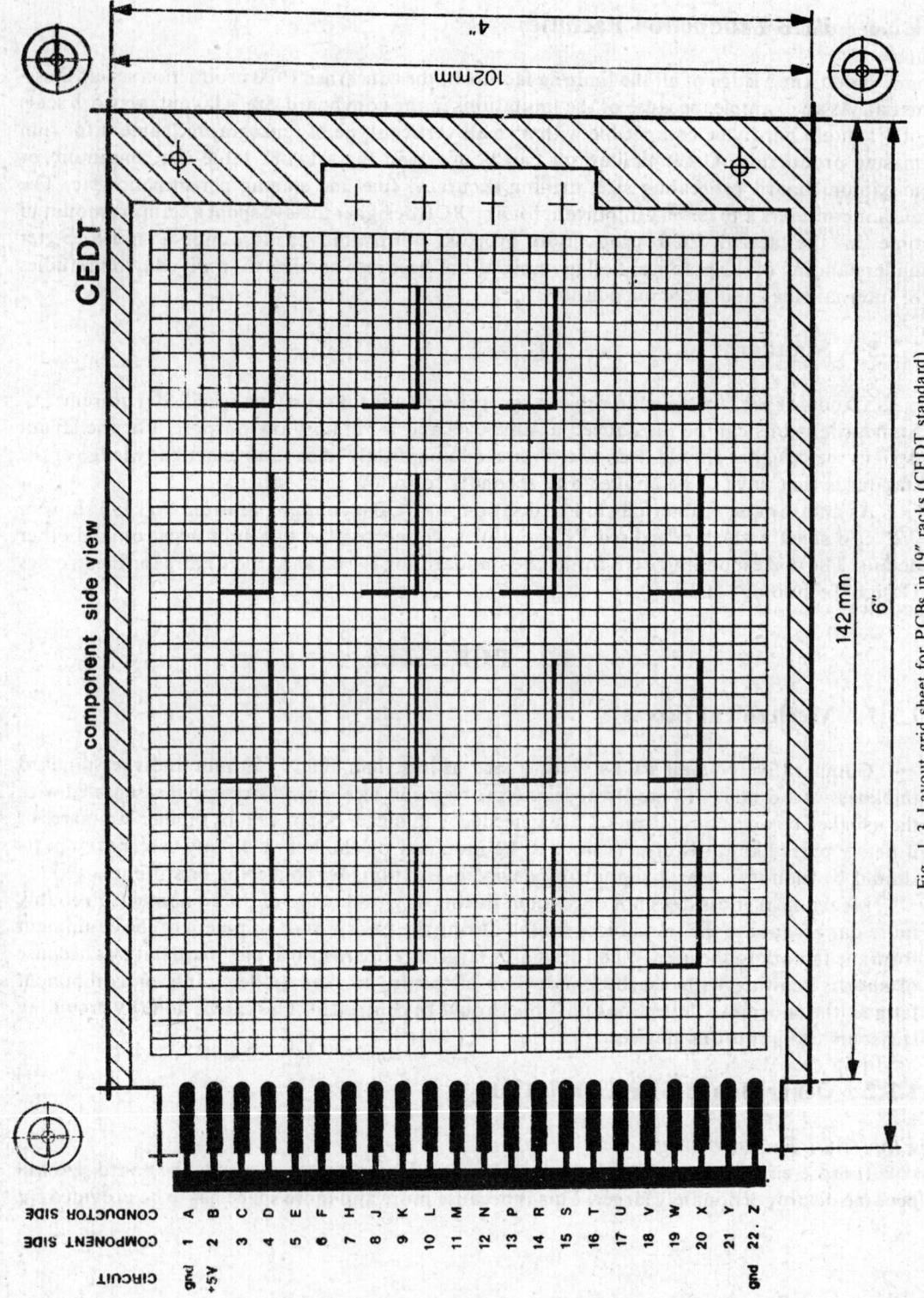

Fig. 1-3 Layout grid sheet for PCBs in 19" racks (CEDT standard)

the interconnections. At the same time, longer interconnections are more susceptible to functional disturbances. This clearly speaks against too large boards. On the other hand, if too many small boards form one complete circuit, other difficulties and higher costs are caused by the external interconnections and connectors required.

Testing and Servicing

To enable efficient testing and repair of PCBs, smaller board sizes should be given preference. In large boards, the isolation of those parts of the circuit that are not working is very difficult because there are usually no means to interrupt or influence the signal flow. If the circuit is realised on different smaller cards, the faulty card is not only easier to detect but also easier to repair or exchange. Another advantage in having an electronic circuit split into subunits on different boards, is the possibility to use the same subunits (e.g. power supply, demodulator) also in other kinds of equipment if it is so designed.

Modifications

It is a good practice to utilise the board area to not more than 95% and to provide at least 5% area for later modifications of the circuit. Where useful, there should be provision to add ICs or other components. With this, a complete redesign of the PCB can in many cases be avoided and it is a known fact that hardly any design works immediately without minor modifications.

Equipment Dimensions

In many cases, the PCB size and number is dictated by the exterior design concept of the equipment: In an ultra-slim stereo-amplifier, only one PCB can be accommodated which is of considerable size. Or, a paging system receiver PCB has to be very small because the receiver is carried in the pocket.

1.3.3 Standard Sizes

From what has been previously said, it is clear that the ideal board size is neither very large nor very small. It seems that an optimum size is given with side lengths in the range of 150–250 mm. However, the constraints on PCB size are manifold. It is not possible to have universally applicable standards. In each case, the optimum solution must be found, by taking account of all the individual constraints.

In industrial electronics, the 19″ rack has been fairly well adapted as a standard rack and PCB sizes have had to follow this standardisation to a certain extent. Even so, no common base in PCB sizes for such racks has emerged till now. Fig. 1–5 shows an industrial electronic equipment in a 19″ rack by using the same standard size as shown in Fig. 1–3. In Table 1.1 the standard PCB sizes are shown as recommended by DIN (German Standards Organisation). The DIN standards have found fairly wide acceptance, especially in Europe.

Fig. 1-4 Slim digital multimeter with the circuit realised on one PCB (Photo CEDT)

LAYOUT PLANNING

Fig. 1-5 19″ rack with PCBs of the type shown in Fig. 1-3. Note the easy access to the test pins (Photo CEDT)

Table 1-1 DIN recommended PCB standard sizes in millimeters, extracted from DIN 41494.

Front Panel	Printed Circuit Board			
		Length		
Height	Height	L_1	L_2	L_3
132.5	100	100	160	220
177.0	144.5	100	160	220
221.4	188.9	100	160	220
265.9	233.4	100	160	220

1.4 LAYOUT APPROACHES

1.4.1 Materials and Aids

Simple Approach with Sketching of Components

The method works with the simplest of means: A sheet of paper with a grid printed on it, a pencil and an eraser are all the materials required. First, the board outlines and connectors are marked, followed by sketching of the component outlines with connecting points and the conductor pattern. In the development of the conductor pattern and component locating, there is a continuous need for erasing, as designing of the layout proceeds on the basis of trial and error. With grid paper, after continuous erasing, the paper base gets damaged and could become rather spoiled towards the end.

An improvement of this method is to use a (transparent) tracing sheet for the development of the layout design. The grid sheet is placed under the tracing sheet which makes all the information drawn on the grid sheet and the tracing sheet simultaneously available to the designer. This method has become very popular as erasing is simple on the tracing paper.

Layout Sketching with 'Puppets'

'Puppets' is a trade-mark of Bishop Graphics Inc. and is used in general for the concept of individually die-cut, transparent layout patterns that represent commonly used electronic components. 'Puppets' show the characteristic of adherence to any clear, non-matte drafting film (Fig. 1-6).

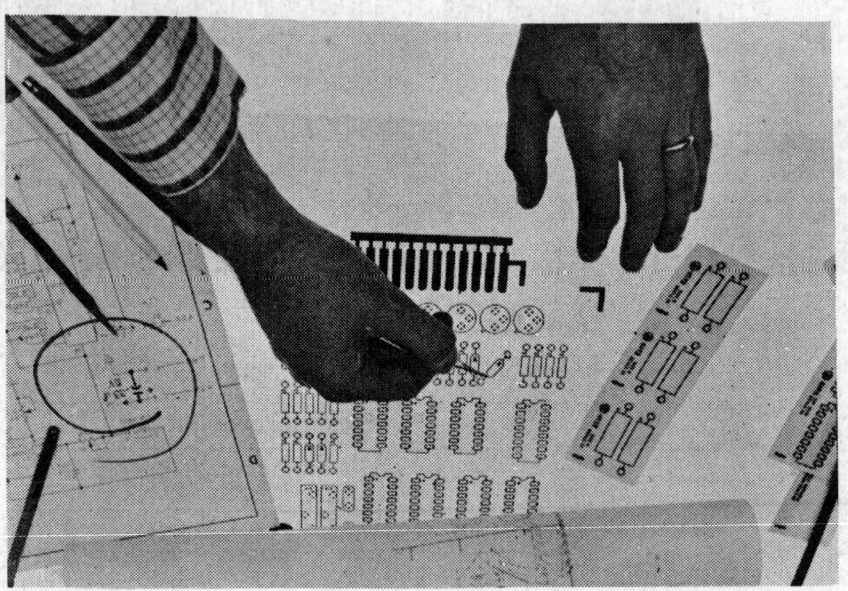

Fig. 1-6 Using 'Puppets' for the component layout (Courtesy: Bishop Graphics Inc., Westlake Village, Ca. 91359)

When using 'puppets', a transparent polyester foil (with the non-matte surface on the upper side) is placed over the grid sheet. Now, the component layout can quickly be done by using the respective 'puppet' for each component. For interconnections, a sheet of tracing paper is placed over the component layout and the conductors can be pencilled on this tracing paper. Changing the position of a component is done by lifting the tracing sheet on one side and simply removing and placing afresh the concerned 'puppet' on the polyester film. Once all the components are allocated and all the interconnections drawn, the component lead holes and outlines are also marked on the tracing sheet to complete the layout.

The use of 'puppets' helps to save time in layout designing: It minimises the time spent in erasing of components located and their redrawing plus redrawing of all the affected interconnections. The use of 'puppets' is recommended particularly where larger and more complicated layouts are designed. It can help to save 20–50% of the layout designing time.

1.4.2 Procedures

The first rule is to prepare each and every PCB layout as viewed from the component side (top side). This rule must be strictly followed to avoid confusion which would otherwise be caused.

Another important rule is not to start the designing of a layout unless an absolutely clear circuit diagram is available, if necessary, with a component list.

Certain designers prefer to complete the component layout first and then start with the interconnections. Others develop the layout of components and interconnections simultaneously. But the common features of any approach should be to develop the layout in the direction of the signal flow as far as possible. Only this way will one achieve the shortest possible interconnections.

Among the components, the larger ones are placed first and the space in between is filled with smaller ones. Components requiring input/output connections come near the connectors. All components are placed in such a manner that desoldering of other components is not necessary if they have to be replaced.

In the designing of a PCB layout, it is very important to divide the circuit into *functional* subunits. Each of these subunits should be realised on a defined portion of the board. This will prove helpful not only in functional reliability but also enables a faster testing and servicing of the board.

If the input/output connections have not yet been clearly defined in the circuit diagram itself, the PCB layout designer must verify that the layout realisation conforms to the requirements of the mother board or the external wiring. Especially, check power connections which are mostly taken from a distinct connector location.

In designing the interconnections, which are usually done by pencil lines, actual space requirements in the artwork must be considered. This again calls for familiarity with artwork practice. It is also a useful practice to draw the conductor lines only on grid lines: Spacing between the conductors will become uniform and it must only be checked when the conductor passes between pads. In addition, the layout can be rather roughly sketched and will still be clear enough for the artwork designer.

To identify the conductor lines on layouts for double-sided PCBs, pencils of different colours can be used. Another possibility is to do all of them with an ordinary pencil, but to draw interrupted lines for the conductors on one of the sides.

1.5 DOCUMENTATION

1.5.1 Scope

In small electronics industry, it is possible that the circuit designing engineer also prepares layout and artwork; the documentation in this case will probably consist of a few loose sheets of paper with some notes. There will hardly be any information or communication problem. However, in normal case, various persons get involved with design and the purpose of documentation now becomes to provide the information to the right person. How this is implemented varies from place to place; smaller companies usually go for the minimum essential while large companies have exact laid down norms, forms, lists and files.

The layout designer on his side, is concerned about the circuit diagram and the component list. These inputs have to be complete and must contain all information which is relevant to the layout design. Similarly, the output of the layout design stage must provide all necessary information for the completion of the artwork and the mechanical drawings.

1.5.2 Circuit Diagram

The usual way of presenting a circuit diagram will not be further elaborated here. Instead of this, a few points have to be touched on, which may be additionally necessary for the layout designer.

Connector: Input and output connections of the PCB which have a predetermined location on the connector or equipment side, must be clearly marked with the necessary particulars.

Critical signals paths: Where the conductor length should be minimum, or conductor should be 'guarded' by ground lines (see **Chapter 5**) or have a distinct width, then these are shown in the circuit diagram, e.g. with dotted lines.

Heat sources: Components producing considerable heat have also to be identified in the circuit diagram so as to avoid concentrated heat on the board or a close neighbourhood to heat-sensitive components.

1.5.3 Component List

For simpler circuits, it may not always be necessary to have a separate component list: The component specifications can be incorporated in the circuit diagram. In all the other cases, a separate component list is prepared by the circuit designer giving all the desired information about the components. Note that the component list must also include heatsinks and mechanical hardware which is mounted onto the PCB.

A component list, depending on the needs, may include the following columns:
—component code
—exact specifications
—supplier of the component
—price
—selection criteria or tolerances

—matching requirements
—mechanical dimensions

For the component code which is also used in the circuit diagram, the following letters have been recommended by the American National Standards Institution (ANSI):

  R = resistor (e.g. R1, R2, R3...)
  C = capacitor
  Q = transistor
  CR = diode
  U = integrated circuit
  T = transformer

1.5.4 Layout Sketch

The end product of the layout designing is the pencil-sketched component and conductor drawing, which is called 'layout sketch' and which contains all relevant information for the preparation of the artwork.

Besides the component outlines, component holes and interconnecting pattern, the layout sketch should also include information on:

—Component holes: Usually, in a given PCB, most of the holes required are of one particular diameter and this diameter is mentioned once in the layout sketch. Holes of a different diameter are shown with a code in the actual layout sketch. The code must be explained outside the layout area (Fig. 1-7A).
—Conductor width: Here also, a code can be used for the conductors with a special width (Fig. 1-7B).
—Minimum spacing to be provided.

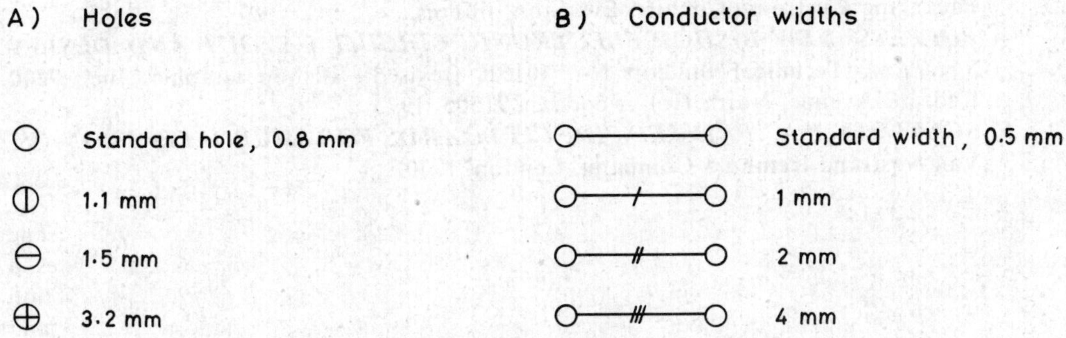

A) Holes

○ Standard hole, 0.8 mm
⊕ 1.1 mm
⊖ 1.5 mm
⊕ 3.2 mm

B) Conductor widths

○——○ Standard width, 0.5 mm
○—/—○ 1 mm
○—//—○ 2 mm
○—///—○ 4 mm

Fig. 1-7 Layout coding example

1.5.5 Mechanical Drawing

Mechanical drawing is an important document for the mechanical fabrication steps of a PCB and it also includes details about processing of the boards. The usual information requirements in the mechanical drawing are as follows:

- Board outlines with tolerances, including connector
- PCB laminate to be used
- Plating specifications
- Registration and mounting holes with locations and diameters.

1.5.6 Assembly Drawing

The assembly drawing, along with the component list, provides all the necessary information for correct assembly of the particular board. The assembly drawing shows board- and component-outlines. The component-outlines are supplemented with the proper component code; special mounting instructions may also be mentioned.

For prototype boards, the assembly drawing may also consist of a blueprint of the PCB artwork into which the component outlines and codes have been added. Note that the artwork would be spoiled in an ordinary blueprint exposure machine because of round bending. But the exposure can be easily carried out in a PCB exposing equipment in which no bending will harm the delicate artwork.

BIBLIOGRAPHY

1. *GRID SYSTEM FOR PRINTED CIRCUITS*, Publication No.97 (3rd edition), International Electrotechnical Commission, Geneva, Switzerland, 1970.
2. LUND P. : *GENERATION OF PRECISION ARTWORK FOR PRINTED CIRCUIT BOARDS*, John Wiley & Sons, Chichester, 1978.
3. *PRINTED WIRING DESIGN GUIDE*, The Institute for Interconnecting and Packaging Electronic Circuits, Evanston, Illinois.
4. *PUPPETS, NEW BISHOP ELECTRONIC CIRCUIT LAYOUT AND DESIGN SYSTEM*, Technical Bulletin No. 1015R (revised), Bishop Graphics Inc., 7300 Radford Avenue, North Hollywood, Ca.91505, 1973.
5. SCARLETT J. A.: *PRINTED CIRCUIT BOARDS FOR MICROELECTRONICS*, Van Nostrand Reinhold Company, London, 1970.

2
LAYOUT, GENERAL RULES AND PARAMETERS

2.1 INTRODUCTION

To design a PCB layout means essentially to solve two tasks: One is to design the interconnections for the components and the other is to simultaneously minimise the magnitude and influence of the parasitic effects connected with the realisation of such interconnections. Parasitic effects influencing the working of an electronic circuit on a PCB can be caused by the resistance or inductance of a conductor or the capacitance between two conductors; furthermore, the heat generated in the circuit may alter its characteristics etc.

It is therefore very important for the designer to have a basic understanding on parasitic effects. His personal experience will be built on this basic knowledge. The author wishes to emphasise the need for sufficient personal experience which is necessary to produce really good layouts: There are many paths possible to realise a layout which leads to a good result and each layout designer has to develop and improve his own methodology by first-hand experience.

In this chapter, an account will be given of common basics to be understood by any layout designer. They are relevant in almost any kind of circuit realisation on a PCB. For more circuit-specific design rules, the following chapters provide some additional guidance.

2.2 RESISTANCE

2.2.1 Resistance in General

The copper conductor tracks on a PCB have a finite resistivity which introduces a voltage drop proportional to the current flowing in that particular conductor.

For practical purposes, it is helpful to know the resistance of a 1 mm wide conductor per cm of length. We assume a *standard copper foil of 35 µm thickness* without any plating:

$$R = \frac{\rho \times l}{A} \quad [\Omega]$$

ρ = resistivity [Ω cm $\times$ 10^{-6}]
l = conductor length [cm]
A = conductor cross-section [cm^2]

For ρ_{Cu} = 1.7241 (at 20°C), l = 1 cm and A = 35 $\times$ 10^{-4} $\times$ 0.1 cm^2, the result will be

$$R = \frac{1.7241 \times 10^{-6} \times 1}{35 \times 10^{-4} \times 0.1} = 0.0049 \ [\Omega]$$

The figure to remember is herewith a resistance of *roughly* 5 mΩ for a 1 mm wide conductor of 1 cm length for standard copper foil of 35 μm thickness.

Example: A 0.5 mm wide conductor of 12 cm length will have a resistance of 5 $\times$ 12 $\times$ 2 = 120 mΩ, since half the conductor width gives double as high resistance.

2.2.2 Resistance and Temperature

The temperature difference within electronic equipment between storage and maximum operating temperature can easily assume values of 60°C difference or more. This will significantly increase the conductor resistance and has to be accounted for in a worst-case design:

$$R_1 = R_0 \ (1 + c_T \ (T_1 - T_0))$$

R_1 = resistance at temperature T_1
R_0 = resistance at temperature T_0
c_T = temperature coefficient of conductivity

Example: The inside temperature in an electronic equipment under operation is 85°C while the outside temperature is 20°C. What will be the resistance of the 0.5 mm conductor line in the previous example at operating temperature? The temperature coefficient for copper conductivity is + 0.0039.

$$R_1 = 120 \ (1 + 0.0039 \ (85 - 20)) = 150.4 \ m\Omega$$

We can herewith see that the resistance has increased by full 25%.

2.2.3 Practice

In most electronic circuits, comparatively small currents are flowing, for which the conductor resistance can practically be neglected. However, if it comes to supply and ground lines, especially of high-speed and certain digital circuitry, much broader conductors than ohmically necessary have to be provided to bring the wave impedance between supply- and ground-lines down to a minimum. This will be treated separately in this chapter.

Exclusively *ohmic* considerations for conductor width determination are necessary for conductors carrying higher currents at low frequencies. For instance, in electronic power supplies and power electronics circuits. For such purposes, Fig. 2-1 gives some guidance to choose the appropriate conductor width. The figure shows also the conductor temperature rises which may be caused due to over-currents. Note that for all practical purposes, no substantial temperature rise should occur on the conductors under normal working conditions.

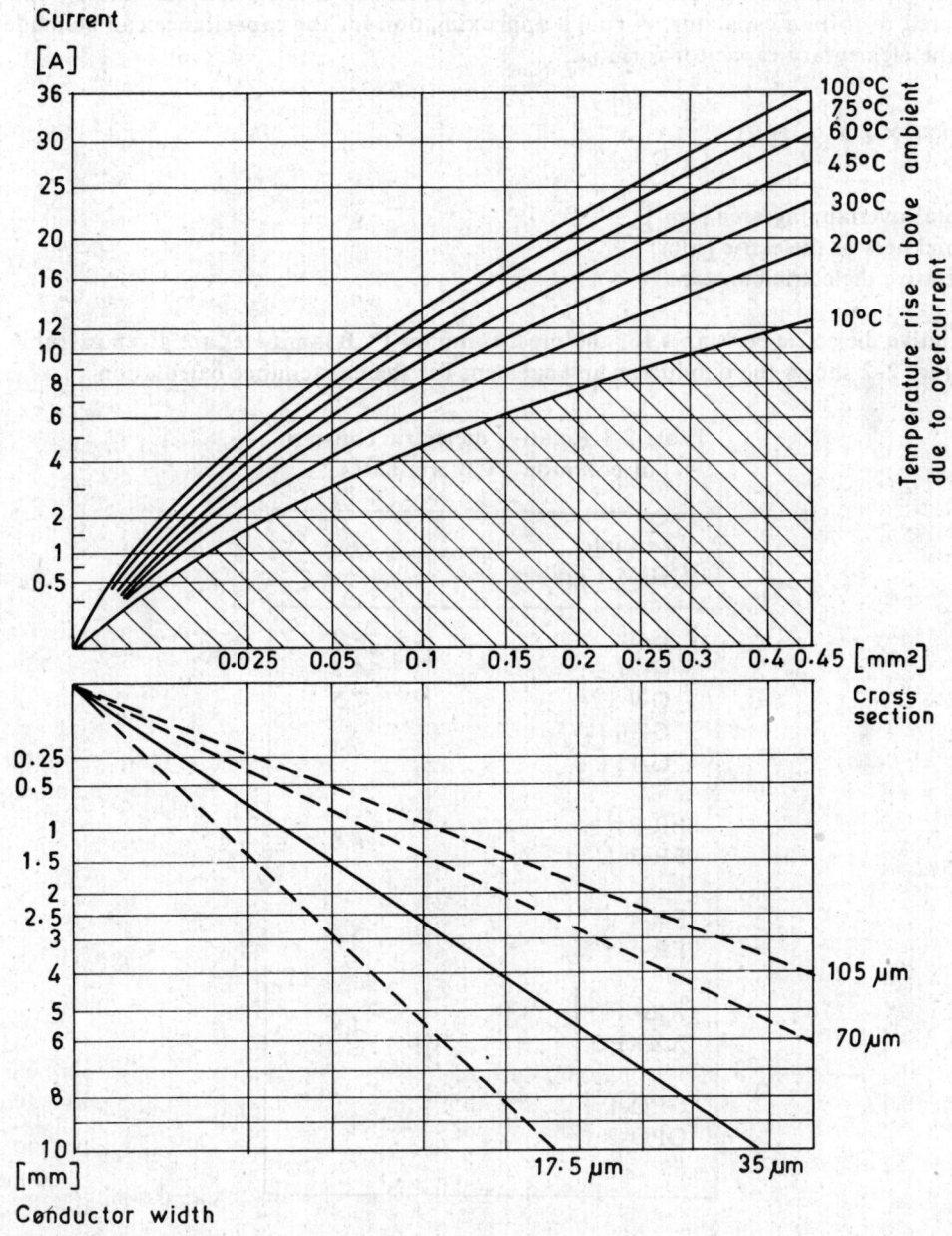

Fig. 2-1 Conductor widths for safe operating temperatures [1]

2.3 CAPACITANCE

2.3.1 Capacitance between Conductors on Opposite Sides of the PCB

Two PCB conductors (flat conductors) separated by a dielectric (laminate), can be considered to form a capacitor. A rough approximation for the capacitance can be made by using the elementary capacitor formula:

$$C = 0.886 \times \epsilon_r \times \frac{A}{b} \, [\text{pF}]$$

A = total overlapping area [cm^2]
b = thickness of dielectric [mm]
ϵ_r = relative dielectric constant

The relative dielectric constants for the most common PCB laminates are given in Table 2-1 while Fig. 2-2 shows the conductor arrangement for the capacitance calculation.

Table 2-1 Relative dielectric constant of common PCB laminates.

Laminate (NEMA grade)	ϵ_r
G-5	8.0
G-7	4.2
G-9	7.5
G-10, G-11	5.4
FR-2, FR-3	4.8
FR-4, FR-5	5.4
XXXP, XXXPC	4.8
GPO-1, GPO-2	4.4

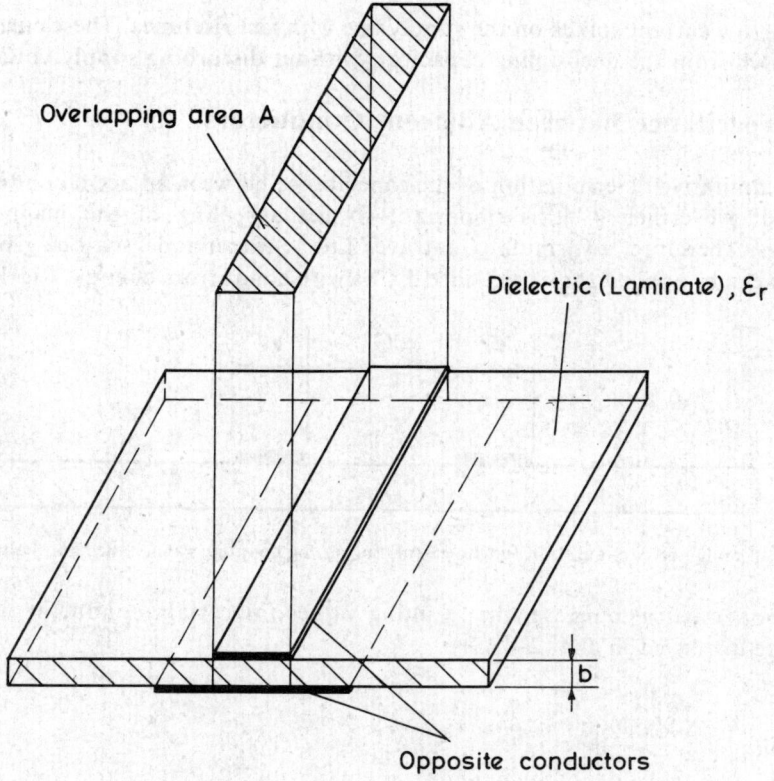

Fig. 2-2 Two conductors separated by a dielectric

In most of the cases, the conductors do not overlap completely and the thickness of the dielectric is of the same magnitude as the conductor width. The result will be an inhomogeneous field. The capacitor formula given gives only an approximate value; practical measurements on the actual PCB will give the exact value and might show a slightly higher value than the calculation.

Example: To get a better idea on the capacitance values to be expected, a simple example will be calculated here: On a G-10 type of laminate, the supply and ground lines are running exactly symmetric on the opposite sides of the PCB, in order to keep the impedance between them low. The total length in common is 250 mm, the supply line width is 2 mm and the ground line width is 4 mm.

$$C = 0.886 \times 5.4 \times \frac{0.2 \times 25}{1.6} = 15 \ [pF]$$

The example shows the usefulness of running broad supply- and ground-lines exactly opposite on the PCB in order to reduce the magnitude of spikes on supply- and ground-lines of fast switching circuits: The supply and ground-lines form a *distributed decoupling capacitor* of 15 pF across the PCB (see also Chapters 3 and 4). Actually, the TTL circuit realised on the PCB will give narrow current spikes on the supply line with fast risetimes. These current spikes can thus be drawn from the decoupling capacitor, without disturbing supply voltages.

2.3.2 Capacitance between Adjacent Conductors

The quantitative determination of the capacitance between adjacent conductors is rather complicated. Nevertheless, it is important to get a feeling of the magnitude of such capacitances. Therefore, a formula (Contraves Ltd., Switzerland), shall be given on the next page which can be applied for single-sided PCBs with conductors of equal widths, as shown in

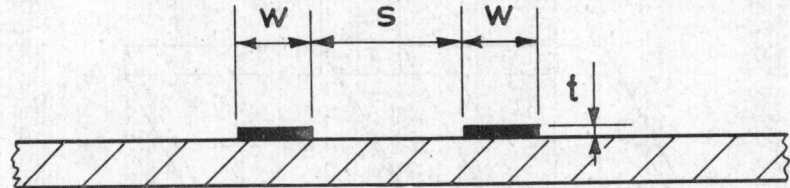

Fig. 2-3 Conductor configuration for the calculation of the coupling capacitance and inductance

Fig. 2-3. For practical purpose, the resulting values for certain conductor widths on G-10 laminate are displayed in Fig. 2-4.

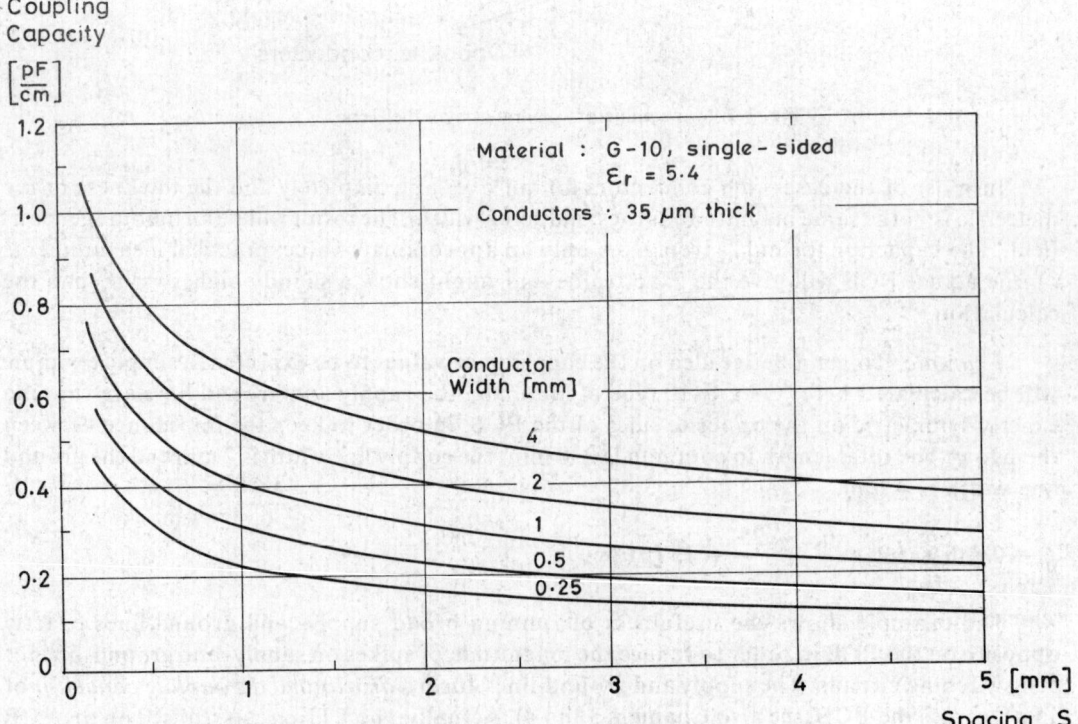

Fig. 2-4 Capacitive coupling between adjacent conductors

LAYOUT, GENERAL RULES AND PARAMETERS

$$C' = 0.122 \times \frac{t}{s} + 0.0905 (1 + \epsilon_r) \times \log_{10} (1 + \frac{2w}{s} + 2\sqrt{\frac{w}{s}} + \frac{w^2}{200}) \, [\frac{pF}{cm}]$$

$$C \doteq C' \times l$$

l = parallel running length [cm]
s = distance between the two adjacent conductors [mm]
t = thickness of the conductors [mm], which is usually 0.035 mm for ordinary copper-clad laminates
w = width of the conductors [mm]
ϵ_r = relative dielectric constant of the laminate

Example: Let us again see in a practical case, how much the capacitive coupling can be: Two conductors of 1 mm width are running parallel with 1 mm spacing for a length of 150 mm. From Fig. 2-4, the coupling capacity is taken as 0.4 pF/cm. This gives a total effective capacity of $0.4 \times 15 = 6$ pF.

Since the capacitive coupling is undesirable in normal circuits, ways to minimise it have to be found where it could assume critical values. This is achieved by observing the following rules which can help to reduce the coupling by a factor of 3 to 10.
— Keep critical conductors narrow and provide sufficient spacing between them (Fig. 2-4).
— If necessary, run a ground line between the critical conductors. The broader this ground line, the better the result will be.
— Where such a ground line has been provided, the two signal conductors should run at a very close spacing to it. This is to keep the capacitive coupling to ground high while the coupling between the signal lines at the same time becomes less.

2.4 INDUCTANCE OF PCB CONDUCTORS

In fast-signal or high-speed logic circuits, the conductors are considered as *transmission lines*. The most relevant characteristic of a transmission line is the wave impedance Z_w. If ohmic losses are neglected, the wave impedance in homogeneous materials can be calculated by using the following formula:

$$Z_w = \sqrt{\frac{L'}{C'}} \, [\Omega]$$

L' = inductance per unit length
C' = capacitance per unit length

With fast-pulse and high-frequency circuits, in designing the conductor pattern, the capacitive and inductive couplings are of a major concern. Even in logic circuits operating at a clock-rate of only 10 kHz, high-frequency components of the rectangularly shaped signals can easily cause problems. The conversion of a known risetime to the corresponding 3dB-frequency can be done with the following formula:

$$f_{3dB} = \frac{0.35}{t_r} \text{ [MHz]}$$

f_{3dB} = 3dB limit frequency [MHz]
t_r = risetime 10 − 90% [μsec]

When fast signals or high-speed logic has to be realised on a PCB, it is important to also know the inductance of a conductor arrangement. The calculation of the inductance of conductors is a rather involving procedure. We shall give here only one formula (Contraves Ltd., Switzerland), which can be used for single-sided boards with two conductors of the same width (Fig. 2–3). The respective values for some conductor widths for a parallel conductor length of 10 cm are shown in Fig. 2–5.

$$L' = 0.00921 \times \log_{10}\left(\frac{s + w}{t + w}\right) + 0.006 - 0.004 \times \left(m + \frac{s + w}{10 \times l}\right) \left[\frac{\mu H}{cm}\right]$$

$$L = L' \times l \text{ [μH]}$$

L = inductance [μH]

L' = inductance per unit length $\left[\frac{\mu H}{cm}\right]$

l = parallel running conductor length [cm]
s = distance between the two adjacent conductors [mm]
t = thickness of the conductors [mm], which is usually 0.035 mm for ordinary copper-clad laminates
w = *width of the conductors* [mm]

m is a factor which depends on the ratio of $\left(\frac{w}{w + s}\right)$.

It can be approximated by the following formula:

$$m = 0.0967 \times \left(\frac{w}{w + s}\right)^{2.082}$$

2.5 CONDUCTOR SPACING

When designing conductors in the layout, the minimum spacing requirements for the final artwork should be known. This is important because the pencil line conductors in the layout do not represent the real space conditions as it will be in the artwork.

Spacing considerations based on the *breakthrough voltage* are treated in section 8.4.3 and the specifications of MIL STD 275-B are given in Fig. 8–11.

Only one rule shall be given here: Minimum permissible spacing should be utilised only where there is no way to avoid it; otherwise higher spacing should be given. This is to minimise the reject rate during the PCB production.

LAYOUT, GENERAL RULES AND PARAMETERS

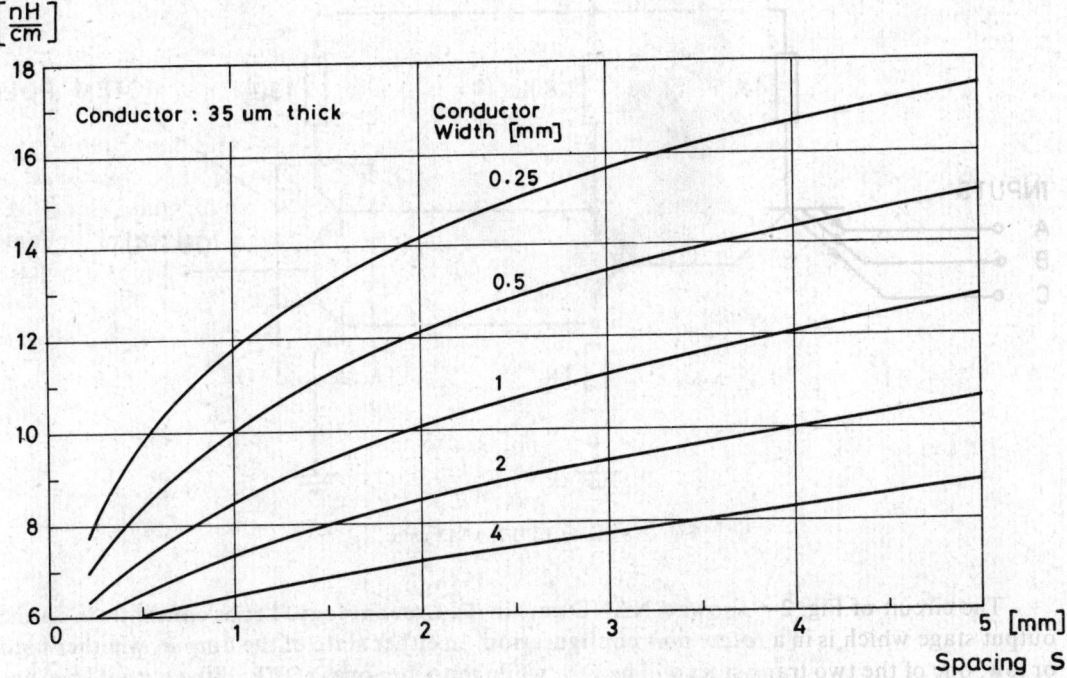

Fig. 2-5 Inductance of parallel running conductors

2.6 REALISING SUPPLY AND GROUND CONDUCTORS

2.6.1 Width

When the supply and ground lines have to be put into the layout, remember that they are not just conductive links. The width of such lines and the pattern in which they are distributed, is of highest importance for stability of the circuit voltages: If the ground potential reference system is unstable, then what other voltage could be stable on such a board?

An unstable supply voltage or ground system can have different reasons. In rare cases only, resistive losses in such conductors is the source. Maximum possible currents on a PCB should always be known to enable the design of conductors with an adequate width. But in most cases, difficulties arise because of the high signal frequencies and fast risetimes in logic circuits.

A special attention must be given to the designing of TTL circuit boards: Short current spikes are introduced when the gates are switching over from one logical state to the other. Since TTL circuits belong to the most common digital circuits in industrial applications, these spikes deserve a closer look in this chapter. But still more details can be found in Chapter 3.

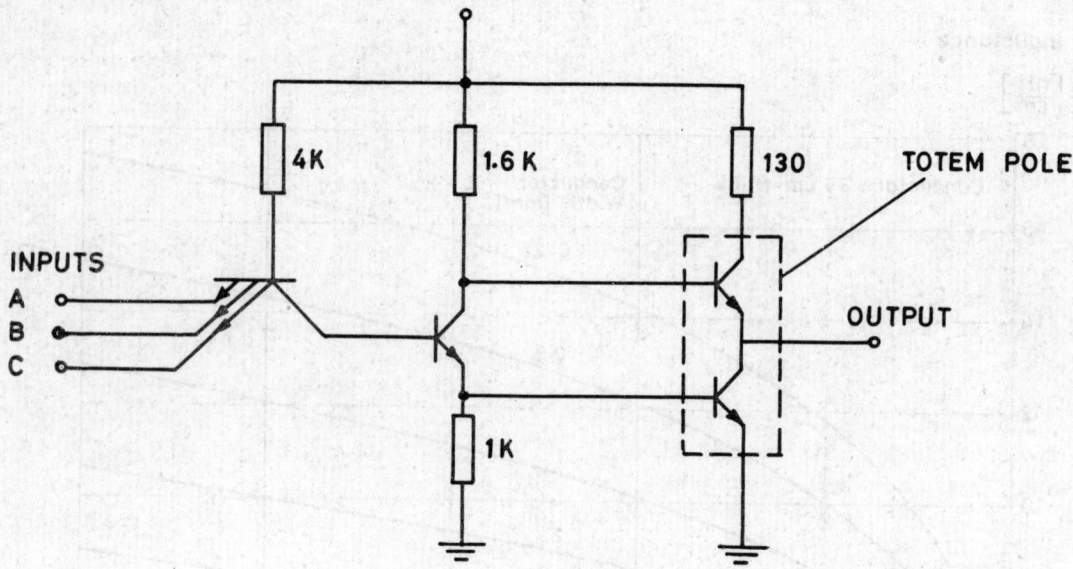

Fig. 2-6 Basic TTL-NAND gate

The circuit of Fig. 2-6 shows a NAND gate in TTL technology. Let us concentrate on the output stage which is in a *totem pole* configuration. In either state of the output, whether high or low, one of the two transistors will be ON, while the other one is OFF. If the output level has to change to the other state, both the transistors will be ON for a short period. This period will typically be of a few nanoseconds duration. During this period, the current in the output stage will practically be limited only by the comparatively small 130Ω resistor. During the change-over to the other state, the power supply must also provide the energy to change the level of all the conductors connected to the gate's output to the new level. This again draws a fairly high current from the source and/or ground because of the capacitive load. Current spikes of roughly 30–60 mA peak current can be expected for changing a 50Ω-line from a TTL output. In a logic system, it can easily happen that a few hundred gates and flip-flops are changing the output state at the same time, as for instance in synchronous counters. In such cases, a heavy current spike will appear on the supply and ground lines. The task of the layout designer is to provide in particular an electrically stable ground line system, in order to avoid unwanted triggering of gates due to a high voltage drop in the ground lines caused by these current spikes.

The fundamental rule for any circuit is:

$$W_{Ground} > W_{Supply} > W_{Signal}$$

W_{Ground} = conductor width of ground line
W_{Supply} = conductor width of supply line
W_{Signal} = conductor width of signal lines

Where several supply voltages are used in a circuit, it must be ensured that the ground line can carry the combined load under worst-case conditions.

LAYOUT, GENERAL RULES AND PARAMETERS

The rule for TTL circuits is:

$$W_{Ground} \geq 2 W_{Supply} \qquad W_{Supply} \geq 2 W_{Signal}$$

In TTL circuits, it is recommended to utilise all the unused board area for ground conductors or a ground plane (see also Chapter 3).

2.6.2 Configuration

In any kind of circuit, a low wave impedance between the supply- and ground- conductors is mostly desirable. Because of this priority, these conductors are usually designed first in the layout. A low wave impedance can be obtained with broad conductors which are capacitively coupled. Fig. 2-7 shows the basic configuration in single- and double-sided PCBs which must be aimed at. It is very clear that these configurations cannot be maintained throughout a layout, especially for single-sided PCBs. In designing a good layout, care should be taken to see that they are implemented to the maximum extent possible.

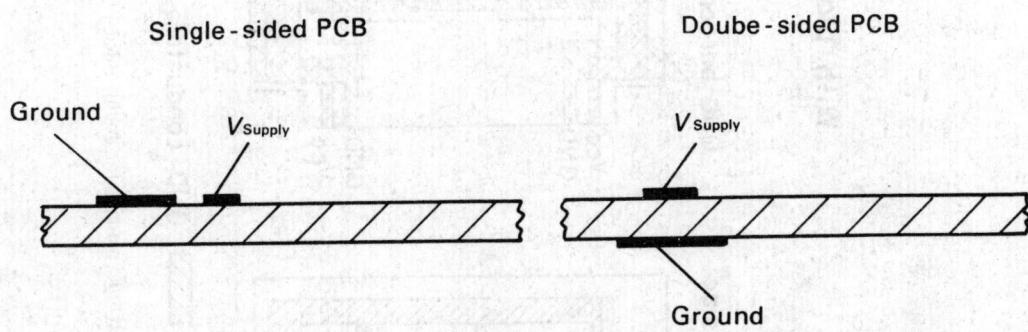

Fig. 2-7 Low-impedance ground/supply conductors

In TTL circuits, the obstructions for running signal lines caused by V_{cc} and GND conductors are quite severe. Fig. 2-8 gives workable solutions for double-sided boards. The signal lines have to be arranged in the X/Y pattern: On one side all the lengthwise interconnections while all the cross direction interconnections on the other side. A solution is shown for both implementations; without plated-through holes as well as with the availability of plated-through holes.

Where analog and digital circuitry have to be accommodated on the same PCB, the complete separation of the corresponding ground conductors is highly essential. This is to avoid any current spikes from the digital part reaching the analog part where they could get amplified and cause errors, oscillations or instability within the whole analog part. The interconnection of the two separate ground conductors can be done on the most stable reference ground point of the system, i.e., at the power supply ground point itself or, if not possible, on the backpanel connector of the PCB. This rule does not only apply for ground conductors but also for supply lines, as long as there are common supply current requirements. Common supply current paths have to be avoided, as they will lead to cross-coupling of disturbances, especially from the digital to the analog circuits.

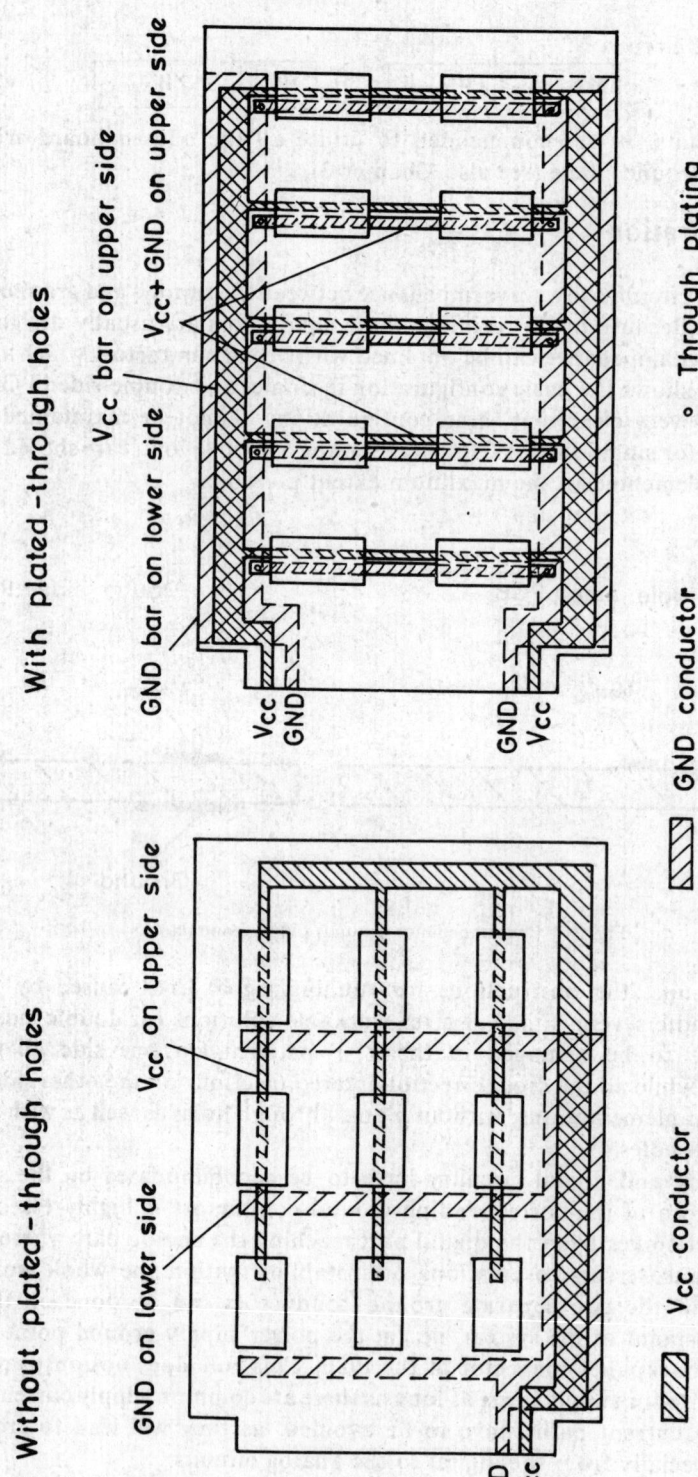

Fig. 2–8 V_{cc}- and GND-line patterns for double-sided PCBs with TTL circuits

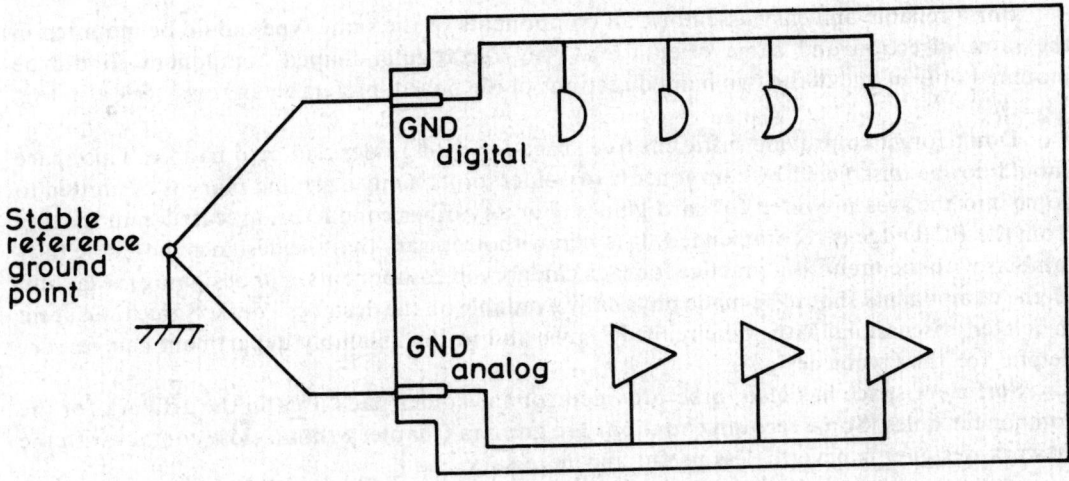

Fig. 2-9 Complete separation of analog and digital ground conductors. The need of separation applies equally also for common supply conductors

2.7 COMPONENT PLACING AND MOUNTING

2.7.1 Component Placing Approaches

The actual location of the components in the layout is responsible for the problems to be faced during routing of the interconnections. It is therefore necessary to analyse the type of circuit and to identify priorities for the particular circuit:

—In a highly sensitive circuit, the critical components are placed first and in such a manner as to require minimum length for the critical conductors.
—In a less critical circuit, the components are arranged exactly in the order of the signal flow: This will result in a minimum overall conductor length.
—In a circuit where a few components have considerably more connecting points than the others (e.g. LSI circuits), these key components have to be placed first and the remaining ones are grouped around them.

From these typical approaches, it becomes evident that a practical solution will always form some kind of a compromise. But the general result to be aimed at is always to get shortest possible interconnections.

2.7.2 Mounting Considerations

The mounting of all components has to conform to accepted practices. Smaller components do not need special provisions: The solder joint provides the mechanical fixation. Bigger and heavier components are adequately secured with clamps or clips and suitable space has to be provided in the layout. A guideline to avoid mechanical overstressing of solder joints is a maximum of 10 g load per solder joint in a board without plated-through holes. The corresponding figure for plated-through hole joints is 25 g.

For a reliable and easy assembly, all components of the same type should be mounted in the same direction and same orientation. Any rectangular-shaped component should be mounted only in one of the two main directions of the board, preferably in rows. (See also Fig. 21-3.)

Don't forget to provide sufficient free space for card guides and card handle. This space should not be obstructed by components or solder joints. Only a ground plane is permitted to come into the area provided for card guides. For any other conductor, at least 5 mm spacing from the PCB edge is recommended. It is herewith necessary that the designer should be fully familiar with the mounting practice for each category of components. For designing the layout, all the components should be made physically available on the designer's desk. Some time spent in related discussions with the circuit designer and in the assembly department can be very helpful for the layout designer.

Sufficient space has also to be provided for the solder pads used in the artwork for the component holes. Some recommendations are given in Chapter 8, but a close contact with the artwork designer is nevertheless useful and necessary.

2.8 COOLING REQUIREMENTS AND PACKAGE DENSITY

2.8.1 Heatsinks

Where heatsinks have to be provided, sufficient free space around them will improve their efficiency. In any case, a free air flow via the heatsink must be achieved and no bulky component nearby should obstruct the air flow. The layout designer should know the exact position the board is to be mounted in the equipment (horizontally, vertically), where in the cabinet, and how the cooling and air flow directions have been planned. These are all questions to be answered by the equipment designer. In a vertically mounted PCB two heatsinks should not be designed one above the other.

Where heatsinks with unidirectional slots are applied, the air flow must always pass the heatsink in the same direction as the slots are made to ensure maximum exchange of heat.

2.8.2 Package Density

A high package density on a PCB looks very attractive and gives the impression of an optimum design. While such attributes may be true in many cases, the high package density can also be less reliable, costly and difficult for servicing of the equipment.

The optimum package density is linked with many constraints, such like
—purpose and use of the equipment: whether airborne, portable or permanent (fixed) installation,
—heat produced and cooling provisions: whether hermetically sealed, natural air flow, or forced cooling,
—type of components on the board: silicon devices have a shorter life, the higher the temperature,
—type of PCB used (interconnection density): whether single-sided, double-sided, or multilayered.

—component technology: whether discrete semiconductor, SSI, LSI
—and many others.

From this, it is evident that there is no simple formula to determine the optimum package density. But let us look at the extremes: If the density is very low, more PCBs will be used to realise the same circuit. The result will also be more volume of the equipment, more PCB connectors and backpanel wiring with more parasitic influences on the working of the circuit; briefly, the reliability comes down while the price goes up. An unnecessarily high package density will give a higher reject rate in the PCB production, higher circuit temperatures, probably more crosstalk, difficult servicing etc., which again brings finally the reliability down and makes the costs higher.

2.9 LAYOUT CHECK

After the layout design is completed, a thorough check is carried out to ensure that all requirements have been considered suitably. Such a check must be done thoroughly; any impatience here will definitely fall back onto the designer.

To make the checking easier, different check lists are given here to cover the most important aspects. But it is impossible to include all constraints in such lists since the requirements in a PCB design are always specific. The author also suggests to wait a certain time after the completion of the layout before starting the final check. This is to gain some distance to the work done. Only after such a change, we should start to analyse, criticise, check and give last improvements to the layout.

2.9.1 General Considerations

—Are the components regularly distributed to give a uniform package density over the entire board?
—Are the components accessible for an easy replacement?
—Is there some free space left for possible circuit extensions or modification?
—Have test points been provided?
—After insertion of the card into the equipment, will there be sufficient clearance around?
—Where necessary, has heatsinking been provided and does a free air flow pass the heatsinks?
—Are heat-sensitive components kept at adequate distance from heat-producing components?
—Is the orientation unisequivocally defined for all components?
—Is the number of jumper wires minimum?
—Do the outermost conductors still have enough distance from the edge of the board?
—Is the access to adjustable components (trimmers, presets, etc.) possible from the correct side?
—Can test equipment easily be connected to the board (e.g., clipping-on of logic probe)?

2.9.2 Mechanical Considerations

—Is the board size optimum and compatible with the PCB manufacturing process?
—Have registration and board mounting holes been provided?
—Are heavy components adequately fixed?

—Has the number of different hole diameters been restricted to the minimum?
—Has the proper hole diameter been provided for component mounting?
—Can the assembled board withstand the mechanical stresses and vibrations occurring in transport and normal use?

2.9.3 Electrical Considerations

—Is the signal flow smooth, with short interconnections?
—Has the circuit been divided into functional subunits on the board?
—Is there a 100% compatibility between circuit diagram and layout?
—Critical signal paths: Have optimum precautions been taken such as minimum length, guard lines, clear separation of input and output lines?
—Has adequate conductor spacing been provided?
—Has a close coupling between supply and ground lines been realised (low wave impedance)?
—Can the supply line conductors withstand a short-circuit on the board?
—Is there sufficient ground line width?
—Have the analog and digital circuit parts independent ground lines?

BIBLIOGRAPHY

1. COOMBS, C. F.: *PRINTED CIRCUITS HANDBOOK*, McGraw-Hill Book Co., New York, 1967.
2. LUND, P.: *GENERATION OF PRECISION ARTWORK FOR PRINTED CIRCUIT BOARDS*, John Wiley & Sons, Chichester, 1978.
3. *PRINTED WIRING DESIGN GUIDE*, The Institute for Interconnecting and Packaging Electronic Circuits, Evanston, Illinois.
4. SCARLETT, J. A.: *PRINTED CIRCUIT BOARDS FOR MICROELECTRONICS*, Van Nostrand Reinhold Co., London, 1970.

3
DESIGN RULES FOR DIGITAL CIRCUIT PCB'S

The four main problems that can affect digital PCBs if not properly designed are:

a) Reflections (causing signal delays and even double-pulsing, i.e., conversion of one pulse into two or more pulses)
b) Crosstalk (interference between neighbouring signal lines)
c) Ground- and supply-line noise
d) Electromagnetic (E.M.) interference from pulse-type E.M. fields (in high-noise situations).

Most of these problems can be fully overcome by using properly designed printed circuit layouts.

3.1 REFLECTIONS

In families of fast rise-time ICs, the conductors cannot be considered as short-circuits, but must be looked at as pieces of transmission lines. These transmission lines are normally mismatched on both sides, so that multiple reflections take place, based on the particular value of wave impedance Z_w of the conductors.

The main point to consider while designing the PCB conductors (and other interconnecting wires) is that they should have the *proper value of wave impedance Z_w*, viz., that value which gives the least reflection problems, for the particular kind of digital integrated circuits (ICs) used. The desired value of wave impedance Z_w will be obtained by choosing the *width* of the signal lines (i.e., the width of the PCB conductors connecting outputs and inputs of the digital ICs) accordingly as well as the distance between signal lines and ground line. Relatively large values of Z_w (such as needed for TTL and, especially, for C-MOS ICs) call for *thin signal conductors*; lower values of Z_w (such as needed for ECL-ICs) call for broader signal conductors. As reflection can cause double-pulsing (i.e., conversion of one pulse into two pulses), they should be kept small, even if we are dealing with a digital circuit having a *low* operating frequency, where delays are not so important (but double-pulsing still cannot be tolerated).

3.1.1 TTL Integrated Circuits

Fig. 3-1 shows the basic situation where 2 gates are connected over a signal line with wave impedance Z_w. Fig. 3-2 shows wave impedance values plotted for a typical PCB situation. Fig. 3-3 shows the reflections for $Z_w = 50$, $Z_w = 100$ and $Z_w = 150\,\Omega$, for the rising edge, in the case of TTL-ICs, type SN 7400.

Fig. 3-4 shows, for the same 3-wave impedances, what happens to the negative-going edge of a TTL waveform. The reflection chart according to BERGERON, which is used to determine graphically the reflections, is also given in Fig. 3-3 and Fig. 3-4 (see [4], [5], [6], [11] for an explanation of this chart). One may note that for TTL a wave impedance of 100 to 150 Ω is desirable. This can easily be obtained on the PCBs by making *the signal lines approximately 0.5*

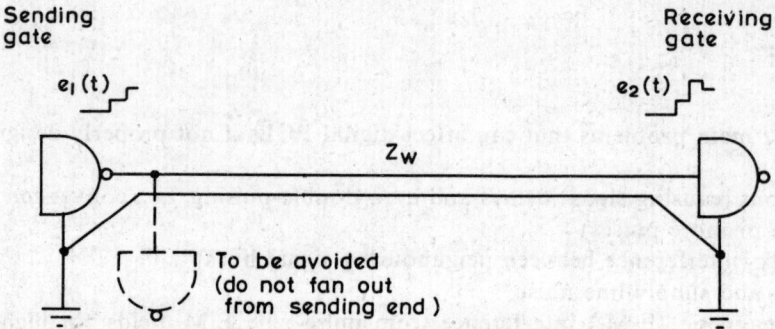

Fig. 3-1 TTL-line: Sending gate driving a receiving gate over a transmission line

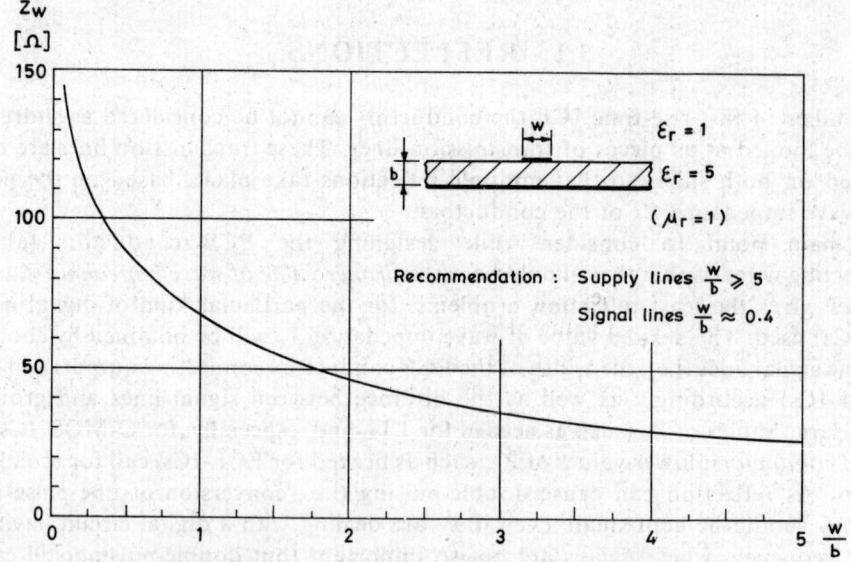

Fig. 3-2 Wave impedance Z_w for PCB conductors [12]

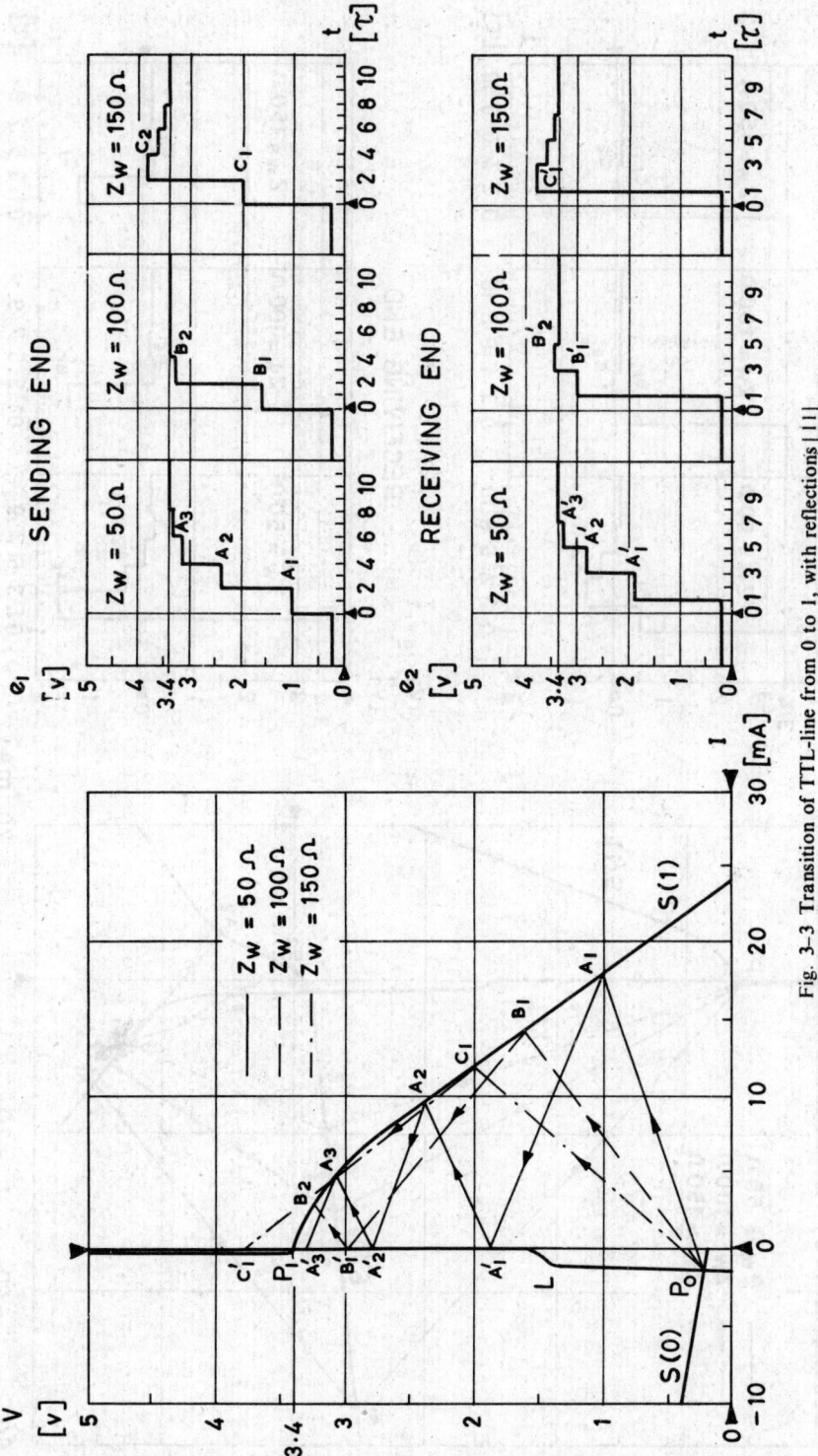

Fig. 3-3 Transition of TTL-line from 0 to 1, with reflections |11|

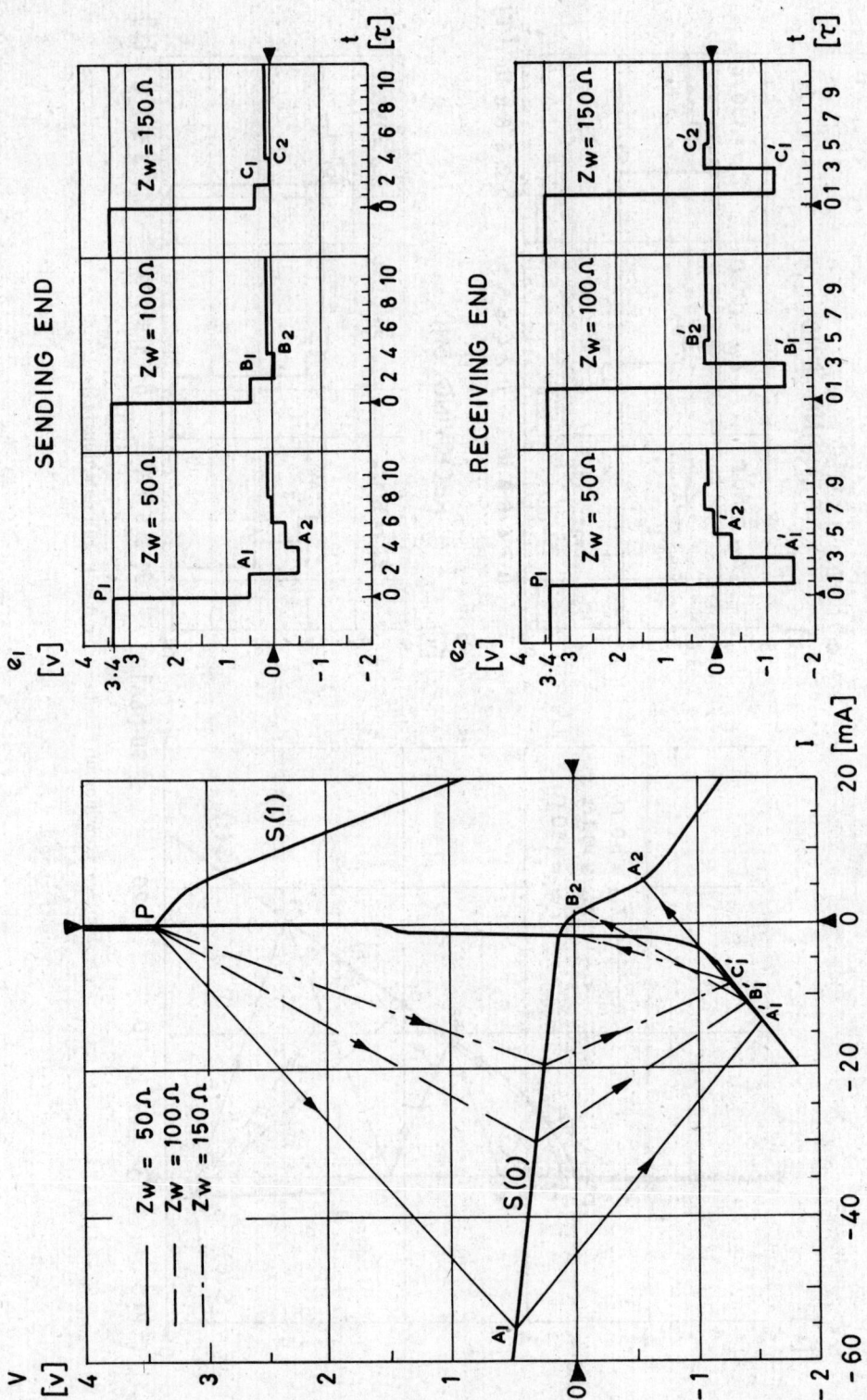

Fig. 3-4 Transition of TTL-line from 1 to 0, with reflections [11]

mm wide (if they are sitting just on the other side of a ground plate and if the PCB-dielectric which separates them is 1.6 mm thick, which is a standard thickness for PCBs; if ground lines are further away, one may have conductor widths for signal lines up to *about 1 mm*).

The use of very broad conductors will give wave impedances of 50 Ω and less. As one can see from Fig. 3-3, the rising edge of a pulse will be often delayed considerably, there will be voltage steps A_1, A'_1 which are situated in the transition zone between a logical 0 and a logical 1 and which can cause dangerous double-pulsing if a bit of additional noise is present. The falling edge, in the case of low impedance lines (50 Ω and less), will possess a negative spike (A_2 and A'_1 in Fig. 3-4) which can, in a *worst-case* situation (the figures show only a *typical* situation), be as large as a few volts of negative voltage and thus harm the IC. Furthermore, currents drawn from or sunk-back into the IC by low-impedance lines will be very large; 20 mA and –55 mA for 50 Ω lines for the rising and falling edges, respectively. These current spikes have to be supplied by the V_{cc} line (rising edge) or fed to the ground-line (falling edge) and will thus cause supply-line and ground-line noise, respectively. We can sum up this all by saying that in TTL-PCBs broad, low-impedance conductors of 50 Ω or less must be avoided. Very high impedance lines ($\geqslant$ 200 Ω wave impedance) will also cause trouble: harmful overvoltages (higher than C_2 and C'_1 on the rising edge, see Fig. 3-3) and double-pulsing (a short pulse higher than point C'_2 on the falling edge, see Fig. 3-4). Actually, very high impedance lines are not commonly obtained with usual PCB layouts; they would, however, occur with open wiring (loose wires, signal wires not twisted with ground line), as might be used between the PCBs. So, one should remember that for TTL-ICs, loose wiring must be avoided and signal line connections *between* PCBs should always run near the ground line and preferably be twisted with the latter.

3.1.2 C-MOS Integrated Circuits

Here the corresponding reflection charts and voltage waveforms are given in Figs. 3-5 and 3-6 for the case of 100 Ω transmission lines, i.e. moderately broad PCB conductors and V_{cc} = 10 V. One can see that considerable delay is involved, especially for the rising edge of the pulse. During the intermediate steps A_3, A_4, A_5 ... (Fig. 3-5), the input voltage of the receiving gate is just within the transition region, and the noise margin gets considerably reduced. For 50 Ω or less, i.e., for broader lines, the situation is even much worse. *So avoid broad PCB conductors for the signal lines of C-MOS circuits*. In fact wave impedance should be kept high and *conductor widths of signal lines as low as possible, i.e., preferably signal lines with a width of 0.5 mm or less should be used*. One can easily guess that with wave impedances of 150 to 300 Ω the transitions will be considerably faster. (This can be easily seen if the corresponding lines with a slope corresponding to Z_w are drawn in the BERGERON-reflection charts of Figs. 3-5 and 3-6.) Such very high wave impedances can best be obtained, if signal conductor widths are kept low, and also if full ground plate is not used (for C-MOS!) and the ground lines are not *too* broad and not kept *too* near the signal lines. Luckily this is generally possible and done in the case of C-MOS PCBs, because C-MOS is not so critical with regards to crosstalk and ground- and supply-line noise and, therefore, does not require so broad ground and supply conductors (see Sections 3.2 and 3.3). In the wiring between PCBs, one should definitely avoid using 50 Ω cables, and also for the wave impedances Z_w to remain high, the ground should not run too near the signal lines.

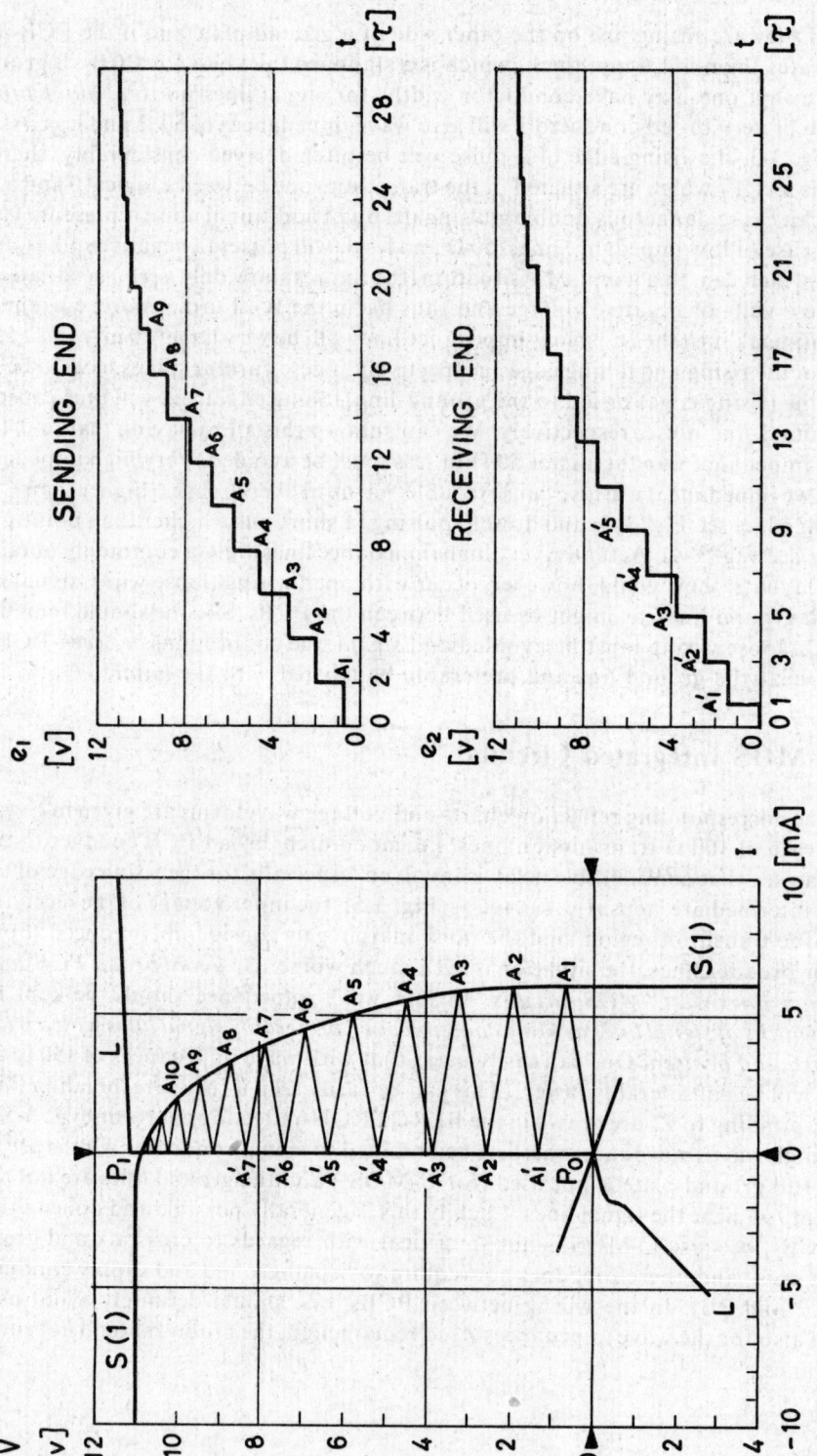

Fig. 3-5 Slow transition of C-MOS line from 0 to 1, with many reflections

DESIGN RULES FOR DIGITAL CIRCUIT PCB'S 39

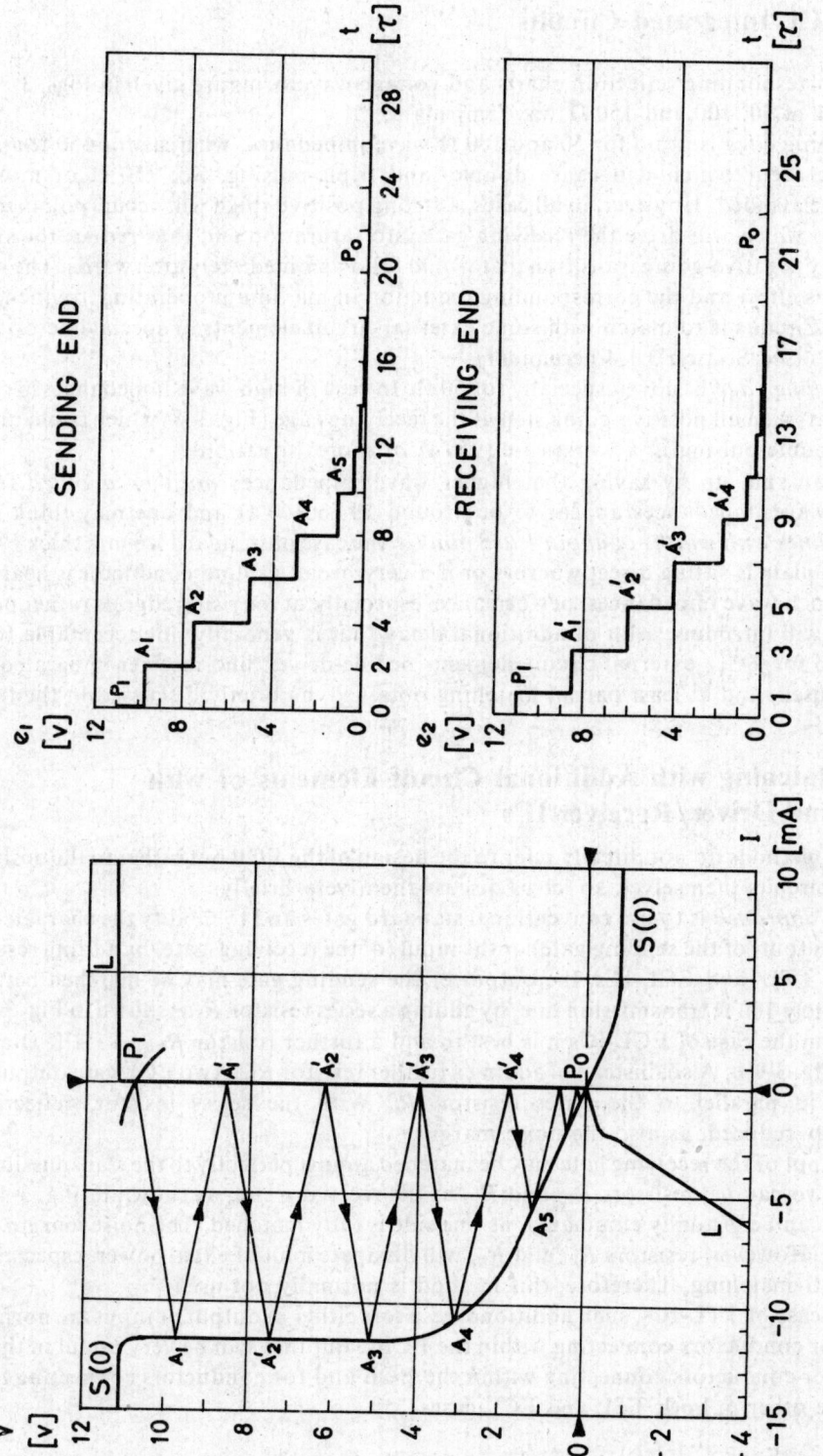

Fig. 3-6 Transition of C-MOS line from 1 to 0, with reflections

3.1.3 ECL Integrated Circuits

The corresponding reflection charts and voltage waveforms are given in Figs. 3-7 and 3-8 for the case of 50, 100 and 150 Ω wave impedance.

The *rising edge* is rapid for 50 and 100 Ω wave impedance, whereas for 150 Ω and more, oscillations occur which will cause double- and triple-pulsing. So, 150 Ω or more, must definitely be avoided. However, in all cases, a strong positive spike will occur (points A_2, B_2, C_2 in Fig. 3-7) which will drive the receiving gate into saturation and thus reduce the switching speed of any negative-going transition that would follow immediately afterwards. The only way to avoid this effect and the corresponding reduction in maximum operating frequency of the Integrated Circuits is to match with some external circuit elements or special line-driver/line-receiver ICs (see Section 3.1.4 hereunder).

The *falling edge* is slow, especially for 50 Ω. In case of high wave impedances ($\geqslant$ 150 Ω), there appears a small positive-going step at the receiving edge (Fig. 3-8) which could ultimately result in double-pulsing in a worst-case (250 Ω or more) situation.

One can sum up by saying that higher wave impedances *must be avoided* for ECL. Optimum wave impedances appear to be around 50 to 100 Ω and one may think of *using conductor lines with widths of about 1 to 3 mm* (in the case of standard 1.6 mm thick PCBs and if a ground plate is sitting directly across or if a very broad ground conductor is nearby). But even with such wave impedances, performance, especially at the rising edge, is rather poor, and reflections will introduce a lot of additional delay that is generally not acceptable for ECL-systems. So for ECL, external circuit elements or line-driver/line-receiver integrated circuits should be used, and at least partial matching obtained, as briefly discussed in the following paragraph.

3.1.4 Matching with Additional Circuit Elements or with Line-Driver/Receiver ICs

These methods do not directly refer to the design of the PCB but rather to the design of the electronic circuits themselves, so let us discuss them very briefly.

A first approach is to use conventional standard gates and to modify the characteristics of either the output (of the sending gate) or the input (of the receiving gate) by adding resistors. In the case of TTL- and ECL-ICs, the output of the sending gate may be matched better to an approximately 100 Ω transmission line, by adding a series resistor R_s as shown in Fig. 3-9 A and B. In fact, in the case of ECL-ICs it is best to add a further resistor R_{EE} of 1.2 KΩ as shown dotted in Fig. 3-9 B. Also, instead of adding a further resistor R_{EE}, two ECL gate outputs can be connected in parallel to the series resistor R_s. With the series resistor, reflections are considerably reduced, as also the noise margin.

The input of the receiving gate may be matched almost perfectly to the transmission line by including two parallel resistors R_{P1} and R_{P2} at the receiving end, as shown in Fig. 3-10. Here, reflections can be virtually eliminated, as one side is fully matched. The noise margin will also be reduced. However, resistors R_{P1} and R_{P2} will dissipate a lot of extra power, especially in the case of 50 Ω-matching. Therefore, this method is normally not used.

In the case of TTL-ICs, such additional resistors either at output or input are normally not required for conductors connecting within the PCBs. But they can be very useful in the case of ECL-ICs for conductors connecting within the PCB and for conductors connecting from one PCB to the other in both TTL and ECL cases.

DESIGN RULES FOR DIGITAL CIRCUIT PCB'S

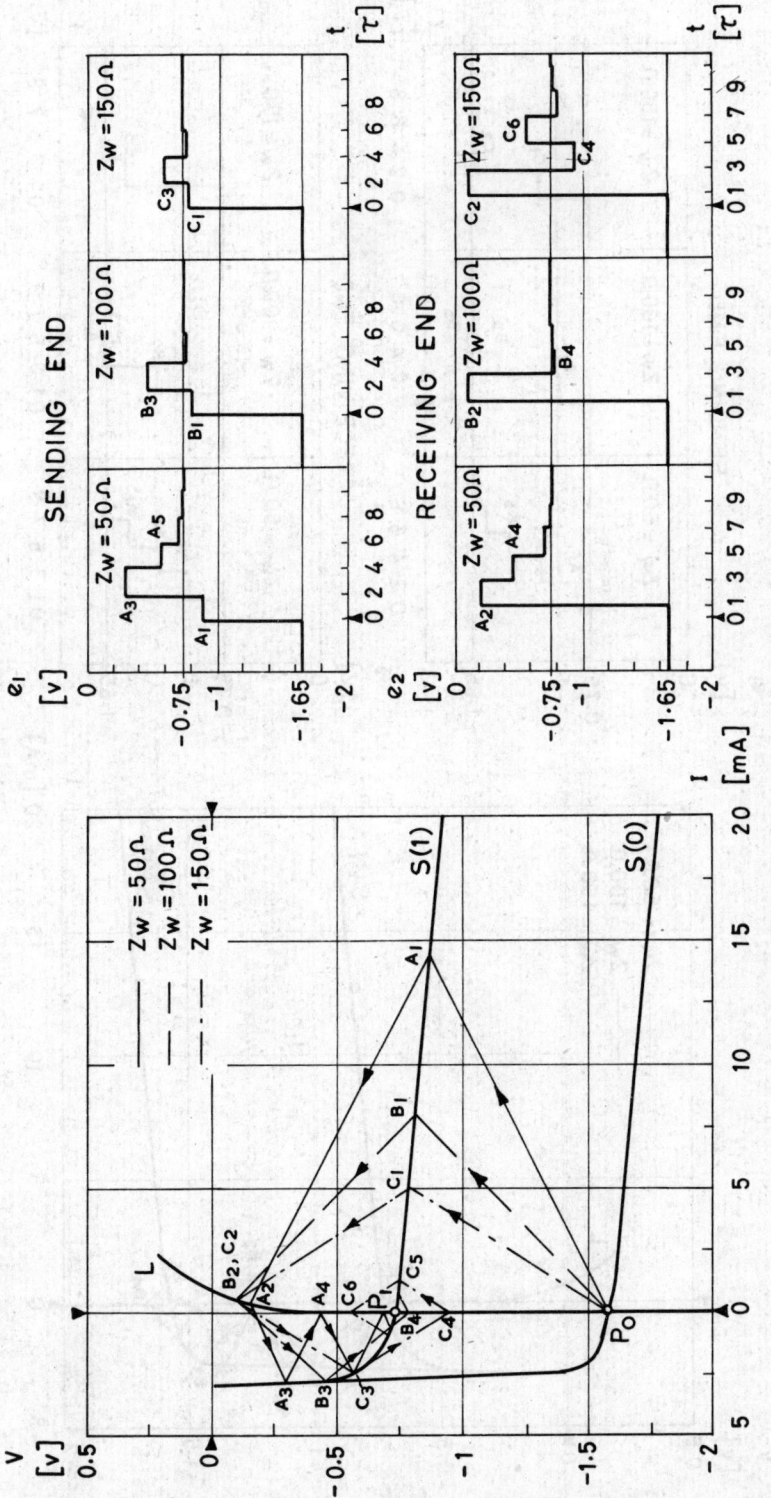

Fig. 3-7 Transition of ECL-line from 0 to 1, with reflections [11]

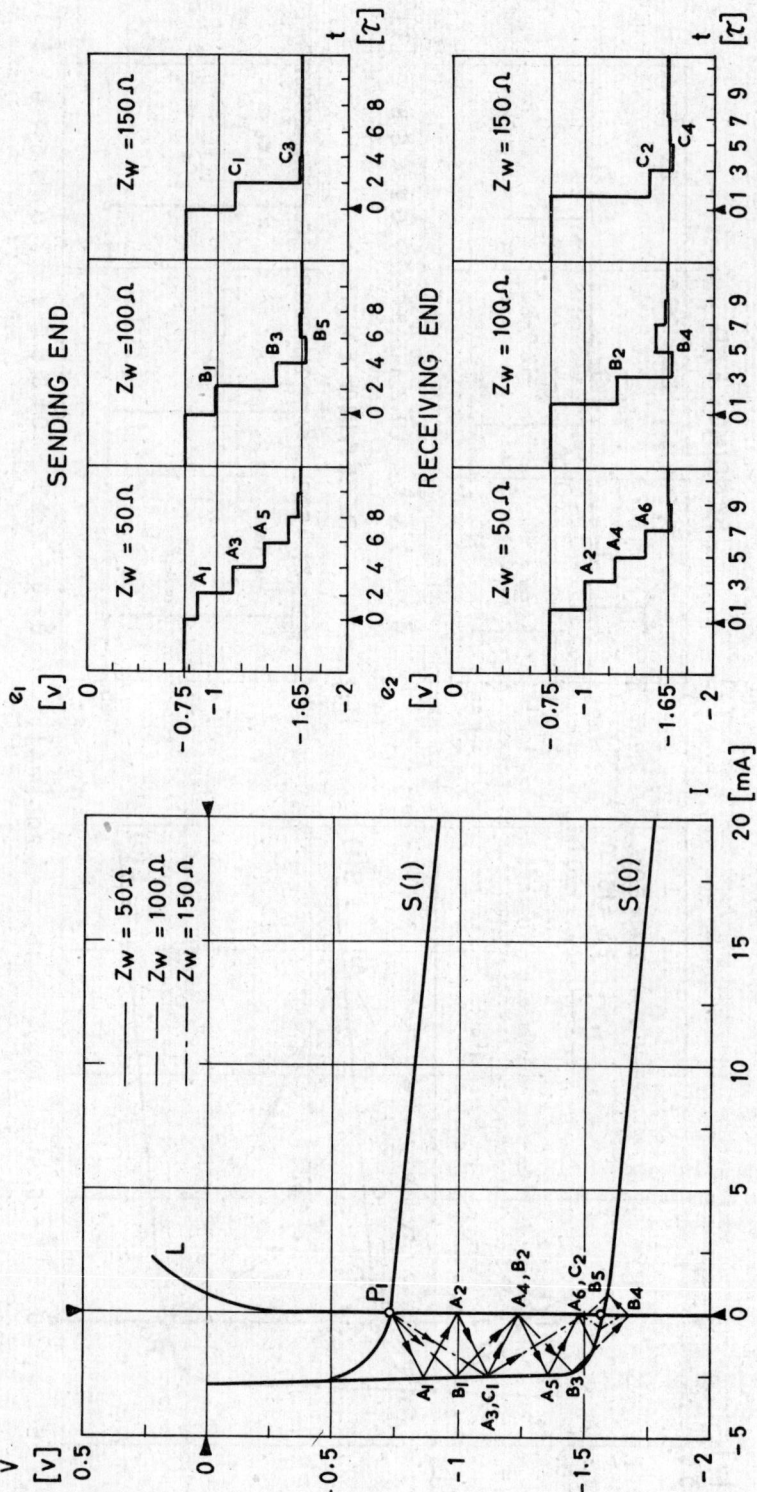

Fig. 3-8 Transition of ECL-line from 1 to 0, with reflections [11]

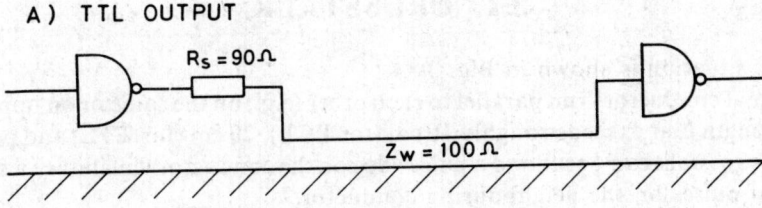

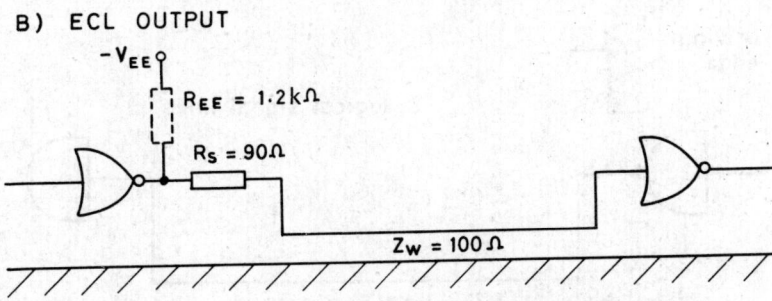

Fig. 3-9 Matching with a series resistor at the sending end:

A) TTL output can be matched for the transition from 1 to 0 (only) by putting a series resistor of value $R_s = Z_w - 10\ \Omega$
B) ECL output can almost be matched for both transitions, by putting a series resistor of value $R_s = Z_w - 10\ \Omega$ and a 'pull-up' resistor R_{EE}

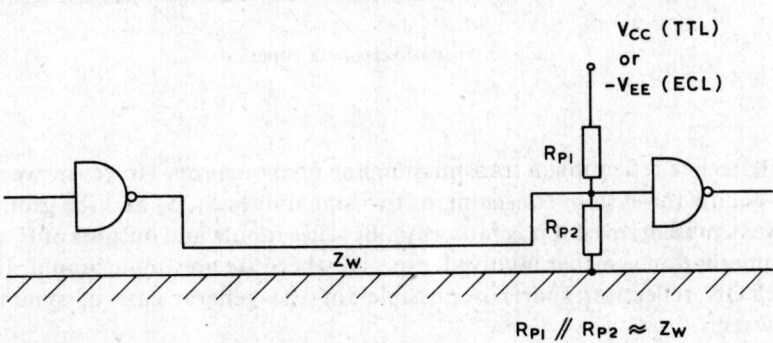

Fig. 3-10 Matching with parallel resistors at the receiving end:
The parallel combination of R_{P1} and R_{P2} should be roughly equivalent to Z_w as the input of the gate has usually a much higher impedance. Note that power dissipation becomes truly excessive (over 100 mW) for $Z_w = 50\ \Omega$

The second approach consists of using special line-driver and line-receiver ICs. These are normally used only in connection with long cables, either with shielded 50 Ω lines or, what is cheaper and almost equally good, with twisted pairs of signal lines operating as differential lines. Both kinds of cables, i.e., the shielded cable as well as the twisted differentially driven cable will have the advantage of being quite immune to external Electromagnetic (E.M.) interference and to crosstalk problems.

3.2 CROSSTALK

The basic situation is shown in Fig. 3-11.

If two signal conductors run parallel to each other (e.g., on the same or on opposite sides of a PCB) for a length that exceeds roughly 10 cm (for ECL), 20 cm (for TTL) and perhaps 50 cm (for C-MOS), crosstalk can occur and a pulse edge on the conductor will induce a short spike or even a train of pulses on the neighbouring conductor.

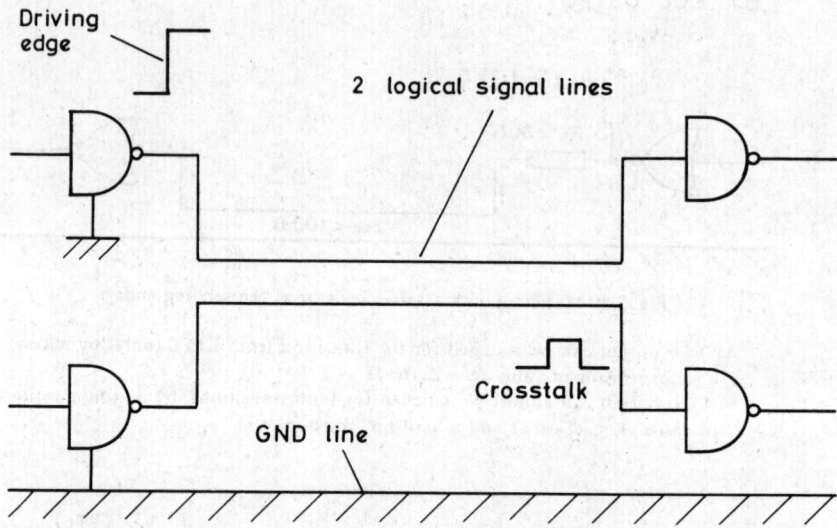

Fig. 3-11 Basic crosstalk situation

Crosstalk is, like reflection, a transmission line phenomenon. However, we are faced here with a three-conductor-system consisting of two signal lines S_1, S_2 and the ground line GND, (which is always present, even if it is far away). Because inputs and outputs of IC gates are non-linear, the computation is rather involved, especially because no simple graphical method (like the BERGERON reflection chart) is possible for the general case of symmetrical three-conductor-systems.

The considerations that follow show that crosstalk can be kept low and uncritical, if a ground line or a ground plate is nearby. A ground line that runs between two signal lines eliminates almost all crosstalk and should be used wherever the situation becomes too critical. A specially problematic situation occurs when signal lines run next to each other with a logical flow (signal flow) on *opposite* directions (as shown in Fig. 3-14 B): Here it will be generally advisable to run a ground line between the two signal lines, specially for the more critical cases of TTL- and ECL-ICs (C-MOS-ICs are much less critical with respect to crosstalk). Actually we may note that *two* wave impedances describe the crosstalk behaviour.

In fact, the basic parameters describing the geometry of the transmission lines (i.e.,

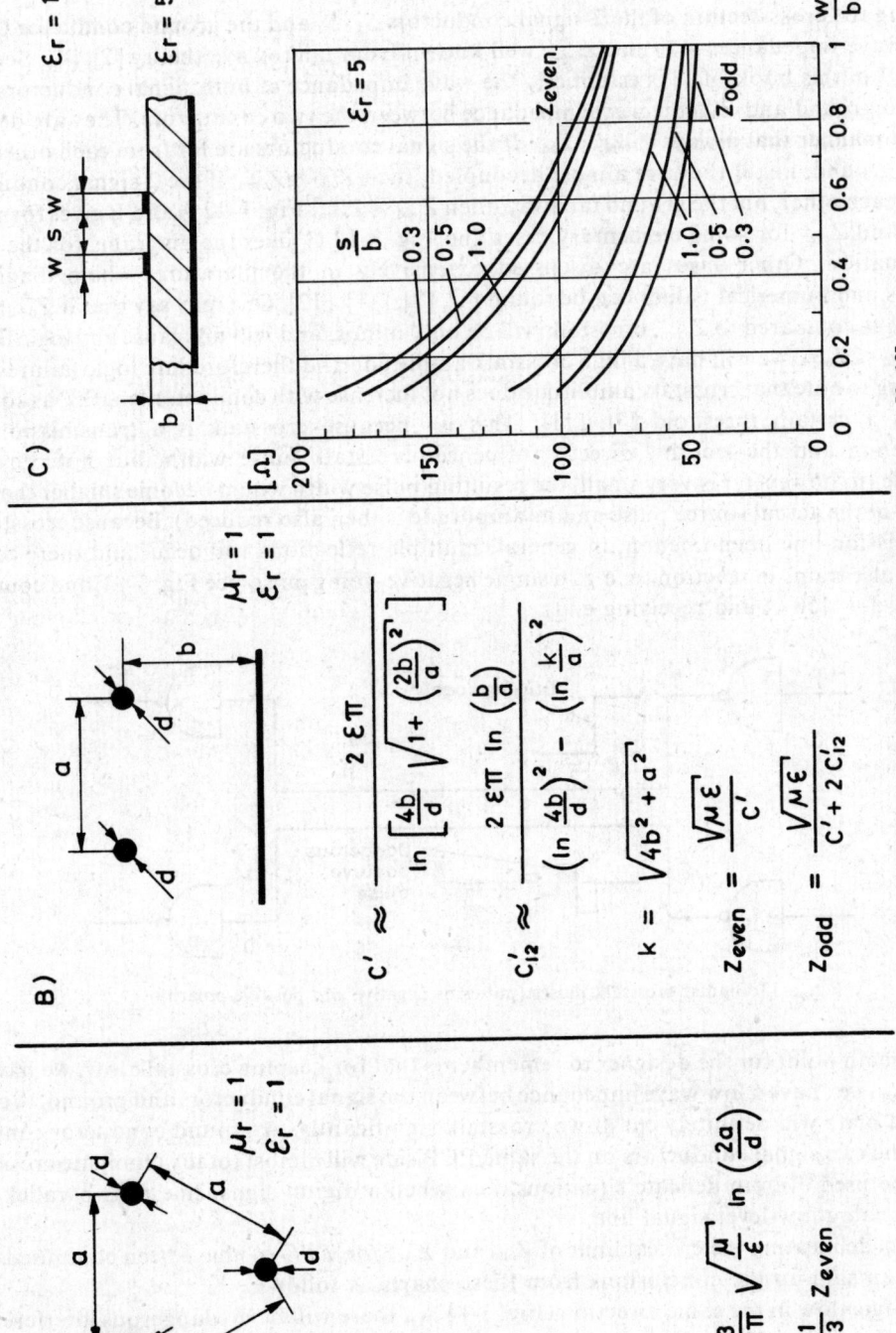

Fig. 3–12 Z_{odd}- and Z_{even}-values for various 3-conductor-systems [3, 12]

describing the cross-section of the 2-signal conductors S_1, S_2 and the ground conductor GND) are the wave impedances Z_{odd} and Z_{even}, well known from microwave theory [2], [4]. See also Chapter 7 of this book. Z_{even} is essentially the wave impedance of both signal conductors with respect to ground and Z_{odd} the wave impedance between the two conductors. They are defined in such a manner that always $Z_{even} \geqslant Z_{odd}$. If the signal conductors are far from each other, but near the ground, i.e., if they are almost decoupled, then $Z_{odd} \approx Z_{even}$. If the 2-signal conductors are near each other, but the ground far away, then $Z_{odd} \ll Z_{even}$. Fig. 3-12 A and B gives formulae for Z_{odd} and Z_{even} for some elementary cases and Fig. 3-12 C gives the diagrams for the basic PCB situation. Other cases are discussed extensively in the literature, where diagrams, equations and numerical values can be found [1], [3], [11], [12]. One may say that if Z_{odd} is not too small as compared to Z_{even}, crosstalk will be only minor, and will not cause any logic faults. But if $Z_{odd} \ll Z_{even}$, we will have a high crosstalk amplitude and therefore sure logic failures. It is interesting to note that crosstalk amplitude does not increase with conductor length l, as soon as we cross a certain threshold [3], [11]. This is because crosstalk is a transmission-line phenomenon and the length l directly influences crosstalk pulse width, but not crosstalk amplitude (if, however, l is very small, the resulting pulse width would become smaller than the rise-time of the actual source pulse and its amplitude is then also reduced). Because crosstalk is a transmission-line phenomenon, in general, multiple reflections will occur and there can be bipolar pulse trains in reaction to, e.g., a single negative-going pulse (see Fig. 3-13; and compare with Fig. 3-7, 150 Ω and receiving end).

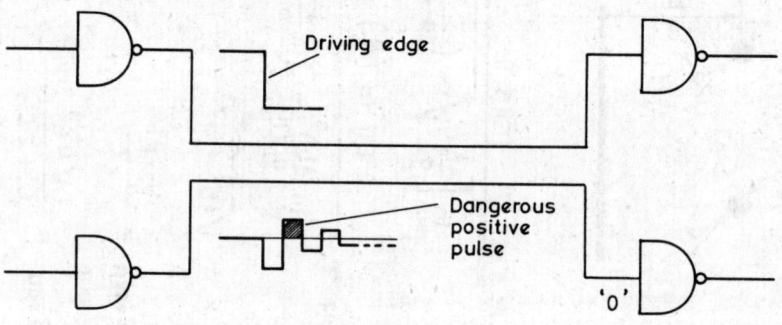

Fig. 3-13 Bipolar crosstalk pulses (pulses of negative *and* positive polarity)

The main point for the designer to remember is that for keeping crosstalk low, we have to reduce Z_{even}, i.e., have a low wave impedance between the signal conductors and ground. A close by ground plate will definitely cut down crosstalk significantly. A ground conductor running *between* the two signal conductors on the same PCB side will almost totally eliminate crosstalk and can be used even in delicate situations, i.e., when a digital signal line runs parallel to a sensitive analog low-level signal line.

Charts delineating the critical limit of Z_{odd} and Z_{even} *for TTL-ICs* have been computed [11] and one can sum up the conclusions from these charts as follows:

For logic flow in the same direction (Fig. 3-14 A), there will be no dangerous interference, as long as $Z_{odd} \geqslant 0.5\ Z_{even}$. However, if the logic flow on the two neighbouring conductors is in opposite directions (Fig. 3-14 B), the situation is much more dangerous and there will be crosstalk, unless $Z_{odd} \geqslant 0.8\ Z_{even}$. The reason for this is that a low-impedance output is sitting, in this case, directly across a high-impedance input and a lot of crosstalk is induced at that end. This reasoning applies to other IC families also, like ECL and C-MOS, so that one should generally avoid running two lines with opposite direction of logic signal flow, close to each other.

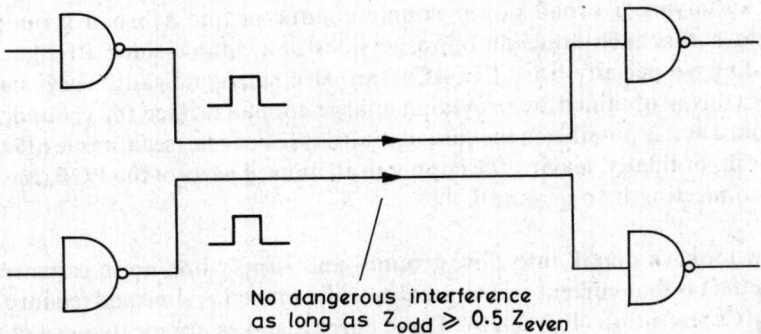

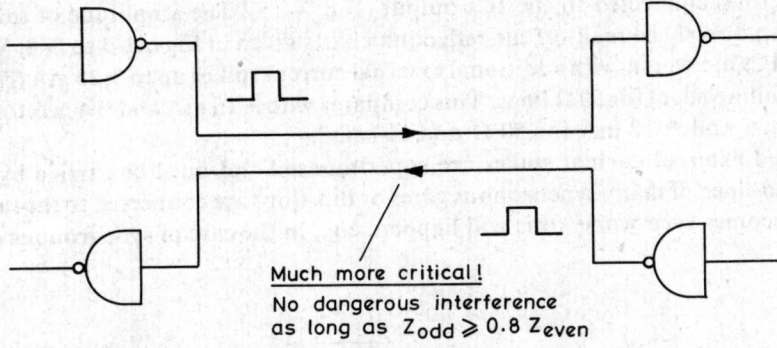

Fig. 3-14 Basic crosstalk configurations (and critical Z-values for TTL)

If one has to run two neighbouring parallel lines with logic signal flow in opposing directions, it will be wise to place a ground line in between, especially if the length of the parallel-running lines is ≥ 20 cm and if one is dealing with TTL- or ECL-ICs.

In the case of C-MOS-ICs, crosstalk is much less dangerous, because of the higher noise immunity of this family (both with respect to amplitude and also with respect to minimal pulse width that will cause a logic fault).

ECL-ICs, on the other hand, have lower noise immunity (in both respects) and will be more sensitive to crosstalk than TTL-ICs.

3.3 GROUND- AND SUPPLY-LINE NOISE

This is perhaps the most serious problem with TTL-ICs, and is present, in reduced form, with ECL-ICs and, in very much reduced form, with C-MOS-ICs. This problem can definitely be solved with proper PCB layout and with a few, very simple measures.

The two principal measures are (as we will shortly see):
1. To have a low impedance Z_{wv} between power supply-line and ground-line (this is obtained by having a broad power supply conductor and a broad ground conductor sitting right across each other on opposite sides of a double-sided PCB);
2. By providing (especially for TTL-ICs) an electromagnetically very stable ground conductor (this is obtained by providing a large copper surface for ground, either a full ground board as it is possible in the case of multilayer boards, (see Chapter 19), or at least a ground mesh, or finally, leaving the copper in all unused parts of the PCB, such as corners, etc., and connecting it to ground).

Let us now look, in detail, into how ground- and supply-line noise is generated.

The main effect is that current spikes are drawn from the V_{cc}-line and fed into the ground-line during the ICs transition (Fig. 3–15). These current spikes are partly needed to charge or discharge the transistor within the IC while they switch. We may refer to this current as the internal current spike. In TTL, the internal current spike, mainly caused by the Totem-pole output can be as high as 20 mA in amplitude and is 5 nsec in duration. Furthermore, we have an additional external current spike which is required to charge or discharge the transmission lines (i.e. the conductors) connected to the ICs output (Fig. 3–15). The amplitude of this external current spike can directly be read-off the reflection charts given in Figs. 3–3 to 3–8. We can see here that TTL-ICs are worst, with additional external current spikes up to + 25 mA (rising edge) and − 60 mA (falling edge) for 50 Ω lines. This compares with + 15 mA and − 4 mA for 50 Ω and ECL and + 6 mA and − 12 mA for 50 Ω and C-MOS.

Internal and external current spikes are superimposed and must be carried by the same V_{cc}- and ground-lines. If many synchronous gates or flip-flops are connected to the same point, the situation becomes even worse (this will happen, e.g., in the case of synchronous counters).

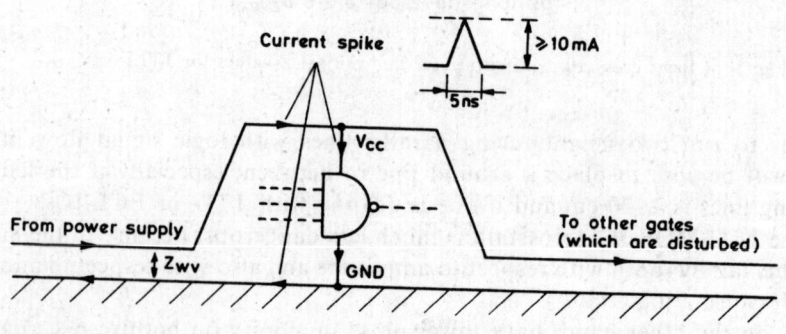

Fig. 3–15 V_{cc} and ground spikes

Remedial Measures

a) To provide the well-known decoupling capacitors (preferably ceramic chips, but never electrolytic capacitors or any wound-type of capacitors). For every 2 to 3 ICs, we may provide about 10 nF in the case of TTL and about 5 nF for C-MOS and ECL. However, decoupling capacitors will only stabilize the voltage difference between V_{CC} and ground and not remove ground noise in any way. They are therefore mainly of use in those transitions, where, as in Fig. 3-16 A, current is *drawn* from V_{CC}/V_{EE} (e.g., $0 \rightarrow 1$ output transitions in TTL and C-MOS and $1 \rightarrow 0$ output transition in ECL). They are not so helpful in the other transitions where current is discharged from the signal line into the ground line, with lesser effect on the V_{CC}/V_{EE} line (Fig. 3-16 B).

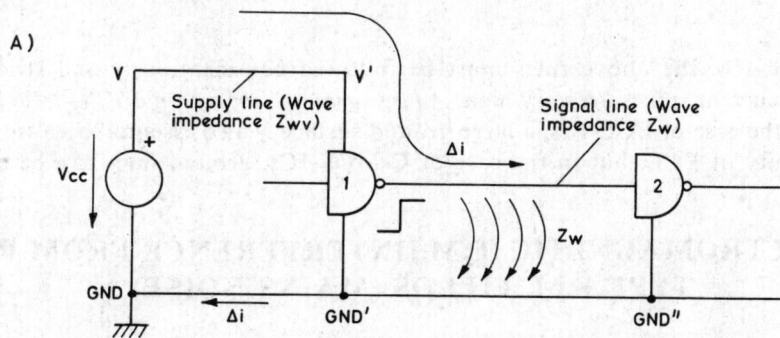

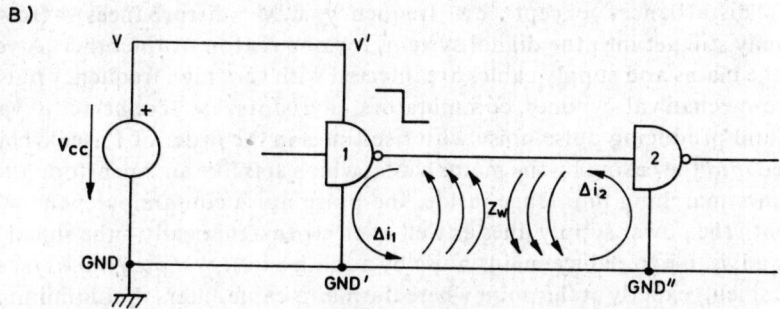

Fig. 3-16 TTL reflections:
A) External current spike Δi supplied from V_{cc} to charge the signal line
(output changing from 0 to 1)
B) Gate output changes from 1 to 0 discharging signal line;
currents Δi_1 and Δi_2 will create ground noise

b) To provide a low wave impedance of 20 Ω or lower (especially low, e.g., approximately 5 Ω for TTL) between V_{CC} and ground. Ideal are broad (5 to 10 mm) V_{CC}- and ground-lines sitting right across each other on a double-sided PCB. This again will stabilize the voltage difference between V_{CC} and ground, but not the ground noise itself.

c) Provide ground with a very large copper surface, so as to give it a good electromagnetic coupling with the 'ideal' grounding point and make it difficult for ground to be 'moved' up on a voltage spike, by a current edge.

In practice, one should not etch away the unused copper surface of a PCB especially with TTL-ICs but *leave it and connect most of it to ground* (and some to V_{CC}) (see Fig. 3-22).
d) A closely-knit grid of broad ground conductors is ideal for a *digital* PCB[1].
e) Conductors which provide ground- and V_{CC}-connections between the power supply and the individual PCBs should be broad[1]. It is recommendable to twist the V_{CC}-line with the ground-line and to provide (for *digital* systems only) multiple ground lines which are running in parallel near each other (this is an efficient way of increasing the 'equivalent ground surface area'; it is in fact the ground *surface* area that counts here and not the cross-section of the ground wires).
f) Strictly avoid using the same ground lines for your digital circuit and for sensitive analog circuits, because a digital ground line always has at least a few mV of ground noise.

For TTL-ICs, the above rules must be followed conscientiously and strictly applied, because, the current spikes are very high. *A large ground surface on a TTL-PCB is absolutely essential*. In the case of ECL-ICs, a large ground surface is also essential because of the lower noise immunity of ECL; but in the case of C-MOS-ICs, ground lines can be considerably narrower.

3.4 ELECTROMAGNETIC (E.M.) INTERFERENCE FROM PULSE-TYPE E.M. FIELDS (MAINS NOISE)

Normally, digital equipment is well shielded (because aluminium casings are used which shield-off all disturbances except low frequency E.M. disturbances). However, some interference may still get into the digital system, for one reason or the other. A very common effect is that the mains and supply cables are infected with very high frequency pulse-type noise (due to electro-mechanical switches, commutators, thyristors, etc., connected at various points to the mains and producing pulse noise with rise-times in the order of 1 nsec). This pulse-type noise is carried *into* the casing by the mains cable (which acts like an 'open door' through which the noise 'comes marching in'). Once inside, the pulse noise contaminates the whole system, influencing both the power supply lines as well as electromagnetically—the signal lines. It is of course highly advisable to reduce mains noise by adding a *well designed mains filter* (having its own separate shield) exactly at the point where the mains cable enters the aluminium casing (see Fig. 3-17; Fig. 3-18 shows examples of wrongly placed mains filters). Even with properly placed mains filters (Fig. 3-17) it is very difficult to filter off mains noise completely and this noise still enters the cabinet although somewhat attenuated and appears as very narrow pulses which can cause considerable havoc in TTL-systems and even more so in high-speed and Schottky TTL and ECL. (In the case of C-MOS and even more so, HTL, the very narrow pulses, which are the only ones which 'survive' the mains filters, are not able to cause logic failures, but in TTL and especially in the other faster logic families they are sometimes able, e.g., to cause a flip-flop to be triggered).

---

[1] But remember that these measures create small ground loops and therefore should *never* be employed for sensitive *analog* circuits.

DESIGN RULES FOR DIGITAL CIRCUIT PCB'S

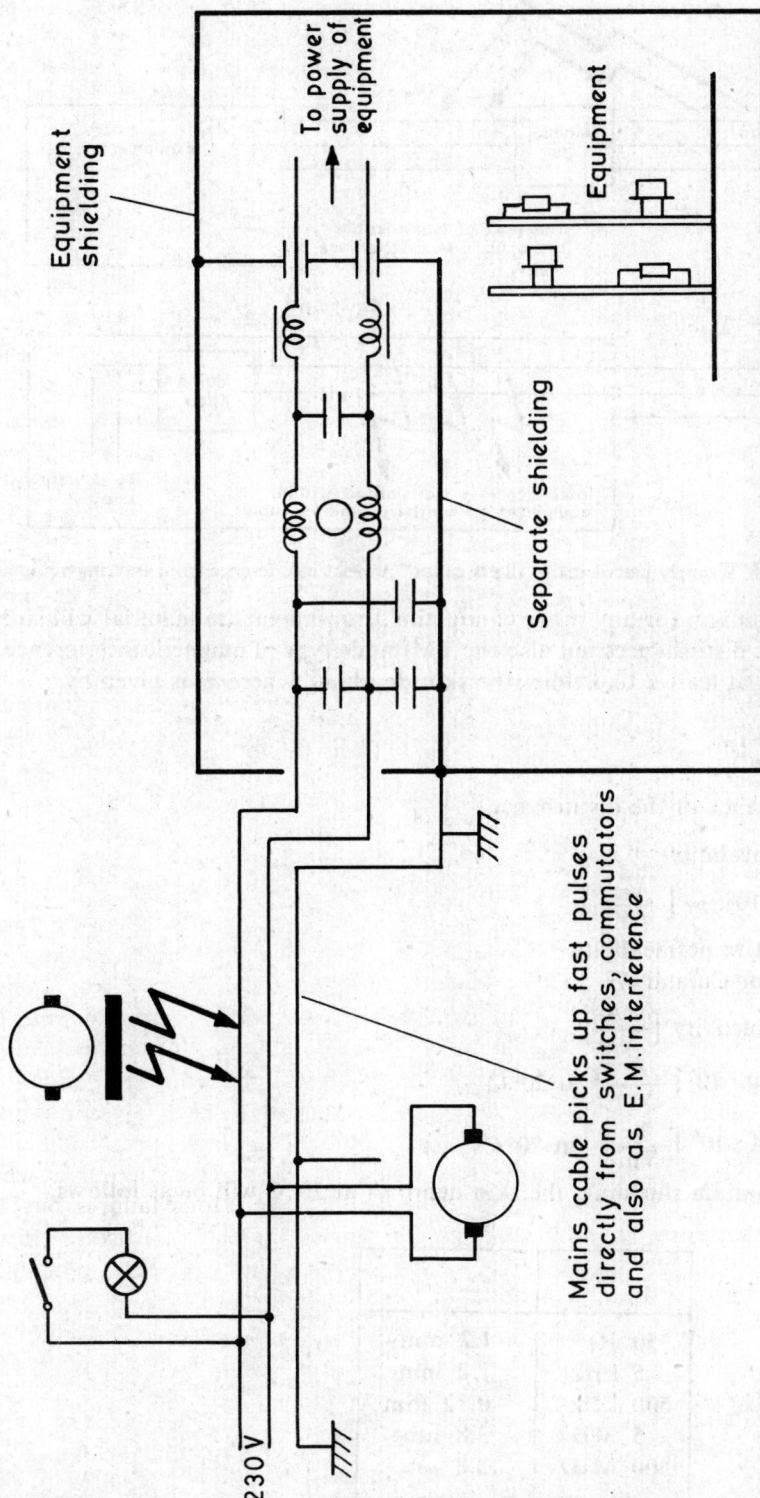

Fig 3-17 Properly placed mains filter prevents pulse-type interference from entering cabinet

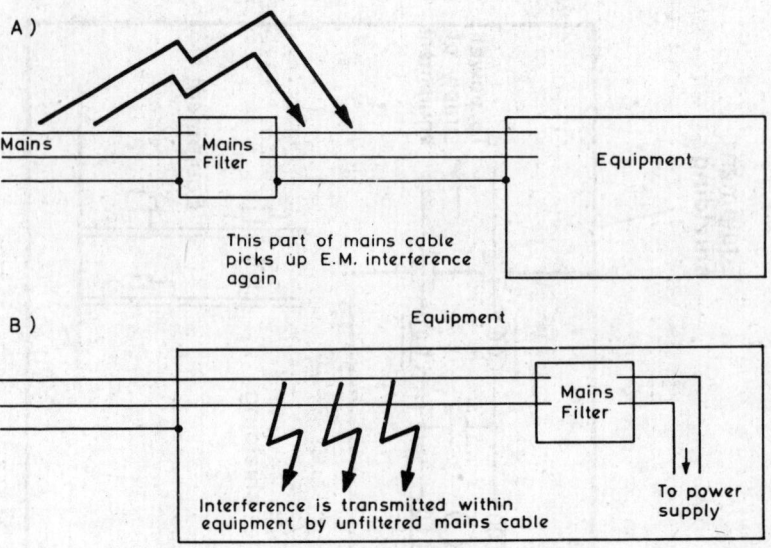

Fig. 3-18 Wrongly placed mains filters cannot prevent interference from entering equipment

A shield of aluminium (or any other conducting, non-magnetic) material will shield-off not only electrostatic disturbances but also the TM-mode type of magnetic interference, provided its thickness d is at least 2 to 3 times the skin depth d_0, where d_0 is given by:

$$d_0 = \frac{1}{\sqrt{\pi \, f \, \mu \, \sigma}} \, [m]$$

where f = frequency of the disturbance $\left[\frac{1}{s} \right]$

μ = permeability

$\mu = 4 \pi \, 10^{-7} \, \mu_r \left[\frac{Vs}{Am} \right]$

μ_r = relative permeability

$\mu_r \approx 1$ for Cu and Al

σ = conductivity $\left[\frac{A}{Vm} \right]$

$\sigma_{Cu} = 58.0 \cdot 10^6 \left[\frac{A}{Vm} \right]$ at 20°C

$\sigma_{Al} = 35.4 \cdot 10^6 \left[\frac{A}{Vm} \right]$ at 20°C

For aluminium shielding, the skin depth d_0 at 20°C will be as follows:

f	d_0
50 Hz	1.2 mm
5 kHz	1.2 mm
500 kHz	0.12 mm
5 MHz	3.8 μm
500 MHz	3.8 μm

For low frequencies f, where d_0 for aluminium becomes very big, magnetic shielding materials (with a high μ) must be used to achieve effective shielding. This is usually very costly.

Digital systems which use high-level signals (a few volts of amplitude and relatively low impedances) are not very sensitive to low-frequency E.M. disturbances. This is why aluminium shielding is usually sufficient for digital systems and no magnetic materials will be used. (Of course plastic cabinets should be avoided.) However, when working with TTL (and with any faster logic) in *a high noise environment,* one should not only use good shielding and a properly placed mains filter, but also see that the *individual PCBs and signal lines are designed in such a way as to minimise the effect of electromagnetic disturbances,* that have entered, in one way or the other, into the interior of the cabinet.

To assess the effect of electromagnetic disturbances on digital signal lines, and to find out some simple design rules (for the PCBs and for the wiring) that would minimise this effect, calculations [7, 8, 9, 10] have been carried out, using a very crude model as shown in Fig. 3.19. A TM mode electromagnetic (E.M.) field was assumed for convenience sake (although in practical cases, especially in the inside of a cabinet, the field will be quite different from a TM field), the E.M. field was assumed to have the shape of a pulse with linear edges, rise-time t_r and amplitude $\hat{B}$, $\hat{E}$ (because of the TM-assumption, we have $\hat{E} = c \cdot \hat{B}$, where c is the velocity of light); the affected signal line was assumed to be running parallel to a ground line at a distance a, both lines having a diameter $d \ll a$. The calculation once again is rather involved; it turns out that the pulse noise first arrives on a signal line, gets reflected on both ends, so that the whole behaviour is governed by transmission line phenomena, multiple reflections etc. Computations can be carried out with the computer or, also, using approximative graphical methods related to the BERGERON reflection chart [8]. Taking into consideration the dynamic noise margin of the ICs used (as well as their non-linear input and output characteristics which play a role for the multiple reflections involved) a critical magnetic flux density B_{crit} can be defined. If the disturbing field exceeds B_{crit}, it will cause logic malfunction.

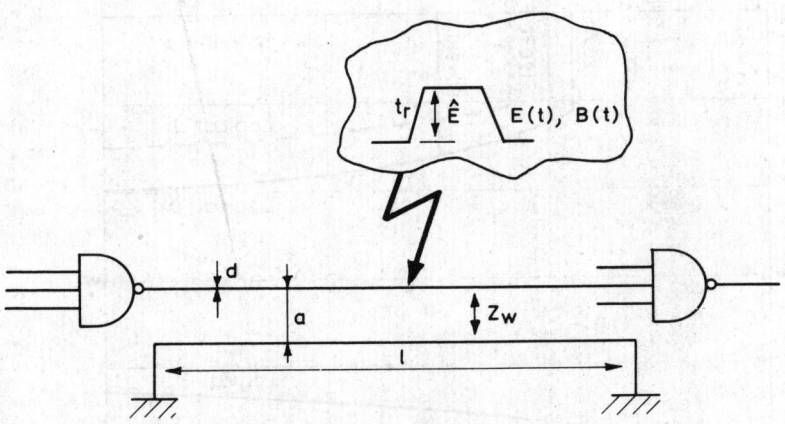

Fig. 3–19 Basic model for determination of E.M. interference

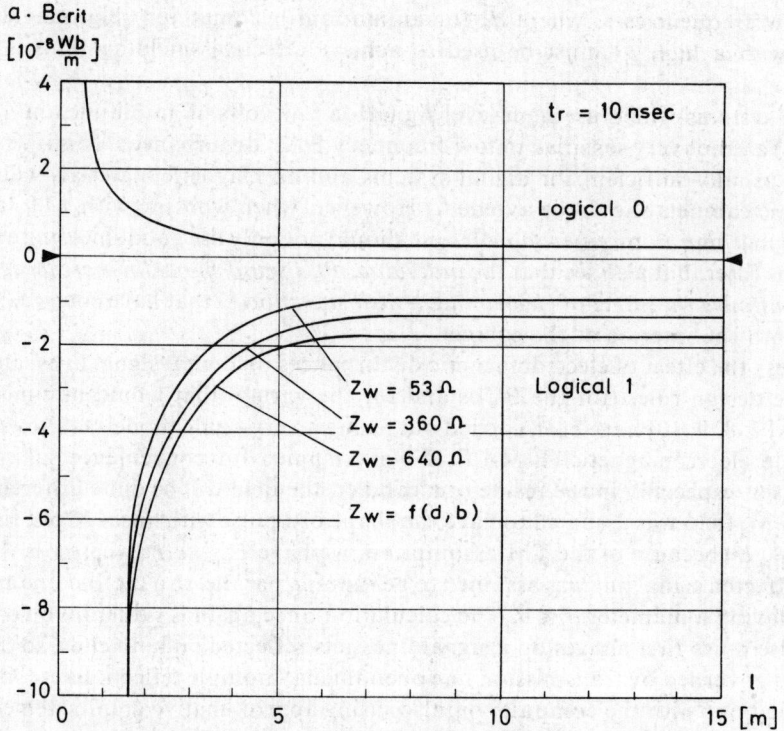

Fig. 3-20 Electromagnetic (E.M.) pulse interference on TTL-lines: Critical E.M. field B_{crit} as a function of line length [8]

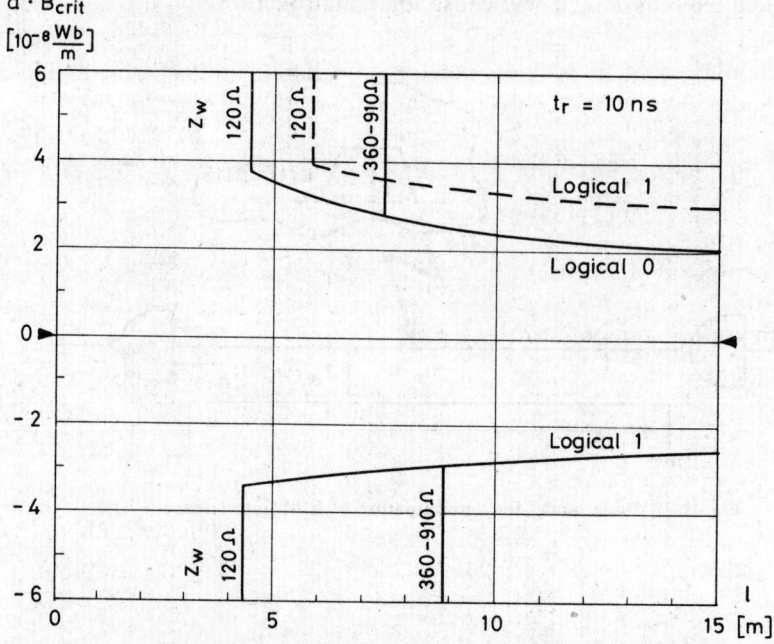

Fig. 3-21 Electromagnetic (E.M.) pulse interference on C-MOS lines: Critical E.M. field B_{crit} as a function of line length [8]

In [8] and [9] B_{crit} has been computed for TTL- and C-MOS-ICs, for a logic 0 and a logic 1 condition. One finds that B_{crit} *is strictly proportional to* $\frac{1}{a}$, where a is the distance between the disturbed signal line and ground line; i.e., that if the distance between signal line and ground line is doubled, then the critical field strength that will cause a logic malfunction is halved. Therefore, the product $(a \cdot B_{crit})$ can be plotted. In Figs. 3-20 and 3-21, the results are given for TTL and C-MOS, respectively, showing $(a \cdot B_{crit})$ as a function of the length l of the signal line; in fact, the longer the signal line, the lower is the critical field, i.e., the easier the signal line can be disturbed. Looking at Fig. 3-20, one sees that TTL signal lines are quite easily disturbed, especially if they are longer than 1 to 2 meters and especially in the logical 0 state. C-MOS-connections, on the other hand, have to have a length of at least $4\frac{1}{2}$ meters, before they can be logically disturbed and then, too, a high field strength is required. (Remember, however, that C-MOS-ICs can quite easily be damaged, or even destroyed, by overvoltages *at their input, even if they are not logically disturbed*!) One can sum up the *practical conclusions* of this problem as follows:

1) In addition to shielding, proper mains filtering (see Fig. 3-17) should always be employed when operating a digital system in a high-noise environment. In the case of TTL and any faster logic family, it is very difficult to filter out all disturbances by shielding and using mains filters.
2) Therefore and because of the low value of $(a \cdot B_{crit})$, as given in Fig. 3-20: If TTL is used in a high-noise environment, keep the distance between the logical signal lines and the ground line small; i.e., *run the signal lines near to ground, both on the PCBs and also when interconnecting PCBs*.
3) C-MOS is much less sensitive than TTL to disturbances and it is not necessary, for C-MOS, to keep the signal lines so near to a ground line.
4) Remember that although C-MOS-ICs may not be logically disturbed, they can be easily *destroyed* by overvoltages. This calls for special protection circuits especially against electrostatic overvoltages[2], if C-MOS is used in a high-noise environment, or even in any application where it is subject to electrostatic voltages (e.g. through touching).

3.5 RECOMMENDATIONS AND SUMMARY

The following two tables sum up the practical conclusions that one draws from the above detailed discussion. In addition, Fig. 3-22, shows an example of a good PCB layout for TTL-ICs. Note the larger ground lines and the thin signal lines. Note also that all unused areas have been left with copper and connected to ground.

---

[2] These special protection circuits have definitely to be added by the user, *even if* the ICs already contain some built-in protection devices themselves; the built-in protection devices are never sufficient.

A) *RECOMMENDATIONS FOR* PCBs

	ECL	TTL	C-MOS
Recommended Z_w [Ω] between signal conductors and GND	50–100	100–150	150–300
Recommended width of signal conductors and distance to GND	(1 to 2)$\times b$; GND plate very nearby	0.5$\times b$; GND plate nearby	0.5$\times b$; No GND nearby
Recommended Z_{wv} [Ω] between V_{CC} and GND	<10	<5	<20
Recommended width of V_{CC}-line	5$\times b$	10$\times b$	About (2 to 3)$\times b$
Recommended width of GND-line	Broad	Very broad	Not excessively broad
General recommendation	—Connect all unused copper surfaces to GND —Use GND–plate or GND–grid		

b = PCB board thickness.

B) *SUMMARY OF EFFECTS*

	ECL	TTL	C-MOS
Reflections	Very critical. Use matching resistances.	Critical. Use thin signal lines.	Critical (but anyway C-MOS is a slow family). Use high-impedance lines (thin and far from GND).
Crosstalk	Critical. Use nearby GND.	Critical. Use nearby GND.	Not critical. Nearby GND not required.
Supply/ Ground noise	Critical. Use nearby GND.	Very critical. GND-plate, GND-grid or large copper surface absolutely essential.	Not critical. Very large GND not required.
Outside E.M. interference	Very critical. Do not use in noisy environment.	Critical. Keep signal-lines near to GND-lines.	Not critical. Nearby GND not required. But external protection of circuitry against over-voltages needed (especially against electrostatic over-voltages).

DESIGN RULES FOR DIGITAL CIRCUIT PCB'S

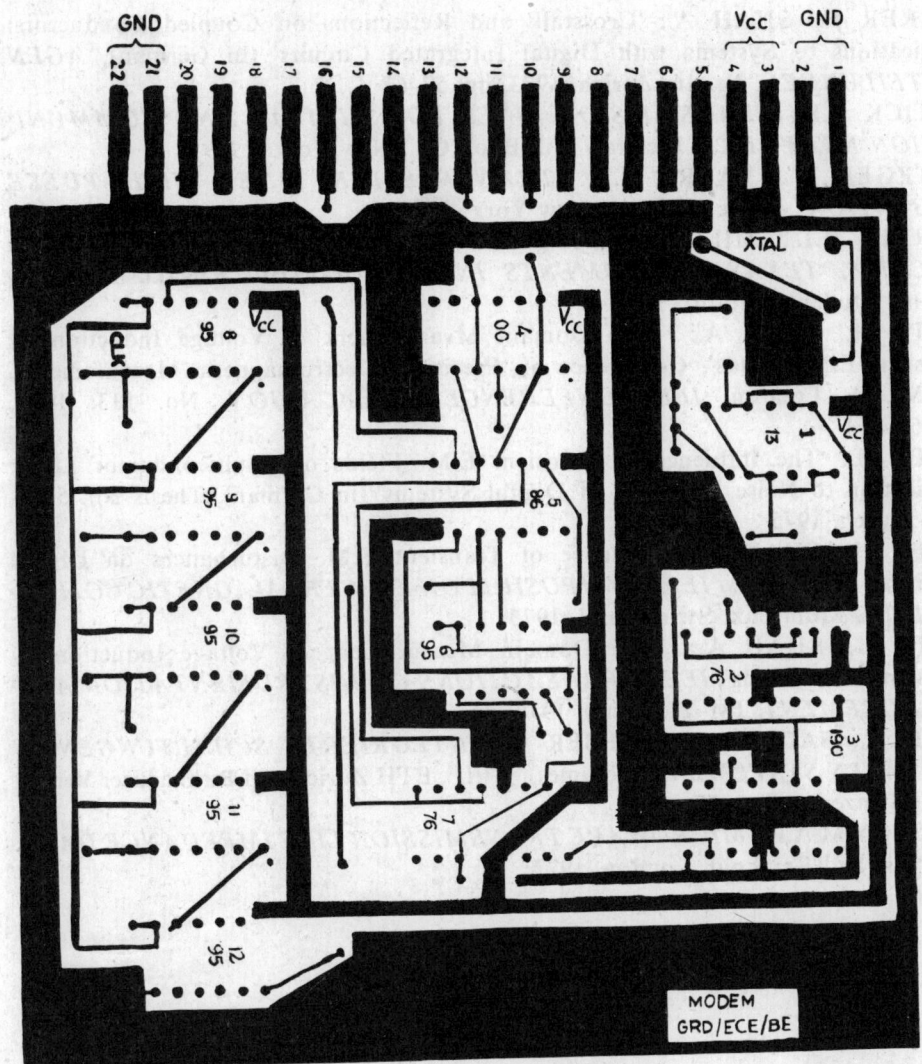

Fig. 3-22 PCB layout for a TTL-system, showing broad V_{CC}-conductors and large ground surface (Photo CEDT)

BIBLIOGRAPHY

1. BRYANT T., WEISS J.: 'Parameters of Microstrip Transmission Lines and of Coupled Pairs of Microstrip Lines', *IEEE TRANSACTIONS*, MTT-16, 1968, pg. 1021–1027
2. COHN S.B.: 'Shielded Coupled-Strip Transmission Lines', *IEEE TRANSACTIONS*, MTT-3, 1955, pg. 29–38

3. FURRER F., SHAH A.: 'Crosstalk and Reflections on Coupled Conductors: Applications to Systems with Digital Integrated Circuits' (in German), *AGEN MITTEILUNGEN,* Nr. 16, Zurich, 1973, pg. 57–67
4. MATICK R.E.: *TRANSMISSION LINES FOR DIGITAL AND COMMUNICATION NETWORKS,* McGraw-Hill Book Co., New York, 1969
5. METZGER P., VABRE J.P.: *TRANSMISSION LINES WITH PULSE EXCITATION,* Academic Press, New York, 1966
6. MORRIS R.L., MILLER J.R.: *DESIGNING WITH TTL INTEGRATED CIRCUITS, TEXAS INSTRUMENTS INC.*, McGraw-Hill Kogakusha (International Student Edition), 1971
7. NAITO H., SHAH A.: 'Time Domain Measurement of Voltage Induction by Transient E.M. Fields', Conference on Precision Electromagnetic Measurements, CPEM 74, London, *IEE CONFERENCE PUBLICATION,* No. 113, 1974, pg. 262–264
8. NAITO H.: 'The Influence of Transient E.M. Fields on Multiconductor Lines; Application to Noise Problems of Digital Systems' (in German), Thesis No. 5505, ETH Zurich, 1975
9. NAITO H., SHAH A.: 'Influence of Transient E.M. Disturbances on Digital Electronic Systems', *1st IEEE SYMPOSIUM ON ELECTROMAGNETIC COMPATIBILITY*, Montreux, Switzerland, 1975
10. NAITO H., SHAH A.: 'Time Domain Measurement of Voltage Induction by Transient E.M. Fields', *IEEE TRANSACTIONS ON INSTRUMENTATION AND MEASUREMENT,* IM-27, March 1978, pg. 38–42
11. SHAH A., SAGLINI M., WEBER C.: *INTEGRIERTE SCHALTUNGEN IN DIGITALEN SYSTEMEN* (2 volumes), AFIF, ETH Zurich and Birkhaeuser Verlag, Basel (Switzerland), 1977
12. GUNSTON M.A.R.: *MICROWAVE TRANSMISSION LINE IMPEDANCE DATA,* Van Nostrand-Reinhold, London, 1972

4
DESIGN RULES FOR PCB'S IN HIGH-FREQUENCY AND FAST-PULSE APPLICATIONS

Note: This chapter contains specialised information that need *not* be studied by readers who are not involved in the design of such circuits.

4.1 INTRODUCTION

For very-high-frequency circuits (around 100 MHz and above) and for very fast pulses (rise- and fall-times of a few nanoseconds and below), even short pieces of conductors on a PCB have to be considered as transmission lines. For longer conductors, if one wants to avoid reflections, matching of transmission lines becomes a necessity. Such reflections are disturbing because they will cause a loss of bandwidth and an increase in rise-times. For *shorter* conductors, matching is very often impossible and the conductors will behave either capacitively or inductively. It is then a question of proper PCB layout to ensure that one obtains a 'capacitive' conductor wherever this will be less harmful, and an 'inductive' conductor wherever that is preferable. Feedback capacitors (Miller effect) are especially harmful and should be avoided by using guard lines. These layout methods are exemplified giving the PCB layout of a cascode-stage for a typical broadband amplifier. Finally, we will have to stress: as in digital PCBs (Chapter 3), proper ground- and power-supply lines are extremely important for fast-pulse and high-frequency applications. The present chapter does not deal with microwave applications (frequencies above a few GHz), where PCBs are actually used as waveguides and where certain functional elements (circulators, strip-line filters etc.) are directly made as printed circuits (see Chapter 7 of this book for such applications).

4.2 MATCHING OF CONDUCTORS; EFFECT OF MISMATCH IN DIFFERENT CASES

4.2.1 Transmission Lines and Wave Impedances

A conductor on a printed circuit should be considered as a *transmission line* (and not as a short circuit) if its length l (in meters) is more or less:

$$l > \frac{t_r}{100 \text{ nsec}} \quad \text{for pulse applications}$$

$$l > \frac{3 \text{ MHz}}{f_{upper}} \quad \text{for high-frequency applications}$$

t_r = rise-time of the pulse [nsec]
f_{upper} = highest frequency in the signal [MHz]

Such a transmission line has a certain *wave impedance* Z_w. A broad conductor will have a smaller wave impedance than a narrow conductor. Similarly, a conductor which is near the groundplate will have a smaller wave impedance than one which is far away. Typical cross-sections for various types of conductors on PCBs and curves showing the corresponding wave impedances are given in Fig. 4-1 (see also the specialised literature [3, 4, 6, 7, 10]). We should always try to match the conductor to the source and to the load. This means that we should have

Source resistance $R_S = Z_w$
Load resistance $R_L = Z_w$

With such a matching, the line will hardly introduce any noticeable distortion of pulses or any noticeable loss of amplitude in the frequency range we are now talking about.

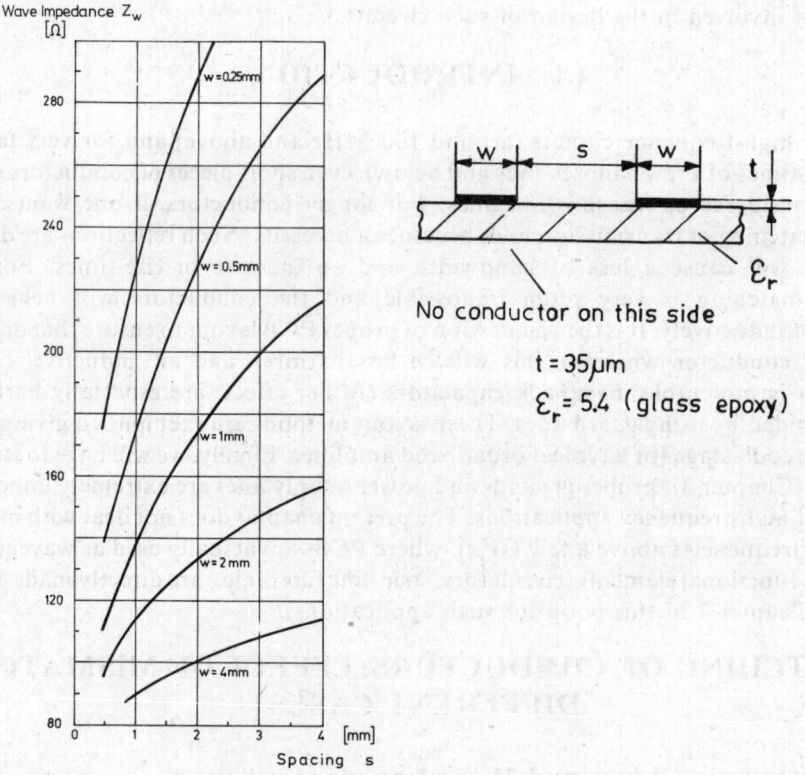

Fig. 4-1a Wave impedance for two parallel running conductor on the same board side (based on data for C' and L' given in Sections 2.3.2 and 2.4 of Chapter 2)

4.2.2 Effect of Mismatch in the Fast-Pulse Case

Let us look at what happens to a fast-pulse travelling over a line, in 3 basic cases:

Case (a) If such a transmission line or interconnection is *matched to both source and load* (i.e., if both the source impedance and the load impedance are equal to the wave impedance Z_w of the transmission line) and if the transmission line is not very long, it will introduce a simple delay (Fig. 4-2a). This delay is approximately 5 nsec/m (3.3 nsec/m in free air or vacuum and $\sqrt{\epsilon_r}$ more when a dielectric with relative constant ϵ_r is present).

Case (b) If such a transmission line is *matched at one end* (either source or load), we will have at the most a single reflection.

If the *source is matched* to the transmission line, but the load mismatched, a pulse fed in from the source will travel along the line, will reach the load, get reflected there (the final voltage step will appear immediately on the load); a reflected wave now travels back to the source where it does not get reflected any more, because the source is matched to the line. The voltage $e_1(t)$ at the source end will have two steps, but the voltage $e_2(t)$ at the load will have just one step (see Fig. 4-2b).

If the *load is matched* to the transmission lines, but the source mismatched, then a pulse fed in from the source will travel along the line up to the load and the final voltage will appear at the load without any reflections at all (see Fig. 4-2c). If on the other hand, a stray pulse had been fed in from the load, then there would have been a single reflection for this case.

Case (c) If the transmission line is *mismatched at both ends*, then multiple reflections will take place (Fig. 4-2d).

These reflections will create disturbances and slow down considerably the rise and fall times of the pulses. Therefore mismatch on both sides must definitely be avoided when dealing with fast pulses. Some examples will be shown hereunder where a graphical method of calculation is used (the so-called 'Bergeron Diagram' or 'Reflection Chart for Pulses' [11]).

Any longer connection (longer than 10-20 cm) between fast-pulse PCBs must definitely be made either with coaxial cables or at least as twisted pairs (where the signal wire is twisted closely with the ground wire). Any longer loose, i.e., open, untwisted wires must be absolutely avoided: Not only will very complex multiple reflections take place, but behaviour of the whole system will become unpredictable. The wave impedances of commonly used coaxial cables are 50 or 75 Ω, the latter for TV applications. Twisted pairs have wave impedances between 100 and 150 Ω, if they are properly (i.e., tightly) twisted.

If the mismatch is only slight (up to $\pm 10\%$ or even $\pm 20\%$), the resulting pulse distortion will not be too bad: It will only cause the pulse to be rounded off, as shown in Fig. 4-3. Due to the finite rise-times of the pulses, and also of the measuring system, sharp edges as drawn in Fig. 4-3a will not appear in practice; the waveforms will always be rounded.

4.2.3 Effect of Matching and Mismatch in the High-Frequency Case

The input impedance seen from the source end of the line according to Fig. 4-2 is given by the expression

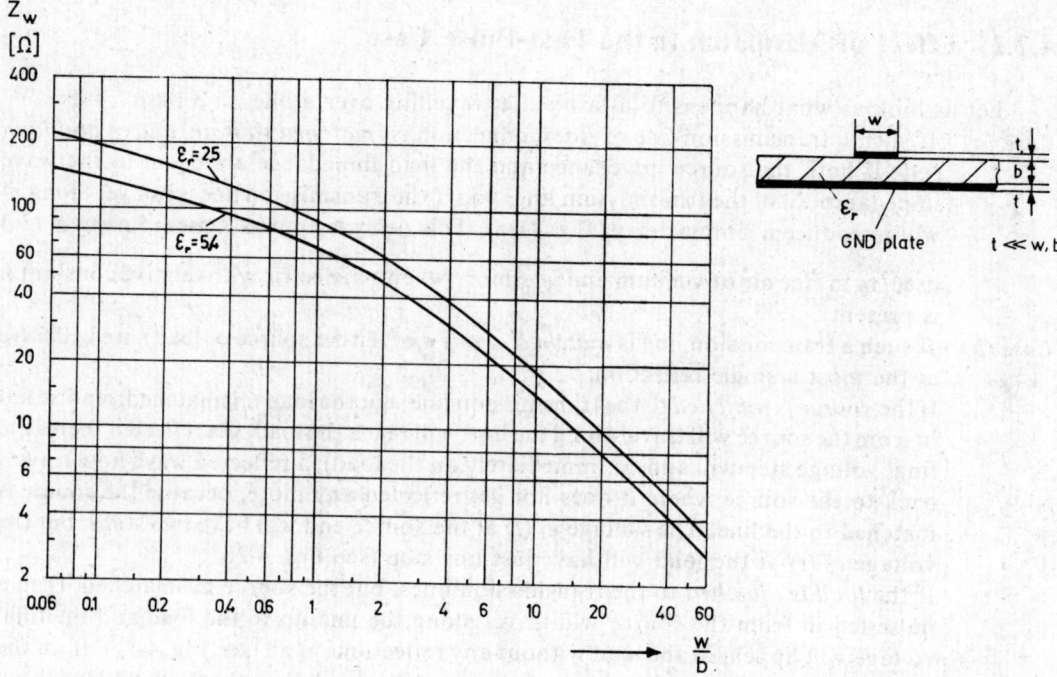

Fig. 4–1b Wave impedance for PCB conductors in microstrip configuration (narrow conductor sitting across a ground plate, according to Fig. 15 of [6c] and taking into account the air-effect given in Fig. 17 of [6c])

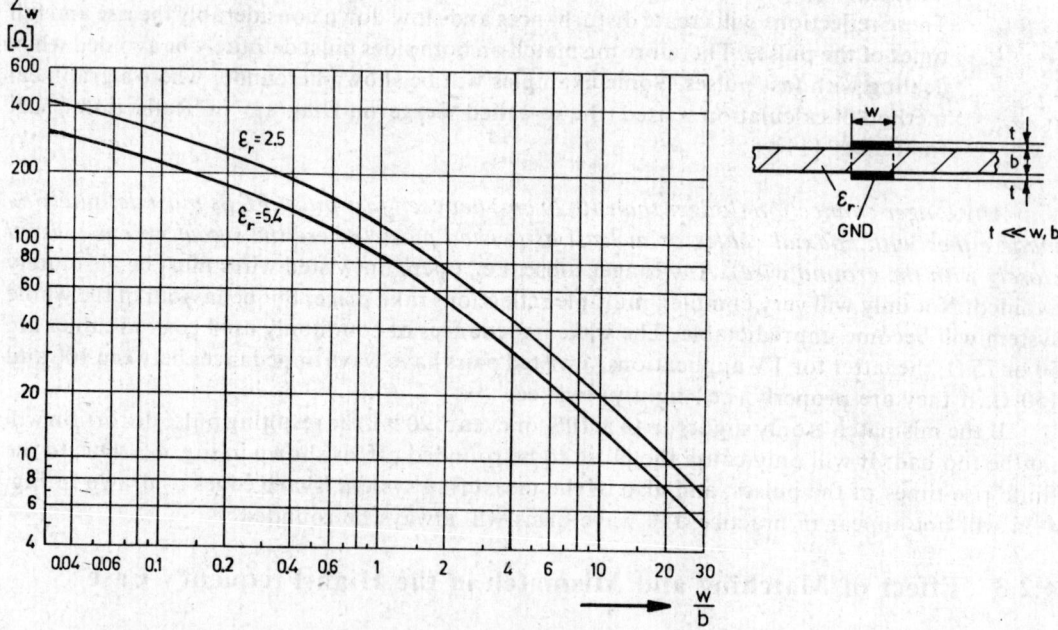

Fig. 4-1c Wave impedance foo PCB conductors in stripline configuration (two narrow strips, on either side of the board, according to Fig. 15 of [6c] and taking into account the air-effect given in Fig. 17 of [6c])

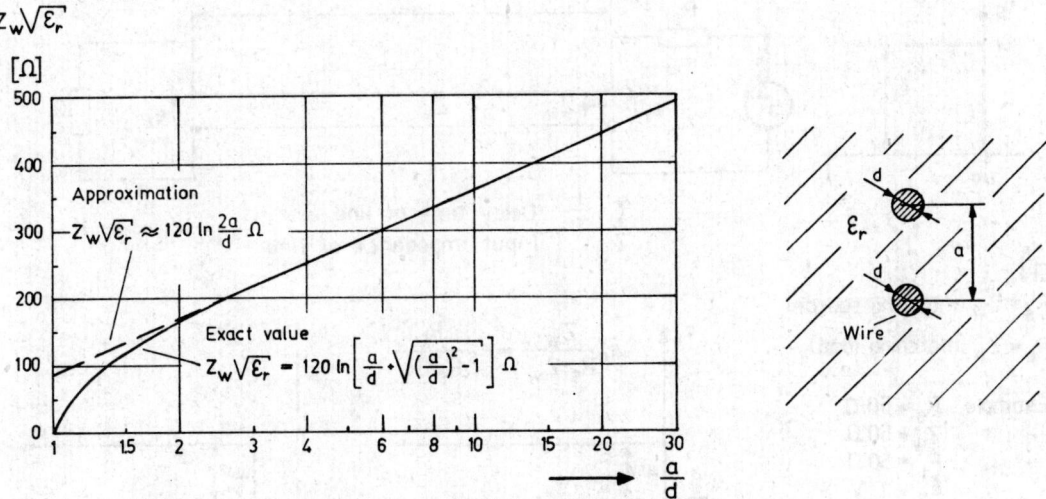

Fig. 4-1d Wave impedance for a pair of two wires with round cross-sections in a dielectric with-relative dielectric constant

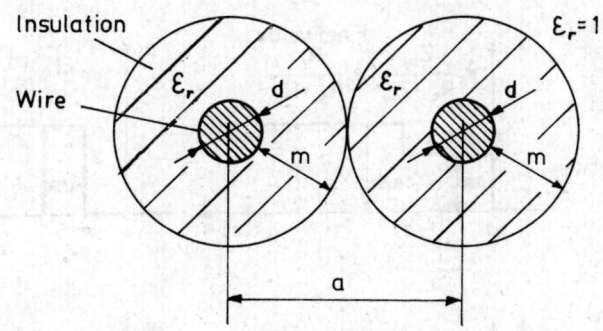

Fig. 4-1e Cross-section of a twisted pair of wires: Its wave impedance can approximately be found from Fig. 4-1d, by setting $a = d + 2m$

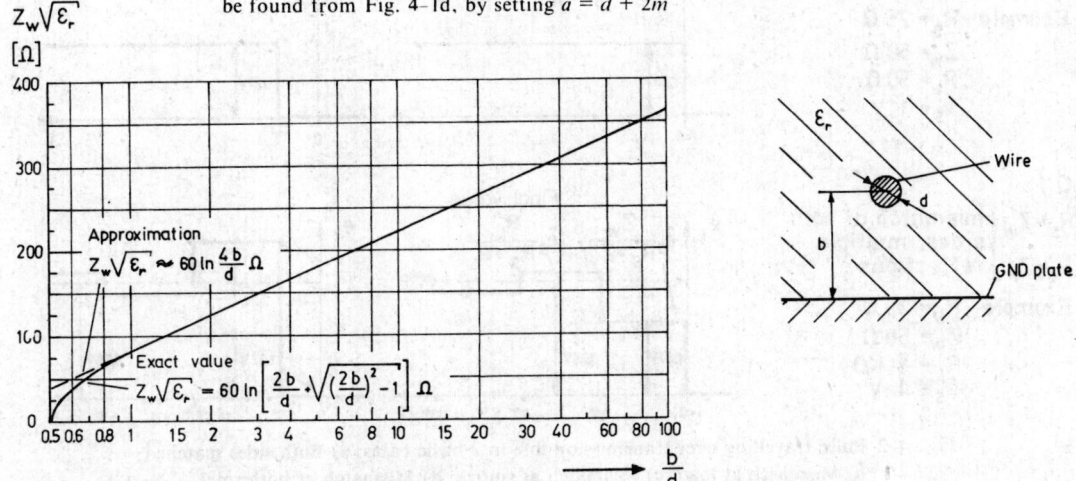

Fig. 4-1f Wave impedance for a single wire above ground plate with free air as dielectric ($\epsilon_r = 1$), according to [10]

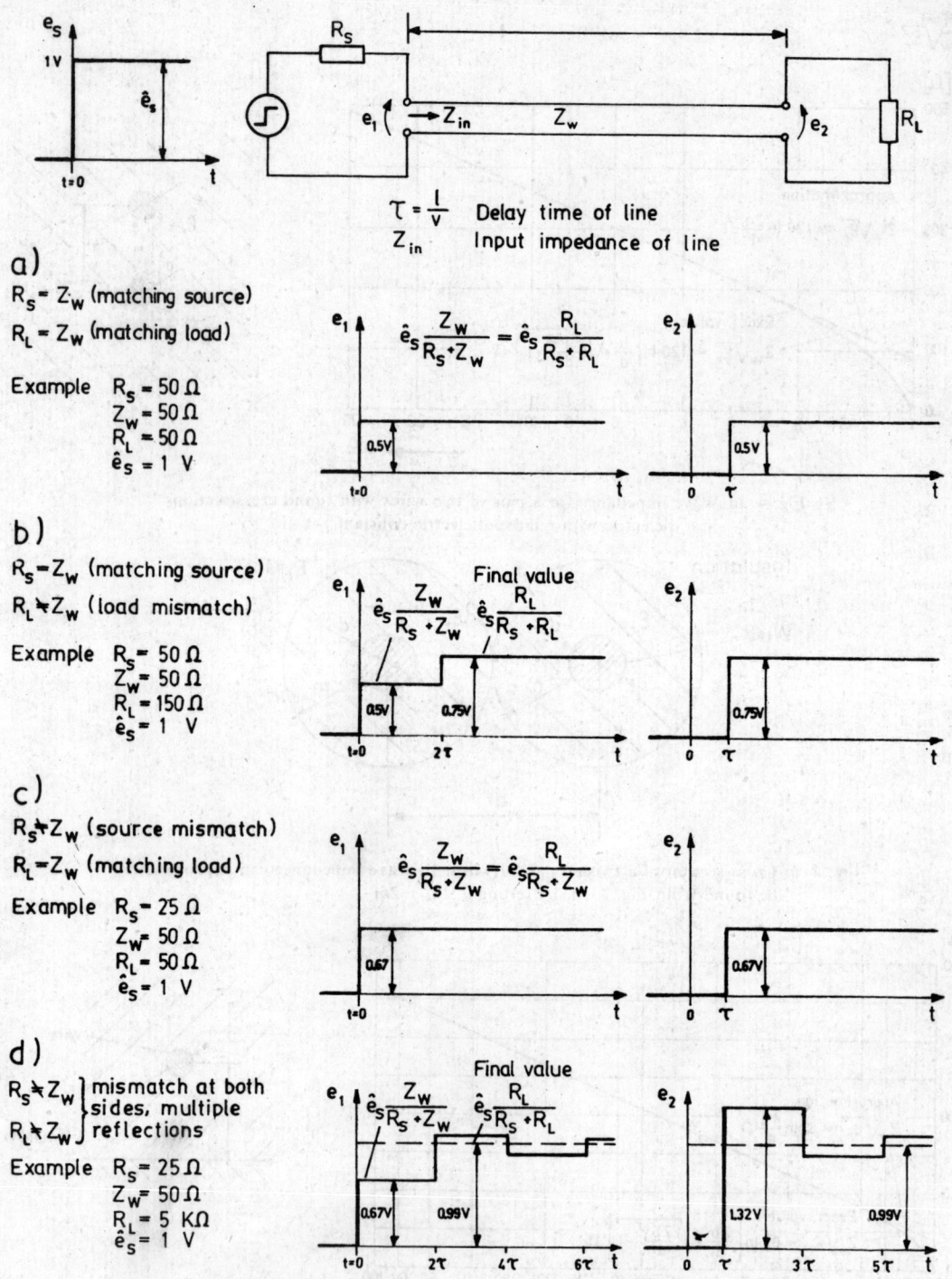

Fig. 4-2 Pulse travelling over transmission line in 3 basic cases: a) Both sides matched; b) Mismatch at load; c) Mismatch at source; d) Mismatch at both sides

DESIGN RULES FOR PCB'S IN HIGH-FREQUENCY AND FAST-PULSE APPLICATIONS

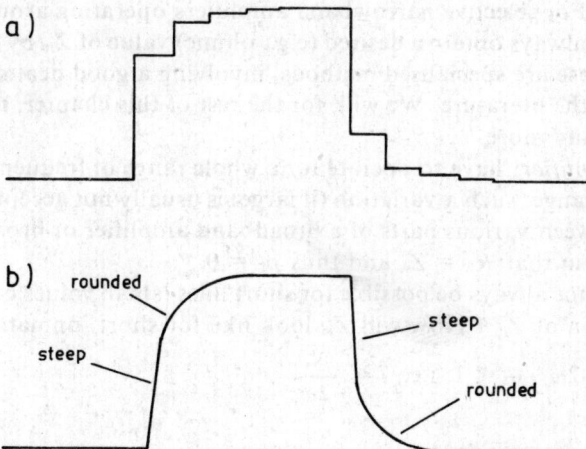

Fig. 4–3 A mismatch of up to about ± 20% will give a sharp (steep) initial step which will be rounded off only for the last portion: a) Ideal staircase waveform due to reflections; b) Actual appearance because of various other rise-time limitations

$$Z_{in} = Z_w \frac{1 + \rho_L \exp(-2j\omega\tau)}{1 - \rho_L \exp(-2j\omega\tau)}$$

where ρ_L is the load reflection factor;

$$\rho_L = \frac{R_L - Z_w}{R_L + Z_w}$$

and $\tau = \frac{l}{v}$ the line delay.

If $\rho_L = 0$ (matched case, i.e., $R_L = Z_w$), then the input impedance of a lossless line is real (ohmic) for all frequencies. If $\rho_L \neq 0$ (unmatched line), then Z_{in} will vary with frequency, becoming ohmic at certain frequency values and being complex (partly inductive or capacitive) in between.

For *Selective Amplifiers* (that only operate at a given frequency ω_0), we can (and will often) use lines that are not matched. Depending upon the value of $(2j\omega_0\tau) = (2j\omega_0 l/v)$ and of ρ_L, we can obtain an input impedance Z_{in} that is either ohmic, capacitive or inductive, as desired by us, at that particular frequency ω_0. A partly inductive line can often be conveniently used to compensate for transistor capacitances (input capacitances), if we are dealing with a selective or narrowband amplifier. An adjustment of the input impedance Z_{in} to the desired value of capacitance or inductance can be achieved by varying *either* R_L (and, thus ρ_L) or the length l of the line (and, thus τ). Smith charts and other specialised techniques are used to determine conveniently the value of Z_{in}. In practical high-frequency cases, not only is $\exp(-2j\omega\tau)$ complex, but even ρ_L becomes complex, because R_L has usually to be substituted, at higher frequencies, by a complex impedance Z_L (containing normally a resistive part R_L and a capacitive part C_L). For selective narrowband amplifiers operating around a fixed frequency ω_0, we can generally always obtain a desired (e.g., ohmic) value of Z_{in} by choosing l and/or R_L correspondingly. These are specialised methods, involving a good deal of calculation and the reader is referred to the literature. We will, for the rest of this chapter, not look at the case of selective amplifiers any more.

Broadband Amplifiers have to operate for a whole range of frequencies. If then Z_{in} varies over this frequency range, such a variation (if large) is usually not acceptable. Therefore, long connecting lines between various parts of a broadband amplifier or broadband circuit should always be matched, so that $R_L = Z_w$ and thus $\rho_L = 0$.

Matching may not always be possible for short lines (small values of l). What will then be the range of variation of Z_{in}? How will Z_{in} look like for short, unmatched lines?

Supposing: $2\omega_1 \tau = 2\omega_1 \dfrac{l}{v} \ll 1$, i.e., $l \ll \dfrac{v}{2\omega_1}$

(where ω_1 is the highest frequency to be amplified by the broadband amplifier)
we can write:

$$\exp\left(-2j\omega \frac{l}{v}\right) \approx 1 - 2j\omega \frac{l}{v} \quad \text{for } \omega \leqslant \omega_1$$

If we take the extreme cases of $\rho_L = \pm 1$ ($R_L = \infty$, $R_L = 0$), we get:

a) *for* $\rho_L = -1$ (short circuit, $R_L = 0$):

$$Z_{in} \approx Z_w \left(j\omega \frac{l}{v} \right)$$

This corresponds to an inductance of value

$$L_{equ} = \frac{Z_w}{v} \cdot l$$

b) *for* $\rho_L = +1$ (open circuit, $R_L = \infty$):

$$Z_{in} \approx \frac{Z_w}{\frac{j\omega l}{v}}$$

This corresponds to a capacitance C_{equ} of value

$$C_{equ} = \frac{1}{Z_w \cdot v} l$$

It is the duty of the designer to see that all unmatched lines are kept so short that these 'parasitic' values of L_{equ} or C_{equ} remain small.

Exactly the same expressions for substituting a short line with equivalent capacitance C_{equ} or inductance L_{equ} will be found in the next section by looking at the case of pulses (i.e., by adopting a time-domain, rather than a frequency-domain in treatment, as adopted here).

In general, a broadband amplifier can always be calculated in the time domain, by looking at pulses and rise-times.

We will therefore, for the rest of this chapter, only deal with pulses, i.e., we will remain in the time domain.

4.3 PULSE CIRCUITS

4.3.1 Inductive and Capacitive Approximations for Lines with Multiple Reflections

TYPICAL CASES
In Fig. 4-4, four typical cases of mismatch are considered:

a) $R_S, R_L \gg Z_w$
b) $R_S, R_L \ll Z_w$
c) $R_S \ll Z_w \ll R_L$
d) $R_S \gg Z_w \gg R_L$

We can see that each of these cases leads to multiple reflections, and step-shaped load voltages $e_2(t)$, which only gradually rise to their final values. Because the time delay between 2 steps is exactly 2τ, where τ is the time delay of the line (approx., 5 nsec/m) and because it takes (if considerable mismatch is there) very many steps to reach even near to the final value, very slow waveforms result. This is why double-sided mismatch must be definitely avoided in fast-pulse techniques.

Not only the interconnections on the circuit should be matched but also wires connecting the circuit to external pulse generators, loads, etc. must be matched preferably on both sides, but at least on one side. Matching the wave impedance to source and/or load impedance to at least about 10% or 20% will render the circuit fast and there will not be too many multiple reflections. Considerable mismatch, on the other hand, will result in rise-times or settling times which are a multiple of the cable's double propagation delay 2τ (i.e., a multiple of 10 nsec for every meter of the cable).

Approximations
There is a convenient rule for assessing the effect of mismatch in cases a) and b):
a) If $R_S, R_L \gg Z_w$ the line appears just like a single, discrete capacitance with where v is the propagation velocity and l the length of the line
$$C_{equ} = \frac{l}{Z_w \cdot v},$$
b) If $R_S, R_L \ll Z_w$, the line appears just like a single, discrete inductance with:
$$L_{equ} = \frac{Z_w \cdot l}{v},$$ where v is the propagation velocity and l the length of the line.

This is the reason why open wires with their high values of Z_w usually appear as inductances.

Theoretical Verification of the Approximations
It is easy to verify *theoretically*, the above approximations. The individual steps $a_1, a_2, a_3, \ldots$ in Fig. 4-4a,b are given by the following expressions:

$a_1 = A\delta (1 + \rho_L)$
$a_2 = A\delta \rho_L \cdot \rho_S (1 + \rho_L)$
$a_3 = A\delta (\rho_L \cdot \rho_S)^2 (1 + \rho_L)$
.
.
.
$a_n = A\delta (\rho_L \cdot \rho_S)^{n-1} (1 + \rho_L)$

ρ_L = load reflection factor = $\dfrac{R_L - Z_w}{R_L + Z_w}$

ρ_S = source reflection factor = $\dfrac{R_S - Z_w}{R_L + Z_w}$

δ = ratio of first wave to source amplitude = $\dfrac{Z_w}{R_s + Z_w}$

Any succession of steps where each following step is ρ times the preceding one (here $\rho = \rho_L \cdot \rho_S$), always has points ($P_0, P_1, P_2 \ldots$) that fit exactly on to an exponential function (see also Fig. 4-4e). This fitting on to an exponential function is even true for cases given in Fig. 4-4 c, d. The exponential function has the form:

$$\exp\left(-\frac{\Delta t}{T_{equ}}\right)$$

where $T_{equ} = \dfrac{-\Delta t}{\ln \rho} = \dfrac{-\Delta t}{\ln (\rho_L \cdot \rho_S)}$

and $\Delta t = 2\tau = \dfrac{2l}{v}$ (duration of the step)

Now, if we have $Z_w \gg R_L, R_S$ or $Z_w \ll R_L, R_S$, then $(\rho_L \cdot \rho_S) \approx 1$ and $A\delta (1 + \rho_L) \ll 1$ so that
1) the mathematical approximation $\ln (\rho_L \cdot \rho_S) \approx - (1-\rho_L \cdot \rho_S)$

   i.e. $T_{equ} \approx \dfrac{\Delta t}{1 - \rho_L \rho_S}$ holds good.

2) individual steps $a_1, a_2, \ldots$ become so small that they can hardly be distinguished any more from the continuous exponential function.

A lumped capacitance C_{equ} (if $R_S, R_L \gg Z_w$) or a lumped inductance L_{equ} (if $R_S, R_L \ll Z_w$) gives the same kind of exponential waveform, for both $e_2(t)$ and $e_1(t)$, provided we now set (see Fig. 4-4a,b):

$T_{equ} = C_{equ} \dfrac{R_S \cdot R_L}{R_S + R_L}$

or $T_{equ} = \dfrac{L_{equ}}{R_S + R_L}$ respectively

DESIGN RULES FOR PCB'S IN HIGH-FREQUENCY AND FAST-PULSE APPLICATIONS

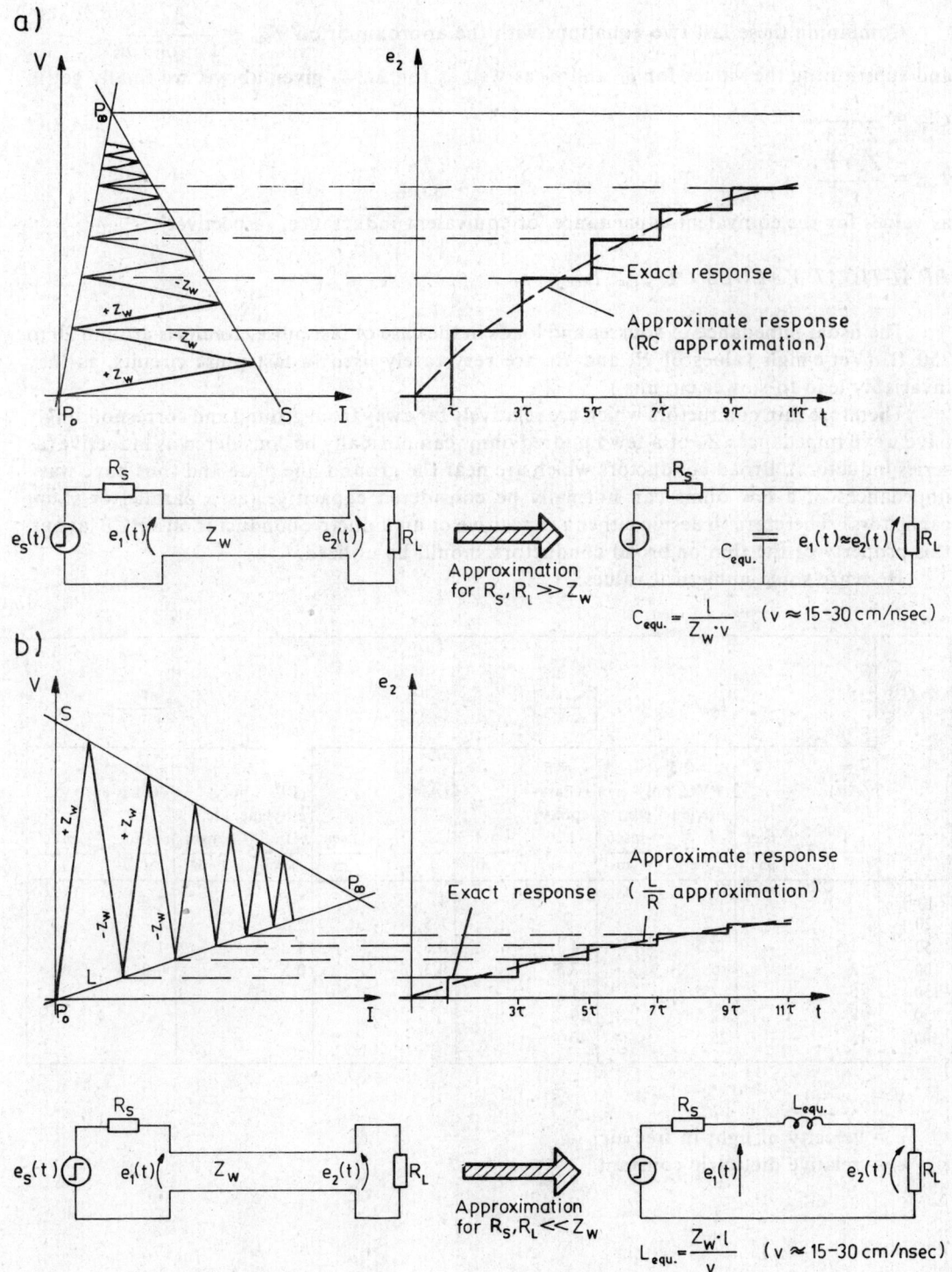

Combining these last two equations with the approximation $T_{equ} \approx \dfrac{\Delta t}{1 - \rho_L \cdot \rho_S}$ and substituting the values for ρ_L and ρ_S as well as for Δt, as given above, we finally get:

$$C_{equ} = \frac{l}{Z_w \cdot v}$$

$$L_{equ} = \frac{Z_w \cdot l}{v}$$

as values for the equivalent capacitance, or equivalent inductance, respectively.

PRACTICAL VALUES OF C_{equ}, L_{equ}

The usual impedance of sources and loads in the case of fast pulse circuits is around 20 to 250 Ω. (Very high values of R_S and R_L are very rarely used in fast pulse circuits, as they invariably lead to slower circuits.)

Therefore thin conductors which are relatively far away from ground and correspondingly have wave impedances Z_w of a few hundred ohms can normally be considered as inductive (as series inductors). Broad conductors which are near the ground line plate and thus have wave impedances of a few ohms can normally be considered capacitive (as a parallel or shunt capacitors). Therefore to design either a capacitive or an inductive conductor on a PCB layout, this property of the thin or broad conductors should be utilized.

Here are some numerical values:

Z_w [Ω]	L_{equ} $\left[\dfrac{nH}{cm}\right]$			C_{equ} $\left[\dfrac{pF}{cm}\right]$		
	$\epsilon_r = 1$ (Air) $\dfrac{1}{v} = 3.3 \dfrac{nsec}{m}$	$\epsilon_r = 2.5$ (PVC, polyethylene, etc.) $\dfrac{1}{v} = 5 \dfrac{nsec}{m}$	$\epsilon_r = 5$ (Glass-epoxy) $\dfrac{1}{v} = 7.5 \dfrac{nsec}{m}$	$\epsilon_r = 1$ (Air) $\dfrac{1}{v} = 3.3 \dfrac{nsec}{m}$	$\epsilon_r = 2.5$ (PVC, polyethylene, etc.) $\dfrac{1}{v} = 5 \dfrac{nsec}{m}$	$\epsilon_r = 5$ (Glass-epoxy) $\dfrac{1}{v} = 7.5 \dfrac{nsec}{m}$
10				3	5	7.5
20				1.5	2.5	4
50	1.5	2.5	4	0.6	1	1.5
100	3	5	7.5	0.3	0.5	0.8
150	5	7.5	11			
250	8	12.5	20			
500	15	25	40			

$v = \dfrac{c}{\sqrt{\epsilon_r}}$

c = velocity of light in free air
ϵ_r = relative dielectric constant

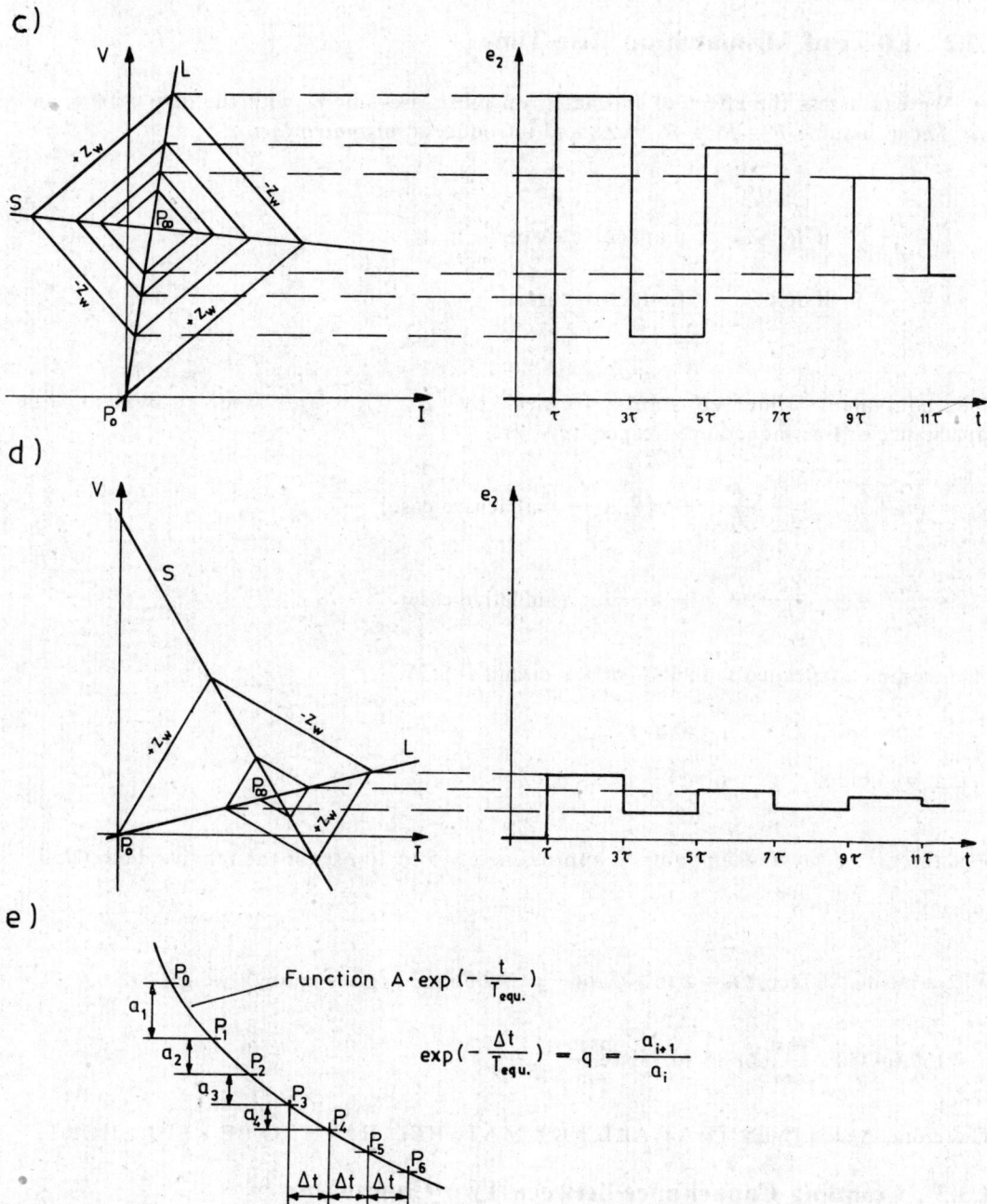

Fig. 4.4 Various cases of mismatch: Load voltage $e_2(t)$, determined with the Bergeron reflection chart [11]:
  a) R_s, $R_L \gg Z_w$ (capacitive approximation for line)
  b) R_s, $R_L \ll Z_w$ (inductive approximation for line)
  c) $R_s \ll Z_w \ll R_L$ (oscillations)
  d) $R_s \gg Z_w \gg R_L$ (oscillations)
  e) Exponential approximation of step-shaped waveform used for cases a) and b)

4.3.2 Effect of Mismatch on Rise-Time

We can assess the effect of mismatch on pulse rise-time t_r, with the help of C_{equ} and L_{equ}: Let us assume $R = R_S = R_L \neq Z_w$ and introduce a *mismatch factor*

$$k = \frac{R}{Z_w} \quad \text{if } R > Z_w \quad \text{(capacitive case)}$$

or $\quad k = \dfrac{Z_w}{R} \quad$ if $R < Z_w \quad$ (inductive case)

The equivalent time constants formed by R_S, R_L and the equivalent line capacitance or line inductance, respectively, are:

$$T_{equ} = \frac{1}{2} R \cdot C_{equ} = \frac{1}{2} \frac{R}{Z_w} \cdot \frac{l}{v} = \frac{1}{2} k \frac{l}{v} \quad \text{(capacitive case)}$$

$$T_{equ} = \frac{L_{equ}}{2R} = \frac{1}{2} \cdot \frac{Z_w}{R} \cdot \frac{l}{v} = \frac{1}{2} k \frac{l}{v} \quad \text{(inductive case)}$$

The rise-time associated with RC- or RL-circuit is [12]:

$$t_r = 2.2 \, T_{equ}$$

so that we obtain: $\quad t_r \approx k \dfrac{l}{v}$

$\dfrac{l}{v}$ is the time delay and amounts to approximately 5 to 10 nsec if the relative dielectric constant ϵ_r is 3 to 9.

With a mismatch factor $k = 2$ to 10, and $\dfrac{l}{v} = 7.5 \, \dfrac{\text{nsec}}{\text{m}}$ (glass-epoxy), we get

$t_r \approx 150$ to $750 \, \dfrac{\text{psec}}{\text{cm}}$ or 15 to 75 $\dfrac{\text{nsec}}{\text{m}}$

Therefore: ALL LINES THAT ARE NOT MATCHED HAVE TO BE KEPT SHORT!

4.3.3 Coupling Capacitance between Two Conductors

The parasitic capacitance between a signal line and ground, or between two signal lines is also given by the expression:

$$C_{equ} = \frac{l}{Z_w \cdot v}$$

Note that the wave impedance Z_w increases only logarithmically with distance. Wave impedances over 500 Ω are very rare. Therefore, the parasitic capacitance C_{equ} between 2 signal lines is not really influenced so much by separating 2 lines (see [16]). However, parasitic capacitances are reduced significantly by shortening the lines, as they are proportional to the length l.

Another method that one can use sometimes to reduce a disturbing parasitic capacitance between 2 conductors is to put a suitable *guard line* in between (see Section 4.7 of this chapter).

4.4 LINE DEFECTS: THE INDUCTIVE AND THE CAPACITIVE APPROXIMATION

Let us take the case of a line (conductor), which is properly matched at both its ends with $R_S = Z_w = R_L$ (see Fig. 4-5a), but has a piece which is defective: The signal line which should run strictly parallel to ground is nearer to the ground on a small stretch of length l_{def} resulting in a wave impedance Z'_w which is lower *locally*. Alternatively, we can also look at the case where the signal conductor is further away from ground ($Z'_w > Z_w$) on a small stretch. If two lines with different wave impedances Z_w, Z'_w are fixed together, they will give rise to reflections in a similar way as a mismatch at the end of the line. Therefore reflections will occur at the boundaries B and C. (Because source and load are matched, no reflections will occur at the boundaries A and D): When a wave arrives at point B from the left (Fig. 4-5a), i.e. in forward direction, a part of it will be reflected back to the source and a part will be transmitted onwards on the defective stretch of line up to point C. Here again a part of this 'new' wave will be reflected back to point B (where it will be again partly reflected and travel forward to C) and a part will be transmitted over the boundary and travel forward towards the load. Complicated multiple reflections will occur and the voltage waveform $e_2(t)$ on the load will have the characteristic step-shape indicated in Fig. 4-5b.

If $Z'_w > Z_w$ the voltage waveforms $e_2(t)$ and $e_1(t)$ will be similar to those introduced if a series inductor had been soldered into the line at the place of the line defect (Fig. 4-5c). The equivalent inductance L_{equ} is a fraction of the value [nH/cm] given in Section 4.3.1, for a mismatched line with, the corresponding value of Z_w and ϵ_r.

If $Z'_w < Z_w$ the voltage waveforms $e_2(t)$ and $e_1(t)$ will be similar to those introduced, if a shunt capacitor had been introduced into the line at the place of the line defect (Fig. 4-5c). The equivalent capacitance C_{equ} is a fraction of the values [pF/cm] given in Section 4.3.1 for a mismatched line with the corresponding value of Z_w and ϵ_r.

It is quite clear therefore that any line defects will further slow down the pulse rise-times, just as if a real inductor or capacitor (corresponding to L_{equ} or C_{equ}) had been there. (This is true provided the relative mismatch in line impedances, viz. $k = (Z'_w - Z_w)/Z_w$, has a larger magnitude than 10 %, otherwise the resulting amplitudes of the reflections due to mismatch are so small, that the line defect has hardly any practical effect. Small values of relative mismatch give waveforms similar to the one in Fig. 4-3.

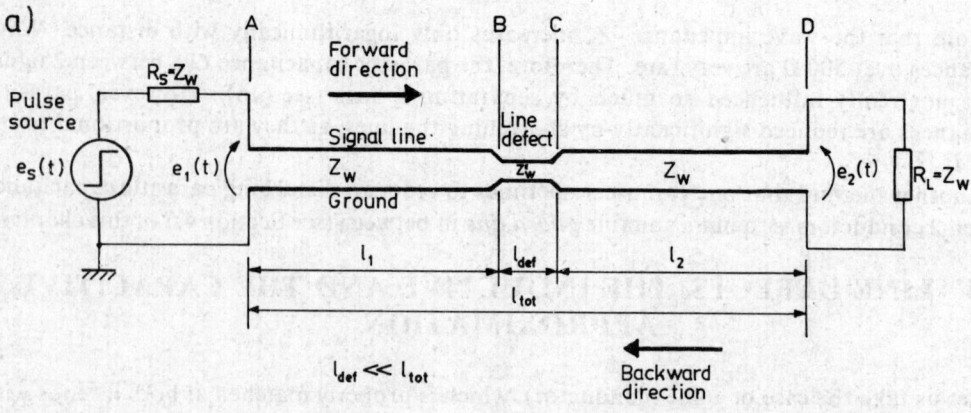
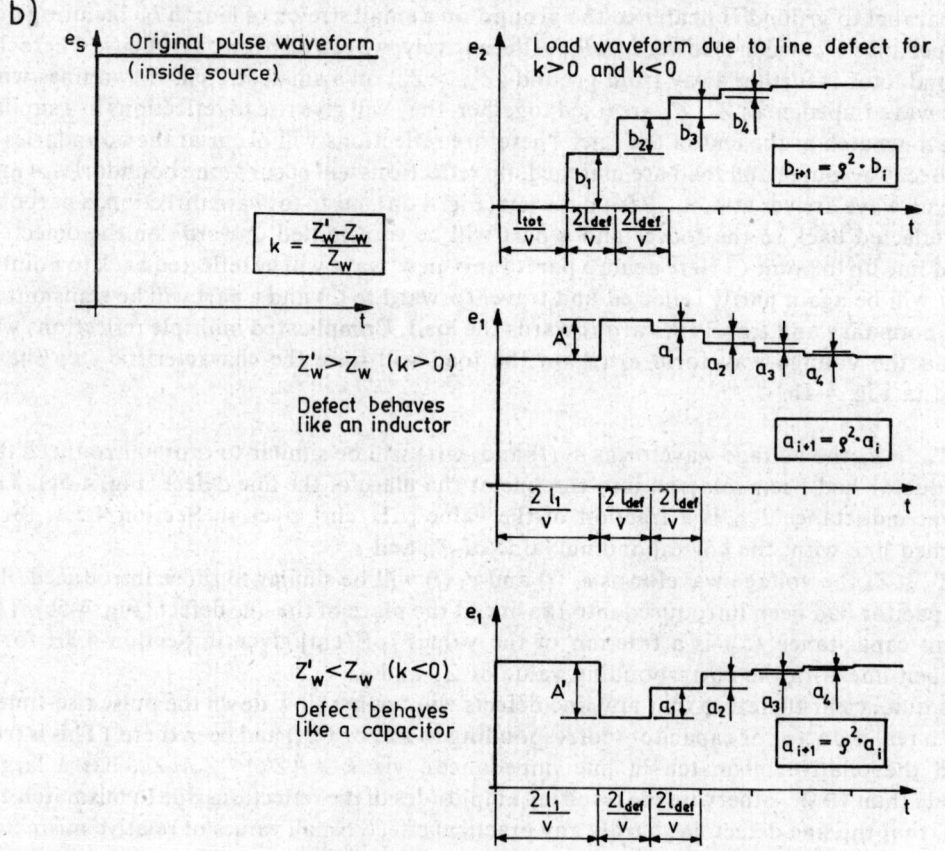

Fig. 4–5 Effect of a line defect on fast-pulse transmission:
 a) Basic configuration
 b) Waveforms $e_s(t)$, $e_1(t)$ and $e_2(t)$

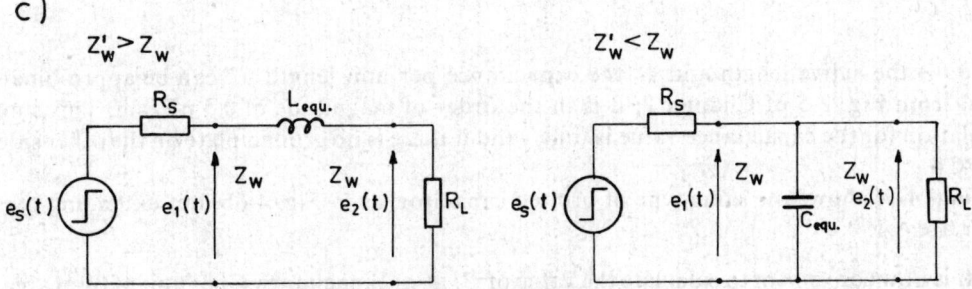

Fig. c) Equivalent circuit diagrams

4.5 PRINTED CAPACITORS AND PRINTED INDUCTORS

It is possible to obtain small values of capacitance and inductance directly as a printed circuit pattern. The advantage is that such printed capacitors have usually very good high-frequency properties. This means printed capacitors have low (parasitic or unwanted) series inductance and series resistance and printed inductors have low (parasitic) series resistance and shunt capacitance. Being intimately and directly connected with the wiring pattern, such printed capacitors and inductors are ideally suited for decoupling power-supply and ground lines (e.g., as additional decoupling capacitances in parallel with other capacitors) and for providing small values of compensation series inductance and compensation shunt capacitance, used to improve the bandwidth of amplifier stages and the rise-time of pulse circuits.

4.5.1 Realisation

Typical configurations for printed capacitors are shown in Fig. 4-6 while Fig. 4-7 shows typical configurations of printed inductors.

Fig. 4-6a shows a printed capacitor using electrodes on either side of a double-sided PCB. It can be calculated in the same way as the classical plate capacitor:

$$C = \frac{A}{b} \epsilon_0 \epsilon_r$$

where A is the surface area (of the smaller electrode, if both electrodes do not have identical areas); b is the PCB thickness and $\epsilon = \epsilon_0 \cdot \epsilon_r$ the dielectric constant of the PCB.

For $b = 1.6$ mm and $\epsilon_r = 5.4$ (glass-epoxy), we obtain:

$$C \approx 3 \cdot A \left[\frac{pF}{cm^2}\right] \quad A = \text{electrode area [cm}^2\text{]}$$

Fig. 4-6b shows a printed capacitor located entirely on one side of the PCB. Its capacitance C can be calculated very roughly by setting:

$$C = l \cdot C'$$

where l is the active length and C' the capacitance per unit length. C' can be approximately found from Fig. 2-5 of Chapter 2; it is in the order of magnitude of 0.5 pF/cm. This simple calculation for the capacitance value is only valid if there is no ground plate on the other side of the PCB.

Fig. 4-6c shows the same type of printed capacitor, as in Fig. 4-6b, but extending over a larger area A.

It is now convenient to calculate the value of C/A, of capacitance C per unit of total area A, using values given in [2] and represented graphically in the figure.

Note In this example the fabrication tolerances play a large role in reproducing the capacitance value exactly: An increase in conductor width w, will at the same time decrease the distance s—both changes will increase C rather than compensate for each other.

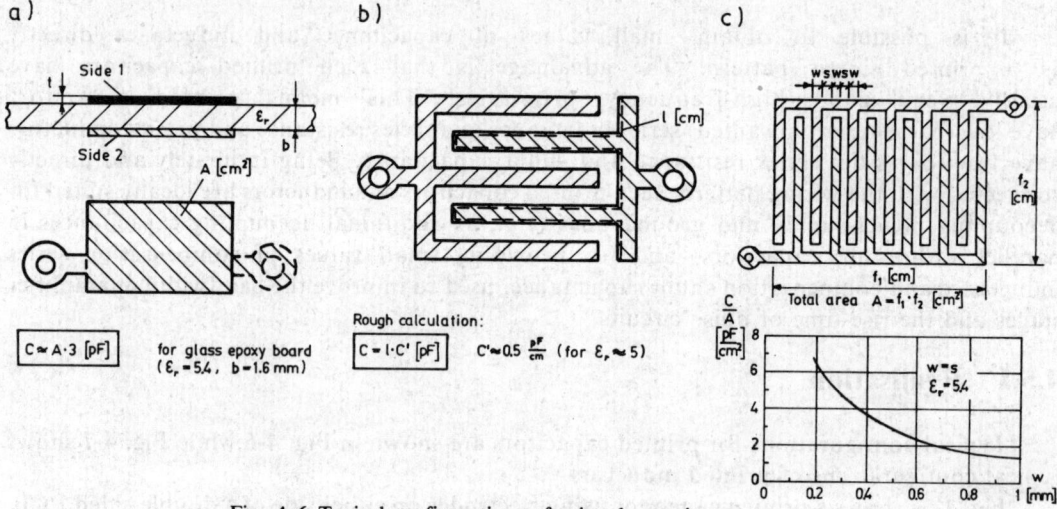

Fig. 4-6 Typical configurations of printed capacitors

Fig. 4-7a shows a round single-turn loop inductor. To determine its value of inductance L, we can use the equation (see [1, 6b]:

$$\frac{L}{f} = 2 \cdot \ln\left(\frac{2f}{w+t} \, 2.5\right) \left[\frac{nH}{cm}\right] \quad \text{for} \quad \frac{f}{w+t} \geqslant 5$$

where f is the mean (or average) circumference of the loop, w the width of the conductor, t the thickness of the copper layer on the PCB, ln denotes natural logarithm (base e). This equation is represented graphically in the figure.

Fig. 4-7b shows a round spiral-type loop with many turns. Here, we can use the formula (see [1, 6b]):

$$\frac{L}{a} = 0.0215 \cdot N^{5/3} \cdot \ln \frac{8a}{c} \quad \left[\frac{\mu H}{cm}\right]$$

where a is the mean or average radius, $a = 1/2 \cdot (R_1 + R_2)$; N is the number of turns and $c = R_2 - R_1$ the 'radical depth of winding'. Where printed inductors are used as frequency-determining elements in resonators or filters, the quality factor Q is important. To get high values of Q, round (rather than square) loops should be used, with value of $c/a \approx 0.8$.

Fig. 4-7c shows a square loop- or spiral-type inductance. It has a slightly higher inductance (about 10% more) than a round loop or spiral, but a lower value of Q. It is also simpler in layout-making.

Fig. 4-7d shows a meander-type conductor that can serve as inductor, if it is sufficiently thin (e.g., $w = 0.5$ mm). As a very rough calculation of inductance value, we can write:

$$L = l \cdot L'$$

where l is the length of the conductor and L' the inductance per unit length. L' can be very approximately found from the data given in Section 4.3.1 of this chapter.

$$L' = \frac{Z_w}{v} \quad (v = \text{transmission velocity})$$

L' is in the order of magnitude of about 3 to 10 $\frac{nH}{cm}$.

4.5.2 General Remarks

If inductors and capacitors with small tolerances ($\pm 5\%$ or less) have to be fabricated as printed elements, the accuracy requirements on the artwork and, especially on the production process (photoresist coating and etching operation) become extremely difficult. We had already hinted at this in the case of printed capacitors according to Fig. 4-6c: If we have a tendency to underetch while producing the PCBs, then w will be increased and s decreased. Both changes will increase the capacitance value. Suppose, $w = s = 0.5$ mm (by no means an extreme case), a change as small as $\Delta w = +10$ μm, $\Delta s = -10$ μm will already increase the capacitance value by a few %, as can be computed from Fig. 2-5 of Chapter 2.

Artwork tolerances are less critical, because one can always proceed by trial and error: Design a PCB — fabricate it — measure capacitance C or inductance L and then modify slightly the artwork — fabricate again—measure again, etc., until you have the exact value of C or L which you want. However, to be able to do this, one needs a fabrication process where the fabrication tolerances are very narrow and very well controlled so that PCBs are obtained in a fully reproducible fashion, and very small modifications in the artwork get translated into reproducible and correspondingly change in the measured values of C and L.

Another remark is that the engineer designing printed capacitors and inductors will gradually have to build up his own design data (charts, tables) that will give him more precise information, for his own type of PCBs and of layout designs, than what we give here.

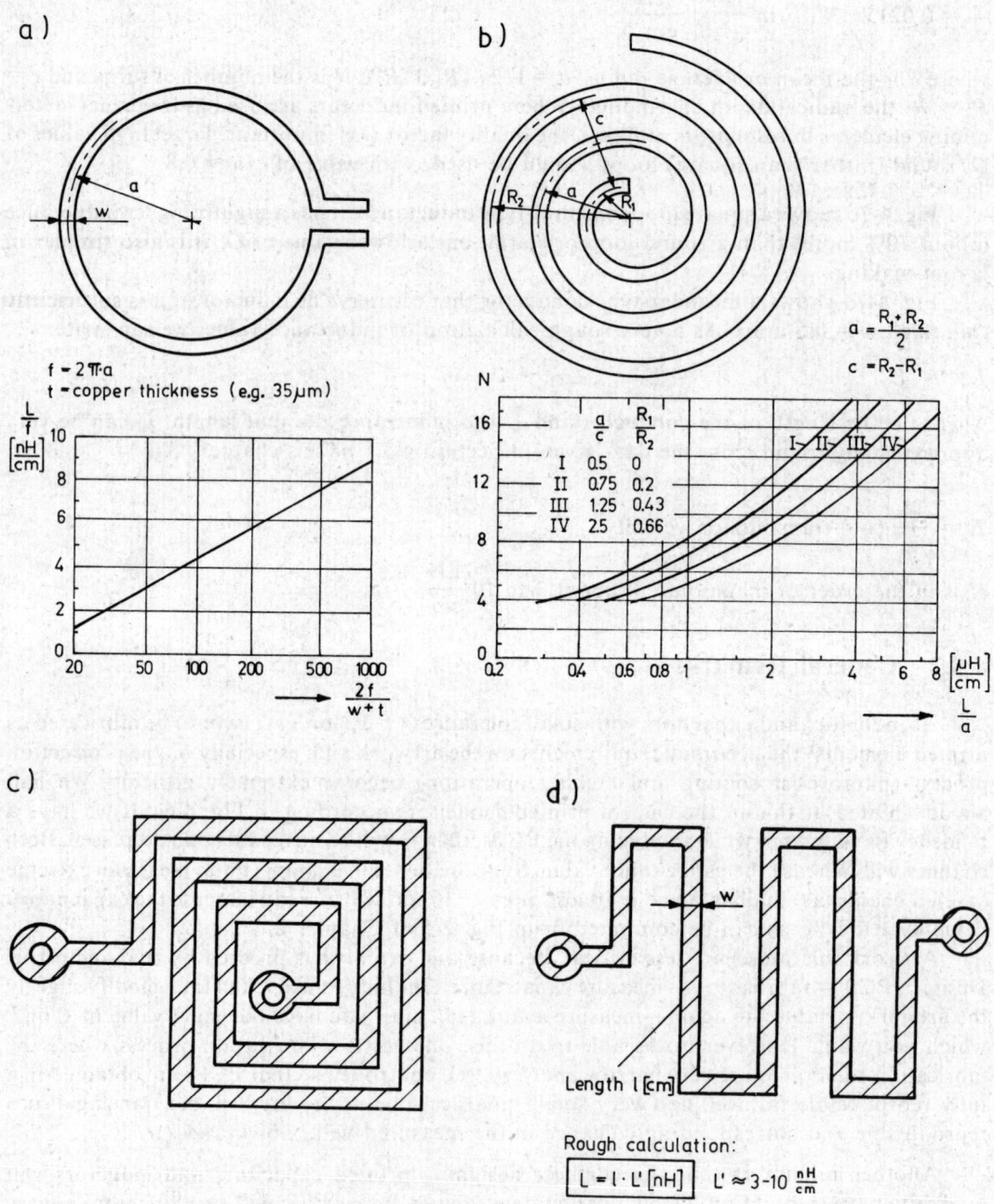

Fig. 4-7 Typical configurations of printed inductors

4.6 GROUND AND SUPPLY LINES

Because the basic function of ground and supply lines is to ensure certain constant DC potentials, the beginner is misled to believe that the high-frequency or fast-pulse properties of these lines are not important. The contrary is true: In fast circuits, the currents drawn from the power supply line and fed back into the ground do not have constant DC currents; rather they have very high frequency components, such as current spikes.

For the proper behaviour of fast-pulse circuits and high-frequency circuits, it is necessary to supply such current-spikes and other fastly-varying currents, and still to keep ground and supply at a constant potential: Otherwise, the behaviour of the circuit will deteriorate significantly (e.g., fast-pulse circuits will become 'slow-pulse' circuits and behave in a strange manner). Therefore, power supply lines just like the signal lines, must be kept as short as possible. The following rules should be followed:

1. Keep power supply lines and ground lines short.
2. Solder at least one decoupling capacitor (value: approximately, 5 to 50 nF) with good high-frequency properties (with low series inductance) on every PCB and between each power supply line and ground, so that fast current-spikes can be supplied by this capacitor and do not need to be drawn from the (more distant) power supply itself. As decoupling capacitor, a *ceramic chip* with short leads is ideal. Wound-types of paper, polyester and other capacitors are not recommended (they have a large series inductance) and electrolytic capacitors should be definitely avoided.
3. For very-fast or very-high-frequency circuits, it is advisable to have a printed circuit capacitor—on the print—in parallel with the ceramic-chip decoupling capacitor. The printed circuit capacitor has the most direct, i.e., the fastest connection with power supply and ground lines.
4. It is advisable to have broad power supply lines and ground lines sitting very near each other, preferably even sitting just across the printed circuit board: i.e., facing each other on both sides of a printed circuit board. This will turn the whole length of the power-supply conductor on the PCB into a *very low impedance line*—a low impedance line can supply high current spikes without having a large momentary voltage drop.
5. The ground line should be the very broadest conductor on the PCB (about double as broad as the power-supply lines)—thus the voltage spikes will be less on the more critical ground line than on the less critical power supply lines. If a lot of copper surface on the PCB is connected to the ground line, it becomes electromagnetically very difficult to change the potential of the ground and the ground will remain very stable and not exhibit any voltage spikes. Therefore, leave the copper in all unutilized parts of the PCB (e.g., in the corners, etc.) and connect all this copper to ground. In many critical high-frequency and fast-pulse applications, one even recommends taking a double-sided PCB and connecting one side entirely to ground (as *ground plate*). Such a ground plate also serves as a shield, to reduce electromagnetic interference.

4.7 CONCEPT OF GUARD LINE: EXAMPLE OF A CASCODE STAGE

Let us take the cascode stage of Fig. 4-8 as example of a typical high-frequency circuit and see how a suitable PCB can be designed. The parasitic elements C^*, L^* are drawn with dotted lines.

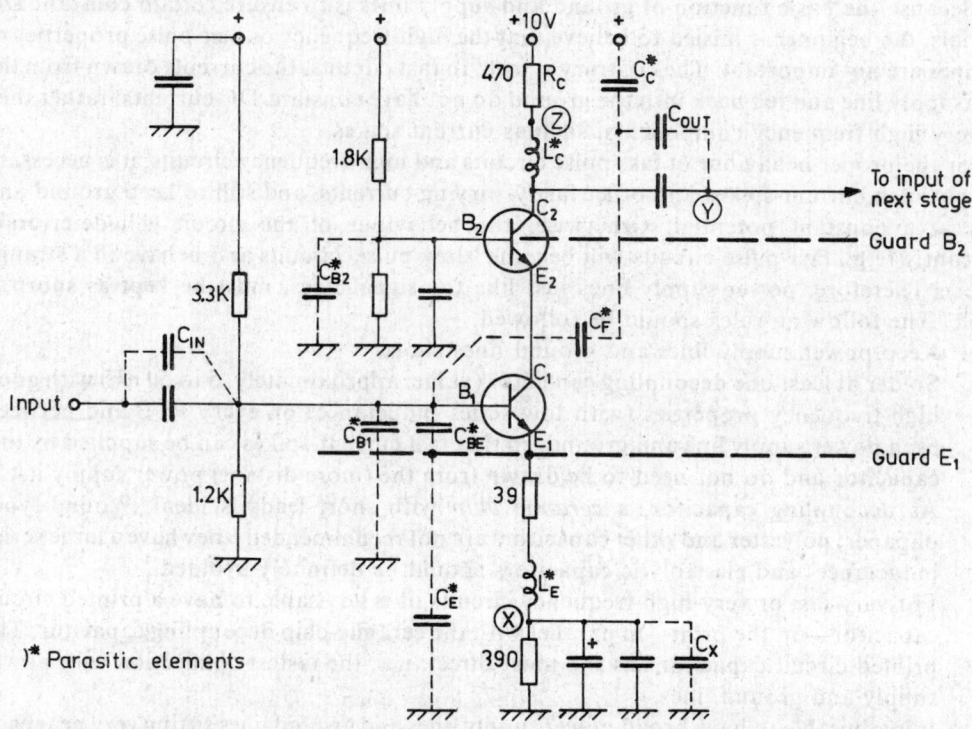

Fig. 4–8 Example of a fast-pulse circuit: Cascode-stage for a broadband amplifier (designed for transistor 2N918 with a gain of 6 per stage and a rise-time of 12 nsec per stage: gain-bandwidth product approximately 180 MHz)

The two-transistor cascode stage is specially suited for high-frequency applications because it has a very low feedback capacitance (Miller capacitance) between output C_2 and input B_1. In fact, any feedback capacitance C_f between C_2 and B_1 would invariably lead to a high dynamic input capacitance $A \cdot C_f^*$, where A is the stage amplification. It is precisely with the object of reducing interior value of C_f that a cascode stage rather than the simpler one-transistor common-emitter stage is chosen.

However, this very purpose of having a cascode stage would be defeated, if a high parasitic feedback capacitance C_f^* is introduced *through the PCB*.

Therefore, the design of the PCB must take into consideration the following points:
a) Conductors connected to point C_2 must be kept far away from conductors connected to B_1.
b) To reduce C_f^* even further, we provide a *guard-line* that is connected to point B_2. Point B_2 is, as far as AC- or high-frequency considerations go, shorted to ground potential through capacitor C_B.
Such a broad grounded guard-line will reduce C_f^* drastically and further decrease any undesired feedback from the output back to the input.
c) In emitter E_1, a small parasitic capacitance C_E^* to ground of a few pF value is

DESIGN RULES FOR PCB'S IN HIGH-FREQUENCY AND FAST-PULSE APPLICATIONS   81

desirable, as it will increase the upper bandlimit (see [10] of Chapter 5), but any inductance L_E^* in series with R_E is not desirable. This is precisely achieved if point E_1 is connected to a broad conductor running near and parallel to the ground conductor.

d) For collector C_2, the opposite is true: An inductor L_C^* of a fraction of 1 µH will actually serve as a compensation inductance, increasing the bandwidth by up to approximately 50% (see [12, 15]). However, a capacitance C_C^* between point C_2 and ground will act as a load capacitance and will therefore decrease the bandwidth. Therefore, point C_2 should be kept far away from ground and the conductors leading to resistor R_C should be very narrow and long, possibly even going in a meander or spiral pattern as shown in Fig. 4-7b, c, d.

e) Capacitance C_{B1}^* between input (point B_1) and ground should also be small. This is obtained by using a guard-line connected to emitter E_1. Because of this guard-line E_1, which is almost at the same potential as B_1, the capacitance C_{B1}^* between base B_1 and ground is drastically decreased. Of course, at the same time C_{BE}^* is increased (this does not matter too much because the emitter E_1 is almost at the same voltage as the base B_1 and therefore very little current flows through C_{BE}) and C_E^*, also is increased (this again does not matter too much because the emitter E_1 is a low-impedance point; in certain cases a capacitance C_E^* may even be helpful, see step c above).

f) The potential between ground and V_{CC} must be stabilised by connecting a decoupling capacitor (ceramic chip) between these two points. In parallel to this ceramic chip, we can have a printed (PCB) capacitor as shown in Fig. 4-6a. Similar PCB capacitors are placed in parallel to soldered capacitors C_{in}, C_{B2}, C_{out}, C_x.

The result of these design steps is shown in Fig. 4-9. For the layout of Fig. 4-9, we can assume a double-sided PCB with a full ground plate on the other side.

4.8 RISE-TIME LIMITATIONS AND HIGH-FREQUENCY LOSSES DUE TO SKIN-EFFECT, DIELECTRIC LOSSES AND RADIATION LOSSES

Note that the considerations under this section are important for very fast pulses only.

If a line is matched (by choosing source and load terminations R_S, R_L equal to Z_w), one would actually be transmitting all pulses in an ideal manner, without changing their rise-time, provided the line had really no losses. However, real lines have losses (due to skin-effect or dielectric losses or radiation losses) and so even a line that is perfectly matched will increase the rise-time of the pulses that it carries. Rise-time limitations due to line losses are almost always much less severe than those due to *line mismatch*. They will only be felt for long lines (length of at least 1 or 2 m), i.e. for the wiring and cables between PCBs and for PCB delay lines, or otherwise for very fast pulses ($t_r < 0.1$ nsec). They will be briefly discussed in the following; orders of magnitude rather than exact values will be given for the resulting rise-times.

Let us look at the cases of *longer wiring* (cables) and long PCB delay lines first: For long matched high-frequency cables with good quality dielectrics, the skin-effect is the greatest limitation (see [9, 11, 16]), whereas if one uses matched cables with ordinary dielectrics, the dielectric losses become much higher and often limit the rise-time at a much larger value than if only the skin-effect was there. The same happens with PCB delay lines. PCB delay lines are used, e.g., in oscilloscopes, to delay a whole signal waveform by a constant value, say 100 nsec;

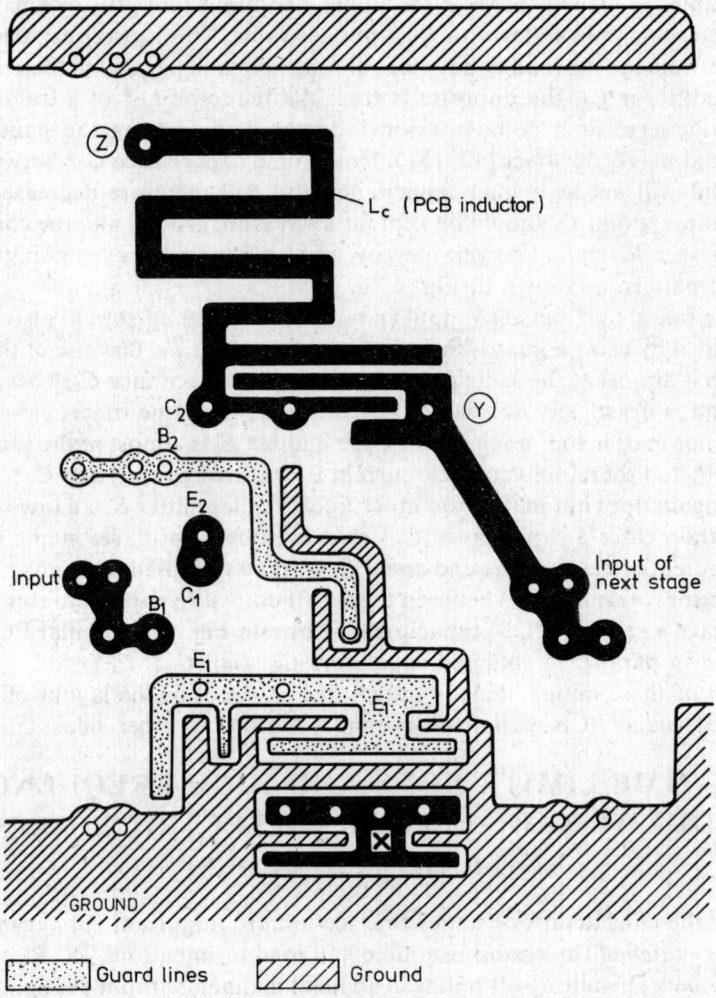

Fig. 4–9 PCB layout for the circuit of Fig. 4–8: Schematic (not actual practical) layout showing, with some exageration, various possible specific design measures that can be used to improve performance (e.g., guard lines E_1 and B_2 to reduce input capacitance; use of printed inductors and capacitors). Actual optimal working PCB will need still further modifications by trial PCB fabrication and measurement

they are transmission lines with a total length of several meters, fabricated on a PCB either in a meander form (similar to Fig. 4-6c) on a rigid PCB, or as a straight line on a flexible but folded PCB.

Looking now at *shorter conductors* on PCBs (length < 20 cm), we find that we often have a mixture of dielectric losses (especially with glass-epoxy and paper-phenolic PCBs), radiation losses and, to a certain extent, of skin-effect losses. These losses are only felt if the pulses have rise-times in the picosecond range; it is therefore not easy to predict their behaviour.

In the following, we shall look at the various effects separately and give very rough, approximate values (orders of magnitude) that describe their individual contributions to the pulse rise-time t_r.

4.8.1 Skin-Effect Losses

The skin-effect losses are sometimes also called 'conductor losses'. First we will consider skin-effect limitations only. The skin-effect gives rise to losses that are proportional to the square root of frequency ($\sim \sqrt{f}$) and therefore this leads to a rather unusual pulse-response waveform.

Theoretical considerations show the following interesting conclusions for a high-frequency cable with skin-effect losses but without dielectric losses (see [9, 11]).

1) If a matched cable having skin-effect is driven by an ideal step, then the response $e_2(t)$ at the load will always have the form of a so-called 'error function' as shown in Fig. 4-10. Because the second half of the response $e_2(t)$ is very rounded, the application of the standard rise-time definition (the rise-time t_r is usually defined as going from 10 to 90% of the response) to this case would give rather unrealistically high values. For step response due to skin-effect, one therefore uses a rise-time $T_{0\ 50\%}$ which is measured between 0 and 50% of the response rather than the usual rise-time $t_r = T_{10\ 90\%}$. One finds in fact for the case of skin-effect:

$$t_r = T_{10-90\%} \approx T_{0-90\%} = 30 \cdot T_{0-50\%}$$

2) The skin-effect rise-time $T_{0\ 50\%}$ for a matched high-frequency cable driven by an ideal step increases with the square l^2 of the cable length, according to the equation[1] (see [11]):

$$T_{0-50\%} = \frac{\alpha^2}{\pi \cdot f} \cdot l^2; \quad f\,[\text{Hz}], l\,[\text{m}]$$

---

[1] In deriving these relationships, a certain *approximation* has been made. Therefore, the error function waveform, the l^2-dependence and the corresponding equation for $T_{0-50\%}$ given here is only valid, as long as the condition is fulfilled: $t \ll \frac{1}{4} \cdot \frac{f}{\alpha^2} \cdot \frac{1}{v^2}$ where $t = T_{0-50\%}$ for this consideration.

(See [11], Section 5.4 where the condition is given in the form of $t \ll \frac{L^2}{\pi K^2}$, and substitute therein $\frac{1}{K^2} = \frac{\pi f}{4 \alpha^2} \cdot \frac{C}{L}$; $v = \frac{1}{\sqrt{LC}}$ = propagation delay).

Using now the relation $T_{0-50\%} = \frac{\alpha^2}{\pi f} \cdot l^2$ itself and solving the inequality for l, one obtains the condition in the form:

$$l \ll l_{\text{crit}} = \sqrt{\frac{\pi}{4} \cdot \frac{f}{\alpha^2} \cdot \frac{Np}{m}} \cdot \frac{1}{v} = 70 \cdot \sqrt{\frac{f}{\alpha^2} \cdot \frac{1}{v} \cdot \frac{dB}{m}}$$

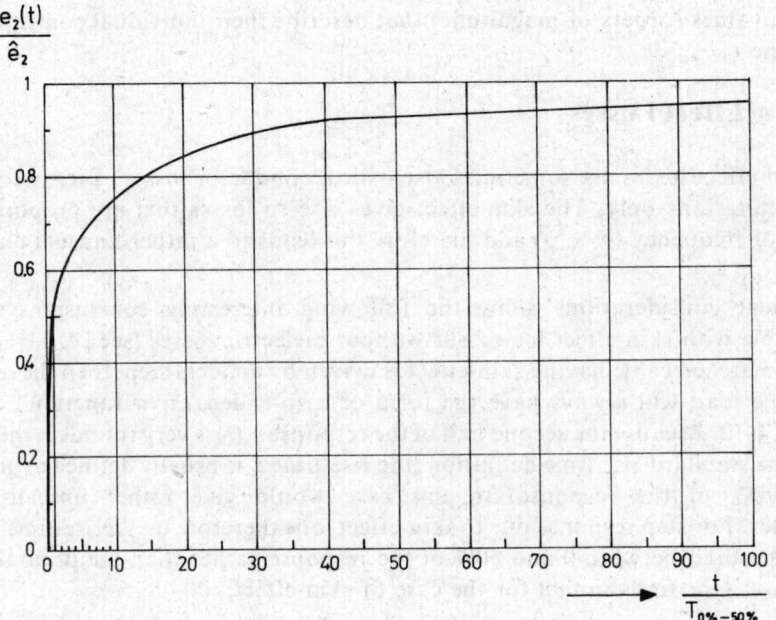

Fig. 4-10 Normalised response curve showing output waveform $e_2(t)$ of matched transmission line over normalised time, if skin-effect is predominant

where α is the attenuation of the cable in Neper/meter at a given frequency f. This can be rewritten as

$$T_{0-50\%} = 0.0042 \frac{\alpha^2}{f} l^2$$

where α is given in decibel/meter.

Because of this law, it is impossible to transport very fast pulses by cable over a long distance (see also [9]).

Remarks

a) The l^2-*dependence* of $T_{0-50\%}$ as well as the equation given above are *only valid for short lengths* $l \ll l_{\text{crit}}$. The condition of validity of the corresponding expressions for

$$t_r = 30 \cdot T_{0-50\%} \text{ is } l \ll \frac{l_{\text{crit}}}{\sqrt{30}}.$$

b) *PCB-microstrip* rise-time limitations due to skin-effect can be conveniently determined with the help of Fig. 4-11, which in turn is based on data given in [6c, 13].

c) *Twisted pair of wires* has been calculated, using $Z_w \sqrt{\epsilon_r}$ values from Fig. 4-1d, f as well as the expressions

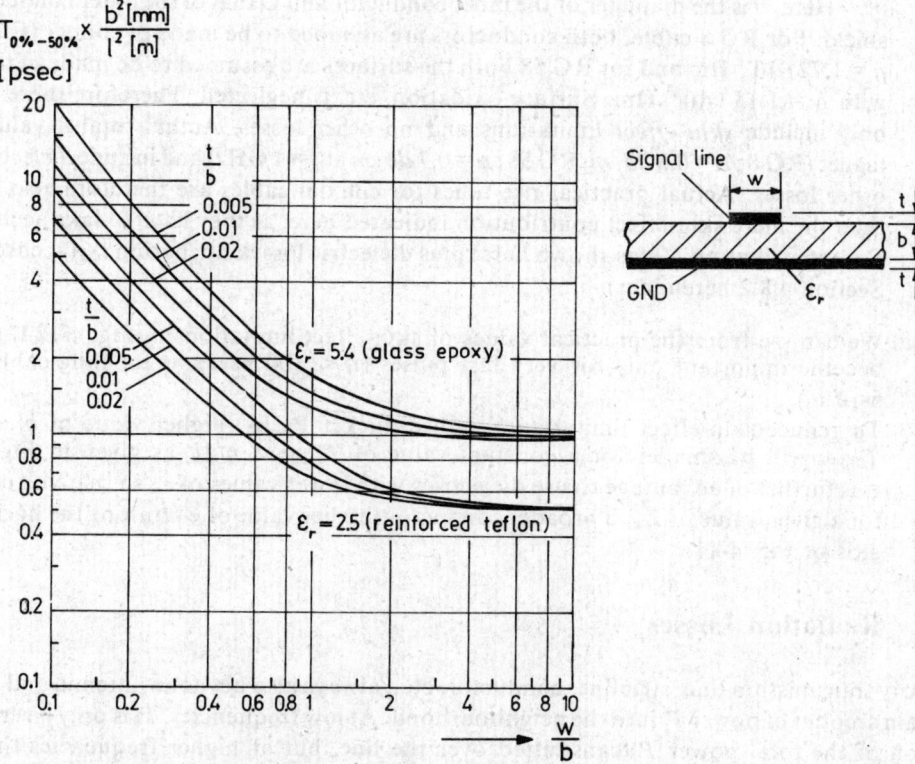

Fig. 4-11 Rise-time limitations by skin-effect in microstrip: Graph giving $T_{0-50\%}$ in function of microstrip dimensions w/b based on Fig. 19 or [6c]

$$\alpha = \frac{K}{2 Z_w} \sqrt{\pi f} \quad ([11], \text{Section 5.3})$$

$$K = \frac{\sqrt{\rho \mu}}{\pi} \cdot \frac{2}{d} \cdot \frac{1}{\sqrt{2\pi}}$$

The last equation is taken from the one given in [6a] for coaxial cables (equation 35) and adapted for the case of twisted wires. The factor $1/\sqrt{2\pi}$ has to be introduced because we have defined here the skin-effect constant $K = R_C/l \sqrt{\omega}$, in accordance with [11], rather than $K = R_C/l \sqrt{f}$ as used in [6a]. R_c stands for skin-effect resistance; ρ for specific resistance ($\rho = 1.7 \cdot 10^{-8}$ Ωm for copper); $\mu = \mu_0 \cdot \mu_r$ is the permeability ($\mu_0 = 4\pi \cdot 10^{-7}$ H/m, μ_r = relative permeability = 1 for copper).

d) *Coaxial cables* have been calculated using the approximate equation:

$$K = \sqrt{\frac{\rho \mu}{\pi}} \cdot \left(\frac{1}{d} + \frac{1}{D}\right) \cdot \frac{1}{\sqrt{2\pi}}$$

This expression for K is taken from [6a] but divided by $\sqrt{2\pi}$ to account for the different definition of K used by us.

Here, d is the diameter of the inner conductor and D that of the outer conductor or shield. For RG 8 cable, both conductors are assumed to be made of copper (Cu) with $\rho = 1.72 \cdot 10^{-8}$ Ωm and for RG 58 both the surfaces are assumed to be made of tin (Sn) with $\rho = 1.15 \cdot 10^{-7}$ Ωm. Surface oxidation, etc. is neglected. Therefore these values only include *skin-effect* limitations and no other losses. Actual total α-values are higher (RG 8: $\alpha = 0.3$ db/m; RG 58: $\alpha = 0.7$ db/m at $f = 1$ GHz) and include *dielectric* and other losses. Actual practical rise-times for coaxial cables are therefore also higher than the mere skin-effect contribution indicated here, as they have to take account of skin-contribution t_{SK} (as shown here) plus dielectric loss contribution t_{die} (according to Section 4.8.2 hereunder).

e) We can see from the practical values of skin-effect limitations in Fig. 4-12 that they become important only for very fast pulses ($t_r < 100$ psec) or for long cables ($l > 5$-10 m).

To reduce skin-effect limitations, choose thicker PCBs (higher value of b, so that $T_{0\,50\%}$ will be smaller for a constant value of $T_{0\,50\%} \times b^2/l^2$ as given in Fig. 4-11. It is further of advantage to use dielectrics with lower values of ϵ_r, so that you can use, for a given value of Z_w, a broader conductor (higher value of w/b along the horizontal axis of Fig. 4-11).

4.8.2 Radiation Losses

Any transmission line, stripline, conductor, etc., always also act as an antenna and radiate a certain amount of power P_r into the neighbourhood. At low frequencies, P_r is only a very small fraction of the total power P transmitted over the line, but at higher frequencies the ratio $\dfrac{P_r}{P}$ increases.

Let us define as cut-off frequency f_c that frequency where $P_r/P = 50\%$, for a given line. We can then very approximately say that the output rise-time of this line will be (for an ideal pulse at its input):

$$t_r \approx \frac{0.3}{f_c}$$

according to a well known relationship between cut-off frequency and rise-time. The wavelength λ_c corresponding to f_c is

$$\lambda_c = \frac{v}{f_c}, \text{ where } v = \text{transmission velocity.}$$

Therefore $t_r = 0.3 \dfrac{\lambda_c}{v}$.

Remember that λ_c is the wavelength at which $\dfrac{P_r}{P} = 50\%$.

For a *wire in free air* ($v = c = 3 \cdot 10^8$ $\dfrac{\text{m}}{\text{sec}}$) at a distance of b over a groundplate and with a wave impedance of Z_w, one finds approximately [6c, 8]:

DESIGN RULES FOR PCB'S IN HIGH-FREQUENCY AND FAST-PULSE APPLICATIONS

			α at f=1 GHz $\left[\frac{dB}{m}\right]$	l_{crit} [km]	$\tau_{0\%-50\%}$ l=20 cm [psec]	l=1 m [psec]	l=10 m [psec]	l=100 m [nsec]	$t_r = \tau_{10\%-90\%}$ l=20 cm [psec]	l=1 m [psec]	l=10 m [nsec]	l=100 m [nsec]
PCB–Microstrip Cu conductor, Fig. 4-1b	ε_r=5.4	Z_w=50 Ω, w=2.5mm, $\frac{w}{b}$=1.5	0.3	6	0.016	0.4	40	4	0.48	12	1.2	120
		Z_w=100 Ω, w=0.5mm, $\frac{w}{b}$=0.3	0.6	1.5	0.06	1.5	150	15	1.8	45	4.5	—
	ε_r=2.5	Z_w=50 Ω, w=4.3mm, $\frac{w}{b}$=2.65	0.2	9	0.008	0.2	20	2	0.24	6	0.6	60
		Z_w=100 Ω, w=1.2mm, $\frac{w}{b}$=0.72	0.3	4	0.016	0.4	40	4	0.48	12	1.2	120
Twisted pair of wires with dielectric, Cu conductor, Fig. 4-1e	ε_r=2.5 Z_w=115 Ω $\frac{a}{d}$=2.33	d=0.15mm, m=0.1mm	0.95	0.4	0.15	3.7	370	37	4.5	110	11	—
		d=0.6mm, m=0.4mm	0.25	6	0.01	0.23	23	2.3	0.3	7	0.7	70
		d=2mm, m=1.67mm	0.07	80	0.0008	0.02	2	200	0.025	0.6	0.06	6
50 Ω coaxial cable ε_r=2.3		d=0.9mm, D=3.6mm, RG 58 low-cost (Sn)	0.59	1	0.06	1.5	150	15	1.8	45	4.5	—
		d=2.25mm, D=8.2mm, RG 8 / RG 213 (Cu)	0.09	40	0.0014	0.035	3.5	0.35	0.042	1.05	0.105	10.5

Fig. 4–12 Practical values for skin-effect limitations (The table gives rise-time values for various types of conductors under the assumption that only skin-effect is present)

$$\frac{P_r}{P} \approx \frac{320\ \Omega}{Z_w} \cdot \left(\frac{\pi h}{\lambda}\right)^2 \quad \text{for } l \gg 40 \cdot b^1$$

By setting $\frac{P_r}{P} = 0.5$, we obtain for $\lambda = \lambda_c$:

[1] This expression has been derived using the condition

$\frac{2h}{\lambda} \ll 1$ (actually $\frac{2h}{\lambda}$ or $\frac{D}{\lambda} < \frac{1}{10}$. see [6c, 8])

This condition is not properly fulfilled in the cases calculated hereunder, especially for $Z_w = 100\,\Omega$: Therefore the rise-time values given in Fig. 4-13 are mere orders of magnitude and not exact values in anyway.

$$t_r \approx \frac{h}{v} \sqrt{\frac{480\ \Omega}{Z_w}} \cdot \sqrt{f(\varepsilon_r)}$$

$$\lambda_c = \pi b \sqrt{\frac{640\ \Omega}{Z_w}}$$

And from there $t_r \approx \dfrac{b}{v} \sqrt{\dfrac{640\ \Omega}{Z_w}}$

For $Z_w = 50\ \Omega$ and air $\left(v = c\right)$: $t_r \approx 3.5 \dfrac{b}{c}$

Note that $\dfrac{1}{c}$ is approximately $30 \dfrac{\text{psec}}{\text{cm}}$, therefore

$$t_r \approx \left(100\ \frac{\text{psec}}{\text{cm}}\right) \cdot b$$

if b is the distance of the wire to ground plate given in centimeters.

If we have a *PCB microstrip* (Fig. 4-1b) instead of a wire in air, the radiation losses are decreased by a factor $f(\epsilon)$ decreasing more or less linearly with ϵ_r, and the constant in the above equation is reduced from $2 \cdot 320\ \Omega = 640\ \Omega$ to $2 \cdot 240\ \Omega = 480\ \Omega$:

This is obtained from the equations given in [6c, 8] with the same reasoning as above; $f(\epsilon_r)$ is the radiation factor given in [6c, 8] under the designation $F_1(\epsilon')$ for an open-circuit microstrip and $F_2(\epsilon')$ for a matched microstrip.

As $v = \dfrac{c}{\sqrt{\epsilon_r}}$, and $f(\epsilon)$ very approximately $\sim \dfrac{1}{\epsilon_r}$, the overall effect is that a higher relative dielectric constant ϵ_r hardly has any influence on rise-time, as long as Z_w and b remain constant.

Fig. 4-13 which is based on data given in [6c, 8], shows the orders of magnitude for rise-time limitations due to radiation losses in function of ϵ_r.

For the case of a *matched microstrip* with glass-epoxy ($\epsilon_r = 5.4$) and $Z_w = 50\ \Omega$:

$$t_r \approx \left(8\ \frac{\text{psec}}{\text{mm}}\right) \cdot b,$$

where b is the distance of the PCB signal strip (signal conductor) from the ground plate in millimeters.

For $b = 1.6$ mm (common PCB thickness): $t_r \approx 12$ psec.

We can see that radiation losses can be generally decreased if signal lines are run nearer to ground plate (lower value of b). On the other hand, radiation losses (and corresponding rise-times) are increased, if

DESIGN RULES FOR PCB'S IN HIGH-FREQUENCY AND FAST-PULSE APPLICATIONS

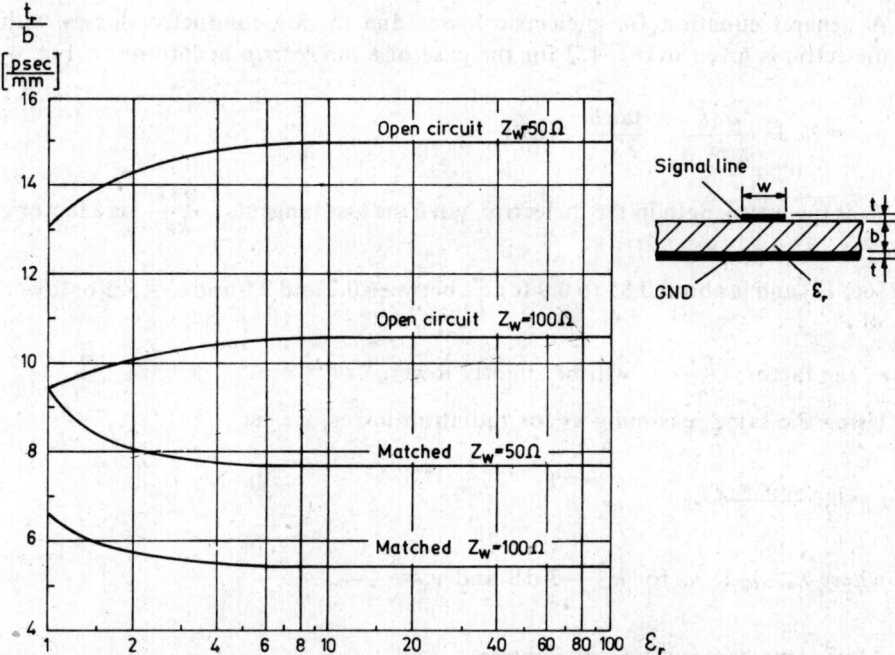

Fig. 4–13 Rise-time limitations by radiation losses in microstrip: Graph giving t_r in function of ϵ_r for matched and open-circuit microstrip, based on data given in [6c], equation 37a, 37b, and 38. The values shown indicate merely orders of magnitude

—no ground plate is used (this drastically increases b to a few centimeters perhaps and therefore also t_r)[1].

—discontinuities, geometric irregularities and mismatch are present.

Radiation losses would limit the rise-times to about 5 to 10 psec if a PCB with ground plate is used (this limitation is, of course, hardly felt, usually). However, if no ground plate is used, rise-times caused by radiation losses can become easily a few hundred psec and will then be felt for very fast pulse systems[1].

Note that radiation losses are approximately independent of the length l of the conductor, as long as $l \geqslant 40 \cdot b$; they will therefore be relatively more important for shorter conductors, such as PCB conductors. For longer conductors, such as cables/wiring between PCBs and PCB delay lines, skin-effect and dielectric losses become relatively more important.

4.8.3 Dielectric Losses

Dielectric losses are more material-dependent than skin-effect losses, as they evidently change very much with the kind of dielectric used. The following remarks can be made about pulse-behaviour in the presence of dielectric losses:

---

[1]If there is no ground plate, the radiation losses cannot be calculated with the equations given here, because the assumption $l \geqslant 40 \cdot b$ is no more fulfilled. Nevertheless, radiation losses are strongly increased if b is strongly increased.

a) A general equation for dielectric losses due to non-conductive losses within the dielectric is given in [6c, 15] for the case of a *microstrip* according to Fig. 4-1b:

$$\alpha_D = 27.3 \cdot \frac{q \cdot k}{k_e} \cdot \frac{\tan \delta}{\lambda_d} \, l \, [\text{dB}]$$

λ_d is the wavelength in the dielectric, $\tan \delta$ the loss tangent, $\frac{q \cdot k}{k_e}$ is a factor given in [6c, 13] and is about 0.85 to 0.9 for $\frac{w}{b}$ between 0.2 and 2.0 and $\epsilon_r = 5$. For lower values of ϵ_r, the factor $\frac{q \cdot k}{k_e}$ will be slightly lower.

Using the same reasoning as for radiation losses, we set

$$t_r \approx 0.3 \, \frac{\lambda_{d \, cut \, off}}{v}$$

where $\lambda_{d \, cut \, off}$ is λ_d for $\alpha_D = 3$ dB and $v = \frac{c}{\sqrt{\epsilon_r}}$.

After some computation, we obtain

$$\frac{t_r}{l} \approx 0.91 \cdot \frac{q \cdot k}{k_e} \cdot \tan \delta \, \sqrt{\epsilon_r} \cdot 10 \, \left[\frac{\text{nsec}}{\text{m}}\right]$$

b) Rise-time due to dielectric losses rises about linearly with conductor length l (and not, as with skin-effect, proportionally to l^2).

c) For small conductor length ($l \leq 20$ cm), as usual with PCBs and for low-cost dielectrics, dielectric losses are more limiting than skin-effect; for longer length ($l > 10$ m) and for high-quality dielectrics, the skin-effect takes over, because of its dependence on l^2.

d) Dielectric losses are the limiting factor with the usual PCB-dielectrics (glass-epoxy, paper-phenolic, etc.: $\tan \delta = 0.01$ to 0.02; see also [5], table 3) where they lead to values of rise-time t_r of a few hundreds of $\frac{\text{psec}}{\text{m}}$.

e) Dielectric losses are also especially limiting for general-quality wires with PVC insulation ($\tan \delta \approx 0.03$)[5]. Resulting rise-times are in the order of magnitude of $0.5 \, \frac{\text{nsec}}{\text{m}}$.

f) To keep dielectric losses low, and thus to increase the speed of the circuit wherever dielectric losses are really the limiting factor, one must use dielectrics with lower losses (lower value of loss tangent $\tan \delta$ at high frequency, e.g., Teflon[1]). These high-frequency dielectrics are very costly therefore they should only be used if they are

---

[1]Trademark of E.I. Du Pont de Nemours & Co., Inc.

really required, i.e., if dielectric losses and not skin-effect (or other effects) are the limiting factors. We can always build and test the same PCB once with a low-cost laminate and once with a laminate with a good high-frequency dielectric. By comparing the measured values of rise-time t_r, we can see whether the quality of the dielectric has any influence or not.

g) High-quality dielectrics have values of tan δ from $2 \cdot 10^{-3}$ to 10^{-5} (see [6c]) and therefore rise-time t_r of 1-30 $\frac{\text{psec}}{\text{m}}$ which are hardly ever felt.

4.8.4 Concluding Remarks on Line Losses

Various types of losses and especially skin-effect and dielectric losses increase their effect on the rise-time t_r as the conductor length l is increased. Therefore the first rule is to keep conductors of fast-pulse circuits as short as possible.

Other measures are selective and depend on which losses are the ones which are more limiting in a given case:
— To reduce *skin-effect losses*: Increase PCB-thickness (i.e., thickness b of the dielectric), decrease ϵ_r and, above all, *keep line length l small* (here, t_r increases with l^2).
— To reduce *radiation losses*: *Use a ground plate on one side of the PCB*, and decrease PCB-thickness b, avoid discontinuities. If no ground plate can be used, at least run signal lines near to ground.
— To reduce *dielectric losses*: Use a PCB laminate with a *good high-frequency dielectric*, i.e., the dielectric should have a low value of loss tangent tan δ.

All these limitations are only felt if these lines are matched. If the lines are mismatched, then the first and most important measure is to match all lines.

4.9 SUMMARY AND RECOMMENDATIONS

4.9.1 Summary of Effects

Matched lines are fastest: Single-sided matching is actually enough, but double-sided is recommendable. Matched lines will have rise-times of usually $< 1 \frac{\text{nsec}}{\text{m}}$. Rise-times t_r of something like 100 psec for a line length of $l = 20$ cm are common and are due to skin-effect, radiation losses, and—if a low-cost dielectric is used—also due to dielectric losses. Note that skin-effect causes t_r to increase with l.

Mismatched lines will give multiple reflections and slow down the circuits: If $R_S, R_L \ll Z_w$, the line behaves as an *inductor* with an equivalent inductance $L = \frac{Z_w}{v} l$ $\left(v = 10\text{-}30 \frac{\text{cm}}{\text{nsec}}\right)$.

If $R_S, R_L \gg Z_w$, the line behaves as a *capacitor* with an equivalent capacitance $C = \frac{1}{Z_w \cdot v} \cdot l$ $\left(v = 10\text{-}30 \frac{\text{cm}}{\text{nsec}}\right)$.

The rise-time increase due to a mismatched line will be a multiple of its transmission delay of approximately 5-10 $\frac{\text{nsec}}{\text{m}}$; as a rough estimate, one can take it to be 10–100 $\frac{\text{nsec}}{\text{m}}$ or $0.1 - 1$ $\frac{\text{nsec}}{\text{cm}}$.

Line defects:—if Z_w is locally reduced (line goes near to ground): 'capacitive' defect
—if Z_w is locally increased (line goes further away from the ground):'inductive' defect.

Printed capacitors and *printed inductors* can be used for low values of C or L; they have good high-frequency properties.

4.9.2 Recommendations for Design

1) Use ground plate or other very large ground surface.
2) Use broad power supply lines.
3) Ground and power supply lines should run near each other and be parallel; if possible put added PCB capacitance between them.
4) Provide decoupling capacitor to be soldered on PCB between ground and power supply lines.
5) Definitely keep all lines which are not matched very short. You will have rise-time increases of up to 1 $\frac{\text{nsec}}{\text{cm}}$ of line length otherwise.
6) Decide which parasitic elements C^* or L^* are more harmful and design conductor layout accordingly.
7) Provide guard-lines in-between (grounded or connected via capacitor to ground) wherever a parasitic capacitance has very dangerous effects (e.g., feedback capacitance in amplifier stages between output and input).
8) For large-size PCBs (long conductor lengths), and very fast rise-times (or very high upper band-limits), sometimes dielectric losses are important. In this case: Use PCBs with suitable high-frequency dielectrics.
9) In other very-fast-pulse cases, skin-effect losses are important: To reduce these, use a ground plate and avoid discontinuities.
10) Because of these losses, we should recommend: For very fast pulses ($t_r <$ 1 nsec), even matched lines have to be kept very short. Rise-times of matched lines increase by 1 to 10 $\frac{\text{psec}}{\text{cm}}$ or 100 to 1000 $\frac{\text{psec}}{\text{m}}$ due to these losses.

Acknowledgement

The author of this chapter wishes to acknowledge the contribution of Mr. Herman Curtins, Institute of Microtechnics, University of Neuchatel, whose assistance was decisive in bringing this chapter and, especially, Section 4.8 (on skin-effect and other losses) to its final form.

BIBLIOGRAPHY

1. DUKES, J.M.C.: *PRINTED CIRCUITS—THEIR DESIGN AND APPLICATION*, MacDonald, London, 1961.
2. 'Gedruckte Schaltungen, Konstruktion und Zeichenpraxis (Datenblatt 10109)', Hans Kolbe & Co., 3371 Gittelde (Germany), 1974.
3. GRIVET, P.: *THE PHYSICS OF TRANSMISSION LINES AT HIGH AND VERY HIGH FREQUENCIES* (translated from the French edition of Masson & Cie.), Academic Press (Vol. I, 1970), London.
4. GUNSTON, M.A.R.: *MICROWAVE TRANSMISSION LINE IMPEDANCE DATA*, Van Nostrand-Reinhold, London, 1972.
5. HAFNER, CH.: 'Bestimmung der dielektrischen Eigenschaften der Ummantelung eines Kabels mit Hilfe eines Optimierungsprogrammes', *BULLETIN SEV,* Zurich, Switzerland.
   Vol. 69, No. 24 (16.12.1978) pg. 1315–1320
   Vol. 70, No. 3 (10.02.1979) pg. 137– 141
6. HARPER, C.A.: *HANDBOOK OF WIRING, CABLING, AND INTERCONNECTING FOR ELECTRONICS,* McGraw-Hill Book Co., New York, 1972.
   6a. Chapter 4: Coaxial Cable and Connector Systems, by Jack Spergel.
   6b. Chapter 8: Rigid Printed Wiring and Connector Systems, by William H. Taylor.
   6c. Chapter 11: Formed High-Frequency Circuits, by Joseph B. Marshall.
7. KUPFMUELLER, K.: *EINFUEHRUNG IN DIE THEORETISCHE ELECTROTECHNIK,* Springer Verlag, Berlin, 1968 (9. Auflage).
8. LEWIN, L.: 'Radiation from Discontinuities in Stripline', *PROCEEDINGS IEEE,* Vol. 107, London, February 1960 (Part C).
9. MATICK, R.E.: *TRANSMISSION LINES FOR DIGITAL AND COMMUNICATION NETWORKS,* McGraw-Hill Book Co., New York, 1969.
10. MEINKE, H. and GUNDLACH, F.W.: *TASCHENBUCH DER HOCH-FREQUENZTECHNIK,* Springer Verlag, Berlin, 1968 (3. Auflage).
11. METZGER, G. and VABRE, J.P.: *TRANSMISSION LINES WITH PULSE EXCITATION* (translated from the French edition of Masson & Cie.), Academic Press, New York, 1969.
12. MILLMAN, J. and TAUB, H.: *PULSE, DIGITAL AND SWITCHING WAVEFORMS,* International Student Edition, McGraw-Hill—Kogakusha, New York and Tokyo, 1965.
13. PUCEL, R.A., MASSE, D.J., HARTWIG, C.P.: 'Losses in Microstrip', *IEEE TRANS. MICROWAVE THEORY TECH.*, Vol. MTT-16, No.6, June 1968, pg. 342–350 (correction in No. 12, December 1968, pg. 1064).

14. SHAH, A.: *FAST PULSE TECHNIQUES,* CEDT Publications, CEDT, Indian Institute of Science, Bangalore-560 012, India, 1978.
15. SHAH, A., PELLANDINI, F., BIROLINI, A: *GRUNDSCHALTUNGEN MIT TRANSISTOREN,* AMIV-ETH Verlag, Zurich, Switzerland, 1972 (2. Auflage).
16. WINNINGSTAD, C.N.: 'Nanosecond Pulse Measurements', *IRE Wescon Convention Record,* paper 23/1, 1961 (excellent article).

5

DESIGN RULES FOR ANALOG CIRCUIT PCB'S

It is assumed that the reader is already familiar with the content of Chapter 2 which applies to the layout design of any analog PCB. However, a few more specific hints shall be given here which will help to achieve a good analog PCB design.

5.1 COMPONENT PLACING

Components which need access from the frontplate, have to be placed exactly according to the instructions of the equipment designer. The same applies for all connectors, receptacles and any other connection to the PCB. Components for internal adjustment, such as potentiometers, trimmers, switches, etc., have to be arranged near the board edge and oriented in the proper direction: The access to these components should be free and not obstructed by other components. Especially, components with a metal case should not be placed very near to the access path that is needed, e.g., for the adjustment screw driver: Otherwise while adjusting, the screw driver might cause a short-circuit between the metal-cased component and the equipment chassis.

Where mounting screws are provided, it should never be forgotten that the nut and in particular the washer need considerable space on the board in which no other conductive track can be run. The placing of heat-producing and heat-sensitive components must be carefully planned:

—The direction of air flow in the equipment has to be known.
—The air flow should have a free and unobstructed path via the heat-producing components. Remember that the free air flow goes always in vertical direction, i.e., from below to top.
—Heat-producing components are regularly distributed over the entire board area, as far as it is feasible. This is to avoid a locally overheated board.
—Hot components have to be separated from the board surface by suitable spacers.
—The locating of heat-sensitive components must avoid a close proximity to hot components as well as a place in a flow of already warmed up air.

Component holes are always placed in grid intersections to take care of standard lead spacings for the components as well as NC drilling requirements.

5.2 SIGNAL CONDUCTORS

5.2.1 In General

Signal conductors in analog circuit PCBs have to fulfill a variety of different tasks, e.g., input, reference level, feedback, output, etc. Therefore, it should not be surprising that a signal line for one application has to be optimised in a different manner than for another application. But there are some common considerations among which minimising the conductor length is the most important.

The length of any signal conductor should be made as short as possible. This is because the magnitude of the undesired inductive and capacitive coupling effects increases more or less proportionally to the length of the particular conductor.

The requirement to keep all signal conductors short, looks somewhat difficult to achieve. But a good approach is to have the most critical signal conductors identified and to put them first in the layout while the remaining ones are placed subsequently. This of course is only possible if the sensitive components have been placed accordingly.

The signal conductor layout has to be done particularly carefully for the following type of circuits:

—High-frequency amplifiers/oscillators.
—Multistage amplifiers especially with high-power output stage.
—Feedback amplifiers/regulators with remote sensing lines.
—High-gain DC amplifiers (thermal effects).
—Amplifiers handling low-level signals.
—Precision differential amplifiers.

In these cases, the preparation of a layout from the schematic circuit diagram is quite crucial for the circuit performance: There are several *invisible* effects due to a PCB conductor resistance,-capacitance,-inductance, solder joints. etc., which, if not properly cared for, will result in poor circuit performance. Apart from making of a good PCB, there are also other techniques such as shielding or guarding with metallic enclosures which have to be employed for a satisfactory operation of the final equipment. Here, only the points relevant to PCB design shall be discussed.

5.2.2 High-Frequency Amplifiers/Oscillators

It is a common experience of electronic circuit designers that at high frequencies (>10 MHz), it is easier to build up an oscillator than an amplifier: You design and wire an amplifier and it will oscillate! To give an example, consider the circuit of an emitter follower as given in Fig. 5-1. At higher frequencies and with a long voltage supply line, the circuit may oscillate because the voltage gain in an emitter follower can be high, i.e., gain = $\dfrac{Z_{collector}}{Z_{emitter}}$

DESIGN RULES FOR ANALOG CIRCUIT PCB'S

Without any parasitic inductance in the voltage supply line, there would be $Z_{collector} = Z_{emitter}$. This gain coupled into the biasing resistors R_1 and R_2 can give sufficient feedback to cause oscillations.

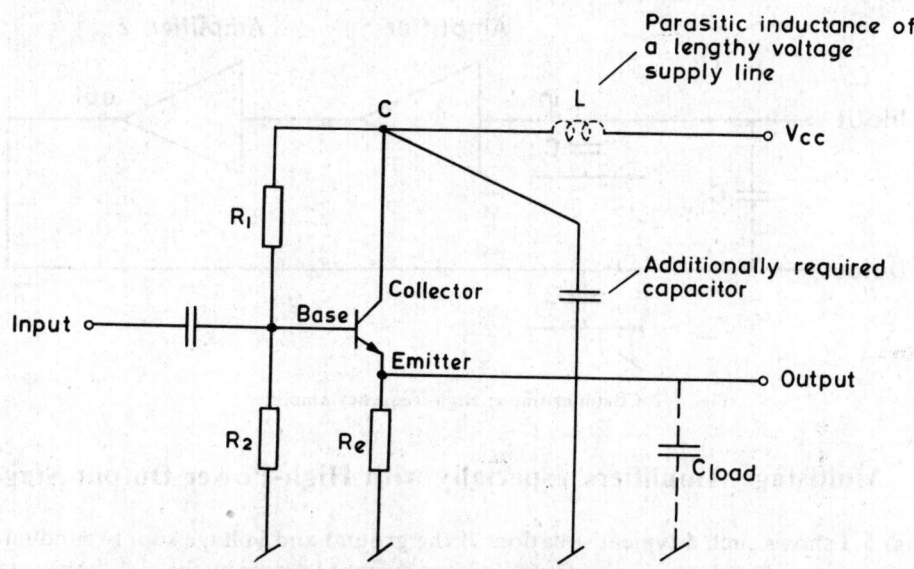

Fig. 5-1 Circuit example: Emitter follower stage

The precautions one has to take in the above case is to put a capacitor between the collector (point C) and ground. The value of this additional capacitor should be greater than the load capacitance of the emitter C_{load}. The length of the voltage supply conductor should in any case be as short as possible.

In designing the layout of an oscillator circuit, similar problems will arise with the result that the oscillator does not oscillate at the desired frequency. To avoid this, the layout has to be very carefully made in order to reduce the capacitive coupling between signal lines to the maximum extent. Alternatively, some allowance can be made during the circuit design taking into account the capacitive coupling effects. But this approach also requires considerable experience.

Especially in the design of large-bandwidth operational amplifier stages (e.g., with OpAmp μA 715), one should never forget to decouple the voltage supply line. Otherwise, the feedback comes in the form of oscillations which can be observed on the oscilloscope screen with the working circuit. Large-bandwidth operational amplifiers, even if used at a low frequency, can oscillate because of their high gain if proper voltage supply line decoupling and the normal compensating network have not been provided.

An improper layout of a high-frequency amplifier can also result in a reduced bandwidth of the amplifier. Fig. 5-2 shows such a situation. If the input and output conductors are close to each other, there can be a feedback resulting in oscillations. Also, if the ground conductors are close to the signal conductors, the higher capacitance can reduce the bandwidth of the amplifier because this capacitance along with the output resistance acts as a lowpass filter. The proximity of input and output conductors can further result in another interesting effect which is called

Miller effect: The capacitance in between gets effectively multiplied by the amplifier gain and makes it to appear at the amplifier input thus further degrading the bandwidth. Herewith, sufficient spacing must be provided between such conductors to avoid this effect.

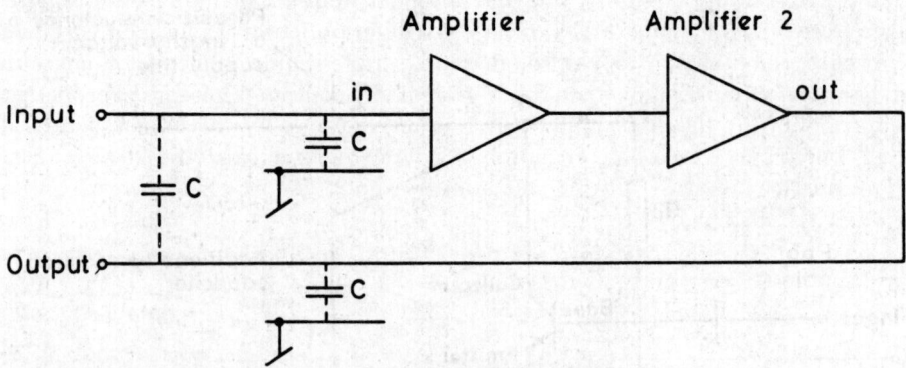

Fig. 5-2 Circuit example: High-frequency amplifier

5.2.3 Multistage Amplifiers especially with High-Power Output Stage

Fig. 5-3 shows such a typical situation: If the ground and voltage supply conductors are long and connected as shown in this figure, a large current I drawn by the high-power stage will flow through the conductors with their own resistivity R. Additional resistance may come also from improper solder joints or from a bad edge connector contact, both for V_{cc}- as well as GND-conductor. This will in effect modulate the V_{cc}- and GND of the input stage resulting in low-frequency oscillations which is called *motorboating effect*. To overcome this effect, decoupling of the power supply conductors with sufficiently large capacitors (as the oscillations are of low frequency) must be done. Alternatively and to the extent possible, separate V_{cc}- and GND-conductors can be provided for the two different stages so that there is no common supply- or GND-path.

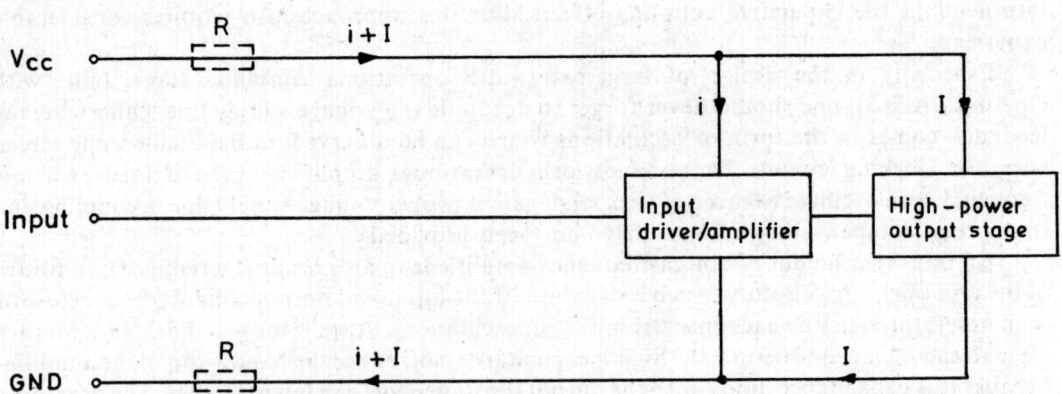

Fig. 5-3 Circuit example: Multistage amplifier with high-power output stage

5.2.4 Feedback Amplifiers/Regulators with Remote Sensing Lines

In designing high-wattage power supplies (e.g., with the specifications 5V/3A; regulation 0.01–0.1%), considerable care in PCB layout design as well as in planning of the remote sensing line to the front panel is necessary to meet the regulation factor. In fact, only a correct PCB layout can solve the situation here although it looks quite simple. If the common ground is not laid out in the star pattern as shown in Fig. 5-4B and if the output is not sensed properly to give a feedback straight from the final supply output point, the resistivity of the conductors, although very small, will cause a very poor regulation. Also, the rectifier/capacitor circuit (not shown in the figure) must be connected to the star point for good circuit performance.

A) Poor circuit interconnecting (ground line resistance, sense point CS not proper)

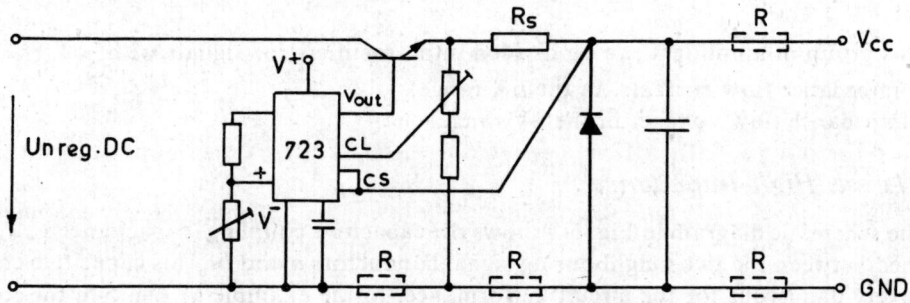

B) Proper layout for a good load regulation

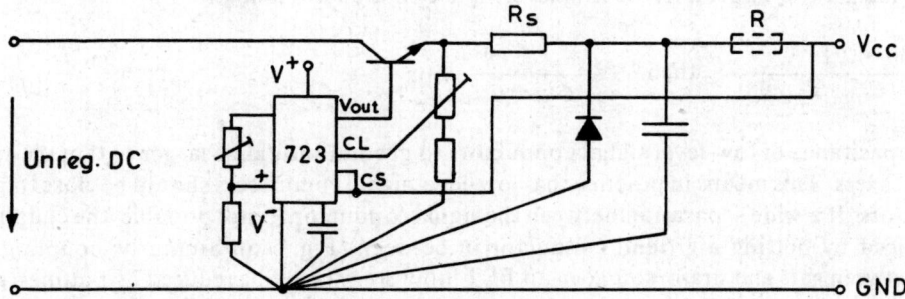

Fig. 5-4 Layout of a power supply PCB

5.2.5 High-Gain DC Amplifiers (Thermal Effects)

High-gain DC amplifiers are generally used to amplify low-level signals. The actual handling of low-level signal conductors is discussed in the next section; here we mainly deal with other points to be considered, like thermal effects, etc.

DC amplifiers have a differential stage to nullify the common-mode effects. Here, besides the layout pattern other points to be considered are associated problems like soldering, temperature gradients, etc. As the amplifier dissipates power, small temperature gradients exist.

When a device (e.g., transistor, DC amplifier) is soldered onto the PCB, a thermocouple junction is formed between copper (i.e., PCB conductor) and the kovar lead of the transistor. If a temperature gradient exists, then different junctions will create different voltages, thus effecting a 'noisy' signal to the amplifier.

Furthermore, this varies with temperature/time. Therefore, all care must be taken in layout design of input stages either to ensure the exclusion of the temperature gradient or at least to have a stable temperature gradient. A possible solution is to put the input stage into a separate enclosure which does not allow a free movement of the surrounding air.

5.2.6 Amplifiers Handling Low-Level Signals

This group of amplifiers can be divided into amplifiers for signals with
— high-impedance (low currents, in the nA-range)
— low-impedance (low voltage, in the nV-range)

Low Level, High-Impedance

The schematic diagram in Fig. 5-5 shows the capacitive coupling (capacitance C_{a-b}) which is formed between the two neighbouring signal conductors a and b. This capacitive coupling can be very dangerous for the circuit performance. In the example in Fig. 5-5, the coupling voltage will be $v_c = j\omega R C_{a-b} v_a$. In high-impedance circuits, R will be very high; disturbances can therefore easily mask a low-level signal. In order to minimise the coupling, it is necessary to provide sufficient distance between the high-impedance conductors and other interfering signal lines. Adequate decoupling can be obtained by keeping the separation distance at least 40 times the signal conductor width.

In fact, if the impedance of low-level signal conductors is high

$$R \gg \frac{1}{j\omega(C_{a-b} + C_{b-\text{GND}})} \quad \text{then } v_c = \frac{C_{a-b}}{C_{a-b} + C_{b-\text{GND}}} v_a.$$

Thus capacitance of low-level signal conductors to ground should be larger so that the coupled voltage is less. This means in practice that low-level signal conductors should be close to ground conductors. If a wide separation between the signal conductors is not possible, the coupling can be reduced by putting a ground conductor in between (Fig. 5-6). Similarly, coupling which exists between gate and drain/source in an FET input stage could be reduced by putting a ground line as shown in Fig. 5-7.

Low-Level, Low-Impedance

If the impedance is low, one has to be careful about induced voltages due to magnetic fields or inductive coupling.

DESIGN RULES FOR ANALOG CIRCUIT PCB'S

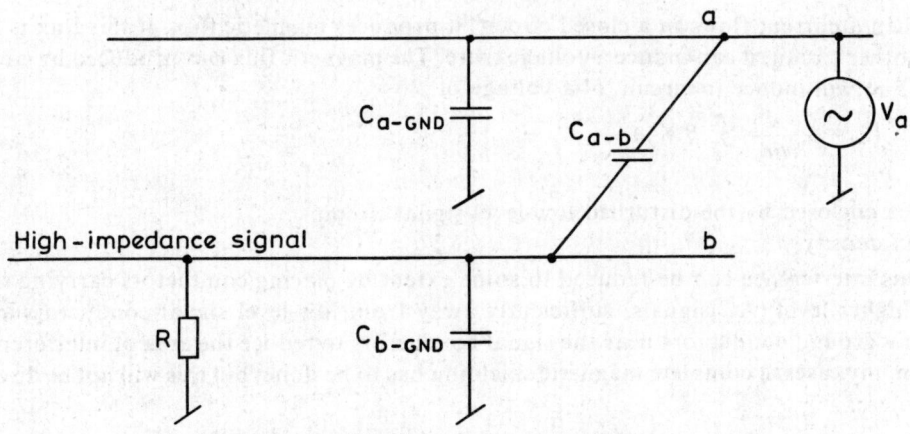

Fig. 5-5 Capacitive coupling between two conductors

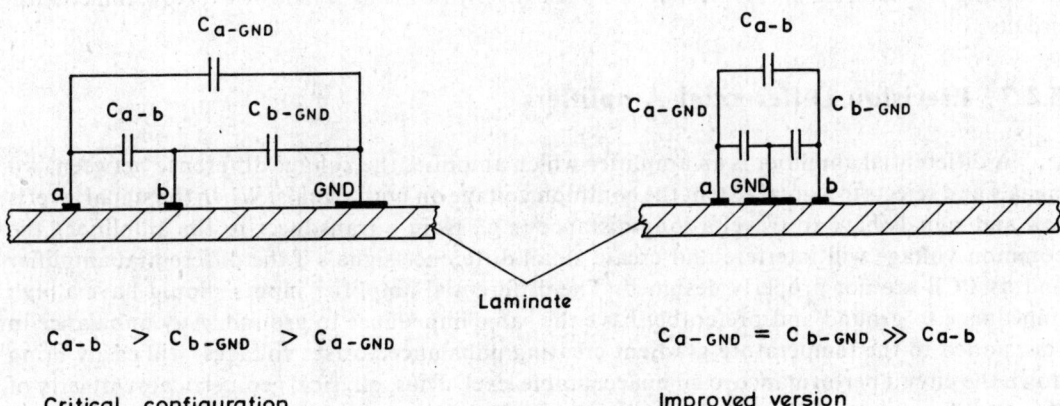

$C_{a-b} > C_{b-GND} > C_{a-GND}$

Critical configuration

$C_{a-GND} = C_{b-GND} \gg C_{a-b}$

Improved version

Fig. 5-6 Crosstalk reduction in parallel running signal lines

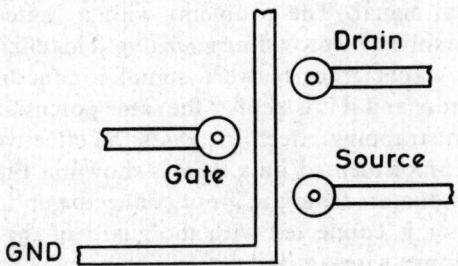

Fig. 5-7 Stopping leakage current in FET input devices

When a current flows in a closed circuit, it produces magnetic flux. If this flux is linked with another circuit, it can induce a voltage there. The magnetic flux ϕ_{a-b} produced by circuit 'a' in Fig. 5-8, will induce in circuit 'b' a voltage of

$$V_m = -\frac{d}{dt} \int_A B \times A$$

A = area enclosed by the disturbed low-level signal circuit
B = flux density.

This interference can be reduced to some extent by placing conductors carrying signals with a higher level (AC signals) sufficiently away from low-level signal conductors and by providing ground conductors near the signal conductors to reduce the area of interference. In fact, in many cases, a complete magnetic shielding has to be done; but this will not be discussed further here.

Also, if an external magnetic field is present where a circuit is operated, a voltage will be induced in this circuit if a ground loop exists. To disable the external magnetic field from disturbing low-level signals, *ground loops must strictly be avoided* in low-level low-impedance circuits.

5.2.7 Precision Differential Amplifiers

A differential amplifier is an amplifier which amplifies the voltage difference between two signals and rejects to a great extent the common voltage on both signals. When the signal level is low and signals have to travel a long distance (e.g., from a transducer to the amplifier), the common voltage will interfere and create small difference signals if the differential amplifier and its PCB are not properly designed. The differential amplifier inputs should have a high impedance to ground and preferably have the same impedance to ground. Any unbalance in this, added to the temperature gradient creating unbalanced offset voltages, will easily bring down the circuit performance to an unacceptable level. Thus, physical geometrical symmetry of the amplifier on the PCB must be taken care of during layout design. In addition to this, the PCB base material chosen should preferably be of glass-epoxy type to reduce leakage currents.

An important thing to remember here is that due to the coupling of signal conductors to ground, a difference signal is produced by the common voltage, which the amplifier cannot distinguish from the actual signal. The problems with a finite impedance of differential amplifier inputs can only be solved by providing *guarding*. This effectively increases the leakage resistance and reduces the capacitance between signal conductors and ground. The guard encloses the signal conductors and if it is kept at the same potential as the 'low-line' of the two signal conductors, the 'bootstrapping effect' increases the effective resistance (Fig. 5-9).

How guarding can be implemented on a PCB is shown in Fig. 5-10: A guard conductor loop encloses the signal conductors from the input connector up to the amplifier input solder joints. The guard conductor is connected with the guard of the equipment. If guarding is carefully applied with the input wires (guard screen), equipment (guard box) and on the PCB (guard conductors), a significant step is made towards reaching the ultimate performance in handling low-level differential signals.

DESIGN RULES FOR ANALOG CIRCUIT PCB'S

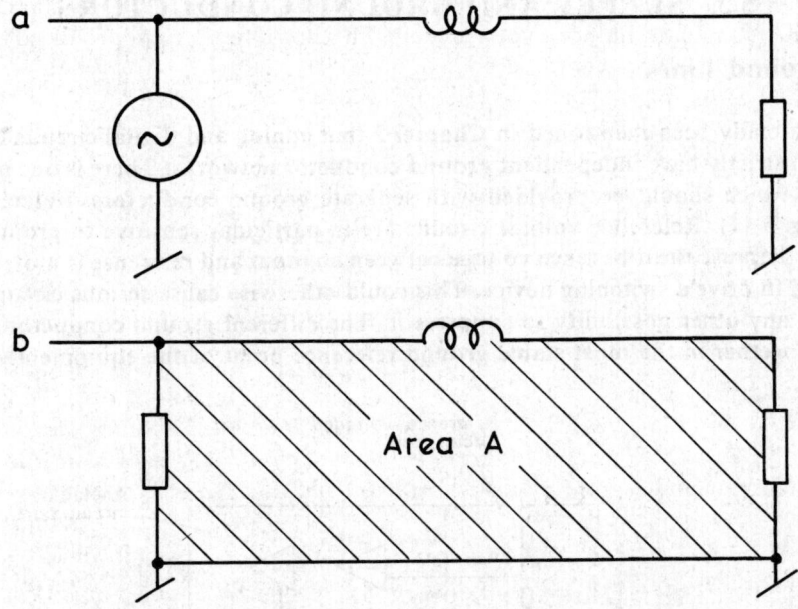

Fig. 5-8 Magnetic coupling

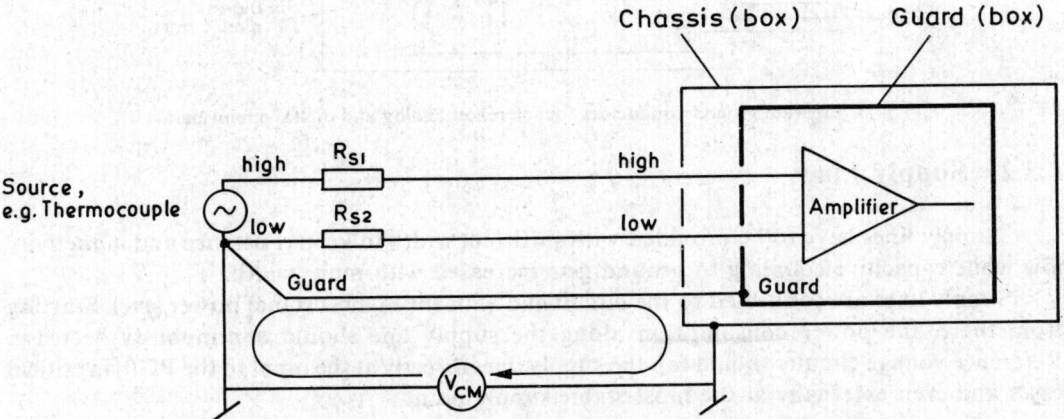

Fig. 5-9 Guarding a differential amplifier (at source end, guard line is at the same potential as low end).

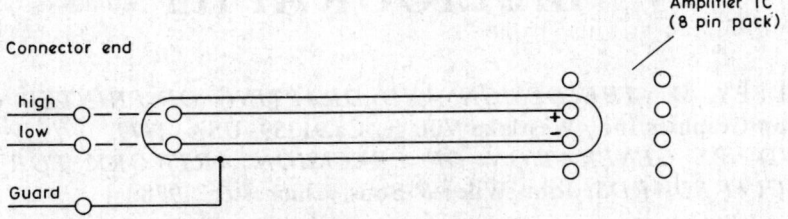

Fig. 5-10 Guarding signal conductors on a PCB

5.3 SUPPLY AND GROUND CONDUCTORS

5.3.1 Ground Lines

It has already been mentioned in Chapter 2 that analog and digital circuits on the same PCB should strictly have independent ground conductor networks. There is one more type of circuit part which should be provided with separate ground conductors: Reference voltage circuits (Fig. 5-11). Reference voltage circuits are in particular sensitive to ground potential fluctuations because the difference voltage between an input and reference is mostly amplified and/or used to drive a switching device. This could otherwise cause serious circuit instability with hardly any other possibility to suppress it. The different ground conductor systems are connected together in the most stable ground reference point of the equipment.

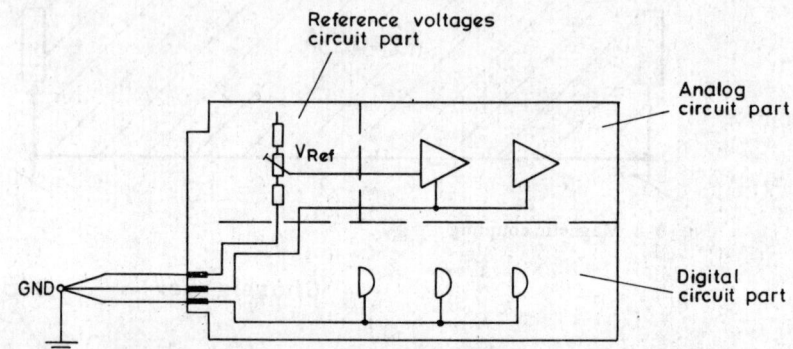

Fig. 5-11 Separate ground conductors for reference, analog and digital circuit part

5.3.2 Supply Lines

Supply lines have to be provided with sufficient width to keep resistance and inductivity low while capacitive coupling to ground gets increased with more width.

Supply lines are connected to the circuit end with the highest signal power level. Starting from there, the power consumption along the supply line should continuously decrease. Reference voltage circuits should tap the supply lines directly at the input to the PCB, in critical cases and even externally at the most stable supply point.

BIBLIOGRAPHY

1. LINDSEY, D.: *THE DESIGN AND DRAFTING OF PRINTED CIRCUITS*, Bishop Graphics Inc., Westlake Village, Ca.91359, USA, 1979.
2. LUND, P.: *GENERATION OF PRECISION ARTWORK FOR PRINTED CIRCUIT BOARDS*, John Wiley & Sons, Chichester, 1978.

6

DESIGN RULES FOR PCB'S IN POWER ELECTRONICS APPLICATIONS

6.1 INTRODUCTION

The design of power electronics PCBs as such has not received much attention in the currently available literature on PCB design. This is not because the subject is of less importance or not worth mentioning but probably because only a small fraction of all the PCB designers have to be careful about it. Therefore, an attempt is made here to highlight at least a few of the most important points.

Many of the design rules mentioned in the previous chapters have to be equally applied in power electronics PCB design. Some rules have become irrelevant. On the other hand, rules which were almost meaningless before, play a key role here. A major difference to be remembered is that comparatively high power is flowing on these PCBs and that a failure occurring under the operational conditions on such boards can easily lead to far more serious consequences, including danger to personnel.

6.2 DIVIDING CIRCUIT INTO HIGH- AND LOW-POWER PART

It is a typical aspect in the application of active high-power electronic components that they are controlled by a control circuit of a considerably lower power level. A simple example: The current flow in the magnitude of 50A at mains voltage through thyristors can mainly be controlled by a TTL circuit which draws less than 1A at a voltage of only 5V. At first one would think that the circuit for power conditioning and for its control would best be continued on one PCB. This, however, would bring the high- and low-power part in very close proximity, resulting in capacitive and inductive coupling between power circuits and control circuits. This may lead to malfunctioning of the equipment. To provide minimum safety requirements, large conductor spacing has to be provided between the two parts of different power level, which again would increase the PCB size.

Because of such difficulties, *it is a functional and safety requirement to clearly separate the high- and low-power part of a circuit onto different PCBs.* High-power part means here: The part working at a considerably higher voltage or higher current level than for instance the levels used in control logic. But also conductors leading to components of the actual high-power part form a part of the high-power part.

As an example, the simplified circuit diagram in Fig. 6-1 shows an SCR control circuit. Note that here even the pulse transformer which provides isolation is not mounted on the control PCB since its secondary winding is connected to the high-power parts.

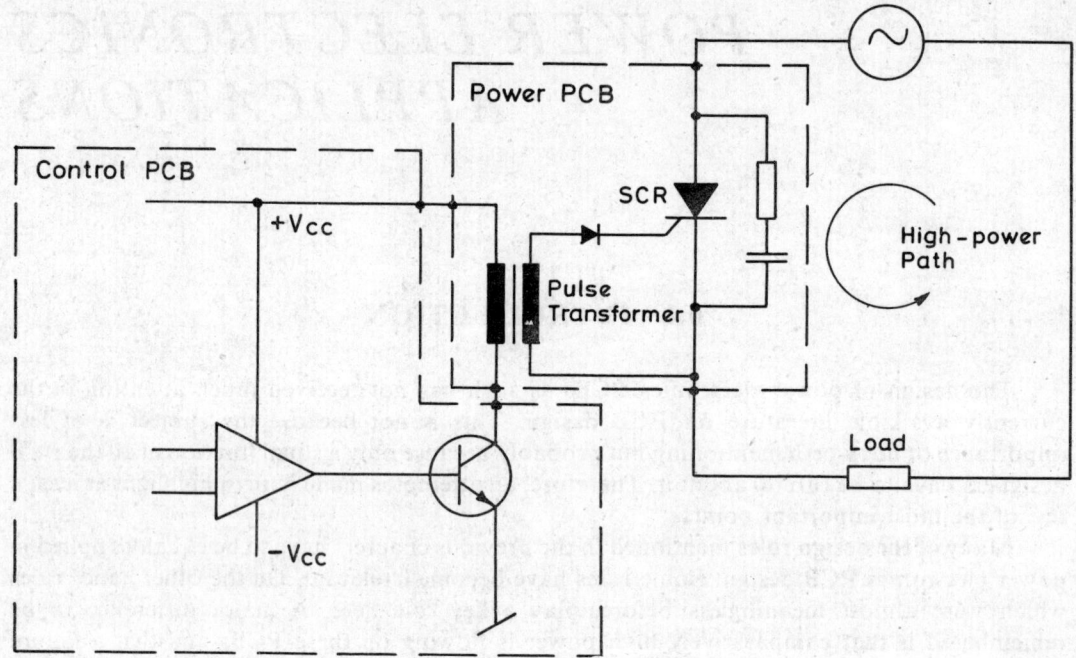

Fig. 6-1 Example of circuit division into high- and low-power parts

6.3 COPPER-CLAD LAMINATES

Base Materials

Power electronics devices dissipate a certain amount of heat which usually calls for suitable heatsinks. If the heatsink is directly mounted on the PCB, the whole board will also be heated up to the same temperature. Therefore, the base material selected has to withstand the continuous operation of the equipment under worst-case conditions.

A very common choice is to go for glass-epoxy laminates. They offer good thermal properties at a modest price.

Base Materials Thickness

Because comparatively heavy components are usually mounted on power electronics

PCBs, the mechanical board strength has to be chosen accordingly. The mechanical strength must also be maintained at the maximum operation temperature of the equipment. The usual laminate with a thickness of 1.6 mm will hardly meet the mechanical property requirements for mounting of heavier components such as pulse transformers, heatsinks, chokes, etc. Therefore, laminates with more thickness are applied. An often preferred standard thickness is 3.2 mm, but others are also used depending on need and availability.

Copper Foil Thickness

Since higher current flow calls for more conductor cross-section, thicker foils than the 35 μm standard thickness are preferably employed. 70 μm thick copper foil is mostly used but even foil with 105 μm thickness can be supplied on request from many copper-clad laminate manufacturers.

6.4 PCB TERMINAL CONNECTIONS AND THEIR ASSEMBLY

Terminal Connections

The ordinary PCB edge connectors can hardly meet the requirements for power electronics PCB connections because of their limited current-carrying capacity and insulation spacing width. A wide range of terminals are available which meet the specific requirements of power electronics.

Fig. 6-2A shows solder terminals which have to be rivetted to the PCB and soldered thereafter. They are available in different designs and diameters and rivetting shaft length for different board thicknesses. Fig. 6-2B shows a few shapes of cable terminals which can be screwed onto the board. The stripped electric cable is inserted into the opened or closed sleeve and crimped with a suitable crimping tool. Only multistranded wire cables are suitable for crimping. Crimping does not require any soldering and gives high-reliability connections. Another type commonly found is the plug-in type connector (Fig. 6-3) popularly known as *Faston* terminal. The multistranded wire cable has again to be crimped into the receptacle. The receptacle can then be inserted into the tab. Such tabs are available in many more varieties than shown and can be soldered or screwed to the PCB, depending on the type chosen. Tongue widths have been standardised and include 2.8 mm, 4.8 mm and 6.3 mm. The 6.3 mm standard has also been internationally accepted in automotive applications. In power electronics applications, the maximum current for the 6.3 mm connections should not exceed 15A and correspondingly less for the smaller connections. Faston connections have been introduced by the US company Amphenol many years ago. Because of their high reliability, they have become one of the most accepted means of interconnection for higher current flows in electronic equipment.

Screw Mounting of Terminals

When the terminals for screw-fastening have to be mounted directly on the PCB for providing electrical contact, some special considerations must be pointed out: In most cases, the pressure required for mounting the cable terminal or tab flat on the PCB should not exceed certain values which are related to the physical laminate properties. This becomes particularly

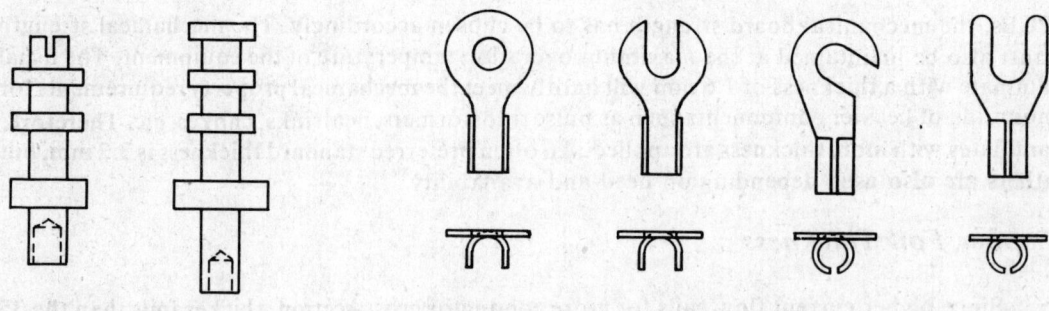

A) Solder Terminals   B) Cable Terminals with open and closed sleeves

Fig. 6-2 Wire and cable terminations

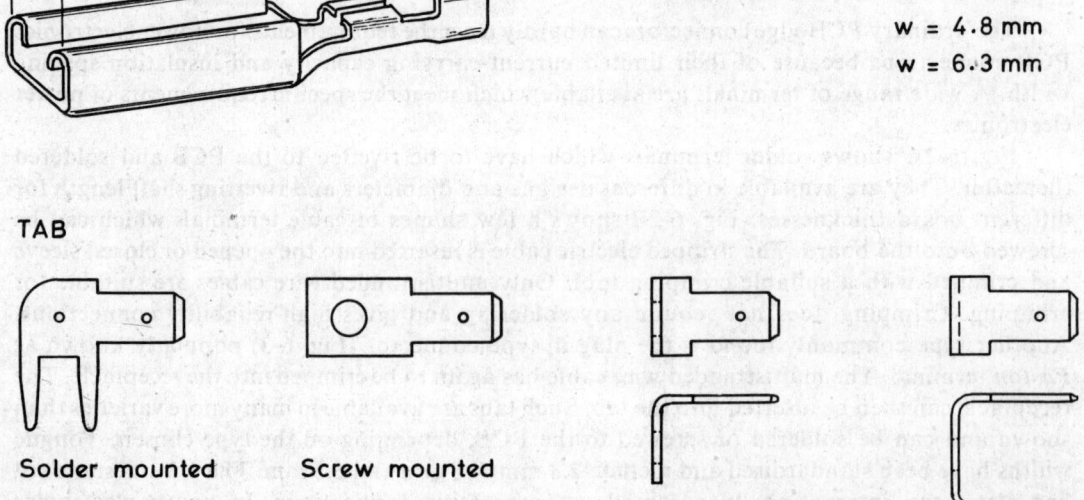

Fig. 6-3 Plug-in type connections

important for bigger terminals meant for a high current flow.

Brass washers must be used wherever pressure is exerted on contact areas and laminates. In addition, split washers have to be used in order to compensate for laminate flow (slow plastic deformation of the laminate under continuous stress) and the resulting contact pressure reduction. To maintain a uniform control within safe limits on the screw-contact pressure, the use of torque wrenches is recommended. Another point of concern is the contact area. It should possess a smooth surface finish and has to be polished (if necessary) prior to the mounting of the terminals. The actual contact area should be symmetrical and well defined: This can be achieved by using washers according to Fig. 6-4.

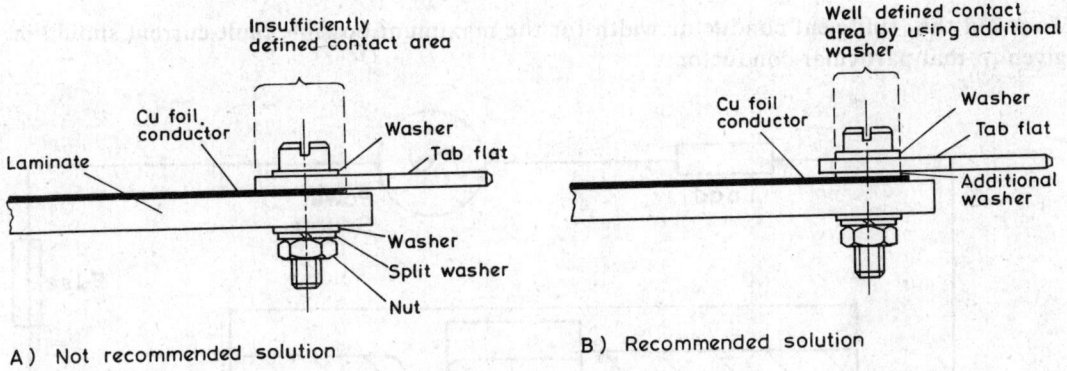

A) Not recommended solution   B) Recommended solution

Fig. 6-4 Contact area definition

Material for Terminals, Tabs and Mounting Hardware

Wherever possible and available, brass parts should be used. They show good conductivity and little corrosion. The electromechanical corrosion between brass and copper is also low; the two metals are relatively closed together in the electrochemical potential order. To enable the terminals and tabs to be easily soldered, a tin, silver or other suitable plating has to be given. Such a plating is also needed for steel parts to prevent corrosion. Where different metal surfaces are mounted face-to-face, the clean and smooth surface finish becomes very important.

6.5 CONDUCTORS

Conductor Width

In the design of power electronics PCBs and their artwork, the copper available on the board surface should be fully utilised because of the larger currents flowing. The following is therefore an accepted practice while designing PCBs for power electronics: First determine the required spacing between the conductors and then allot the remaining copper area to the conductors.

For the conductor spacing, Fig. 8-11 can serve as a guideline during artwork preparation. The mostly irregular shaped conductors in the artwork can be designed as follows: The conductor borders are laid down by taping while the intermediate space is filled with ink. In no place should the conductor cross-section be less than the values obtained from Fig. 2-1 for a conductor temperature rise of 10°C. The worst-case load current must be taken as the base for minimum conductor-width determination.

With a proper conductor layout, it is possible to minimise damages on the PCB in case of circuit failures, which can lead the main current to flow through conductors planned only for small currents. First, it is necessary to know the most probable circuit failures possible. Secondly, the conductors possibly affected on the PCB are determined and a check is carried out whether they can carry the fault current. If not, the widths have to be increased as far as possible. An example is given by the simplified circuit in Fig. 6-5: A short-circuit across the diode on the left side would definitely blow the fuse in the main current path and probably the conductor to the SCR gate on the PCB would also have to be repaired because of over-current.

To avoid this, sufficient conductor width for the maximum possible fault-current should be given to that particular conductor.

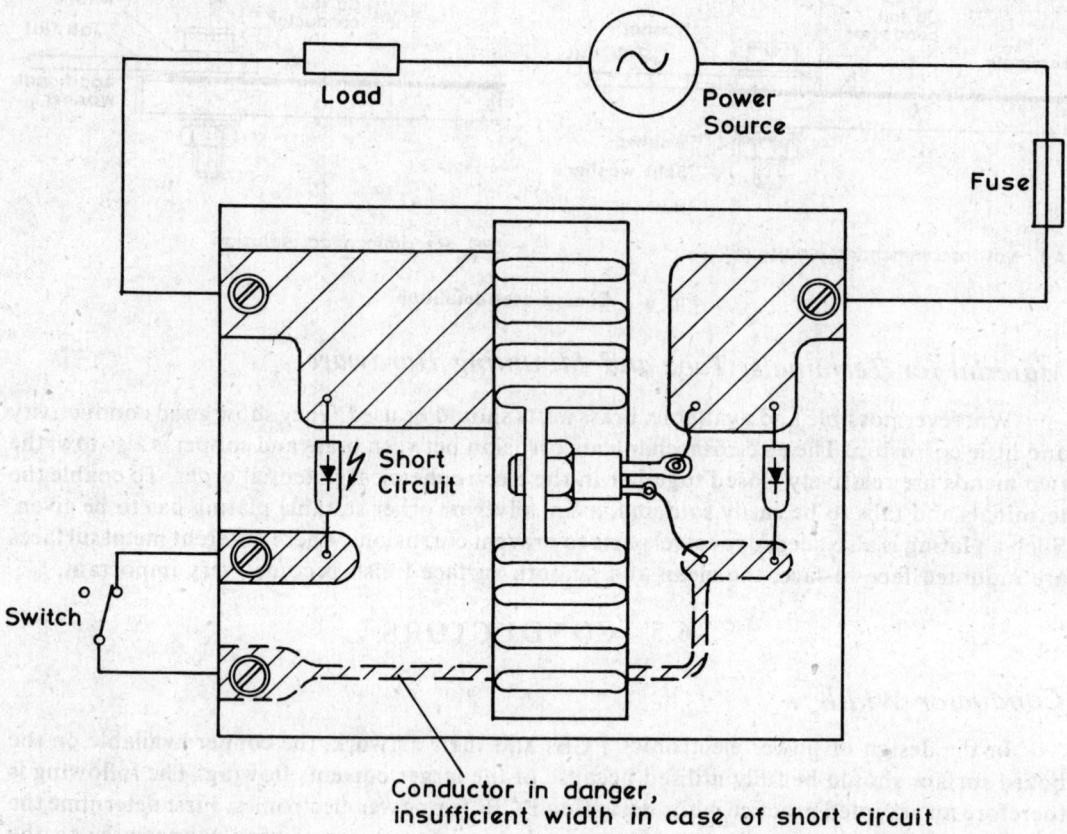

Fig. 6-5 Avoiding of PCB damages through precautions against most probable circuit failures

Spacing

An important aspect of designing highly reliable PCBs is to minimise the possible flash-overs caused by over-voltage or contamination on the PCB surface. Such faults may occur inspite of maintaining conductor spacing within specified safety requirements. Sharply localised high-voltage gradients (i.e., high electric fields), found near low-angle conductor corners must be strictly avoided (Fig. 6-6). Minimum spacing specifications have to be interpreted as guidelines for cases where minimum spacing cannot be avoided. In all other cases, a sufficient safety margin has to be applied.

Resistive Coupling

The high currents usually flowing in power electronic circuits can cause considerable voltage drops if they are carried through *PCB* conductors. These heavy load-currents are

DESIGN RULES FOR PCB'S IN POWER ELECTRONICS APPLICATIONS 111

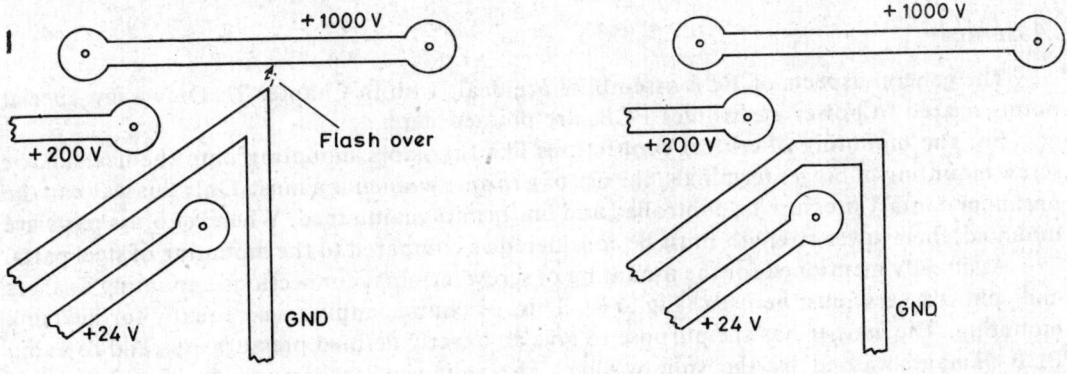

Fig. 6-6 Minimising danger of flash-overs

therefore avoided on the PCB wherever possible. Where the load current cannot be bypassed and has to be carried through the PCB conductor, the voltage drop caused thereby should not have any influence on the functional stability of the circuit. In the simplified circuit of Fig. 6-7, the load current flows unnecessarily through the control PCB, thereby creating a voltage drop which causes circuit instability. The best solution would be connecting *A* to *B* and removing the connection *C-D-B*.

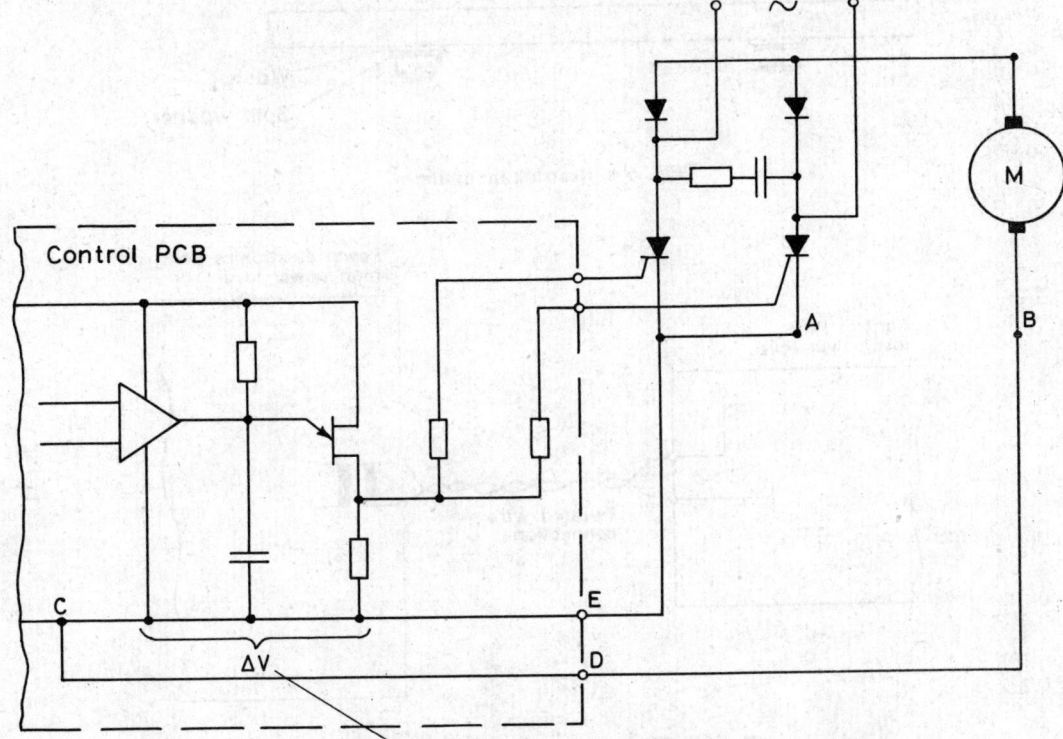

Fig. 6-7 Circuit instability caused by voltage drop in conductor

Assembly

The general aspects of PCB assemblies are dealt with in Chapter 21. Only a few special points related to power electronics PCBs are pointed here:

For the mounting of critical connections like thyristors mounting onto the heatsink or screw mounting of bigger terminals, the use of a *torque wrench* is a must. Only this way can the optimum contact pressure be controlled and uniformity maintained. Wherever brass parts are mounted, their lower strength must be considered as compared to the mounting of steel parts.

As already mentioned for the mounting of screw terminal connections, mounting washers and split washers must be used (Fig. 6-8). This, of course, applies also equally for heatsink mounting. The washer has the purpose to give an exactly defined pressure area and to avoid PCB damages caused by the split washer. The split washer compensates for the plastic deformation of the laminate which flows a bit under the pressure applied. At the same time, it secures the screw connection.

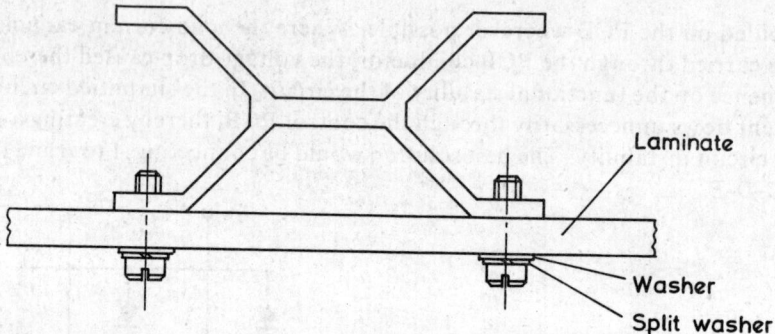

Fig. 6-8 **Heatsink mounting**

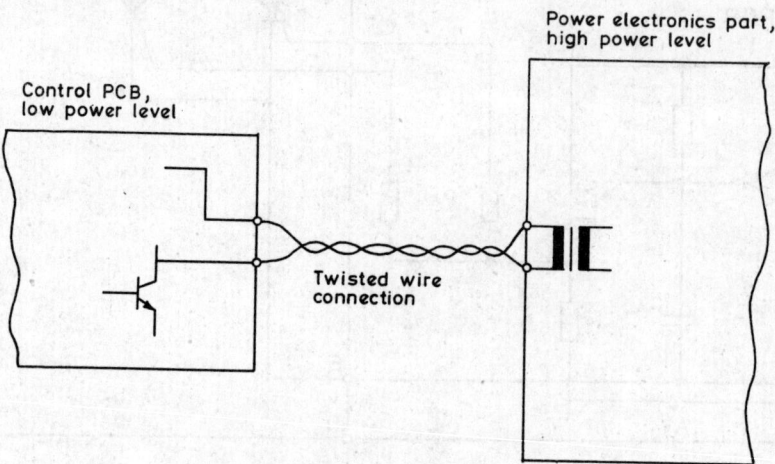

Fig. 6-9 **Minimising EMC interference**

When mounting semiconductor devices on a heatsink, the contact area must be clean and have a smooth surface finish which may be improved by polishing. The application of heatsink compound will give a better heat flow into the heatsink.

The soldering of power electronics PCBs has to be done with utmost care. Any solder splash left could lead to serious consequences. Also, small wire pieces left within power electronics equipment are very dangerous. Special care is needed for multistranded wire stripping where thin strands can become loose.

The assembled PCB is preferably coated with a polyurethane or epoxy lacquer spray for surface insulation. This is to minimise flow of stray surface currents under severe environmental conditions like high humidity, dust and dirt.

The wire connections used between low-power control PCBs and the high-power part of the equipment have to be protected against the interference by electro-magnetic coupling (EMC). Strong electromagnetic fields are usually present in an industrial environment where power electronics equipment is used. A simple and efficient remedy is to twist the wire connections (Fig. 6-9). Besides this, all well known precautions should be taken to avoid control circuit malfunctioning due to EMC. In particular, control PCBs should not be mounted too close to power leads, chokes and transformers carrying currents with fast rise-time waveforms, as commonly found in power electronics circuits.

7

DESIGN RULES FOR PCB'S IN MICROWAVE APPLICATIONS

7.1 INTRODUCTION

At UHF and microwave frequencies, passive components like inductors and capacitors which are used in many circuits such as filters, tuned amplifiers, etc., assume very small design values. Conventional components like wound inductors or parallel plate capacitors, usually cannot be built to yield such small values. Their physical size shrinks to impracticably small dimensions and/or the parasitic effects modify the nature of the component significantly; for example, an inductor may have a too high parasitic capacitance. In such situations, it is possible to use transmission lines of suitable length and with a suitable termination as inductors and capacitors. The impedance Z, as seen at the near end of the transmission line terminals in Fig. 7-1, is related to the far end terminating impedance Z_L and length of the line as follows:

$$Z = Z_c \frac{Z_L + jZ_0 \tan \beta\, l}{Z_0 + jZ_L \tan \beta\, l} \quad [\Omega] \tag{1}$$

where Z_c = characteristic impedance of the line [Ω]

$$\beta = \frac{2\pi}{\lambda} = \text{phase constant (i.e. phase shift per unit length along the line)} \quad \left[\frac{1}{m}\right]$$

λ = wavelength of the wave propagating along the transmission line [m]
l = length of the line [m]
Z_L = far end terminating impedance [Ω]
Z_0 = source impedance [Ω]

For instance, if $Z_L = 0$ (i.e., a transmission line with far-end shorted), then
$Z = jZ_c \tan \beta\, l$ [Ω]. \hfill (2)

By choosing a proper length l, one can realize either an inductor or a capacitor of suitable

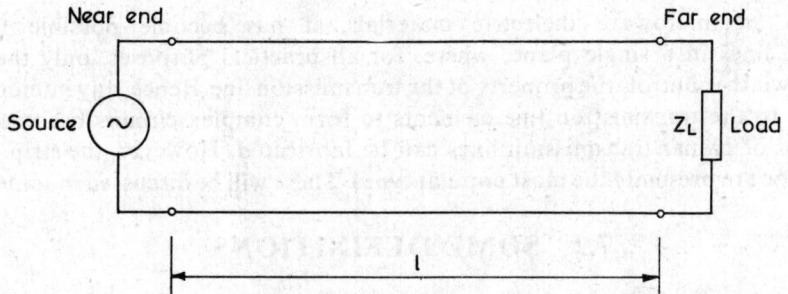

Fig. 7-1 Transmission line connecting a source to a load

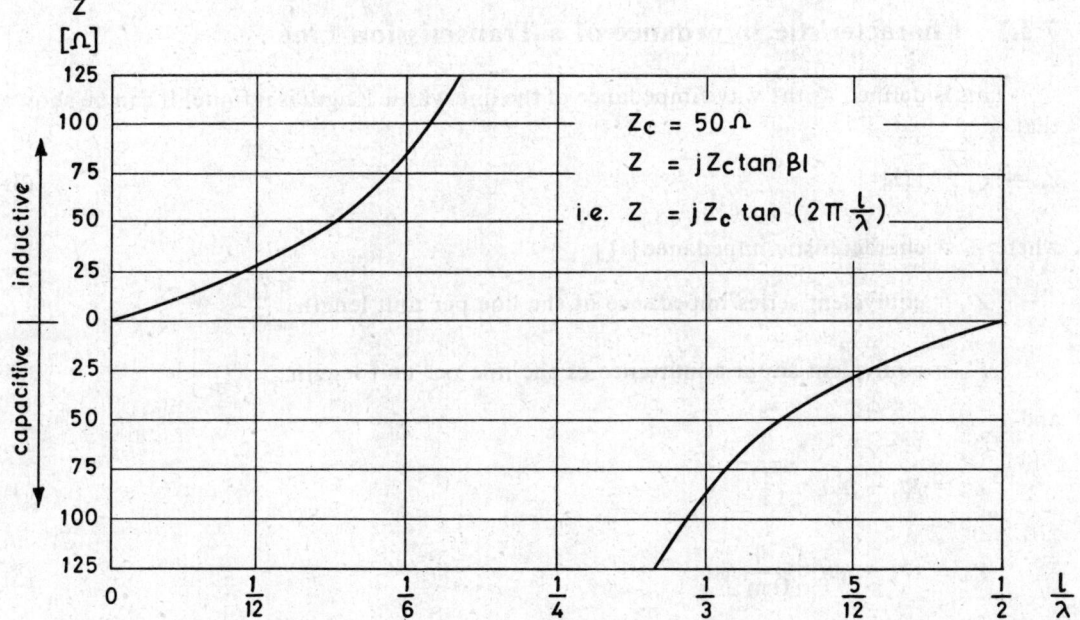

Fig. 7-2 Impedance Z, as seen at the near end of a transmission line with far end shorted

value, at any given frequency. Note that for a given length of transmission line, the value of inductance or capacitance is a function of frequency. Essentially, therefore, UHF and microwave circuitry consist of a number of transmission line elements with suitable lengths and terminations. In case of a purely passive circuit like a filter, these elements are connected accordingly together to form a filter network; and in case of active circuits like amplifiers, these elements form tuned circuits which are suitably connected to active elements like transistors.

Many different types of transmission lines are in use today. Some examples are, coaxial lines, wave guides, etc. In such lines the various properties of the line are usually determined by two dimensions, for instance width and height in the case of a rectangular waveguide. These lines are therefore, referred to as non-planar transmission lines, as the elements forming the transmission line are not in a single plane. With the advent of PCB technology and the development of microwave dielectric materials, it has become possible to fabricate transmission lines in a single plane, where, for all practical purposes, only the dimension (usually the width) controls the property of the transmission line. Hence, any interconnection of components to the transmission line elements to form complex circuits becomes very easy. Several types of planar transmission lines can be fabricated. However, the strip line and the microstrip line are presently the most popular types. These will be discussed in some detail here.

7.2 SOME DEFINITIONS

Before proceeding with the discussion on planar transmission lines, a few transmission line terminologies are described below:

7.2.1 Characteristic Impedance of a Transmission Line

This is defined as the wave impedance of the line whose length is infinite. It can be shown that

$$Z_c = \sqrt{\frac{Z'_s}{Y'_p}} \; [\Omega] \tag{3}$$

where Z_c = characteristic impedance $[\Omega]$

Z'_s = equivalent series impedance of the line per unit length $\left[\frac{\Omega}{m}\right]$

Y'_p = equivalent shunt admittance of the line per unit length $\left[\frac{1}{\Omega m}\right]$

and

$$Z'_s = R'_s + j\omega L' \; \left[\frac{\Omega}{m}\right] \tag{4}$$

$$Y'_p = G'_p + j\omega C \left[\frac{1}{\Omega m}\right] \tag{5}$$

where R'_s = equivalent series resistance of the line per unit length (due to conductor resistance, skin effect etc. $\left[\frac{\Omega}{m}\right]$

L' = equivalent series inductance of the line per unit length $\left[\frac{H}{m}\right]$

G'_p = equivalent shunt conductance of the line per unit length (due to dielectric losses) $\left[\frac{1}{\Omega m}\right]$

C' = equivalent shunt capacitance of the line per unit length $\left[\frac{F}{m}\right]$

For most practical transmission lines, the losses are negligible, i.e., $R'_s \approx 0$ and $G'_p \approx 0$ and Z_c becomes

$$Z_c = \sqrt{\frac{L'}{C'}} \; [\Omega] \tag{6}$$

7.2.2 Propagation Constant

When a power source is connected to a transmission line, the electrical wave is said to 'propagate' along the line. This wave, however, gets attenuated because of the R'_s and G'_p defined above. This attenuation takes place exponentially along the line and can be expressed as $(e^{-\alpha l})$ where α is a constant and 'l' is the distance from the near end. α is called *attenuation constant* and gives the attenuation per unit length. Similarly, the wave also undergoes a phase shift along the line which can be expressed as $(e^{-j\beta l})$ where β is a constant. β is called the *phase constant* and gives the phase shift per unit length along the line. Combining the effects of the two, one can express the decay of the wave along the line as $(e^{-(\alpha+j\beta)})l$ where

$$\gamma = \alpha + j\beta. \tag{7}$$

γ is called the *propagation constant* of the line and is a complex number.

7.2.3 Reflection Coefficient

When a power source is connected to a transmission line which is not terminated by its characteristic impedance, the energy gets reflected from the termination. The reflection is usually given as a fraction of the incident wave as follows:

$$\rho = \text{Reflection coefficient} = \frac{\text{Reflected wave amplitude}}{\text{Incident wave amplitude}} \tag{8}$$

7.2.4 Voltage Standing Wave Ratio (VSWR)

In Section 7.2.3 above, a pulse was applied to the transmission line. However, even if sinusoidal voltages are applied, reflections as described above take place.

The voltage at any point along the transmission line can be got by adding the incident and reflected voltages. In case of sinusoidal inputs, if such an addition of voltages is carried out and if the length of the line is greater than half the wave length corresponding to the input frequency, voltage maxima and minima are established along the line (Fig. 7-3).
VSWR, the voltage standing wave ratio, is then defined as

$$\text{VSWR} = \frac{\text{Maximum value of the voltage along the line}}{\text{Minimum value of the voltage along the line}} \tag{9}$$

VSWR is related to ρ by the relation

$$\text{VSWR} = \frac{1+\rho}{1-\rho} \tag{10}$$

7.2.5 Modes of Propagation

When a power source is connected to a transmission line, the wave propagation along the line can be described completely by describing the direction with respect to propagation direction, magnitude and time variation of electric and magnetic waves. These two waves (E and H respectively) may have many possible orientations with respect to the propagation direction. Each possible orientation is referred to as mode of propagation. One of the most

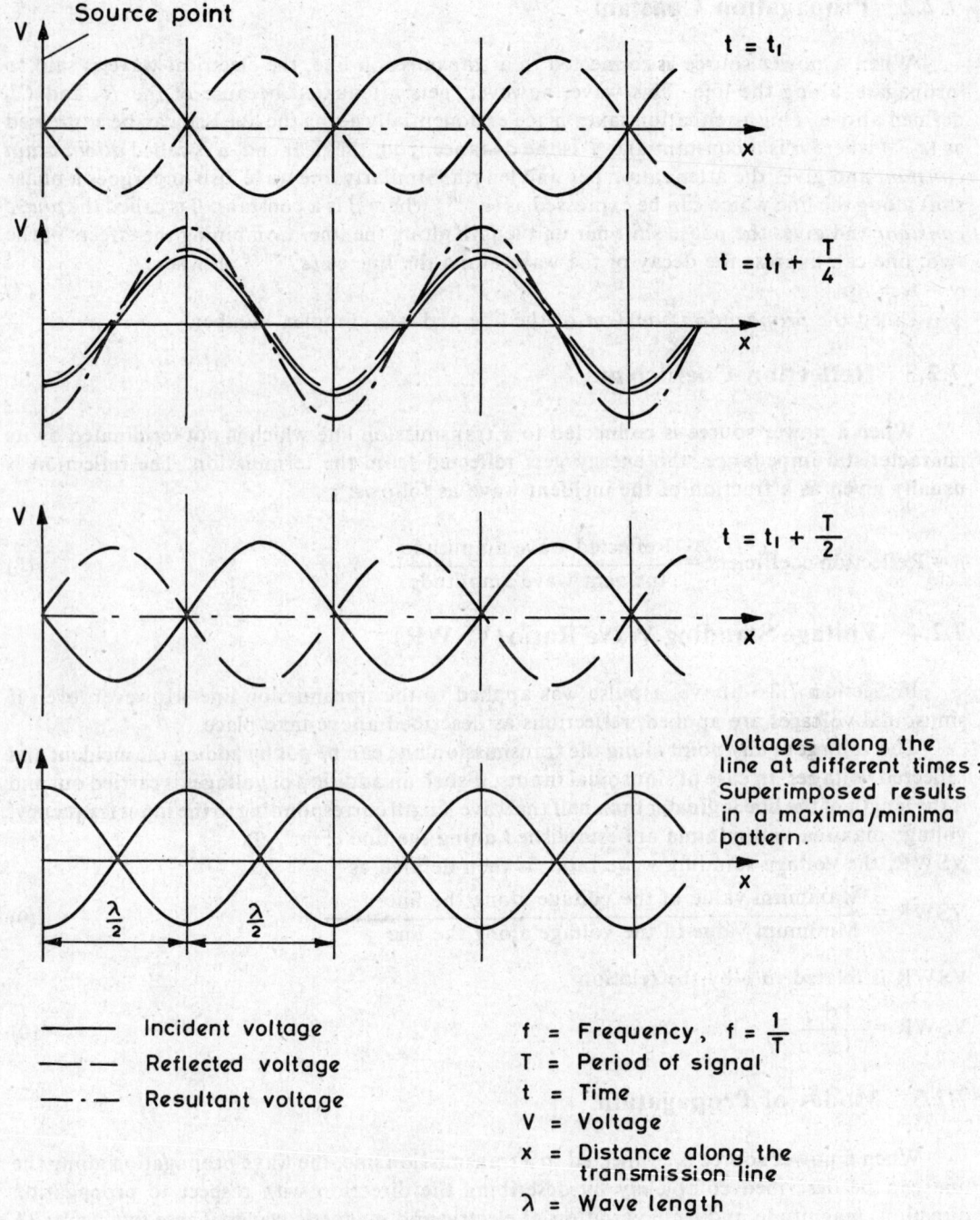

Fig. 7-3 Voltage maxima and minima along a transmission line

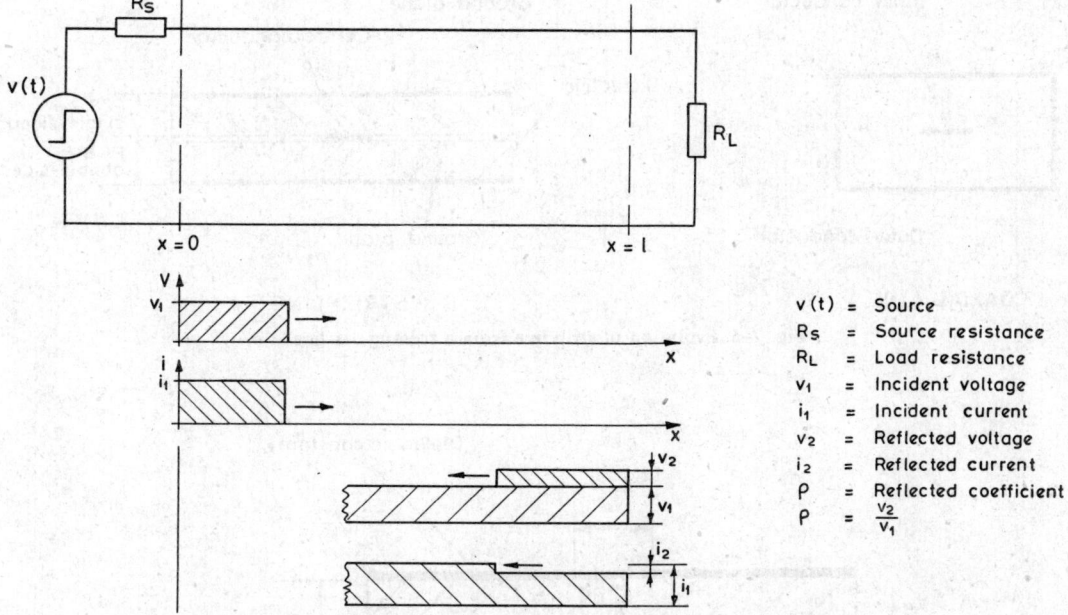

Fig. 7-4 Reflections along a transmission line

common types of propagation is the TEM (*T*ransverse *E*lectric and *M*agnetic). In this mode, the electric and magnetic fields are perpendicular to the direction of wave propagation.

The assumption of TEM mode of propagation is the basis for the transmission line parameters defined above. In most practical cases of microstrip and strip line transmission lines, the propagation mode is almost TEM. Whereas in other types of microwave transmission lines like waveguides, TE and TM modes of propagation appear predominantly.

7.3 STRIP LINE AND MICROSTRIP LINE

7.3.1 Strip Line

The strip line is derived usually from the coaxial transmission as shown in Fig. 7-5.

In Fig. 7-5, the central conductor of the coaxial line is shrunk to zero thickness and the outer conductor is broadened in a direction parallel to the central conductor to infinity. As can be clearly seen from this figure, a strip line is nothing but a sandwich of two PCBs: One double-sided PCB with the transmission line on one side and a ground plane on the other, and one single-sided PCB with ground plane over its entire area.

The mode of propagation in a strip line is nearly TEM.

The following equations determine the characteristic impedance, which is the most important characteristic of the transmission line of such a line.

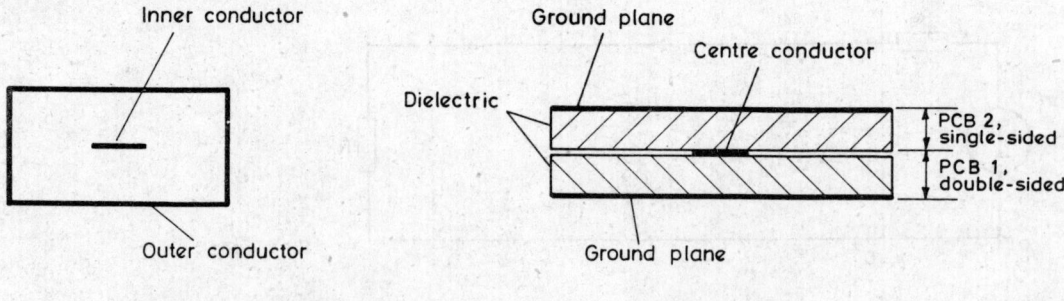

COAXIAL LINE STRIP LINE

Fig. 7-5 Evolution of strip line from a rectangular line

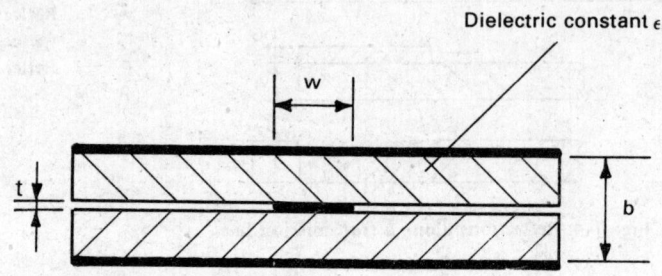

b = Thickness of dielectric

t = Thickness of centre conductor

(t = 0 for ideal strip line)

w = Width of centre conductor

Fig. 7-6 Practically realised strip line

According to [1], the characteristic impedance Z_c of the stripline is given by

$$Z_c \sqrt{\epsilon_r} = 29.976 \, \pi \, \frac{K(k)}{K^1(k)} \; [\Omega] \tag{11}$$

assuming that $\dfrac{t}{b} \leqslant 0.01$

here $K(k) \displaystyle\int_0^{\pi/2} \frac{d\phi}{1 - k^2 \sin^2 \phi}$ (elliptic integral of the first kind) (12)

$$k = \text{sech}\left(\frac{\pi}{2} \cdot \frac{w}{b}\right) \tag{13}$$

DESIGN RULES FOR PCB'S IN MICROWAVE APPLICATIONS

$K^1(k) = K(k^1)$ where (14)

$k^1 = \sqrt{1-k^2}$ (15)

Usually this equation is approximated by the following expressions for $\dfrac{w}{b} \leq 0.5$

$$Z_c \sqrt{\epsilon_r} = 29.976 \ln 2 \left(\frac{1+\sqrt{k}}{1-\sqrt{k}} \right) \;[\Omega] \quad (16)$$

for $\dfrac{w}{b} > 0.5$

$$Z \sqrt{\epsilon_r} = \frac{29.976 \times \pi^2}{\ln 2 \left(\dfrac{1+\sqrt{k^1}}{1-\sqrt{k^1}} \right)} \;[\Omega] \quad (17)$$

where $k = \text{sech} \left(\dfrac{\pi}{2} \cdot \dfrac{w}{b} \right)$ (18)

and $k^1 = \tanh \left(\dfrac{\pi}{2} \cdot \dfrac{w}{b} \right) = \sqrt{1-k^2}$ (19)

Table 7-1 gives directly the characteristic impedance of lines as a function of $\dfrac{w}{b}$ for $\dfrac{t}{b} \leq 0.001$.

Table 7-1 Characteristic impedance of strip line

$Z_c \sqrt{\epsilon_r}\;[\Omega]$	$\dfrac{w}{b} \left(\dfrac{t}{b} \leq 0.001 \right)$
150	0.18936
120	0.32942
100	0.47894
80	0.70931
50	1.40774
40	1.87388
16	5.37003

7.3.2 Microstrip Line

The microstrip line is derived from a 'conductor above ground' line as shown in Fig. 7-7. The conductor above ground is flattened into a plate of zero thickness and the ground is extended to infinity.

As can be seen from Fig. 7-7, a microstrip line is nothing but a double-sided PCB with a conductor line on one side and a ground plane on the other side. The mode of propagation in a microstrip line is not strictly TEM because of a discontinuity in the dielectric and the absence of symmetry of ground plane with respect to the line conductor. The mode of propagation is usually referred to as quasi TEM and calculation of characteristic impedance is quite complex. Fig. 7-8 gives the diagram of a practical microstrip line with various parameters defined as in strip line.

Fig. 7-7 Evolution of the microstrip line from a 'conductor above ground' transmission line

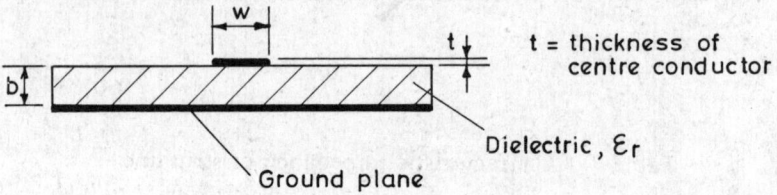

Fig 7-8 Practically realised microstrip line

According to [1], the approximate characteristic impedance of the microstrip line is given by

$$\text{for } 0 \leqslant \frac{w}{h} \leqslant 1: \quad Z_c \sqrt{\frac{\epsilon_e}{\epsilon_r}} = 59.952 \ln\left(\frac{8h}{w} + \frac{w}{4h}\right) \tag{20}$$

$$\text{for } 1 < \frac{w}{h} \leqslant 10: \quad Z_c \sqrt{\frac{\epsilon_e}{\epsilon_r}} = \frac{119.904\pi}{\frac{w}{h} + 2.42 - 0.44\frac{h}{w} + \left(1 - \frac{h}{w}\right)} \tag{21}$$

$$\text{where } \epsilon_e = \frac{\epsilon_r + 1}{2} + \frac{\epsilon_r - 1}{2} \cdot \left(1 + \frac{10\,h}{w}\right)^{-\frac{1}{2}} \tag{22}$$

Alternatively, the following procedure described in [3] can be used to find the approximate characteristic of a microstrip line:
w = width of strip
h = thickness of dielectric = b

Step 1: Compute the width correction due to thickness of the conductor strip as follows:

$$\Delta w = \frac{t}{\pi}\left(\ln\frac{4\pi w}{t} + 1\right) \text{ for } \frac{w}{h} \leq \frac{\pi}{2} \qquad (23)$$

$$\Delta w = \frac{t}{\pi}\left(\ln\frac{2h}{t} + 1\right) \text{ for } \frac{w}{h} > \frac{\pi}{2}$$

where t = thickness of the conductor.
Then correct width $w^1 = w + \Delta w$ \hfill (24)
Let $w^1 = 2a$ and $h = b$ \hfill (25)
Refer to Fig. 7-9 and determine the characteristic impedance of a 'symmetrical strip' i.e., two identical strips placed symmetrically with respect to a ground plane and let this be Z_s.

Thus Z_c, the characteristic impedance of a microstrip, without taking into account the discontinuity in the dielectric, is

$$Z'_c = \frac{Z_s}{2} \qquad (26)$$

From Fig. 7-10 determine 'per cent of air effect' on the dielectric constant and let this be p. Then compute ϵ^1 as

$$\epsilon^1 = \sqrt{\epsilon_r}(1-p) + p \qquad (27)$$

Then characteristic impedance of the microstrip line is

$$Z_c = \frac{Z_c^1}{\epsilon^1} \qquad (28)$$

7.3.3 Attenuation in Strip- and Microstrip-Lines

The wave which propagates along the strip and microstrip transmission line is attenuated due to the following reasons:
a) dielectric loss
b) loss in the conductor of the line
c) radiation loss (mainly in the microstrip line).

Calculation of attenuation constant is quite cumbersome and can be found in the references [2, 3] cited at the chapter end.

7.4 APPLICATIONS OF STRIP-AND MICROSTRIP-LINES

These transmission lines can be put to various uses. As given earlier, these lines with

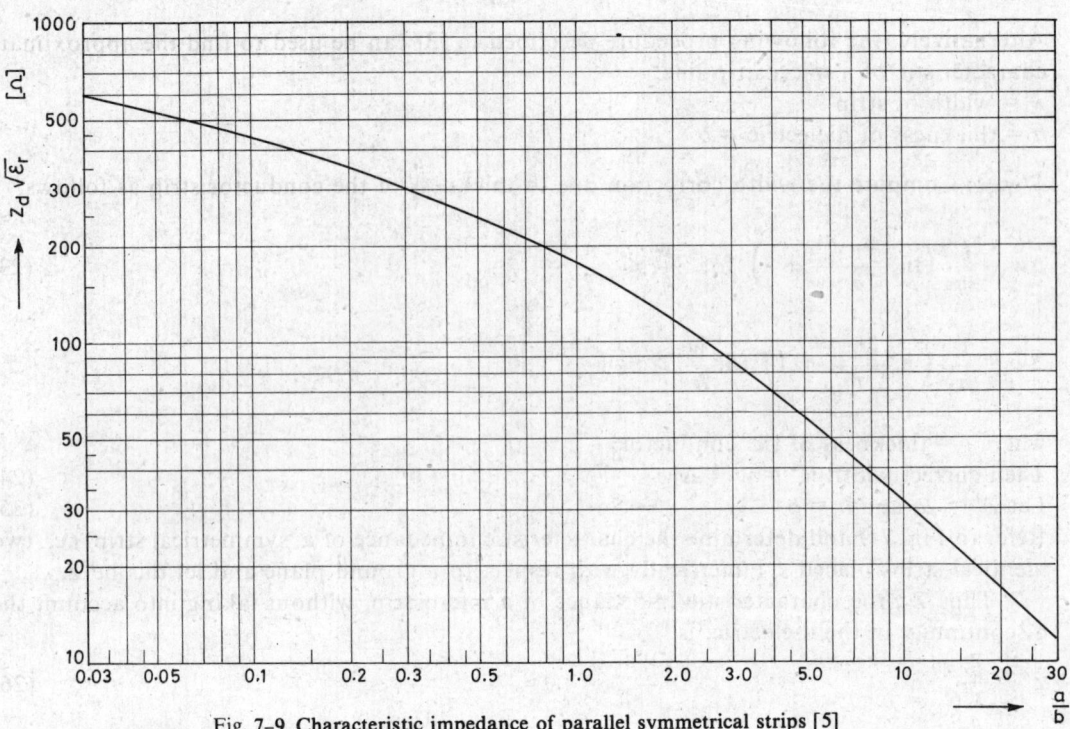

Fig. 7-9 Characteristic impedance of parallel symmetrical strips [5]

Air Effect

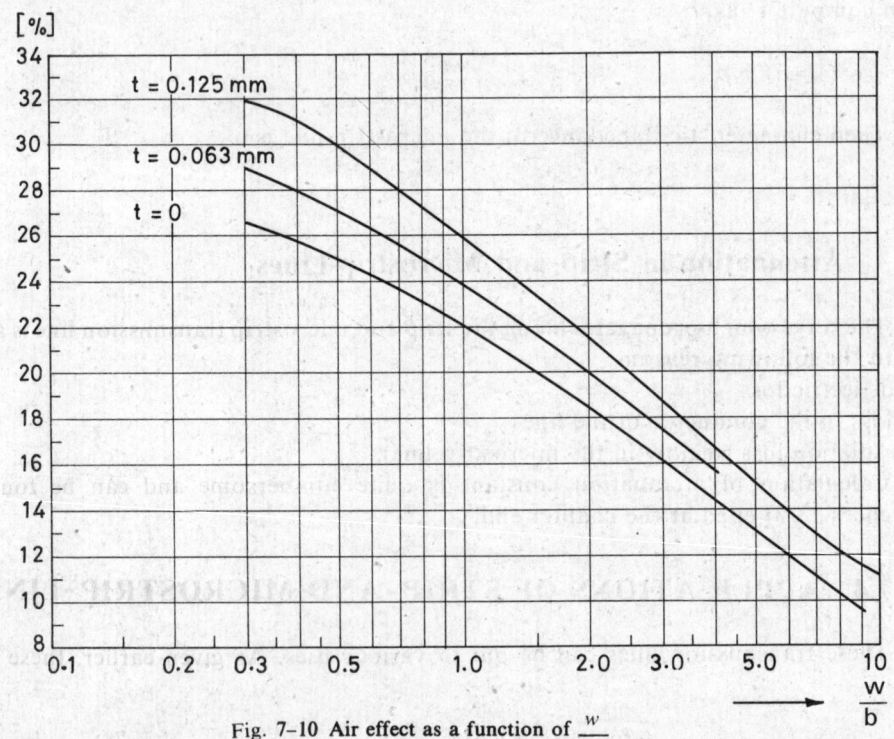

Fig. 7-10 Air effect as a function of $\frac{w}{b}$

suitable terminations can be used as passive elements. A line which is short in length compared to the wave length of the signal transmitted, can be approximated as a *lumped* passive element such as an individual inductor or an individual capacitor. A thin line gives an inductance and a thick line a capacitance.

Transmission lines can also be used as impedance transformers, matching devices, etc.

7.4.1 Filters

A lowpass filter is illustrated in Fig. 7-11. The inductor and capacitor sections of this circuit can be built using transmission line segments to yield lowpass filters with a cut-off frequency in the range of a few hundred megahertz and above.

In a similar manner it is possible to achieve other types of filters.

Design of such filters follows the conventional methods and can be found in the reference [4]

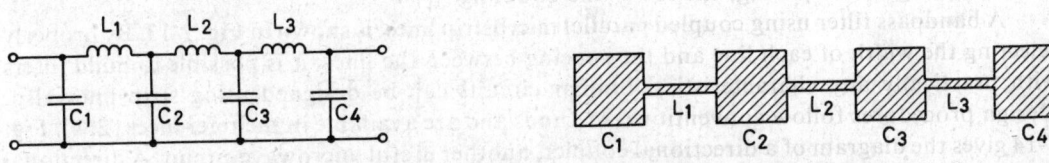

Fig. 7-11 Lowpass filter circuit and its microstrip form of realisation

7.4.2 Coupled Transmission Lines

When two transmission lines are placed close to each other, they are said to be coupled, as there is an interaction between these lines. This coupling can be used to build bandpass filters, band elimination filters, directional couplers, etc.

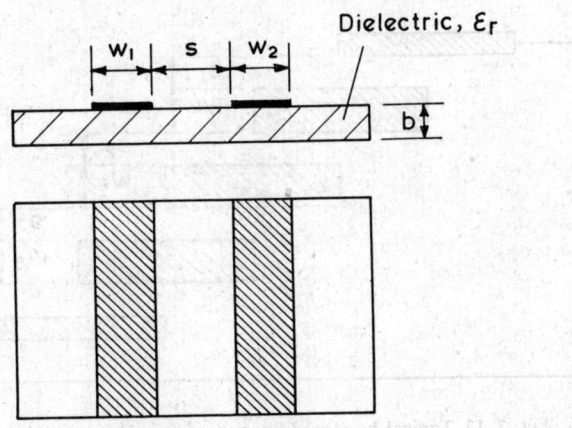

Fig. 7-12 Coupled transmission line

Fig. 7-12 shows two lines placed close to each other. If I_1 and I_2 are the currents flowing through these lines, they can be considered to be made up of two currents I_e and I_o such that

$$I_1 = I_e + I_o, \quad I_2 = I_e - I_o \tag{29}$$
where
$$I_e = \frac{I_1 + I_2}{2}, \quad I_o = \frac{I_1 - I_2}{2} \tag{30}$$

According to (29), I_e is a current which has same direction of flow along the two lines and I_o is a current which flows in opposite directions along the two lines.

I_e is known as the even-mode current and I_o the odd-mode current. The characteristic impedance components, designated as Z_e and Z_o, associated with these currents, are known as the even-mode and odd-mode impedances respectively. The relative values of Z_e and Z_o give an idea of the coupling between the line.

If $Z_e = Z_o$, there is no coupling between the two lines and usually $Z_e > Z_o$ for coupled cases. Greater Z_e (as compared with Z_o) means a tighter coupling between the lines.

As can be readily seen, coupling between lines is a function of spacing between the two lines: The larger the spacing, the lower the coupling.

A bandpass filter using coupled parallel microstrip lines, is shown in Fig. 7-13. By properly selecting the width of each line and the spacing between the lines, it is possible to build filters with practically any characteristics. Similar circuits can be designed using strip lines also. Design procedures follow conventional methods and are available in the references [2, 4]. Fig. 7-14 gives the diagram of a directional coupler, another useful microwave circuit. A directional coupler is a device which couples a small fraction of the power which is flowing in the 'main path' to a port if the power is flowing in one direction and couples negligible power to the same port if the power is flowing in the opposite direction. The amount of power coupled depends upon the distance between the lines, the length of overlap between the lines, terminating conditions on the other port, etc. Such couplers are used to sample power flow in high-power circuits, in microwave instrumentation, etc.

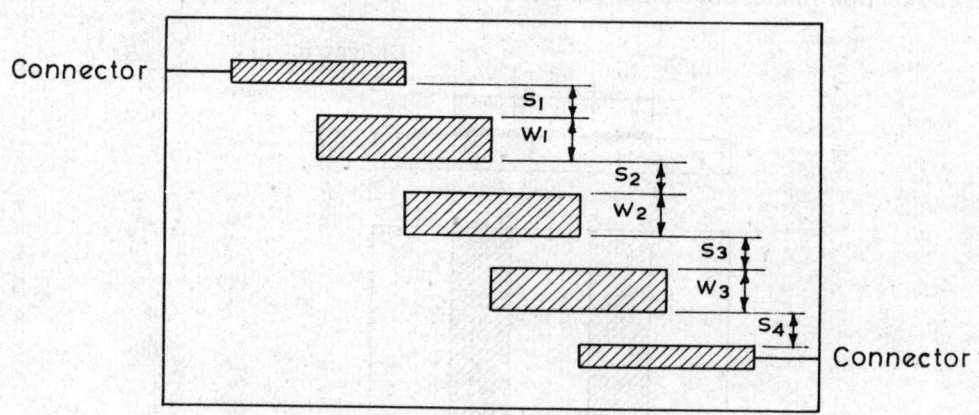

Fig. 7-13 Typical bandpass filter using coupler microstrip lines

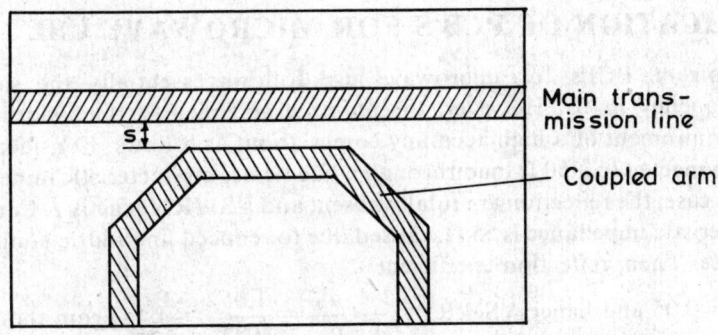

Fig. 7-14 Directional coupler

7.5 MATERIALS FOR MICROWAVE PCB'S

One main problem with strip lines and microstrip lines is the high attenuation per unit length. PCB materials for microwave should therefore be selected so as to yield minimum loss. For this a low-loss dielectric material, which reduces the dielectric loss, is required. Also, a higher dielectric constant yields lower radiation loss. It is also important to have materials whose thickness is very uniform and the dielectric constant stable. Otherwise, the characteristic impedance of the line varies along the line length, which gives rise to unacceptable VSWR values even with a perfect termination.

A high dielectric constant also reduces the size of the microwave circuit and leads to miniaturization and extends usefulness of microwave PCBs to higher frequencies. Table 7-2 gives some typical dielectric materials used at high frequencies and their relevant characteristics.

Table 7-2 Microwave dielectric materials

Dielectric material	ϵ_r	$\tan \delta$	Useful range
Rexolite 1422 (cross-linked) polystyrene	2.53	$2 \cdots 7 \times 10^{-4}$	$2 \cdots 4$ GHz
Cross-linked polystyrene with ceramic filling (custom high K)	$3 \cdots 15$	$7 \cdots 15 \times 10^{-4}$	up to 10 GHz
Silicon resin with ceramic powder filling	$3 \cdots 12$	$2 \cdots 5 \times 10^{-3}$	up to 10 GHz
Teflon fibre glass	2.5	2×10^{-3}	up to 4 GHz
99.6% Alumina	9.6	—	used in microwave ICs

7.6 FABRICATION OF PCB'S FOR MICROWAVE USE

Fabrication of PCBs for microwave use follows essentially the same technology as for low frequency PCBs. However, the required accuracy for line width, etc., is usually higher. The requirement of a high accuracy comes about as follows: Consider, for instance, a 50 Ω source connected to a 50 Ω load through a line whose characteristic impedance is exactly 50 Ω. In such a case, the reflection are totally absent and VSWR is exactly 1. Consider now a line whose characteristic impedance is 55 Ω, caused due to reduced line width, connecting the same load and source. Then, reflection coefficient

$\rho \approx \dfrac{55 - 50}{55 + 50} \approx 0.05$ and hence VSWR $= \dfrac{1 + \rho}{1 - \rho} = \dfrac{1.05}{0.95} = 1.1$. From these results, it can be clearly seen that VSWR becomes poorer if the line width is not exactly what has been calculated.

If strip line is used for the above case on a dielectric with $\epsilon_r = 2.56$, a 50 Ω line would correspond to a line width of 2.36 mm. A decrease of 10% in width of this line (i.e., a decrease of 0.236 mm) would lead to a characteristic impedance 53.4 Ω and hence to a VSWR of 1.06.

The problem of accuracy is still further aggravated when using lines with higher impedances where Z_c is a more sensitive function of $\dfrac{w}{b}$.

The artwork, therefore, is usually made at 4 to 16 times the actual size. Also, because of odd line widths required, one resorts to inked artworks on a stable base sheet (usually 100 to 125 μm thick polyester).

7.7 USE OF STRIP AND MICROSTRIP LINES IN PULSE CIRCUITS

For the transmission of pulses with very sharp rise and fall, e.g., of the order of a few hundred picoseconds to a few nanoseconds, one has to resort to strip or microstrip transmission lines. Such lines, when used for pulse transmission, introduce signal delays and also reflections if the termination at the far end of the line is not proper. Further, if two lines carrying fast pulses are close to each other, they cause interference with each other due to crosstalk. While designing fast-pulse circuits, all these effects have to be taken into account.

BIBLIOGRAPHY

1. GUNSTON M. A. R.: *MICROWAVE TRANSMISSION LINE IMPEDANCE DATA*, Van Nostrand Reinhold Co., London, 1972.
2. HARLAN HOWE: *STRIP LINE CIRCUIT DESIGN*, Artech House Inc., Dedham, Massachusetts, 1974.
3. HARPER C. A.: *HANDBOOK OF WIRING, CABLING AND INTERCONNECTING FOR ELECTRONICS*, McGraw-Hill Book Co., New York, 1972.
4. MATHEI G. L. et al: *MICROWAVE FILTERS, IMPEDANCE MATCHING, NETWORKS AND COUPLING STRUCTURES*, McGraw-Hill Book Co., New York, 1964.
5. WHEELERS H. A.: 'Transmission line properties of parallel strips separated by a dielectric sheet', *IEEE TRANS. MICROWAVE THEORY TECH.*, *Vol.* MTT-13, March 1965, pg. 172–185

8
ARTWORK

8.1 INTRODUCTION

The generation of PCB artwork should be considered as the first step of the PCB manufacturing process. The importance of a perfect artwork should not be underestimated: Problems like inaccurate registration, broken annular rings or too critical spacings are often due to bad artwork. And even with the most sophisticated PCB production facilities, no PCB can be made better than the quality of the artwork used.

8.1.1 Personnel

There are different possibilities to produce an artwork. But a common necessity for all of them is the need for a clean and exact working method, which means taking care even of smallest details, hardly visible to an untrained eye. Skills and patience are the basic assets for artwork designers. Since these prerequisities are hardly taught in teaching institutions, generally the artwork designers get suitable in-house training by working under the guidance of more experienced colleagues. A very important aspect of such a training is the familiarisation with the PCB *fabrication* techniques applied: The artwork designers, mostly with a background in technical drawing, have to experience *personally* the implications and consequences triggered off by unsatisfactory artwork.

8.1.2 Rooms

The artwork generation, as a highly specialised job, is carried out in separate offices which take more care of the special requirements than general drawing offices could do: Individual workplaces are equipped with sufficiently large transilluminated working areas, big enough also for the maximum artwork size. Comfortable chairs increase the capability to concentrate on the artwork and enable an easy reach of all the materials and tools required.

Artwork materials are made available in a central place, e.g., in a cupboard with suitable compartments. Stock checking should be made easy and carried out regularly; this is to avoid a sudden exhaustion of a particular pad or item.

8.1.3 Artwork Scale

For PCBs with plated-through holes or integrated circuits, the generation of 1 : 1 scale artwork would not meet the dimensional accuracy required for the reliable production of PCBs.

There are various sources of inaccuracy contributing to the overall result. A few very common shall be mentioned here as they arise in the artwork taping method on a transparent base foil:

— Pads are usually placed on a grid intersection. The centre hole of the pad is meant to enable an easy centering. The eye of an experienced artwork designer will be able to recognise asymmetries in pad placing down to 0.05–0.1 mm. Because the pads and tapes used have a certain thickness, an additional parallax error is introduced. Depending on the grid location, whether on the top or bottom side of the artwork base foil, additional parallaxes might be introduced. The final average in positional accuracy will therefore be in a range of 0.2–0.25 mm.
— The registration inaccuracy for the different conductive pattern layers itself can be more than 0.2 mm, depending on the artwork approach chosen and the care taken.
— Where minimum permissible spacing is utilised, it must be considered that the tapes tend to creep also in addition to all the above mentioned inaccuracies which cannot be avoided in manual taping. Creeping may well be in a range of 0.1–0.2 mm.

There are still more factors which have not been listed. But it has become obvious that the influence of these factors cannot be tolerated on a direct (1 : 1) scale for professional PCBs with their more and more dense patterns and widths of annular solder pad rings as low as 0.3 mm. Most of the artwork is therefore generated at a 2 : 1 scale which gives an artwork of 4 times the actual PCB area. For demanding fine-line PCBs, only a 4 : 1 scale can provide the final accuracy required. 4 : 1 artwork becomes fairly large (16 times the actual PCB area) and the material costs will also be more. This scale is therefore only used where the inaccuracy reduction factor of 4 is really required. At the same time, the reprographic camera used must also have sufficient copy-board size to accommodate the large artwork for the production of the film masters

8.2 BASIC APPROACHES

Under this section, approaches are discussed with special reference to double-sided PCB artwork. In single-sided PCB artwork, the registration problem does not arise and this makes the situation less critical with respect to overall artwork precision.

Among the five approaches discussed, the *black taping on transparent base foil* must be considered as the most versatile method; it has been widely accepted in industries for all kinds of professional artwork requirements. The *red/blue taping on one transparent base foil* and the *black taping on diazo film* are improved versions of the initial black taping method. Their advantage is in providing a perfect registration between the two artwork patterns for a double-sided PCB. There are however various handicaps in using the red/blue taping method. In the view of the author, for artwork with highest registration requirements, the black taping on diazo film should be given preference.

Both *ink drawing method* as well as *cut-and-strip artwork method* approaches are followed mainly for consumer electronics boards. But, under certain conditions, they are also used for industrial electronics PCBs. These two methods will also be explained here in order to give a complete picture of all the methods practically used today.

8.2.1 Ink Drawing on White Cardboard Sheet

This method is the earliest one used for PCB artwork design. Although it has lost its importance due to the availability of faster and more precise methods, there are still many applications where ink drawings offer the best solution. As for instance, in the design of radio receiver PCBs or for high-frequency PCBs where the conductors require special shapes.

Materials

From the material side, it is an extremely cheap solution: White cardboard paper (Bristol board), good quality Indian ink and an ink pen set are the minimum requirements. Where available, cardboard paper with a blue 0.1" grid is preferably used. The blue lines will not be reproduced in the photographic process for the film master production.

Drawing Practice

The drawing procedure is very time consuming and needs good drawing skills. The circumferences of the solder pads and the centre holes are drawn with a drawing compass and the space in between is filled with ink. A very useful aid for drawing of the solder pads is given with the templates which are available at 1 : 1, 2 : 1 and 4 : 1 scale (Fig. 8-1) depending on the artwork reduction scale chosen. Conductors are drawn with an ink pen either by directly giving the desired width or as double lines which are thereafter filled with ink.

For artwork of double-sided PCBs, the pattern of one side is finished first. The solder pad centres are then transferred onto the other side by pricking and the second pattern can be drawn on the backside of the cardboard sheet.

A major disadvantage of ink-drawn artwork is the difficulties encountered if pattern rectifications or modifications have to be made. The erasing operation will cause a rough surface where later on the lines can become smeared or ragged.

Precision

Compared to other artwork generation methods currently used, the precision which can be obtained by ink drawings is rather on the lower side. A main reason is in the drawing inaccuracies themselves. Conductor widths vary at least by 0.1–0.2 mm and solder pad locations and conductors can easily be displaced by 0.3–0.5 mm. One solution to this can be to choose a higher artwork reduction scale which basically reduces such inaccuracies by the same scale factor.

But a more critical limitation to the precision is given by the dimensional instability of the cardboard sheet which itself can vary in different board directions. Typical figures are 0.01%/°C with respect to temperature changes and 0.005%/%RH with respect to changes in the relative humidity.

Improvements

The basic process, as it has been described, is in practice very often modified in order to eliminate or reduce some of the inherent handicaps:
—The use of a thicker variety of transparent drawing paper instead of cardboard sheet gives

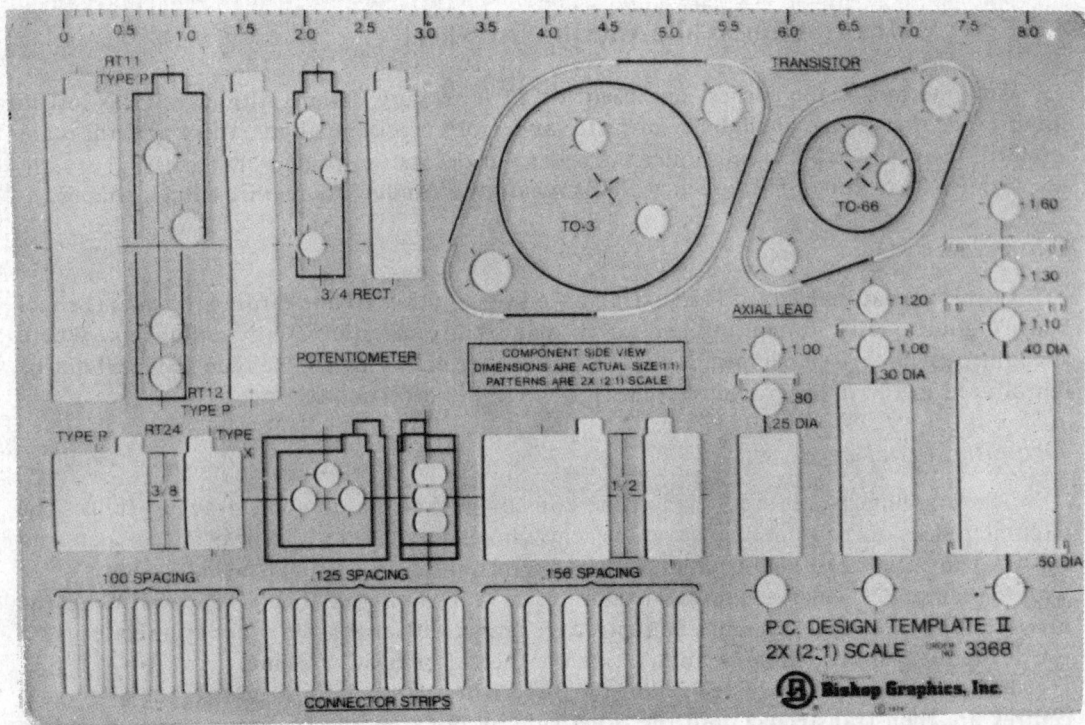
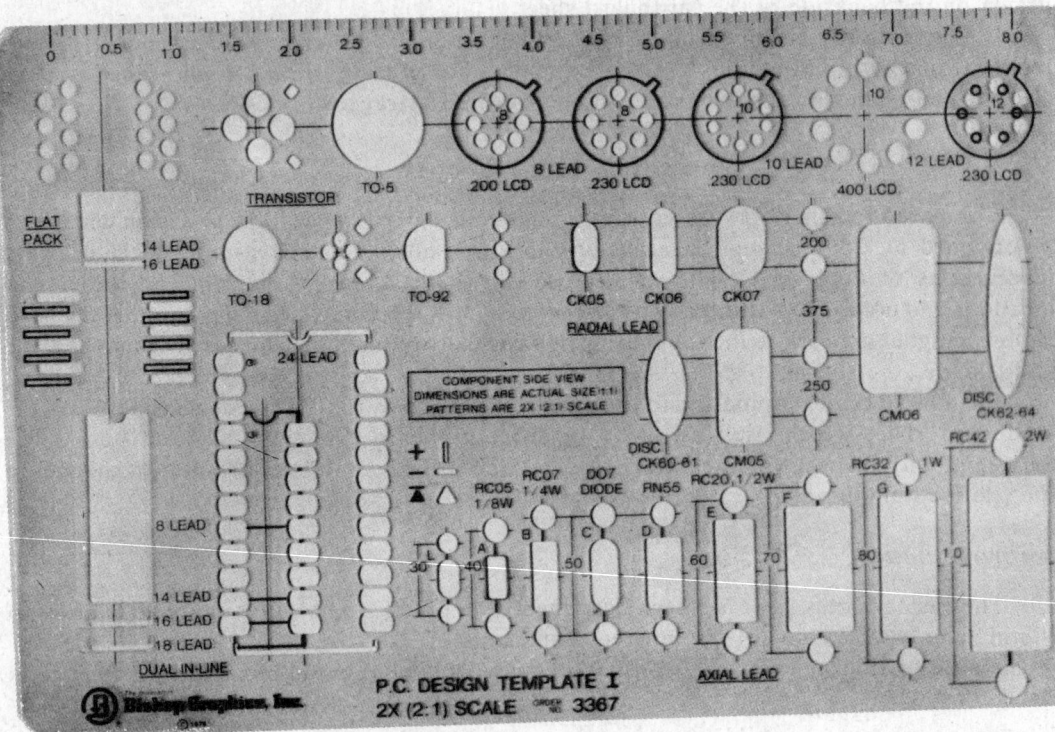

Fig. 8-1 Template set for ink drawing of artwork (Photo CEDT)

more convenience in inking. If the layout sketch is properly scaled, for instance on a grid paper, it can be put underneath the transparent drawing sheet and inking can be carried out faster with the guidance of the layout sketch. At the same time, erasing operations are simplified. The ink artwork on transparent drawing sheet is also suitable for photography on a camera with a through illuminated copy-board. On the other hand, one separate artwork drawing will be required for either conductor side of double-sided PCBs which introduces registration inaccuracies.

The ink pattern may also be drawn on suitable polyester foil which is dimensionally considerably more stable than cardboard paper or transparent drawing sheet. Such polyester foil should be highly transparent and have at least 100 μm thickness. One side must provide a matt/rough surface finish so that the ink will stick to it. The ink drying takes more time on the polyester foil; patience and time are the main requirements for the drawing personnel.

8.2.2 Black Taping on Transparent Base Foils

With the availability of self-adhesive or transfer-type pads and adhesive precision tapes (Fig. 8-2) to the artwork designer, the artwork generation procedures have rapidly changed. Artwork generation does not depend any longer on high drawing skills and endless patience; the artwork is produced in a much faster and more precise way with such pads and tapes.

Pads

They are available in basically two different varieties which are the self-adhesive type and the transfer-type.

Self-adhesive pads: These pads are supplied sticking on a backing paper. If such a pad has to be fixed on the artwork base sheet, a knife blade is slipped under an edge of the pad to peel it off from the backing sheet. The pad is held against the knife blade with a finger and can thus be conveniently positioned on the artwork base. The pad is then slightly pressed down and the position verified. If the pad position has to be shifted, the pad can be peeled off again. The pad is firmly pressed down only after the exact position has been assured.

Transfer pads: The pads are printed on a thin adhesive film of typically 10 μm thickness. The thin film is mounted on the top side against a transparent carrier strip. The pad can be transferred from the carrier strip onto the artwork base by rubbing with a wooden stick on the carrier strip while the pad is exactly positioned on the artwork base. The carrier strip can thereafter be lifted from the artwork base leaving behind the pad. The paper strip, which is usually supplied to protect the adhesive side of the pads, is now placed on top of the pad and rubbed with the wooden stick to improve pad adhesion on the artwork base.

Since the pad does not adhere on the artwork base unless it is rubbed, exact positioning is easily possible. However, transfer pads can easily get damaged because of the very thin layer. Once placed, they can be removed only by destroying them (scratching off with knife blade or sand eraser).

Tapes

Self-adhesive precision tapes are available in a wide range of widths. They are supplied in rolls and have a width tolerance of as little as 0.05–0.1mm, depending on the manufacturer. The very flexible, non-transparent material used, permits artwork with round-bent corners.

However, where round-bent corners are used, it must be considered that the tapes tend to contract a little in the course of time thus changing the curvature which could lead to critical spacing.

All the self-adhesive or transfer-type pads and also the self-adhesive tapes have a limited shelf life which may go up to 3 years. The expiry dates are sometimes printed on the packages.

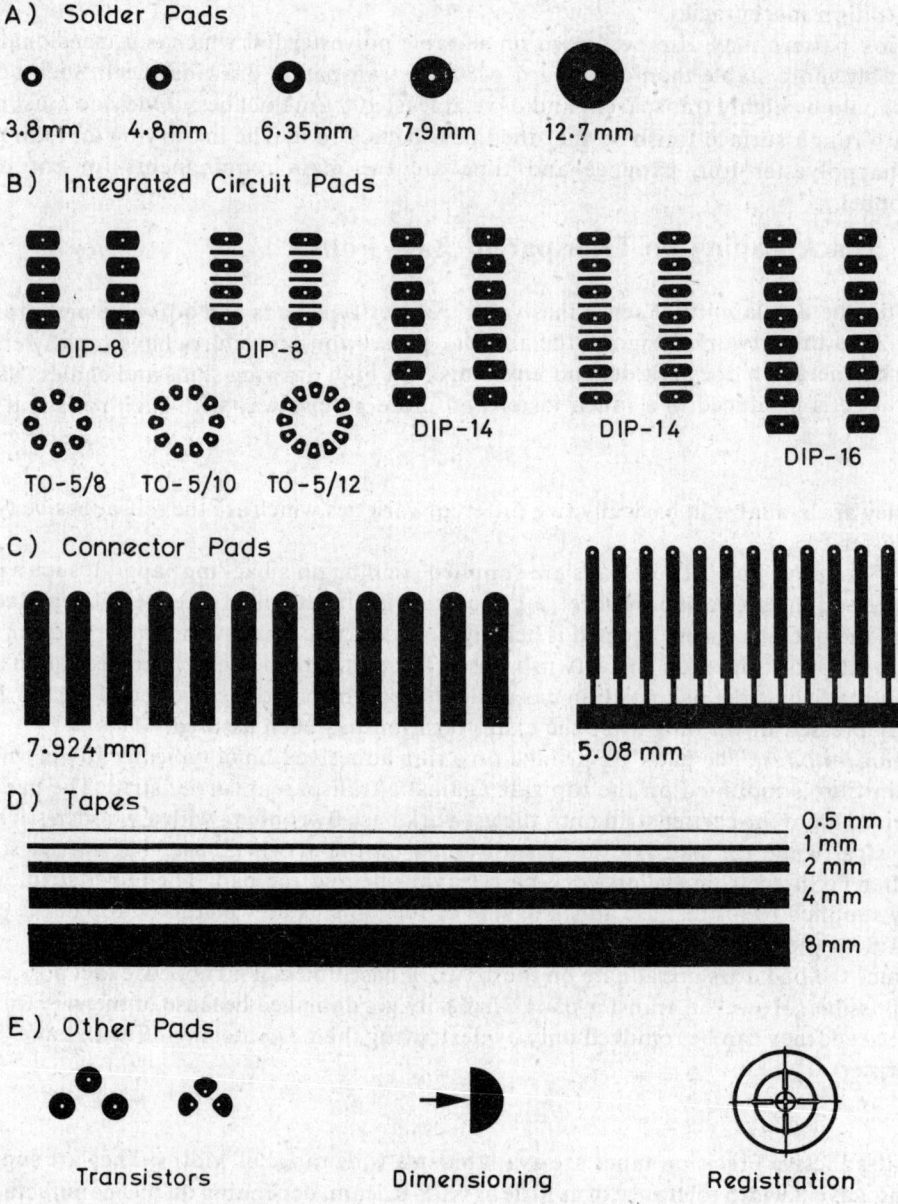

Fig. 8-2 Self-adhesive pads and tapes as commonly used in 2 : 1 artwork generation

Storage has to be done under appropriate conditions: Temperature 5–25°C and humidity below 75% RH.

Artwork Base Foil

The artwork base foils mostly used today are *polyester films* which provide an excellent dimensional stability. Typical values are 17 ppm/°C with respect to temperature changes or 11

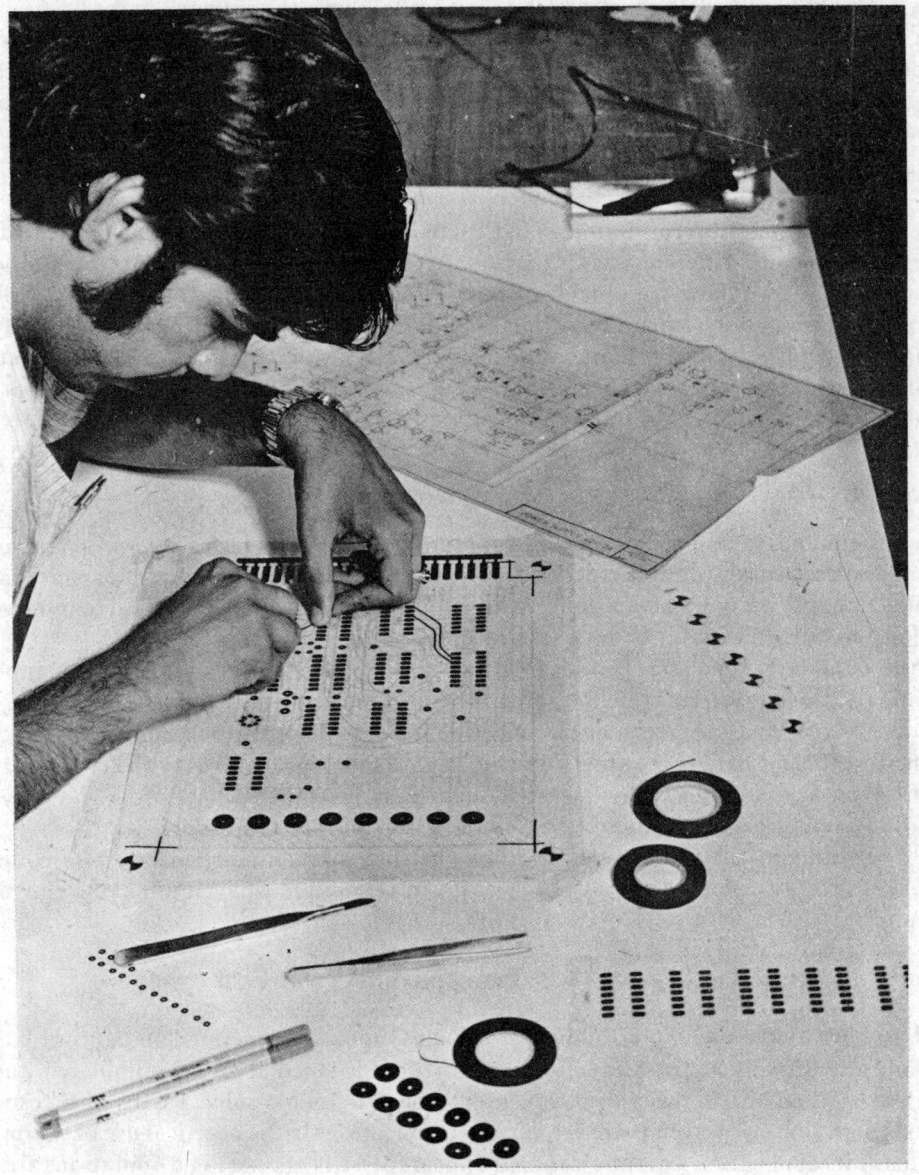

Fig. 8–3 Student preparing artwork (Photo CEDT)

ppm/%RH with respect to changes in relative humidity (ppm means 'parts per million'; RH = Relative Humidity).

Polyester films are available in many thicknesses: 100 μm is considered as the minimum for artwork in order to give sufficient mechanical stability against wrinkling due to shrinkage of the conductor tapes. The surface quality may be matt or glossy. A matt surface is essentially required for ink adherence. Where polyester foil with a printed grid is used, one has to be sure about the accuracy of the grid. Many artwork designers prefer not to use preprinted foil but to employ a highly precise grid sheet which is put underneath the artwork base.

In applications where the high dimensional stability of polyester film is not absolutely required, as for instance for single-sided or small double-sided PCBs, it is also possible to use a thick quality of *transparent tracing sheet* which is considerably cheaper. But the quality should be atleast 90 g/m^2.

2-Layer Artwork

In artwork generation for double-sided PCBs, it is possible to accomplish the 2 different circuitry patterns on 2 different artwork layers. After completion of the first layer, all the pads put on the second layer have to coincide exactly with the ones on the first layer. Practically, it is an extremely difficult job to place hundreds of pads in a manner to give an accurate registration between the two layers. The tolerances depend mainly on the artwork scale chosen and the annular ring widths of the solder pads used. Because of such difficulties, especially in artwork for highly complex circuits with a high density and fine pads, alternative methods are given preference.

3-Layer Artwork

In this method, the first layer prepared contains all the pads which are *common* to both the conductor patterns of the double-sided PCB. This common layer includes solder pads, pads for integrated circuits, via holes and mounting holes. (A via hole is a hole just used to interconnect one side to the other without being used for component mounting.)

The second layer thereafter includes the additional pads and the taping of the component side while the third layer cares of the same for the solder side (Fig. 8-4).

In the 3-layer method, the registration problem for the common pads has been solved. Even then, sufficient care is required to achieve an exact registration for the two conductor pattern layers. Special registration systems have been developed which provide a very high registration accuracy and which are even suitable for multilayer artwork [4].

A disadvantage of the 3-layer method is that for any artwork modification, a simultaneous rectification of two or even three layers is required which is not a very convenient procedure.

8.2.3 Red/Blue Taping on One Transparent Base Foil

Here, the complete artwork for double-sided PCBs can be generated on only one base foil. Therefore, the artwork registration problem for double-sided PCBs is eliminated and this makes the method to be the most convenient one for double-sided PCBs with respect to artwork generation. However, there are various implications in the practical use of this method which make its application less simple than it appears at a first glance. The limitations are in the

ARTWORK

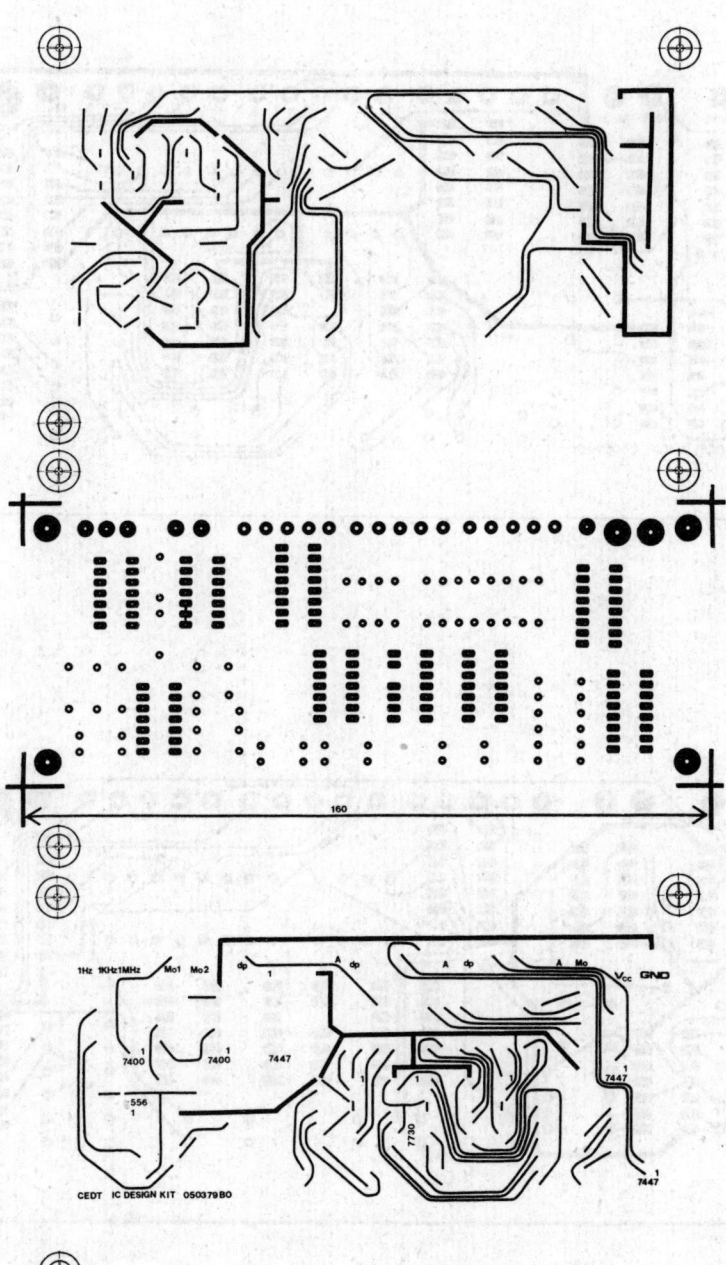

Fig. 8-4 3-layer artwork for double-sided PCB (Photo CEDT)

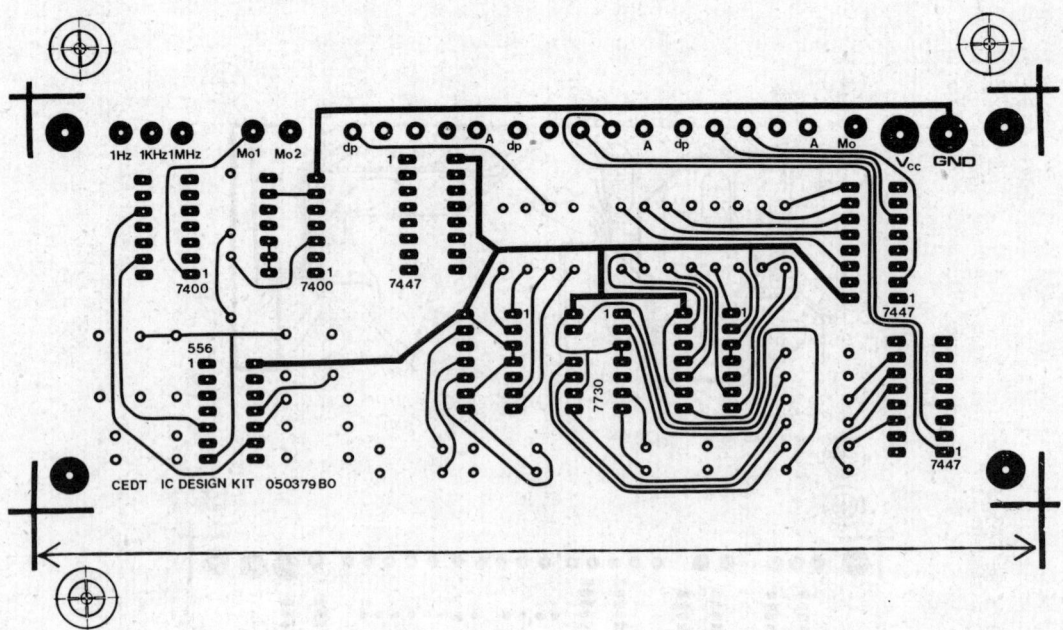

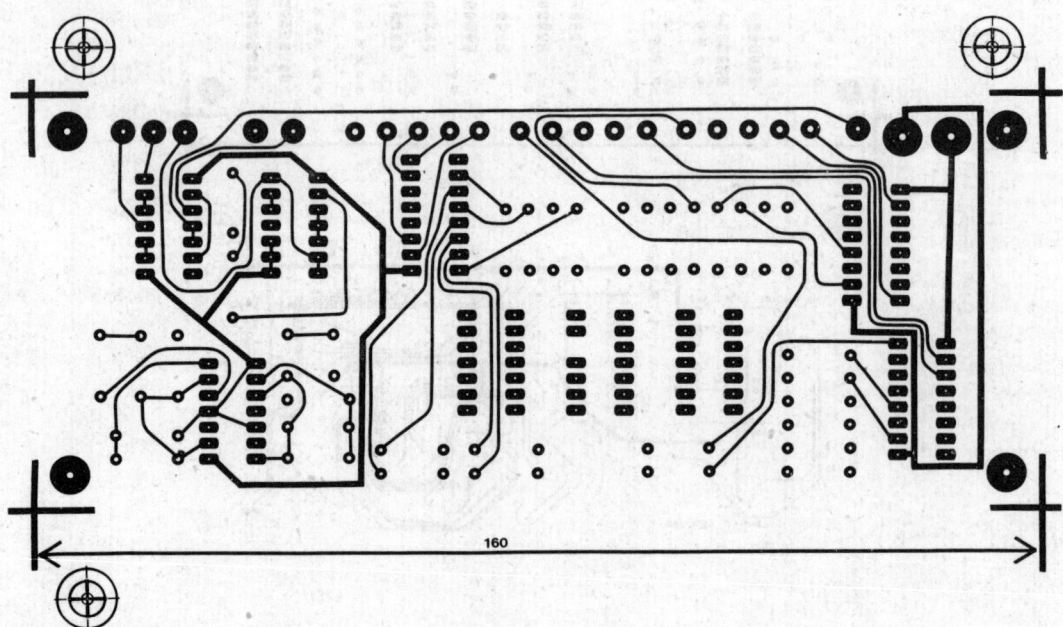

Fig. 8–5 Film masters obtained from the 3-layer artwork shown in Fig. 8–4 (Photo CEDT)

red and blue tapes which cannot be bent, the need for panchromatic film materials for the film master production and in the need to stock more artwork items.

Principle

The component pads, which have to appear on both sides of the PCB, are generated by using black pads as in the previous methods. The conductor pattern, however, is done with red transparent tapes for one conductor side and with the blue transparent tapes for the other conductor side. While producing the two film masters, colour separation by using special filters is applied. A red filter put before the camera lens will filter off the red taped conductors while only the blue tapes and black pattern will appear. Similarly, a blue filter will filter out the blue taped conductors leaving only the red tapes and black pattern visible.

Practice

When producing the artwork, the black pads are first put down on the base foil. It is very convenient to do the red and blue taping on the same side. However, it becomes rather complicated and tiresome if modifications have to be made. The method of putting the red tapes on one and the blue tapes on the other side permits easy modifications but the artwork becomes much more susceptible to damage in handling.

Limitations

Although the red/blue taping offers no chance for a misregistration, there are various other important disadvantages connected with its use. The major drawback is the need of two different type of film materials for the film master production. The usually used orthochromatic film emulsions are not sensitive in the red light spectrum. Therefore panchromatic film must be used for one of the film masters. Another disadvantage is in the stiff red and blue tapes which are made, for transparency reason, of polyester or polyvinyl material. These tapes cannot be stretched and cannot be used to make round corners. Special elbow patterns have to be used for round corners which complicates the taping process. In addition, where tape ends are superimposed, a maximum of only two layers can be tolerated because the tapes are not monochromatic and with more layers, a too different colour would result.

From all these points, it can be seen that the use of the red/blue taping method is not so unproblematic. Still more difficulties might be encountered in the photographic process to obtain the film masters: Red and blue tapes from different manufacturers are different, thereby creating filter problems; the colour corrections of some lenses used in reprographic cameras is such that the image size varies slightly with different colour filters and also a small narrowing of the conductors occurs in the photographic process at the crossings of the red and blue tapes.

The red/blue taping method, even though it looks attractive, will not replace the black taping method as it is the more expensive artwork generation method: It simplifies the artwork procedures on one hand but introduces some new difficulties on the other.

8.2.4 Black Taping on Diazo Films

Principle

First, an artwork of the pattern common to both PCB sides is made in black on polyester foil, similar to the 3-layer method. But now 2 copies of this common layer are made by contact copying on dimensionally highly stable diazo film and the respective conductor patterns are thereafter taped on these copies. The result is two artwork layers which can be perfectly registered.

Diazo Film Material

Diazo film is basically a polyester based reprographic film. Its *emulsion* is sensitive in the blue-violet-ultraviolet range and subdued daylight does not affect it. Diazo film is developed in ammonia vapour and the emulsion gives a grainless image with a clear edge definition and a high resolution. The theoretical resolving power is above 500 lines/mm. The *base material* is polyester which is already known for its excellent dimensional stability. Diazo film is supplied in rolls of 1.015 m width and 10 m length in different thicknesses (Kanva Industries, Mysore, India). Working with diazo film materials does not need special skills and no darkroom is required.

Diazo Film Processing

The contact exposure is made in a vacuum frame with an actinic light source as almost similarly done for the exposure of photoresist coated PCB base material. The development is performed by means of ammonia vapour. No water is involved in processing, therefore no changes occur with respect to dimensions. Over-development is not possible which gives much freedom in the diazo copying process. The copy is immediately ready for further use after the development, which takes only a few minutes.

Further Advantages

Because of the high photographic quality of diazo films, no differences in the image can be detected in 2nd and 3rd generation copies (copies made of copies) if properly carried out. This opens new possibilities for artwork storage: Instead of the taped artwork (which is prone to handling damages and pattern/tape creeping) diazo contact copies can be stored. Where the artwork has to be mailed to the PCB manufacturer, it is definitely better to send diazo copies (Fig. 8-6).

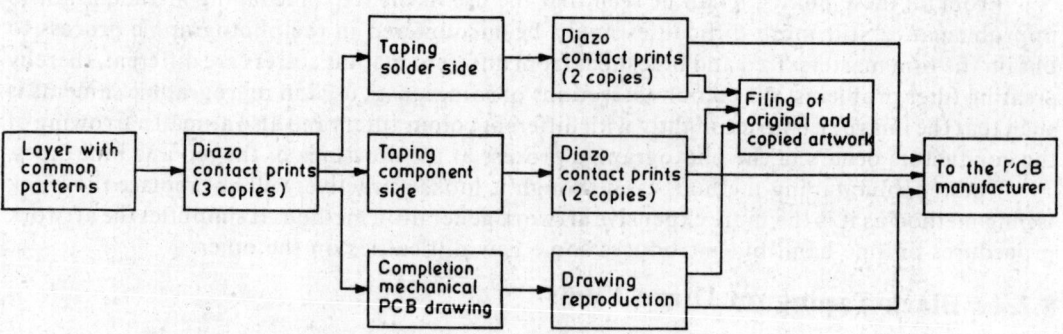

Fig. 8-6 Artwork procedure with diazo copies

8.2.5 Cut-and-Strip Artwork

Principle

The cut-and-strip method depends on the use of a special foil consisting of a clear, stabilised polyester film base of a thickness of 100–125 μm with a red coating which is transparent to the eye but opaque to ultraviolet light. The red coating has only little adhesion to the base and is cut through by means of a knife blade. It can then be peeled off, leaving the negative pattern behind on the transparent base. The artwork with this method is usually done on a 1 : 1 scale and can be used directly as a master for exposure of the photoresist coated PCB material.

Equipment

Depending on the needs, the equipment can be as simple as a transilluminated table and a few hand cutting tools like surgical knife, square, drawing compass with cutting blade and a metal straight edge. However, in most cases coordinatographs are preferred for an accurate and convenient artwork generation (Fig. 8-7). Coordinatographs are known since the 19th century and have been used as precision drawing machines. The coordinatographs available today offer typically an X/Y accuracy of ± 0.025 mm and are precision mechanics instruments. The cutting depth can be adjusted in order not to cut through the transparent base foil but only the red coating.

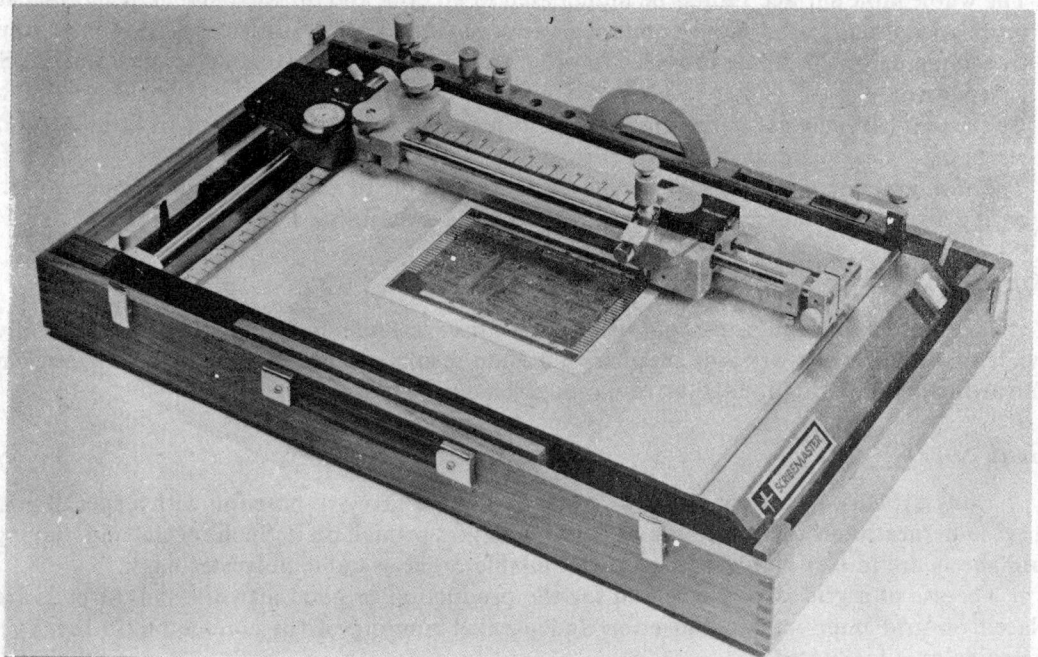

Fig. 8-7 Coordinatograph used for the generation of cut-and-strip artwork (Courtesy: Linton Laboratories Limited, Sawston, Cambridge)

Applications

Since the artwork is normally done at a 1 : 1 scale and is directly used as a master, the method is applied in making prototype boards on a laboratory scale. However, if the board is slightly complicated, this method of artwork generation is too slow, fatiguing to the operator and not sufficiently accurate at the 1 : 1 scale. In the opinion of the author, many of the typical PCB requirements with a cut-and-strip artwork, could probably be satisfied easily with the mechanical PCB milling process as described under Section 18.9.

8.3 ARTWORK TAPING GUIDELINES

8.3.1 Equipment

Transilluminated Table

Transilluminated tables are available in various designs from different manufacturers. It is also possible to get them built according to one's own specifications, with the availability of simple in-house workshop facilities. Since a transilluminated artwork table is more convenient to work with it will probably lead to better artwork results. A few desirable features are given below:

—The whole table surface should be illuminated in an even and diffused manner. Fluorescent tubes are normally used with a combined wattage of up to 100W. Sufficient air flow inside the transilluminator should be provided in order to minimise the heating up of the transparent cover plate.
—Opalescent acrylic glass is preferably used as a working plate. Because of its little flexural strength, a glass plate has to support it.
—The tilting of the working surface should be adjustable to cause minimum fatigue to the sitting artwork designer. Also, adjustable height is a desirable feature.

Knife

The most important hand tool used is the artwork knife. Suitable knives are generally available from the artwork aids suppliers but good results can also be obtained with surgical knives fitted with straight-edge or round-edge blades.

Grid Sheet

Many artwork designers prefer to work on a plain artwork base foil with a special grid sheet underneath rather than to use base foil with grid printed on it. Such special underlaying grid sheets are in fact highly precise and available on very stable polyester base.

The use of a grid is very essential for the production of good artwork. All the pads are placed on grid intersections and clean and parallel running of the conductors in the X/Y coordinates is facilitated.

Grid sheets are available with the following standard increments:

Artwork scale	Increments		
1 : 1	2.54 mm (0.1″)	1.27 mm (0.05″)	0.635 mm (0.025″)
2 : 1	5.08 mm (0.2″)	2.54 mm (0.1″)	1.27 mm (0.05″)
4 : 1	10.16 mm (0.4″)	5.08 mm (0.2″)	2.54 mm (0.1″)

Most of the commonly used electronic components for PCB mounting have a 2.54 mm (0.1″) lead spacing or a multiple thereof. The grid sheet for an artwork on 2 : 1 scale must have grid increments of 5.08 mm (0.2″).

8.3.2 Practical Hints

Order of Artwork Generation

After fixing the artwork base foil on the table, the PCB outlines are identified by suitable corner marking with 1mm tape (Fig. 8-8). The tapes are fixed outside the actual PCB area. The

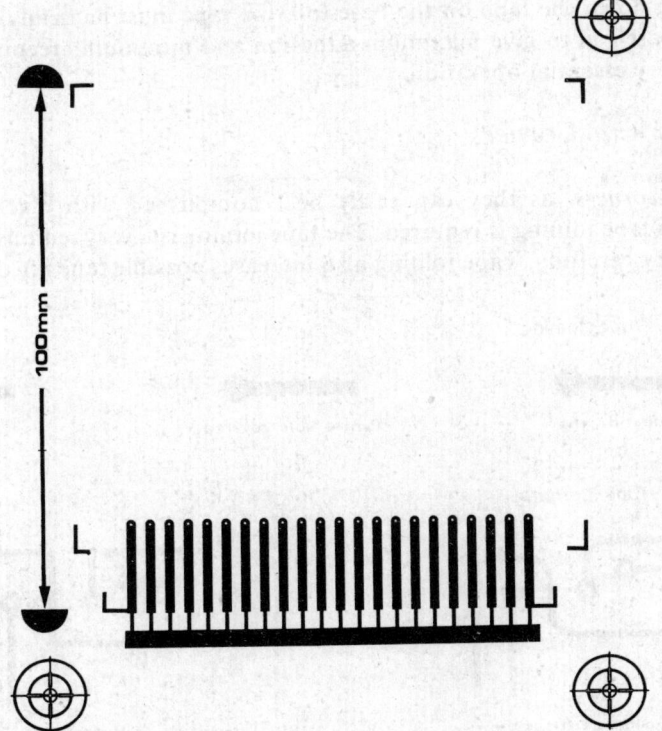

Fig. 8-8 PCB outline identification and dimensioning

inner tape edge represents the cutting edge. The final PCB dimension is shown with special arrow pads for the longest side of the PCB. The dimension must show the actual PCB size value and is used for proper setting of the reduction scale by the artwork photographer.

The next step is fixing of all the pads. The pads placed are always exactly and properly centred on grid intersections: this is enabled by the centre holes in the pads.

The taping of the conductors has to follow now. Where different layers are used for the artwork, the registration between the different layers is accomplished by 3 special registration pads put outside the PCB corners on each artwork layer.

Tape Handling

As a general rule, all the unused tape rolls and artwork pads should always be kept in their original package. This is necessary to protect them from dirt, dust and deterioration.

The cutting of tapes should never be done against the pads of artwork which would leave behind slicing marks. The proper practice is to pull the unwanted portion of the tape against the cutting knife edge. The resulting cut will be clean and without frayed edges.

When applying tapes, they should never be stretched. They would otherwise creep back to a shorter length thereby producing hair-line gaps between the tape end and the pad.

The pad centre holes are always kept free; do not allow them to be covered by tapes. That is how they fulfill their purpose as a registration aid in artwork design and as a centering aid in the drilling operation (Fig. 8-9A).

After putting down the tape on the base foil, the tape must be firmly pressed on its full length onto the base foil to give maximum adhesion and minimum creeping. Use your own fingers for this very essential operation.

Round-Bent or Cut Corners

Round-bent corners, as they can easily be accomplished with black tapes, have the advantage that no tape joining is required. The tape joining is a very tedious procedure which has to be done very carefully. Tape joining also increases possible tape lift-offs in the artwork

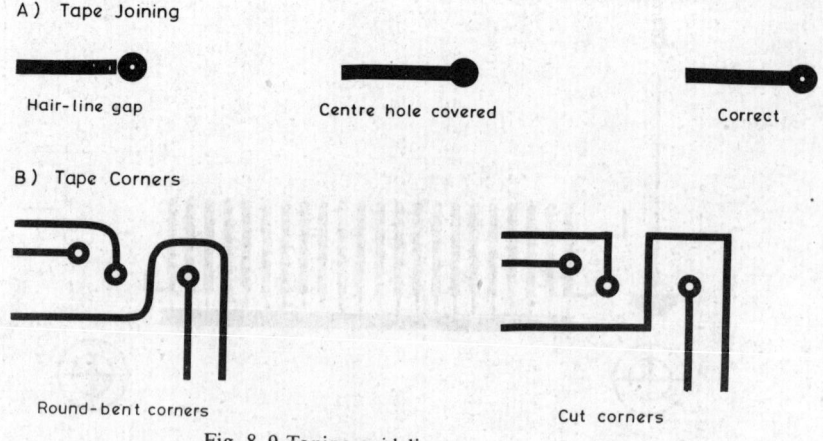

Fig. 8-9 Taping guidelines

which might not be detected. On the other hand, round-bent corners, especially if they are very tight, introduce the danger of creepage which can easily be a few tenths of a millimeter within a few months. In dense patterns with minimum conductor spacing, creeping cannot be tolerated.

In any type of conductor bends, tape stretching has to be carefully avoided and the bends are made rather wide than tight.

Cut corners are preferred in dense patterns to avoid creeping of the tapes. At the same time, best space utilisation is obtained by going for the minimum conductor spacing.

More artwork preparation time may be required because of tape joinings applied. Careful attention and checking is required to avoid tape lift-offs when taking the film master photographs.

8.4 GENERAL ARTWORK RULES

8.4.1 Conductor Orientation

In PCB artwork with a higher conductor density, it is an established practice to run the conductors basically on one side in the direction of X-coordinate and on the other side in the direction of Y-coordinate. Following such a rule, via holes are minimised and the conductor pattern gets fairly regularly distributed.

Where conductors have to be placed in other directions, preference should be given to the 45 degree direction or to the 30/60 degree directions. The strict observation of this rule helps in an optimum utilisation of the space available and gives a well-organised appearance.

In simpler circuit patterns with plenty of space available, the conductors can be run in any direction so as to give the shortest interconnection length. This is specially important in high-frequency PCBs.

8.4.2 Conductor Routing Practice

It is a good rule to begin and end the conductors in a solder pad or in another pad. In certain cases, however, where this would increase the length of the conductor, it can also be terminated by joining another conductor (Fig. 8-10A).

Conductors forming sharp internal angles of less than 60 degrees must be avoided (Fig. 8-10B). This is of particular importance for boards which have to be wave soldered; excess solder would otherwise get deposited in such corners.

Where one or several conductors have to pass between pads or other conductive areas, the spacing has to be equally distributed. Maximum spacing is obtained if conductors are put perpendicular to a narrow passage (Fig. 8-10C). Since closely-spaced parallel running conductors can give manufacturing problems and also difficulties in electrical functioning of the circuit, the conductors should be as widely spread as possible over the area available.

8.4.3 Spacing

When specifying the minimum spacing requirements, there are various factors which have to be kept in view. But in any case the rule holds that minimum spacing is applied only where it cannot be avoided. The yield in PCB fabrication will otherwise come down with minimum spacing. A brief account is now given on the different considerations which influence the design of the PCB.

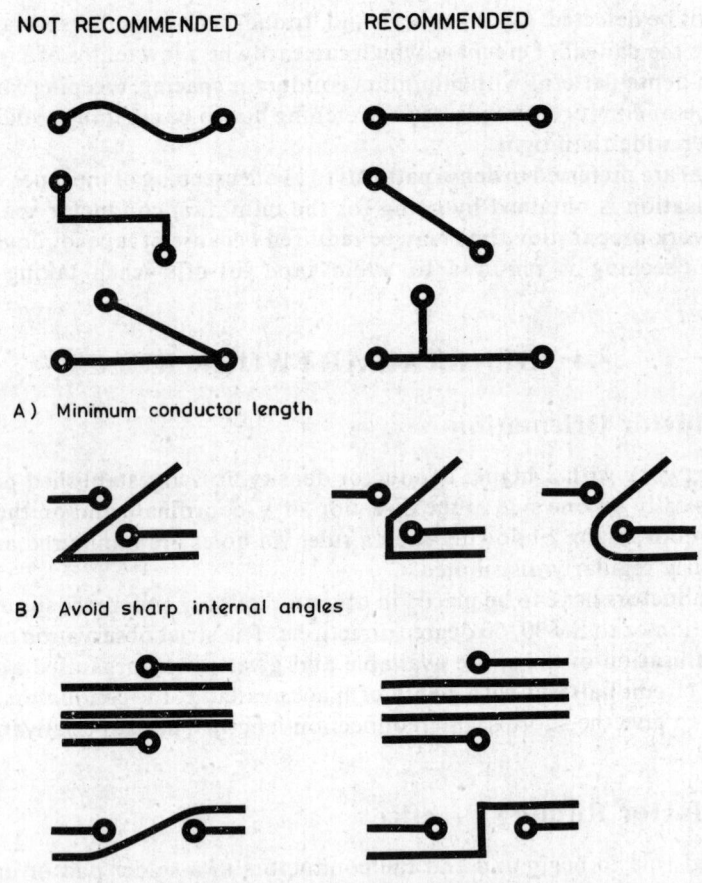

Fig. 8-10 Conductor routing rules

PCB Manufacturing Process

The first concern in spacing is the performance of the PCB fabrication set-up used. The minimum spacing chosen in a PCB design should never reduce the production yield in a significant way (e.g., because of conductor bridging). Here, the experience of the PCB production department can give the best guidelines for the various types of board.

Another consideration will be the PCB process specification for the particular design. Mainly the choice of the plating process has to be mentioned. In pattern plating, a conductor width increase will result (overhang) which can be as much as 125 μm. Panel plating gives width decreases which can go up to the same magnitude (under-etching). The width change itself depends mainly on the thickness of the copper foil in subtractive PCB processes and also on the image transfer method applied (wet-film resist, dry-film resist, screen-printing). Since again many variables are involved, best guidelines are obtained from the experience in the previous production of similar types of PCBs.

Artwork

Minimum spacing specifications must also take care of the overall precision of the artwork prepared. The following factors contribute to the result:
— Artwork method and how many layers involved
— Dimensional stability of artwork base foil
— Artwork aids used
— Individual accuracy of the artwork designer
— Artwork scale.

Voltage Considerations

In order to rule out a voltage flash-over between conductors due to insulation failure, minimum spacing requirements with respect to voltage are specified. The specified value has to be maintained under all circumstances including the worst-case tolerances of artwork generation and PCB processing.

Minimum spacing specifications have been issued by IEC (International Electrotechnical Commission), MIL (US Military Standard), UL (Underwriters Laboratories) and others. They differ partially from each other if compared. As a guideline, the details of MIL Standard 275B are given here (Fig. 8-11). The specifications are divided for PCB application in altitudes below and above 10,000 ft (3,048 m) and whether a protective coating is applied on the PCB or not. Protective coatings are understood to be coatings which form an insulating layer on both sides of the PCB. Very often, epoxy lacquers are employed for such a purpose. (The MIL specifications mentioned are considered to be rather conservative specifications: Especially in high-density boards, lower spacing is often applied).

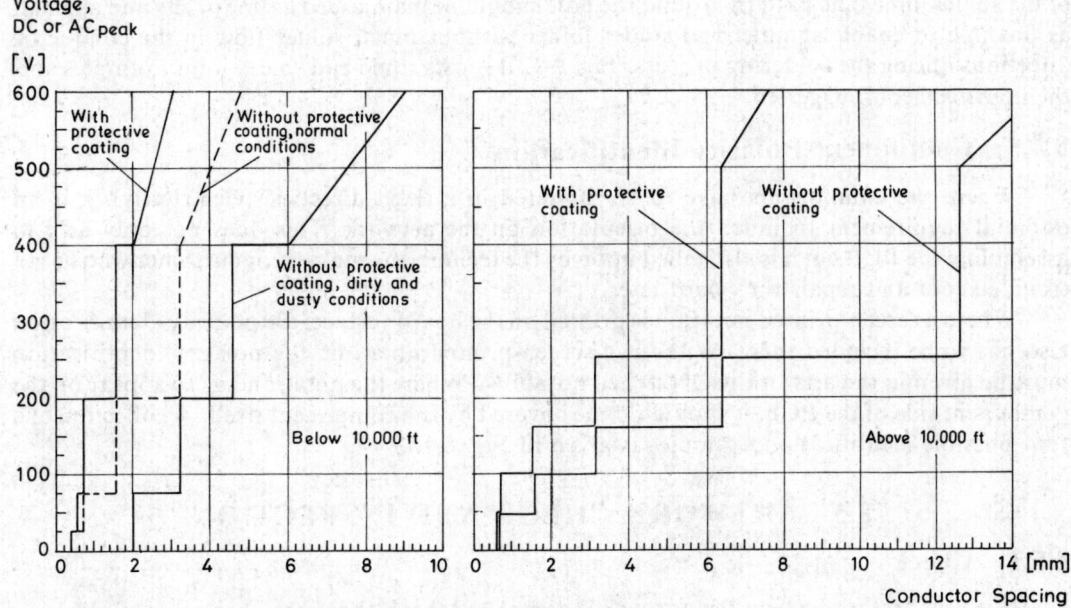

Fig. 8-11 Minimum spacing requirements (MIL Std. 275B)

8.4.4 Hole Diameter, Solder Pad Diameter and Solderability

Hole Diameter

In the interest of a rationalised PCB fabrication, the number of different hole diameters on a PCB has to be kept minimum. On the other hand, any component lead should be fitted only into the hole of an appropriate diameter. Therefore, a certain optimising must be done. Satisfactory soldering results are usually obtained if the diameter of the finalised and plated holes gives about 0.2–0.5 mm clearance as compared with the nominal diameter of the component lead.

Solder Pad Diameter

The diameter of the solder pad in relation to the finalised hole diameter is very important for reliable solder joints. In PCBs with plated-through holes, the widths of the annular ring should be atleast 0.5 mm but preferably more. In PCBs without plated-through holes, the solder pad size must be more because there is no through-plating to give mechanical strength to the solder pads. A rule here is a solder pad diameter of approximately 3 times the drilled hole diameter. However, under all circumstances, sufficient solder pad size must be provided to avoid broken annular rings because of drill position tolerances. Another constraint is the relation between the solder pad size and the width of the joining conductor. The conductor width should always be less than the solder pad diameter but preferably about one third (Fig. 8-12A).

Joining Conductor and Solder Pad

The way of joining the conductors into the solder pads is very important for the reliability of the solder joint. The pattern around the hole should be maintained as uniformly and as small as possible to enable symmetrical solder joints without much solder flow in the conductor directions during the soldering process. Fig. 8-12B shows some bad solder joint examples and the improvements suggested.

8.4.5 Component Polarity Identification

Where the components have to be mounted in a fixed direction (electrically), it is an essential requirement to index the orientation on the artwork. This does not only help in assembling the PCB but it is also relied upon by the technician, engineer or customer who has to trouble-shoot and repair the board later.

Where a screen-printed notation is printed on the board surface, the polarity identification also has to be included there. Without a screen-printed notation, the polarity identification must be given in the artwork itself (etched notation). Where the notation has to appear on the component side of the PCB, it should not be covered by the component itself. A collection of a few possible identification examples is given in Fig. 8-13.

8.5 ARTWORK CHECK AND INSPECTION

8.5.1 Check

It is very essential where the artwork is completed that a thorough check is carried out to prove its exact correspondence with the circuit diagram. Even very experienced artwork

ARTWORK

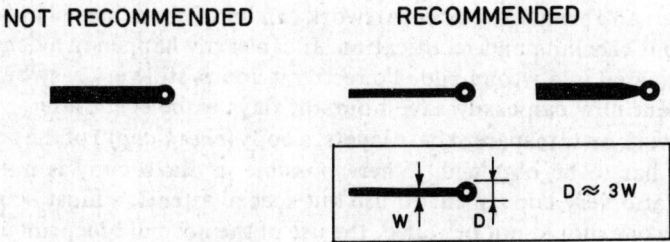

A) Solder pad diameter and conductor width

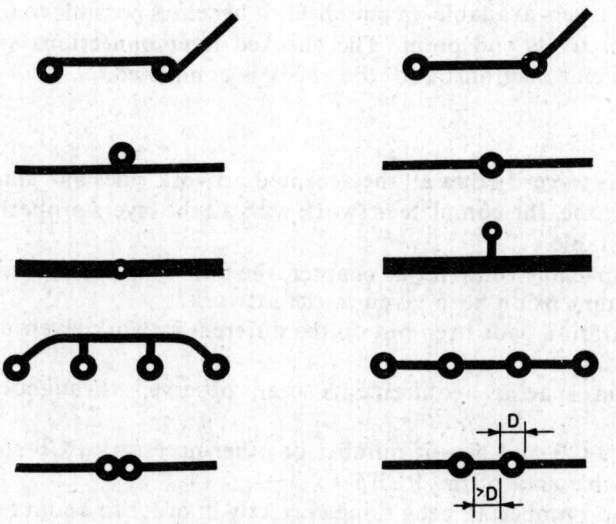

B) Solder joint design examples

Fig. 8-12 Guidelines for the design of reliable solder joints

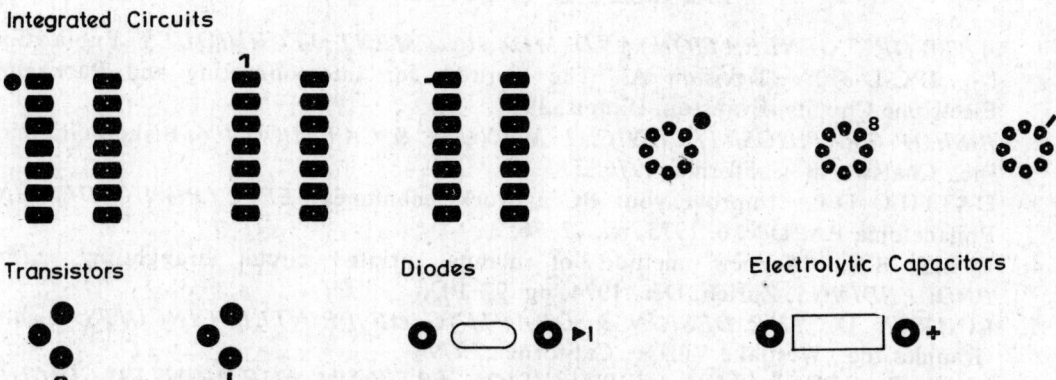

Fig. 8-13 Component polarity identifications

designers know that an absolutely correct artwork can hardly be achieved in a straight-forward manner i.e., without checking and rectification. It can easily happen in a dense artwork that a conductor is terminated in a wrong pad. To rectify it now is still very easy but later will be too late. One hour spent now can easily save hours or days to be spent later.

To check artwork on transparent base layers, a copy (*check copy*) of the combined artwork with all its layers has to be obtained. Where possible, a diazo copy is made. Otherwise, a blueprint copy is also very convenient to use but special attention must be given to the copy process. Since artwork should not be rolled, the use of the normal blueprint exposure machine is ruled out. Instead of this, PCB exposure units can as well serve the purpose to obtain the prints.

By means of such a check copy, the check can easily be carried out with all the information of the various artwork layers available on one sheet. It becomes possible to follow a conductor from its starting point to its end point. The checked interconnections are marked in the schematic diagram (circuit diagram) until the check is completed.

8.5.2 Inspection

The inspection has to verify that all the accepted artwork rules and standards have been followed. For this purpose, the complete artwork with all the layers properly registered is put on a transilluminated table.

Besides all the applicable rules in this chapter, the following points have to be examined:
— Has the final PCB dimension been given in the artwork?
— Have suitable registration pads been put on the different artwork layers and do they match exactly?
— Have the minimum spacing requirements been observed throughout the completed artwork?
— Is the name of the PCB and circuit number or other necessary PCB specifications given in an easily visible area of the PCB?
— Has all lettering and numbering been done correctly in order to be finally readable on the particular side of the PCB?

BIBLIOGRAPHY

1. *ARTWORK GENERATION AND MEASUREMENT TECHNIQUES*, Publication No. IPC-D-310A (Revision A), The Institute for Interconnecting and Packaging Electronic Circuits, Evanston, Illinois, 1977.
2. *BISHOP GRAPHICS, TECHNICAL MANUAL & CATALOG 106*, Bishop Graphics Inc., Chatsworth, California, 1976.
3. DATTILO D.P.: 'Improve your PC artwork techniques', *ELECTRONIC DESIGN*, Philadelphia PA, Dec. 6, 1975, pg. 72–76.
4. ELLIS B.N.: 'A new method of manual printed circuit draughting', *EIPC PROCEEDINGS*, Zurich, Dec. 1974, pg. 92–102.
5. LINDSEY D.: *THE DESIGN & DRAFTING OF PRINTED CIRCUITS*, Bishop Graphics Inc., Westlake Village, California, 1979.
6. LUND P.: *GENERATION OF PRECISION ARTWORK FOR PRINTED CIRCUIT BOARDS*, John Wiley & Sons, Chichester, 1978.

9

AUTOMATION AND COMPUTERS IN PCB DESIGN

9.1 LIMITATIONS OF MANUAL DESIGNING

After completion of the electronic circuit design, one wants to implement fast the related PCBs, for an early prototype testing. But after the prototype testing, modifications on the PCB design are normally required for which again time must be provided. Last but not least, all the production documentation has to be continuously updated in every detail. All these operations are carried out by skilled and experienced technicians. But even with the most reliable personnel, it takes weeks after finishing the circuit design until the film master is available for PCB productions.

A problem quite often encountered in electronics industries is the nonavailability of sufficient and experienced technicians for layout design and artwork generation. Even today, there are hardly any special training schools existing for the professional formation of this category of technicians although such technicians are needed whereever PCBs are to be designed. The usual practice still is to take people with some background in draughting or electronics and to give them specialised in-house training.

In a time analysis of the layout design and artwork preparation (taping) the total time spent for these two operations is typically 40% for the layout sketch design while 60% is spent on the artwork preparation. It is therefore natural that the first step in automation will be the elimination of the manual artwork draughting: The layout sketch is digitised and the information fed to a photoplotter which directly produces the 1:1 artwork on film. Going still further in automation, the next step is the use of computeraided layout design which still needs the active interference of the designer. For this purpose, minicomputers with interactive CRT displays are employed. The last step in automation is to make even the layout design fully automatic and independent of any interference from the designer. Only larger computer systems have the capacity to perform such a task.

The purpose of this chapter is to highlight to some extent the possibilities with each one of these automation steps, and also to indicate their limitations so as to take away some of the "magic" usually attributed to them.

9.2 AUTOMATED ARTWORK DRAUGHTING

When automatic artwork facilities are made use of, there is a significant shortening of the turn-around time in PCB design. The artwork preparation step has been eliminated and is replaced by the considerably faster digitising of the circuit pattern (Fig. 9-1). The digitiser provides the paper tape or the punched cards to the photoplotter as well as the paper tapes required for NC drilling and NC routing (NC stands for *N*umerically *C*ontrolled).

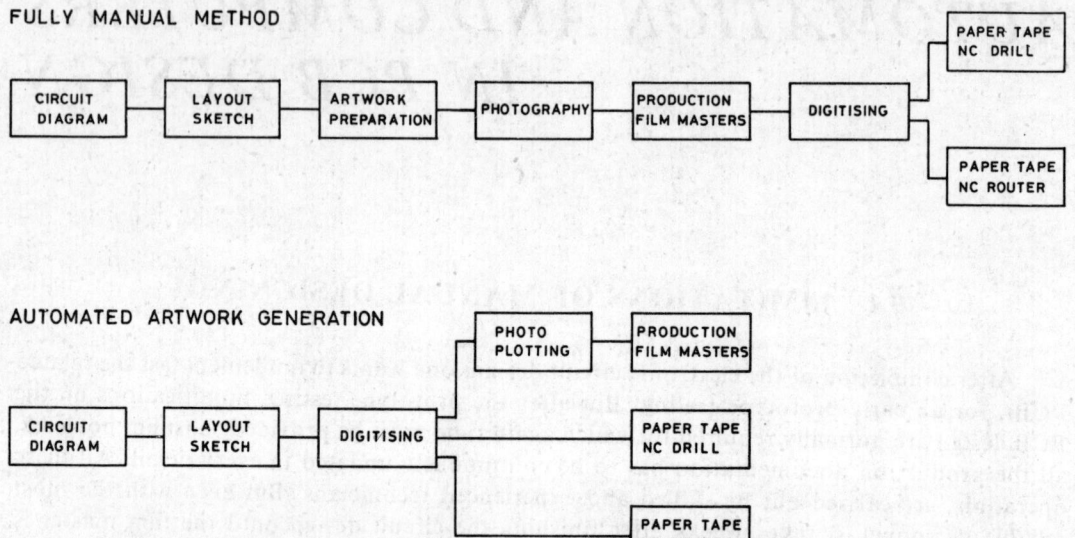

Fig. 9-1 Layout and artwork generation processes: Fully manual and manual layout/automated artwork process

9.2.1 Layout Sketch

The layout sketch can be prepared in an almost similar way as used for the manual taping method. The layout sketch can be a hand-drawn pencil sketch on a suitable grid sheet. However, it is preferably done on an enlarged scale, e.g. 2 : 1 or 4 : 1, to obtain maximum accuracy in the conversion process. Only straight conductors can be accepted for digitising. Depending on the equipment, 45° conductor slopes or even slopes of any other angle are also permitted. The absolute positional accuracy provided by the layout sketch is not very critical: Grid rounding capability of the computer will automatically take care of the nearest grid intersection.

9.2.2 Equipment for Automated Artwork Draughting

Digitiser

Digitising is a graphic-to-digital conversion of all the relevant information incorporated in the layout sketch such as component location, solder pad size, conductor configuration, etc.

A digitiser consists usually of a cursor-operated digitiser, a control console, a disc storage system and a fast pen plotter which are all linked to a minicomputer.

With the hand-held cursor, all the relevant positions on the layout sketch are sensed while the cursor movement gets resolved into X/Y coordinate pulse trains. These pulse trains are stored in the storage system. Additional information on hole diameters, conductor widths, etc., can be entered via the keyboard on the control console.

Fig. 9-2 Large-area coordinate digitiser (Courtesy: The Gerber Scientific Instrument Company, Hartford, CT 06101, USA)

Functioning: There are different working principles applied in the digitisers currently employed. Some designs make use of a coordinatograph which gives the cursor position via precision gears to optical encoders. In another principle, the cursor is not mechanically linked but it produces an inductive field which is detected and followed by a servo-driven carriage below the working table. A further principle applied, without a mechanically moving part, makes use of the electromagnetic interaction between a coil in the cursor and arrays of electrical conductors under the working table surface.

Accuracy: Resolution capability of the digitisers available is mostly in the range of 0.025–0.1mm. The absolute positional accuracy varies between 0.075–0.25mm. Further deviations might be introduced in the digitising process by the operator which is still a human being. For highly precise results, therefore, a recommendation is to work at a 2:1 or even a 4:1 scale. The errors will herewith be proportionally reduced.

Computer

The pulse trains from the digitiser are temporarily stored in the storage system of the mini-computer. The data are then processed by the computer in OFF-line mode and the output is usually on paper tape as required for the photoplotter and the NC routing equipment. A fast pen plotter can make an artwork drawing for control purposes.

Depending on the hardware and software, data modifications and also editing facilities are made use of. Some possibilities are indicated here:

—Grid rounding: Point locations are shifted to the nearest grid intersection, thus improving the overall accuracy of the digitising process.

—Grid base selection: Grid base can be selected according to layout sketch precision and PCB requirements (e.g., 0.1mm, 0.254mm, 0.5mm etc.).

—Scaling factor selection: This is to take care of layout sketches at an enlarged scale.

—Pad patterns: For standard pad configurations (e.g. DIP-16) only one pad position has to be entered while the rest can be recalled from the library (Computer memory) via the keyboard.

—Pad size and type: To be specified via the keyboard.

—Conductor digitising: Only change points to be entered. Width specifications via the keyboard.

—Coordinate system: Either related to one fixed point (datum point) or incremental measurements with zero setting of coordinate register at a new point.

Photoplotter

In the photoplotter, the processed digital data are converted back to graphics by photographic means. The drawing head which is a light spot projector changes its relative position towards the drawing medium which is the photographic film. Depending on the design of the photoplotter, the drawing head or drawing table or both of them are moving parts.

Flat bed plotters: The film is kept flat on the drawing table. Either the drawing head or the table is moved by a highly precise mechanical driving system consisting of digital stepping motors and a gear box with rack and pinion output.

Drum plotters: The Y coordinate movement is carried out by the drawing head while the X coordinate movement occurs by rotation of the cylindrical drum with the film fixed on it. To control the accuracy of the drum rotation is more difficult than to control a linear movement. Drum plotters are therefore not used where the highest possible accuracy is aimed at.

Exposure: Solder pads are exposed by flashing of the light spot projector. The pad diameter depends on the aperture chosen for the exposure. The apertures for solder pads and conductors are selected by the input program. Conductors are exposed with the light spot projector switched on, while the projector or table is moving

Accuracy: Various factors have an influence on the accuracy, but mainly size of drawing area and drawing speed should be mentioned here. Positional accuracy can be expected to be within 0.02–0.04mm while repeatability has approximately 0.005–0.025mm limits. The resolution or step size is typically around 0.0125–0.025mm for precision plotters but is very much interrelated with the plotting speed.

Plotting duration is somewhere between 15 min and 4 hours, depending on the size of the plotting area, the plotting speed, the circuit complexity, etc. The maximum speed for plotting of

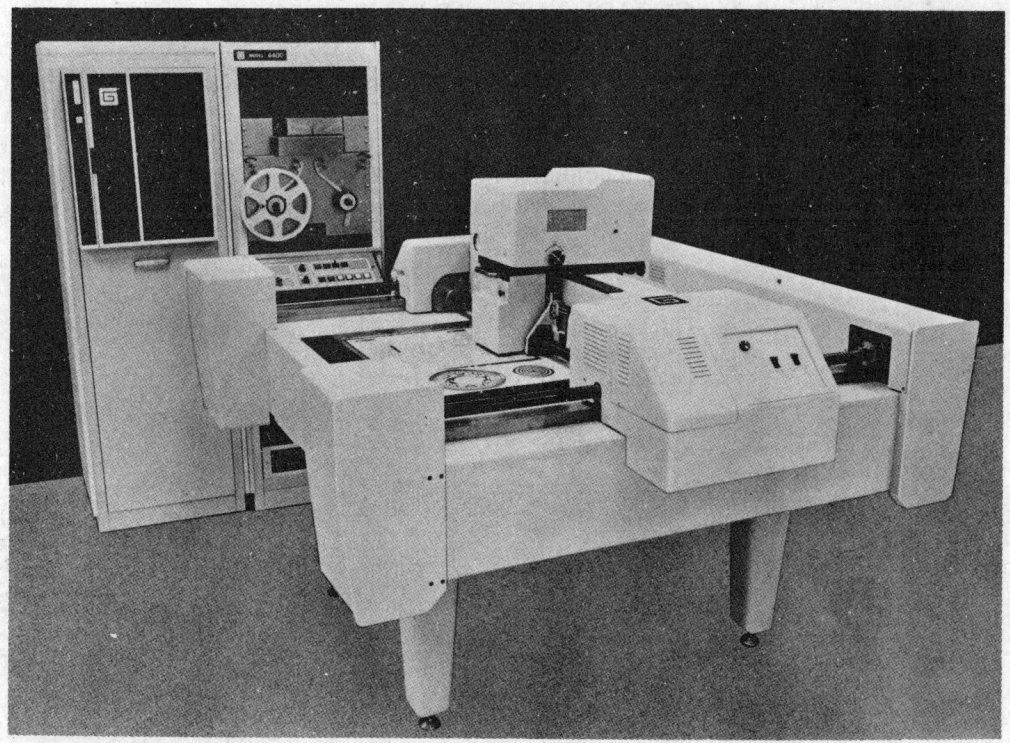

Fig. 9-3 Photoplotter (Courtesy The Gerber Scientific Instrument Company, Hartford, CT 06101, USA)

conductors is typically 100mm/sec. After plotting is over, the exposed piece of film has to be processed as recommended by the film manufacturer.

9.2.3 Scope

The necessary *capital investment* of $ 50,000–$ 100,000 into automated artwork generation facilities can be justified only in larger companies or specialised service bureaus where an extensive use of facilities is guaranteed. It must be pointed out that for manual artwork generation, there is practically no investment required. The equipment costs have therefore to be carefully weighed against the advantage obtained by the automation.

The *advantages* of automatic artwork generation are basically given by the considerably faster artwork preparation. But it also offers a higher artwork precision which is constant and independent of personnel. Very often, also higher package densities are achieved. Paper tapes for NC machining operations are obtained as a by-product. Further advantages are provided where circuit modifications have to be done: The digitised layout on magnetic disc can be loaded again into the computer and modifications are carried out. All the new paper tapes will be 100% updated after the modifications and therefore avoid further confusion in the production.

9.3 COMPUTER-AIDED DESIGN

In computer-aided design (CAD) of layouts, the component placing and interconnecting patterns are developed with the assistance of computer facilities. It is not a fully automated procedure but it makes optimum use of the designer's ingenuity while the tedious routine operations are taken over by a minicomputer. It is a team work between man and machine, where the designer plays his role in a feed-back loop, thus being the controlling element (Fig. 9-4).

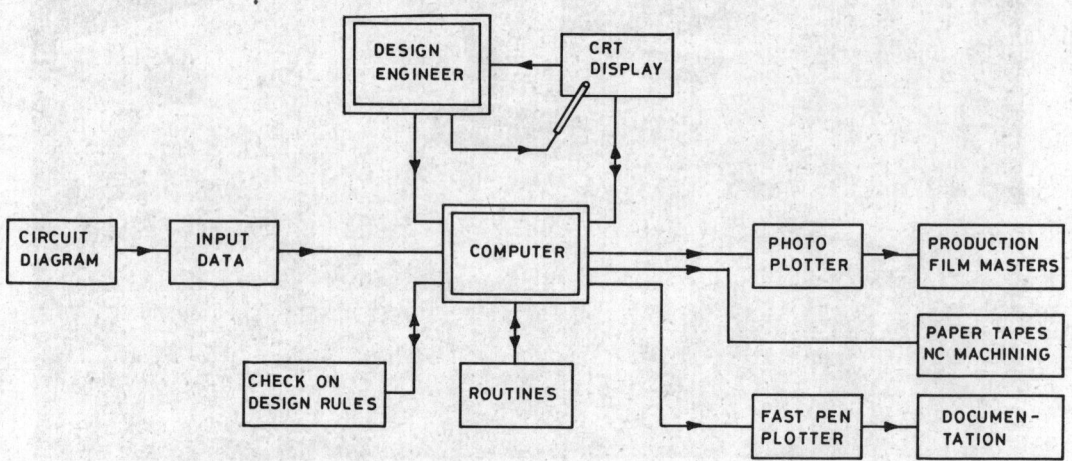

Fig. 9-4 Typical equipment configuration for computer-aided layout design with automated artwork generation

9.3.1 Input Data

Initial Data Requirements

Before starting the layout procedures, complete and detailed specifications on the electronic circuit have to be provided. This includes the following material:
—Schematic diagram with component details, interconnections, edge connector specifications.
—Component list including component name, type specifications, type number, manufacturer.
—Mechanical specifications such as board size and shape, mounting holes, identification of restricted areas with respect to component height, edge connector location, etc.
—PCB specifications, whether single- or double-sided or multilayer board, plated-through holes.
—Pattern specifications on pad type and size, conductor widths, spacing.
—Electrical specifications like restrictions on component placement because of heat production, capacitive or inductive coupling, ground planes, critical-length interconnections, etc.

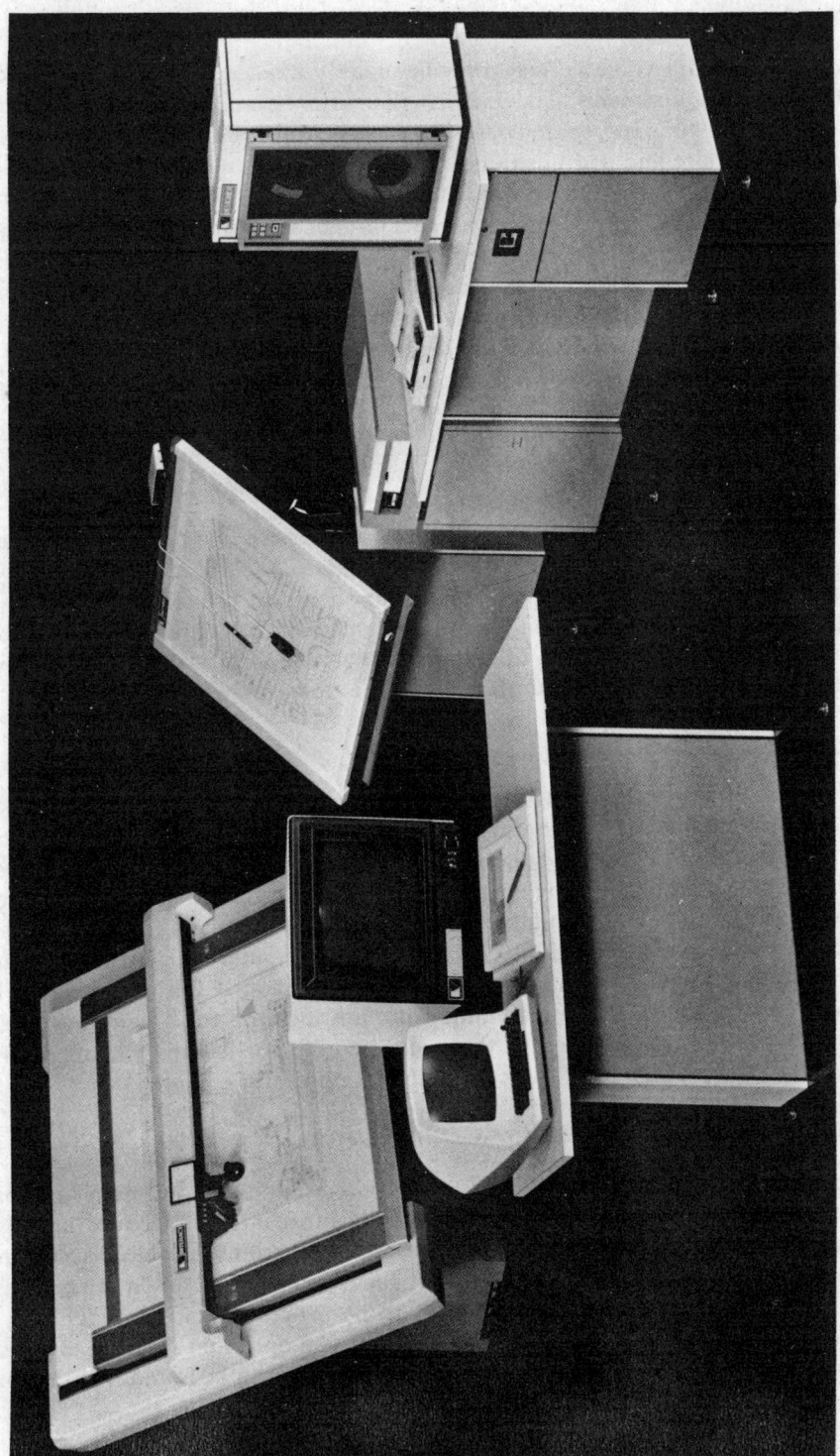

Fig. 9-5 Complete CAD system for PCB design (Courtesy: Computervision Corp., Bedford, Ma. 01730, USA)

Data Preparation

The full lot of initial data available has now to be suitably arranged and modified in order to fit the computer's input requirements. The data can thereafter be entered, e.g., via the teletype.

First, a *component library* is prepared describing each component package type used with its outline, pad type and size and pad position. In the complete component list, package type, component name orientation and side of interconnection on the PCB are identified for each component used. The *connection list* gives an exact account of all the interconnections in point-to-point form. The *list of board details* finally contains various board informations and the X/Y coordinates of the board corners. The absolutely correct entering of these data and in particular of the connection list is utmost important. In many CAD systems, there is provision to enter two connection lists for the same circuit but prepared by two independent operators. Any further processing of the data is possible only after sorting out the discrepancies between the two connection lists.

9.3.2 Component Placement

After the input data have been fully accepted by the computer, component placement is carried out. Those components for which the position has earlier been determined by the designer, will get their place accordingly. The other components are usually placed in a manner to give minimum overall conductor length. Two placement procedures are possible: Either the complete placement can be fully guided by the operator, or it can just be attempted by the automatic placement routine with the interference of the operator wherever necessary.

Manual Placement

The CRT display (CRT = *C*athode *R*ay *D*isplay) shows the board contour as specified in the board list. At the same time, all the components are shown with their outlines in one corner of the display including the straight pin-to-pin interconnections: A picture of wild confusion.

With the help of the light pen, one by one, components can be called up and moved across the display to the position found most suitable. The light pen is a hand-held electronic device. If held close to the display, its position can be detected by the computer with a special program. When moving the components, the interconnections simultaneously follow in straight lines like elastic bands. An important help in finding optimum location for a component is the simultaneous indication, on the display, of the overall conductor length. The operator can herewith see whether the placement of the component is optimum or not.

Automatic Placement with Operator's Interaction

With the computer's placement routine, it is possible to attempt the optimum component placement by automatic means. The components having most interconnections are placed together in groups while the components with predetermined locations are placed accordingly. The component allocation occurs now by minimising the overall conductor length and can be observed on the display. The huge amount of data to be processed and considered by the automatic placement routine makes it very difficult to find the optimum solution. Because of the possibility of the designer to interact, an acceptable solution can be found with combined efforts.

Fig. 9-6 Component placement on the display screen by means of a pen (Courtesy: Computervision Corp., Bedford, Ma. 01730, USA)

9.3.3 Conductor Routing

After the component placement, the interconnections are still shown as straight pin-to-pin connections in all directions across the display. The automatic routine has now the task of finding an interconnection pattern which is feasible on the PCB. This procedure is executed by giving stress to the following priorities:
1. Interconnections are sorted according to length.
2. Conductor routing begins with the shortest interconnections.
3. All the existing obstacles like pads and copper area are sorted out.

4. Via holes are minimised by transferring conductors or portions thereof to the optimum board side. (A via hole is a plated-through hole of usually a smaller diameter which serves as an interconnection between circuit pattern on the two sides of the board but it is not used for component lead mounting.)
5. All the information in the computer memory and on the display is continuously updated.

On the display, pattern of the two different board sides are distinguishable according to different light intensity. The routine will attempt to place all the conductors in X direction on one board side while conductors in Y direction are placed on the opposite side.

The computer as such will hardly be able to place all the conductors onto the board or it might also find solutions with very long tracks which are too risky because of possible unwanted coupling effects. Unfinished interconnections (usually about 20%) are displayed as dotted lines. Since the operator has a better overview on the pattern, he can interfere and finish the conductor routing.

9.3.4 Checking

After completion of the conductor routing, checking routines are used to check that the design takes full care of specified standards. Such check parameters can be rules on spacing, minimisation of the number of plated-through holes, special design rules, current carrying capacity of conductors, opened or disconnected pins, positional errors, etc. The errors are indicated on the display by flashing and can be corrected by the designer's interaction.

After completion of the design, automated artwork generation is applied with the same equipment as dealt with under Section 9.2.

9.3.5 Scope

Performance

A CAD system with the possibilities as described here, will definitely offer various advantages over a fully manual designing. An important consideration is the shortened time for the layout procedure. Also in many cases, the capability to make circuit modifications simple and simultaneously provide a completely updated production documentation, is even more relevant. (Under production documentation, we understand, not only film master and paper tapes for NC machining, but also solder mask generation, reference printing mask, assembly drawings, paper tapes for automatic component assembly, assembly inspection masks, schematic diagrams, etc., depending on the software available.)

With the assistance of CAD, higher package densities can be achieved and complex circuitry with a larger number of ICs per board are realised which could hardly be arranged by a manual design. The resulting patterns will constantly be of the same high precision and of a consistent quality. Especially in multilayer board design, interactive CAD plays an important role.

Computer

When selecting CAD facilities, a careful consideration has to be given to the minicomputer with its storage system. The computer's hardware capacity should not be utilised to more than

60% in *normal* applications. It would otherwise cause an enormous rise in programming costs (software) for the solving of the more difficult problems which utilise the computer capacity to its limits (Fig. 9-7).

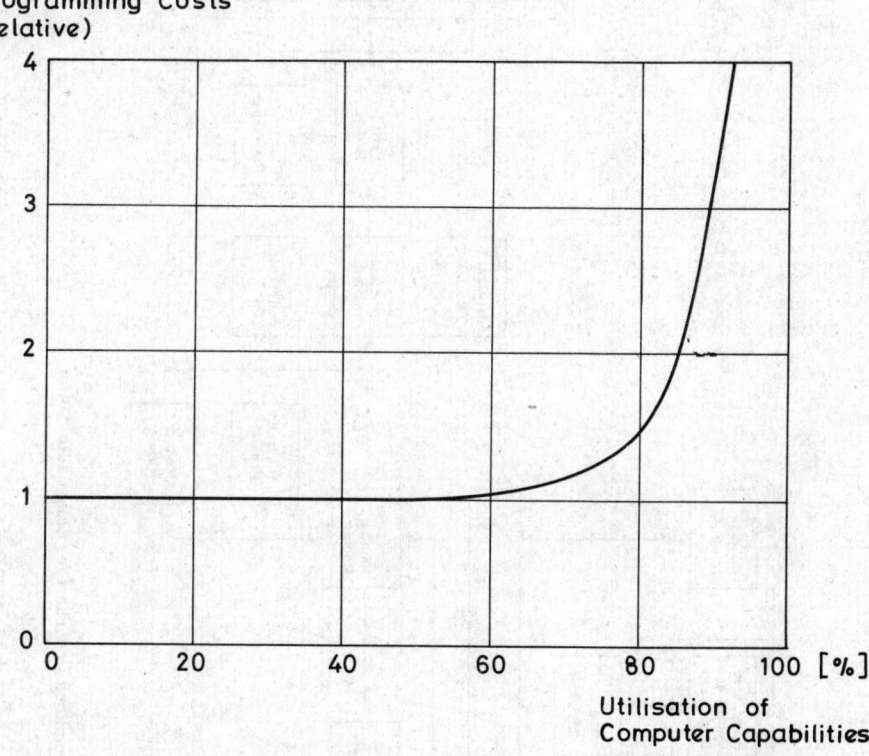

Fig. 9-7 Rise of programming costs with high utilisation of the computer capabilities [1]

Economy

Computer-aided layout design is definitely beyond the reach of smaller companies. The necessary capital investment of approximately $150,000–$400,000, depending on sophistication, calls for full utilisation by the company to recover the investment within a reasonable number of years. The biggest economic advantage of such facilities is in many cases the possibility to provide continuously and rapidly fully updated production documentation along with simple modification procedures, rather than in the fast turn around time for PCB designs. In countries with low labour costs, however, in the view of the author, there is no economic justification of the CAD layout systems.

9.4 DESIGN AUTOMATION

The highest level of PCB design engineering sophistication is given in the full design

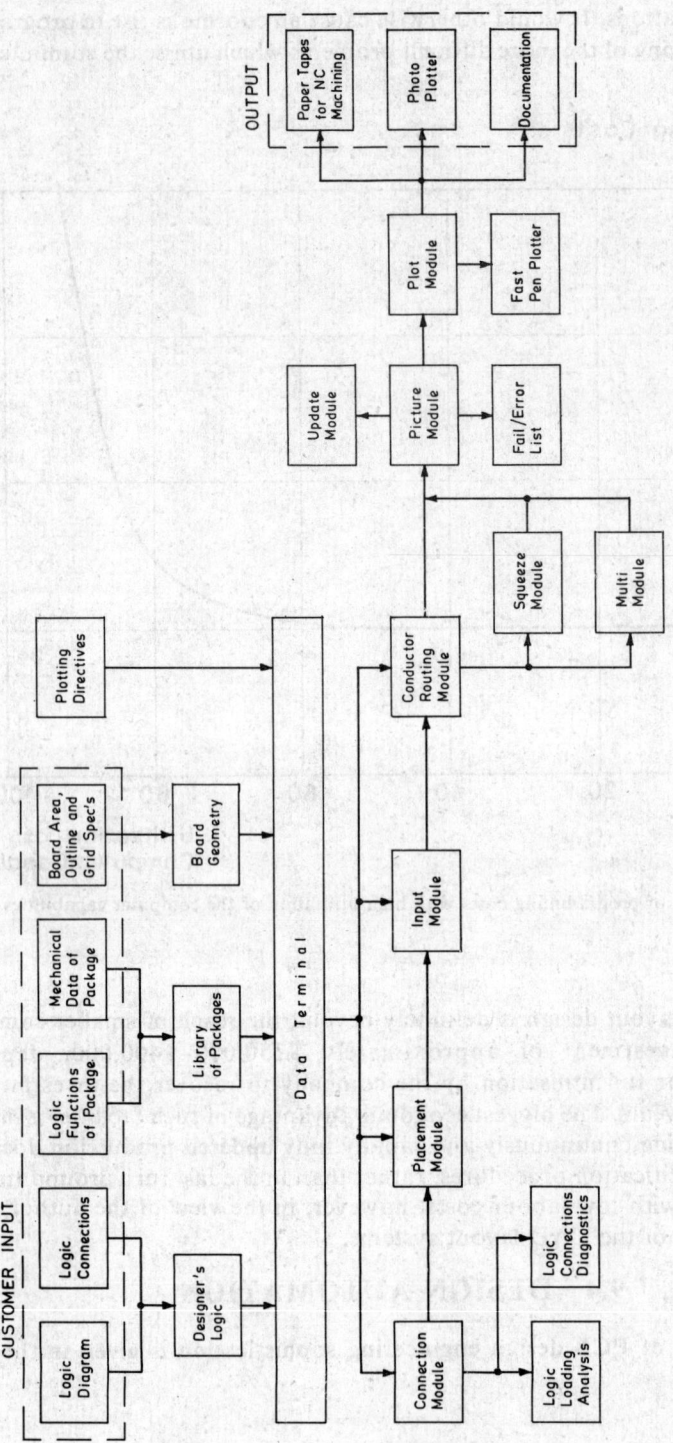

Fig. 9–8 Simplified flow chart for design automation

automation where the computer acts as the designer rather than just as an aid to the design engineer. Because of the large amount of data to be processed, mainframe computers are utilised for this purpose in the time-sharing mode. The software routines or modules employed have much similarities with the ones used in CAD. However, in design automation, they are automatically called by the main program as and when they are required, thus giving far-reaching independence to the designer.

To make full use of the computer capabilities, sometimes even a program routine giving the individual gate assignment to the most optimally located IC package is employed (see Section 9.4.2). The input specifications contain therefore usually a logic diagram and a list of the logic interconnections along with all the necessary details of the available IC packages in the package library (Fig. 9-8).

9.4.1 Conductor Routing

While it is typical in designing with CAD systems that only about 80-90% of the conductor routing is automatically accomplished, design automation has to result in 100% complete conductor routing. This can be achieved by iterative routing programs which also replace previously routed conductor tracks and via holes. Iterative routing requires considerable processing time since a new step can only start after the previous run has been completed. Hence only mainframe computers with their high processing capacity are made use of in design automation.

In the first approach, all the conductors follow a staircase pattern because of the grid system specified. To ensure easy manufacturability of the boards and to provide minimum length for the interconnections, the conductors are straightened in another step thus forming diagonal tracks. Simultaneously, conductor widths and spacings are increased over the minimum, wherever possible.

9.4.2 Component/Package Placement

Many of the digital IC packages used house two to eight similar gates within the same package. The software modules introduced do not only optimise the location of the package on the board but can also move gates from one package to another in order to keep the overall length of interconnections and via holes at a minimum. An automatic check certifies that the fan-out capacity of the gates is taken care of.

Latest software additions provide conductor routing and component placement programs which develop the board design simultaneously on all the conductive layers of the board. This eliminates a dense pattern on the first layer and a far more sparse pattern on the final layer. This is of special importance to the multilayer design.

9.4.3 Performance and Scope

Design automation as such is an extremely efficient tool in the design of highly complex double-sided and multilayer boards with large number of integrated circuits. If compared with CAD, the *overall system flexibility in design automation is more limited* since it does not utilise human ingenuity which is still superior to computer algorithms. Design automation can be successfully applied where this high flexibility is not required and the tasks follow a certain standardisation as for instance in digital system design. Standardisation in this context is understood with respect to mechanical board outlines and types of components/packages used. For standard designs, design automation can usually complete the design job without human interference. But now and then, the need for an interaction by the designer is still required, particularly for more complex tasks.

Design modifications are also possible with design automation; compared to CAD, they will need more time and are costlier because the whole design gets involved. In CAD, the modifications are manually carried out on the display which works more straightforward.

The advantages of CAD in providing production documentation can be still further enhanced in design automation. It may even include engineering report facilities giving account of spare gates and unutilised pins or thermal placement. Design automation requires access to a mainframe computer, usually available only in larger companies. Since these computers are also utilised for a wide range of other tasks, economic constraints in design automation get largely concentrated on the software required. There is less economic pressure for a continuous utilisation of design automation facilities than with CAD where the whole system is exclusively used for PCB design with the need to pay back mainly the large hardware investment.

9.5 SUMMARY AND LIMITATIONS OF AUTOMATION IN PCB DESIGN

Whether to go in for automation in PCB design and artwork generation and to which extent, depends on a wide range of factors and each method has its areas where it can be justified as the best method.

Manual design and artwork generation: This is and will remain the general method for simpler boards like single-sided or double-sided PCBs. The method works also successfully for higher circuit complexity in single or low-volume production quantities. Digital boards of an extremely high sophistication or with more than 100 ICs are very difficult to design manually. Other limitations with respect to quality, time and personnel requirements have been pointed out in section 9.1.

Worldwide, a majority of the PCB design and artwork generation is still done manually. The fully manual method, without any investment required, will remain the major method for some more time although its share gets continuously reduced in particular for digital PCB designs. Manual designing offers the highest flexibility and gives all possibilities for human ingenuity.

Design automation: The full automation of PCB design and artwork generation, like the automation of any other process, is a valuable tool where the input constraints can be standardised and reduced to a smaller number of straightforward-approach rules. Any standardisation is based on a reduction in flexibility which itself defines the limits of the applications. Automation is usually connected with a substantial investment into hardware and software: This calls for a high utilisation of such facilities, or in other words, for a large number of different PCBs to be designed per month, in order to keep an economic balance. Since any economic balance is directly related with the cost factor of labour, an economic justification for design automation in a country with a low labour cost factor seems almost impossible.

Design automation, however, may be a desirable tool for the design of highly sophisticated, digital boards with more than 150 ICs and for complicated multilayer boards where it is extremely difficult, even for an experienced design engineer, to keep an exact overview on the design. The total design time can here be reduced from several weeks to a few days while giving near-optimum results.

Design automation is normally not used for analog PCBs because it is extremely difficult to reduce the various design constraints for a wider range of analog circuit boards to a list of a few straightforward-approach rules as it can be done with digital circuit boards.

Automated artwork draughting and CAD: Being between the extremes of fully manual procedures and complete automation, there are mainly economic reasons for partial automation. Where a tight delivery schedule is important for a considerable number of PCB designs while simultaneously the need for debugging and rectifications should be minimum, CAD is often a preferred choice. Automated artwork draughting provides also a more precise artwork than the manual draughting or taping. It forms usually the last step in CAD and design automation.

CAD can hardly be justified for boards with less than 20 digital ICs, with a discrete component content of more than 50% or where only a very small number of PCBs is normally required of such designs. In all such cases, manual design is given preference. The large capital investment into CAD facilities will always call for a full utilisation of the system: This it does already in countries with a high cost factor for labour and even much more in such countries where labour costs are low.

BIBLIOGRAPHY

1. FASSINI M.G.: 'Computer aided design of printed circuits', *EIPC PROCEEDINGS*, Zurich, March 1978, pg. 1.1-1.16.
2. HEWLETT R.: 'Software partners hardware for efficient and faster CAD for printed circuit board design', *PROCEEDINGS OF THE WORLD PRINTED CIRCUIT CONVENTION*, Volume II, London, 1978.
3. HOSKING K.H. et al.: 'The addition of computer aided design software to a commercially available digitization system for printed circuit board layout', *PROCEEDINGS OF THE WORLD PRINTED CIRCUIT CONVENTION*, Volume II, London, 1978.
4. HOULBROOKE D.J.: 'Tailoring a PCB design system', *PROCEEDINGS OF THE WORLD PRINTED CIRCUIT CONVENTION*, Volume II, London, 1978.
5. KESSELMARK J.D.: 'Artwork: Manual to automated', *PROCEEDINGS OF THE WORLD PRINTED CIRCUIT CONVENTION*, Volume II, London, 1978.
6. LARSON R.: 'Computer-engineer partnerships produce precise layouts fast', *ELECTRONICS*, New York, Jan. 19, 1978, pg. 102-107.
7. LUND P.: *GENERATION OF PRECISION ARTWORK FOR PRINTED CIRCUIT BOARDS*, John Wiley & Sons, Chichester, 1978.
8. VERHAAG E.: 'Computer aided layout of printed circuits', EIPC PROCEEDINGS, Zurich, March 1978, pg. 2.1-2.19.
9. VERHAAG E.: 'Discussion on time-cost analysis', *EIPC PROCEEDINGS*, Zurich, March 1978, pg. 4.1-4.9
10. WEINERT J.: 'INKA, a system for computer aided layout of two-side PC boards', *EIPC PROCEEDINGS*, Zurich, March 1978, pg. 3.1-3.41.

Part B

Part B

TECHNOLOGY OF PRINTED CIRCUIT BOARDS

In the field of PCB technology, the issuing of straight-forward rules on 'how to do it' is even more difficult than in PCB designing where there are some basic laws to guide us towards an optimum design. The multitude of possible paths of technology leading to a finished PCB is clearly visible when visiting different PCB set-ups which turn out more or less similar PCBs: There will never be two PCB set-ups alike. The technology applied in production of PCBs must always form an individual solution for that particular place taking into account all the important local factors.

The field *technology of printed circuit boards* is a very large one; it is impossible to cover its entire width in full detail in one book. Furthermore, where views and experiences are expressed, they usually reflect the views of the author of that particular chapter, based on his personal experience and background. Nevertheless, an attempt to include all the relevant topics here has been made in order to provide a sufficiently complete insight into this field. Many practical hints have been included which take care of the typical problems faced in a country like India.

10
FILM MASTER PREPARATION

10.1 INTRODUCTION

The film negative or film positive which is finally used for the direct exposure of the photoresist-coated PCB or the light-sensitised screen is called *film master*. The film master as such has to be considered as a high-precision tool for etching, comparable to a complex punching tool in mechanical manufacturing. Parameters like dimensional accuracy, sharpness and wear-out resistance will equally apply for a film master. This means that the materials used and the production of the film master need very careful considerations, as it gets materially reflected in each PCB produced afterwards with an imperfect film master. To establish and maintain a reliable fabrication of PCBs on a high quality standard, it is helpful to get a certain familiarity with film emulsion, film base materials, equipment and processing, in order to know, in sufficient details, about their importance in practice.

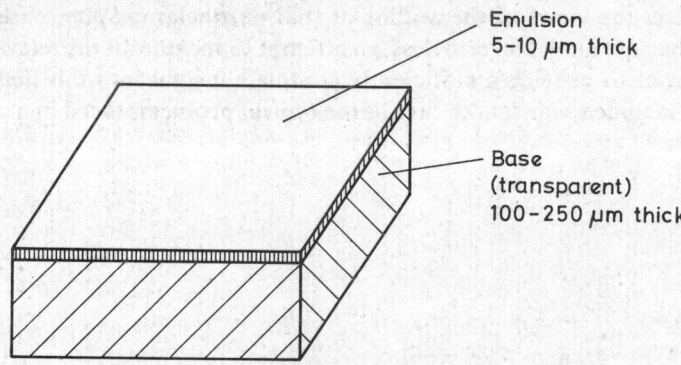

Fig. 10-1 Layers of a photographic film

Any photographic material consists basically of two layers, the emulsion and the base (Fig. 10-1). The emulsion has an approximate thickness of 4–8 μm and counts mainly for the photographic characteristics of the film. Under unexposed conditions, it contains the light-sensitive silver halides stabilised in gelatine or in a synthetic suspension. The material of the base can be acetate, polyester or glass, just to name the ones most often used. The material of the base is the prime factor for dimensional stability under changing environmental conditions. The emulsion has also got its own dimensional characteristics which will affect the overall dimensional characteristics of the film due to the intimate bond between the two layers. However, these dimensional effects through the emulsion will be much less than those caused by the comparatively much thicker base layer. The thickness of base layers is typically 100 - 250 μm.

In PCB technology, the use of polyester-based films has become a standard: Polyester offers the best compromise between dimensional stability, easy processing and handling convenience. Glass bases are used in applications where dimensional stability gets highest priority like in microelectronics and semiconductor fabrication (fabrication of semiconductor devices and integrated circuits themselves). Acetate-based films can be used where dimensional stability is not much of a concern (single-sided PCBs or double-sided PCBs of small dimensions) or where polyester-based films are not easily available or because of price considerations (e.g., in student exercises).

10.2 EMULSION PARAMETERS

Before coming to the emulsions practically used in PCB technology, a few characteristic parameters for the definition of photographic materials have to be explained in a more detailed way. However, for an in-depth study of these theoretical aspects, you are advised to refer to [6] and [8].

Density

The density D, sometimes also called 'transmission density', is defined as the logarithm of the ratio of the illuminance density I_0 falling on the developed image to the illuminance density I_t transmitted by the developed image (Fig. 10-2). The maximum achievable density of an emulsion depends mainly on grain size and emulsion thickness.

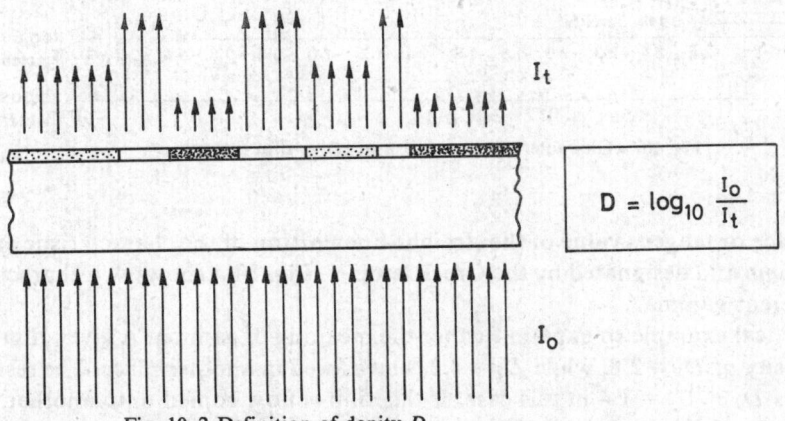

Fig. 10-2 Definition of denity D

A practical example: The illuminance density on a transilluminated inspection table is measured with the densitometer and set to 100 lux. The film master is put on this table and the transmitted illuminance density by the dark portions of the image gives a value of say 0.1 lux. The density D of the film master will be in this case

$$D = \log_{10} \frac{I_0}{I_t} = \log_{10} \frac{100}{0.1} = \log_{10} 1000 = 3$$

Characteristic Curve

The measured values of various densities on the perfectly developed strip of film material are plotted against the exposure logarithms from which they are derived. The resulting diagram will show the characteristic curve of the particular film material. A typical characteristic curve of a negative film material is shown in Fig. 10-3.

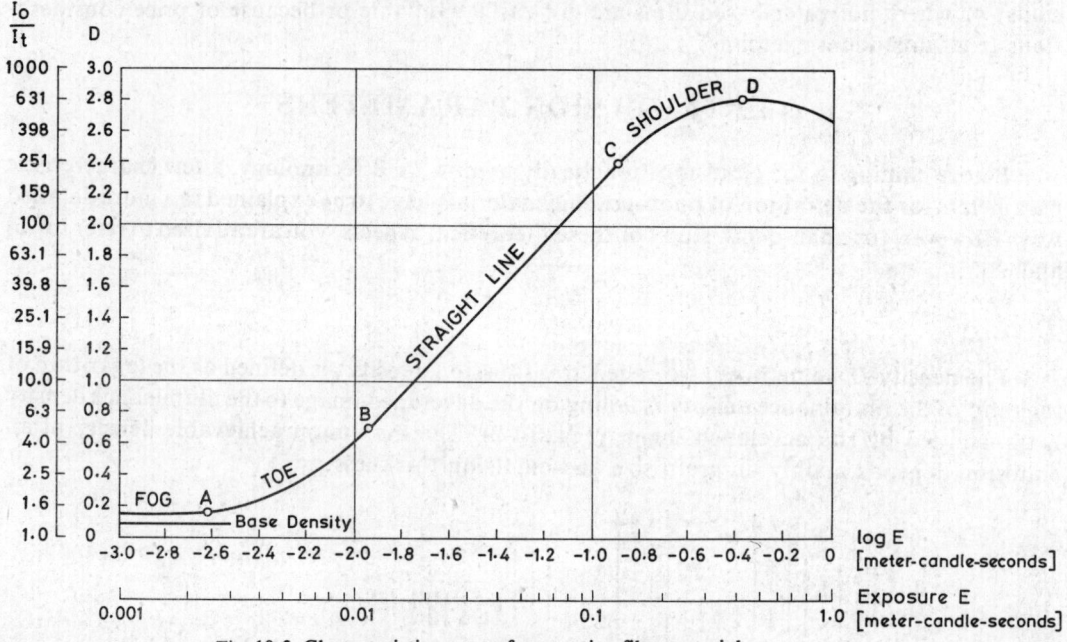

Fig. 10-3 Characteristic curve of a negative film material

Gamma

The slope or tangens value of the straight-line portion of the characteristic curve has been termed *gamma* and designated by the Greek letter γ. Fig. 10-4 shows two characteristic curves with a different gamma.

A practical example to explain further the meaning of gamma: A given film shows in one place a density of $D_1 = 2.8$, while $D_2 = 4.2$ is measured in another place. The resulting density difference is $D_2 - D_1 = 1.4$ in this case. If this film is now copied onto another film material which produces the same density difference (e.g., $D_{1x} = 3.1$ and $D_{2x} = 4.5$ in the corresponding places) then the film material must have a gamma of 1.0 which means the contrast remains unchanged. A film material with a gamma of 8.0 would produce an eightfold density difference

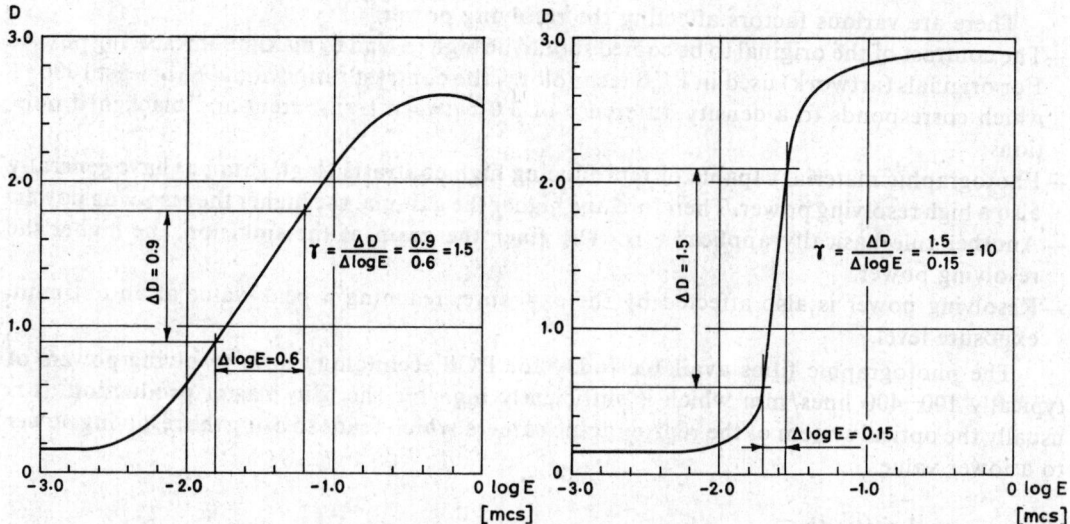

Fig. 10-4 Characteristic curves with a different gamma

(e.g., $D_{1Y} = 0.8$ and $D_{2Y} = 12.0$). In other words, a film material with a gamma above 1.0 enhances the contrast while a film material with a gamma below 1.0 reduces the contrast. The same contrast will be reproduced only by a film material with a gamma of 1.0.

Resolving Power

The resolving power is defined as the ability of a photographic material to maintain in the developed image the separate identity of parallel bars which are very close together. The resolving power is commonly expressed in lines/mm. A line is considered as the sum of the width of a single black and an adjacent clear line. A typical pattern to test the resolving power is given in Fig. 10-5.

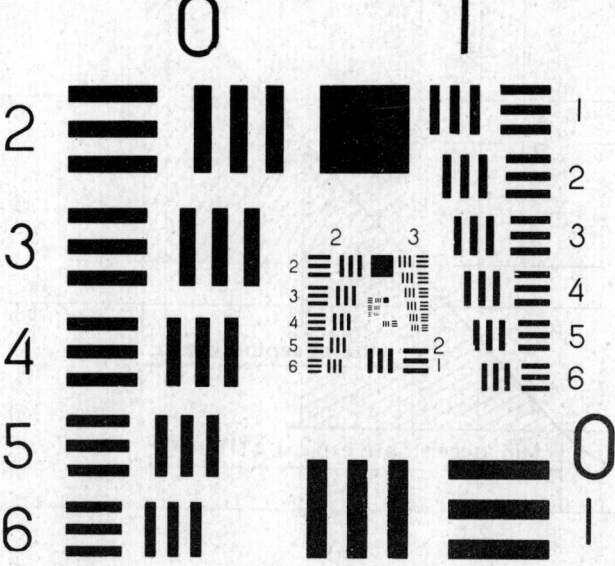

Fig. 10-5 Typical test pattern for resolving power (Courtesy: Itek Corp., Lexington, Ma. 02173, USA)

There are various factors affecting the resolving power:
— The contrast of the original to be copied should be high to lead to maximum resolving power. For originals (artwork) used in PCB technology, the contrast ratio should be at least 1000 : 1 which corresponds to a density difference of 3.0 between transparent and blackened portions.
— Photographic material capable of reproducing high contrasts (high gamma) have generally also a high resolving power. Therefore, the higher the gamma, the higher the resolving power.
— Another rule basically applicable is: The finer the grain of the emulsion, the higher the resolving power.
— Resolving power is also affected by the exposure, reaching a peak value at an optimum exposure level.

The photographic films available today for PCB technology have resolving powers of typically 100–400 lines/mm which is sufficiently high for the film master production. It is usually the optical system of the reprographic camera which tends to bring the resolving power to a lower value.

Exposure Latitude

Exposure latitude is a measure for the freedom we have to vary the exposure without causing loss of details in the result. The exposure latitude is expressed as the factor by which the minimum exposure can be multiplied to give the maximum exposure which still does not reduce the density difference in the result. The example in Fig. 10-6 gives a latitude of $10^{(1.44 - 0.93)} = 10^{0.51} = 3.2$.

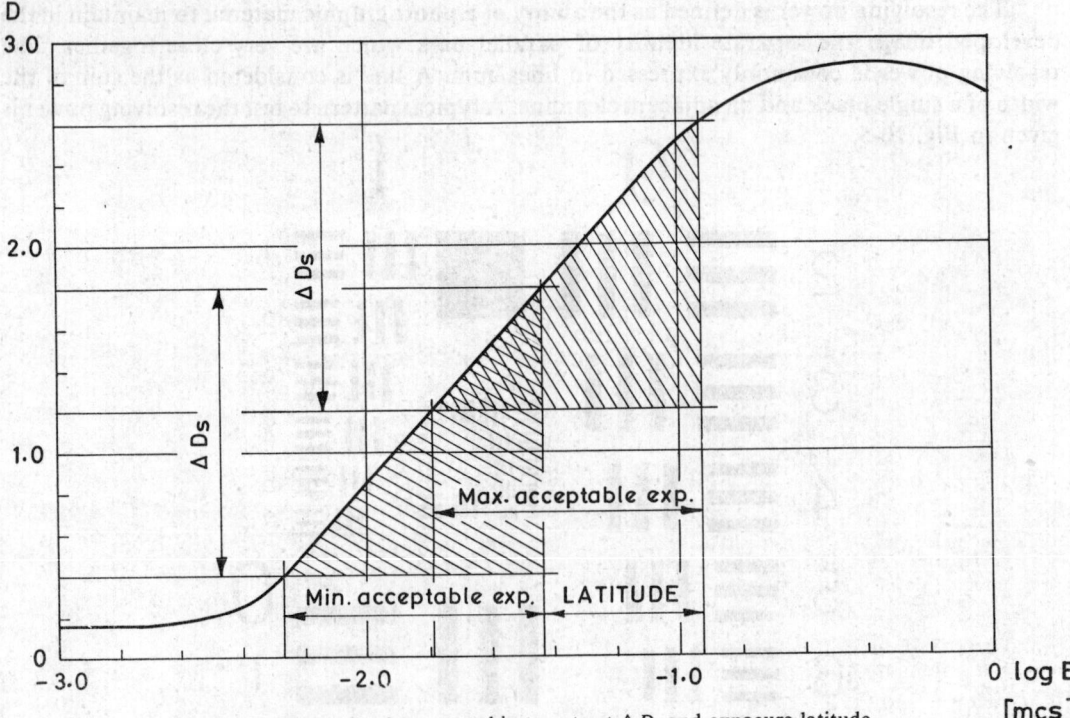

Fig. 10-6 Relation between subject contrast ΔD, and exposure latitude

It is evident from Fig. 10-6 that an emulsion with a high-gamma (steep slope) will give little exposure latitude. This is in fact a handicap with the high-gamma emulsions used for high-contrast work like in PCB technology. It calls for an exact control of exposure times.

Desirable Features of Films used for PCB Film Masters

After this introduction to the characteristic parameters of film emulsions, let us see now what is desirable for making PCB film masters:
— Contrast: The photographic work required in PCB technology is entirely of line-type. Line photography means that only fully black and fully transparent portions have to be reproduced; there are no grey shades and tones needed as in landscape photography. Therefore the films used must have a high gamma which means also a high contrast capability. The gamma of suitable film materials varies typically between 4 and 12.
— Density: The minimum density required is 2.5 which corresponds to a contrast ratio of approximately, 320 : 1 ($\log_{10} 320 = 2.5$).
— Scratch resistance: The emulsion should be scratchproof to a high degree. Scratches are a big enemy of film masters and turn them quickly into an unusable condition. The synthetic suspensions currently used to stabilise the grains give good resistance even for the wet film masters during processing.
— Grain: Since a high resolving power is aimed, it makes fine-grain emulsions necessary. Fine grain goes usually at the cost of film sensitivity (film speed) which is not a very critical parameter in this kind of applications.
— Processability: The film materials are used in an industrial environment and the necessity for special considerations in processing should be minimised. In most of the cases, emulsions with an orthochromatic response to light are preferred i.e., emulsions which are not sensitive to red light. This enables a limited red light illumination inside the darkroom. Furthermore, the exposure latitude and development timings should not be very critical so that constant results can be achieved on a regular scale.

There are of course many more desirable features for film emulsions, but let us look now at the typical emulsions which have found wide acceptance in PCB technology.

10.3   FILM EMULSIONS

Lith Emulsions

The most typical characteristic of lith emulsions is a high gamma, usually in the range of 8-12. Such a high contrast capability makes lith emulsions an excellent tool for line-type photographic work such as PCB film master production. Simultaneously, a high resolving power is obtained. A typical representative is the Kodalith Ortho Film, Type 3, 2556, on Estar base of 100 μm thickness, for which a resolving power of more than 225 lines/mm is claimed [10]. The high gamma, however, calls for strict exposure tolerances because of the resulting low exposure latitude.

Another, and in many cases undesired attribute of lith emulsions is the so-called *infection effect*: Extended developing will cause all silver halides to get reduced (oxidised) as they would have been if exposed. The development, therefore, must be done under close observation of development time and developer temperature.

The fine-grain lith emulsions have an emulsion thickness of typically 6-8 μm which permits the reproduction of details as small as 8 μm in size.

Line Emulsion

Line emulsions have a gamma which is typically in the range between 4 and 8. Compared to lith emulsions, there is less contrast with line emulsions but still sufficient to meet the general requirements of PCB technology.

Line emulsions are gaining more and more acceptance in PCB film master fabrication because they do not have, to the same extent, the disadvantage of a low exposure latitude like lith emulsions. Furthermore, the infection effect does not exist and the development time is therefore uncritical. It is also reported that less pinholes are found which reduces retouching operations.

The thickness of fine-grain line emulsions is typically 4-5 μm. They provide at least the same or slightly higher resolving power as compared to lith emulsions. Details of a size down to 5 μm are reproduced.

It is expected that line emulsions will replace lith emulsions in PCB technology to a large extent due to their advantages which make them better suited for applications in an industrial environment.

10.4 DIMENSIONAL STABILITY OF FILM MASTERS

The absolute dimensional accuracy of a film master is determined by
— accuracy of the original artwork
— dimensional stability of the artwork
— reduction scale of artwork to actual PCB size
— camera accuracy in producing the film master
— dimensional stability of the film master.

There are again various factors influencing the dimensional stability of a film master such as
— temperature variations
— changes in relative humidity
— processing effects
— ageing.

Temperature Variations

The dimensional changes caused by a temperature variation are comparatively easy to deal with: They are fully reversible and independent of film thickness. Higher temperatures cause an extraction while lower temperatures cause a contraction of the film. For all practical purposes in PCB film master fabrication, it is sufficient to consider only the temperature effects on the base; the emulsion with its comparatively little thickness can be neglected here.

Temperature expansion coefficients of the commonly used film base materials are [5]:

Tri Acetate Film   63 ppm/°C
Polyester Film   27 ppm/°C
Photographic Glass   4.5 ppm/°C
Acrylic Glass Plate   70 ppm/°C

The dimensional changes are usually expressed in ppm (parts per million). An example: The temperature expansion coefficient for polyester film is 27 ppm/°C. A temperature variation of 1°C will change the dimensions by 27 millionth of their original value. A temperature rise of 15°C will therefore bring a film of originally 250.000 mm length to 250.000 $\times (1 + 15 \times 27 \times 10^{-6})$ = 250.101 mm.

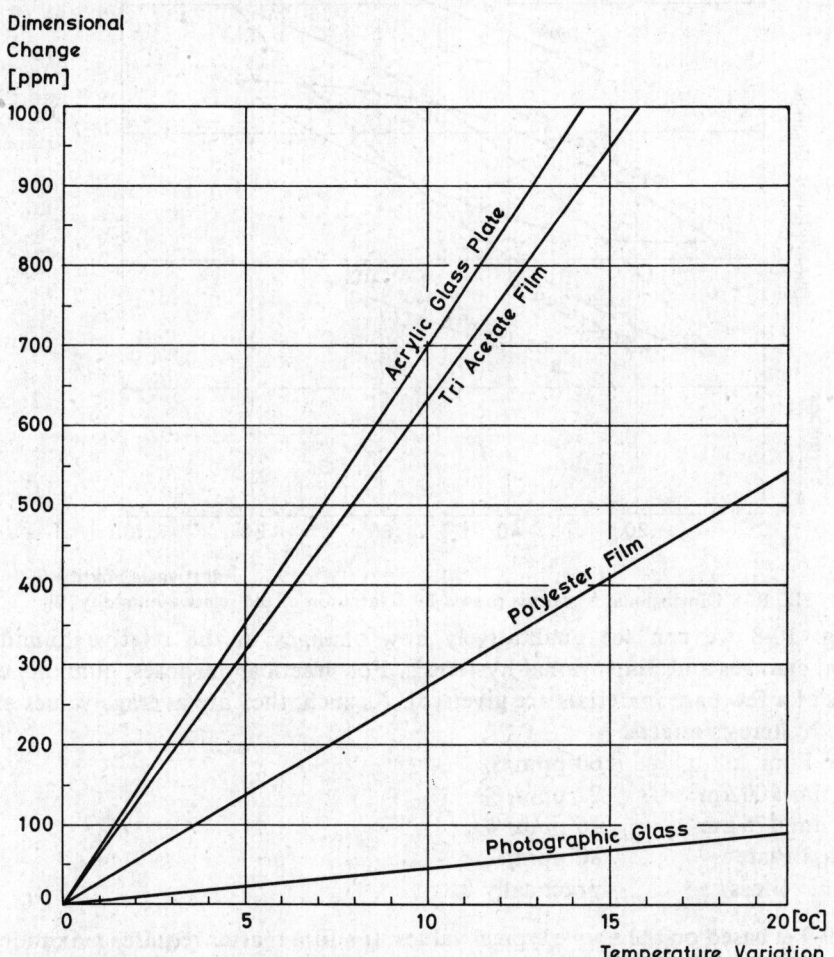

Fig. 10-7 Dimensional changes due to temperature variation [5]

Change in Relative Humidity

Humidity expansion or contraction is also a reversible change and is caused by the gain or

loss of moisture in the air in contact with the film. There are a few factors to be considered for an exact determination of the resulting dimensional changes which are
— thickness and material of the base
— type, thickness and state (processed or unprocessed) of the emulsion
— direction in which the change in relative humidity occurs (dimensional hysteresis).

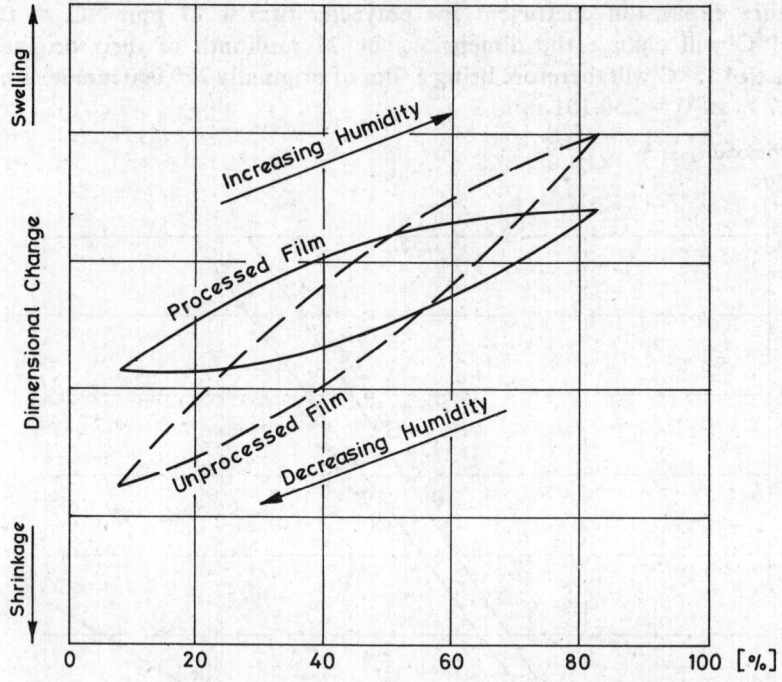

Fig. 10-8 Dimensional hysteresis caused by a variation of the relative humidity [2]

In Fig. 10-8 we can see qualitatively how changes in the relative humidity cause dimensional changes and dimensional hysteresis. For practical purposes, humidity expansion coefficients of a few base materials are given [5]. As such, they are average values and do not include the hysteresis effects.

Tri Acetate Film	60 ppm/%
Polyester Film, 100 μm	21 ppm/%
Polyester Film, 175 μm	16 ppm/%
Acrylic Glass Plate	80 ppm/%
Photographic Glass	practically 0

Fig. 10-9 is based on the above typical values. If a film master requires maximum possible accuracy, exact data and expansion coefficients for that particular film have to be requested from the film manufacturer.

Processing Effects

During development of the film, there will first be the conversion of the exposed silver

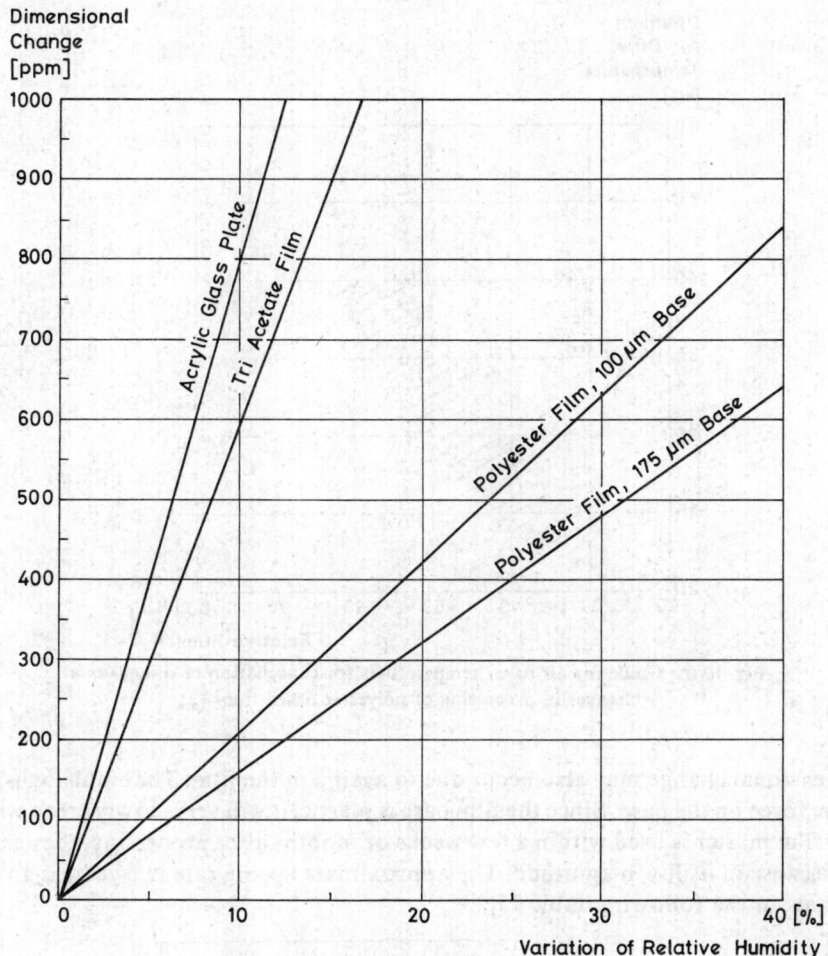

Fig. 10-9 Typical values for dimensional changes caused by a change in the relative humidity [5]

halides to metallic silver, while the bromides are washed away. The washing process itself removes some gelatine through dissolution. Each one of these steps represents a certain irreversible emulsion volume decrease.

If the equilibrium humidity before and after processing is the same and has a low value, a slight increase in film size can be noticed. The increase will be more for films with little base thickness. At a high equilibrium humidity before and after processing, a slight decrease in dimensions will occur. According to this, there must be somewhere in between an ideal humidity where no dimensional change occurs. It is now possible to choose the proper air drying temperature for a given relative humidity, so as to cancel the dimensional change caused by film processing. For polyester based films, a guide in finding the optimum drying temperature in an air drying cabinet is given in Fig. 10-10.

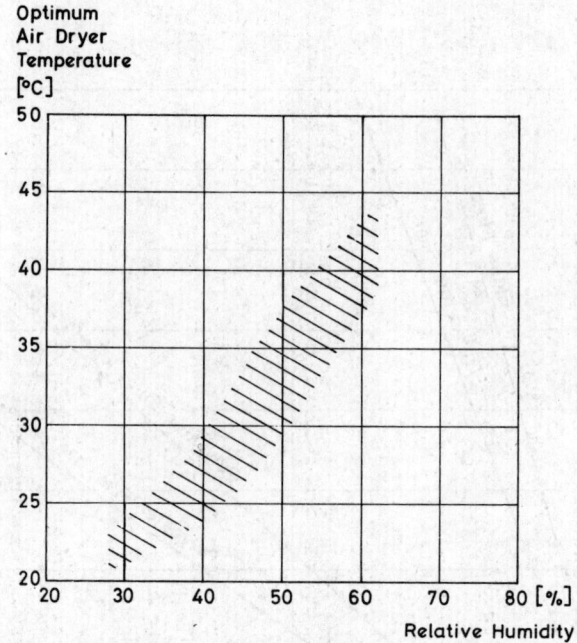

Fig. 10-10 Guide for air dryer temperatures for cancellation of dimensional changes in processing of polyester based films [2]

Ageing

A dimensional change may also occur due to ageing of the film: The emulsion is exerting a compressive force on the base. Since the film base is plastic, it will very slowly creep with time. If a polyester film master is used within a few weeks or months after processing, the ageing can be neglected because of its low magnitude. The approximate ageing rate in 5 years is given for two base materials in the following table [5]:

	Ageing rate in 5 years	
	at 25°C 60% RH	at 32°C 90% RH
Tri Acetate Film	- 300 ppm	- 4700 ppm
Polyester Film	- 250 ppm	- 100 ppm

Conclusions

While each one of these factors causing dimensional changes has been discussed separately, in practice they will simultaneously influence the dimensions of the film master. Under certain conditions, they may be additive or they may cancel each other. For example, a rising air temperature normally causes a film expansion. But this expansion may be offset by a lower relative humidity and vice-versa. However, where dimensional accuracy has to be maintained within narrow limits, suitable measures like temperature- and humidity-control are unavoidable, and films with the best dimensional characteristics have to be chosen (e.g., polyester base of 175 μm instead of 100 μm). If dimensional changes within 300 ppm are acceptable, as a thumb rule, there is no need for further control of temperature and humidity than what is needed for personal comfort (temperature variation not more than 10°C and relative humidity between 30 and 60%).

10.5 REPROGRAPHIC CAMERAS

The reprographic cameras can be put into two categories: the horizontal type (Fig. 10-11) and the vertical type (Fig. 10-12). The horizontal camera basically permits bigger artwork-and film sizes and the room with the camera may be divided into two sections separated by a thin wall (Fig. 10-13). This permits the handling of film materials to be restricted to a small room which can easily be kept clean and dustfree.

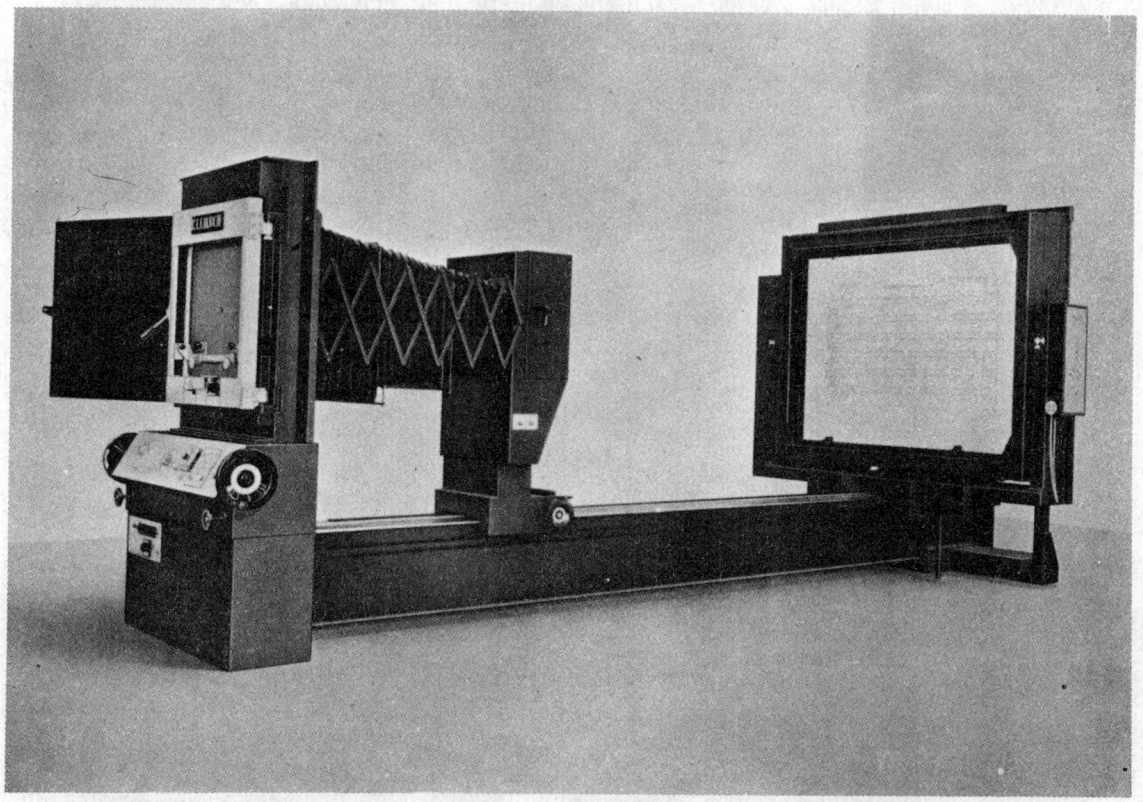

Fig. 10-11 Horizontal-type reprographic camera (Courtesy: Klimsch & Co., Frankfurt/M., Germany)

The vertical cameras need comparatively little floor space, however, the copyboard size is limited (typically 45 × 60 cm) in order to enable a normal standing person to operate the camera with reasonable convenience.

The physical size of a camera is governed by the maximum reduction ratio required, the

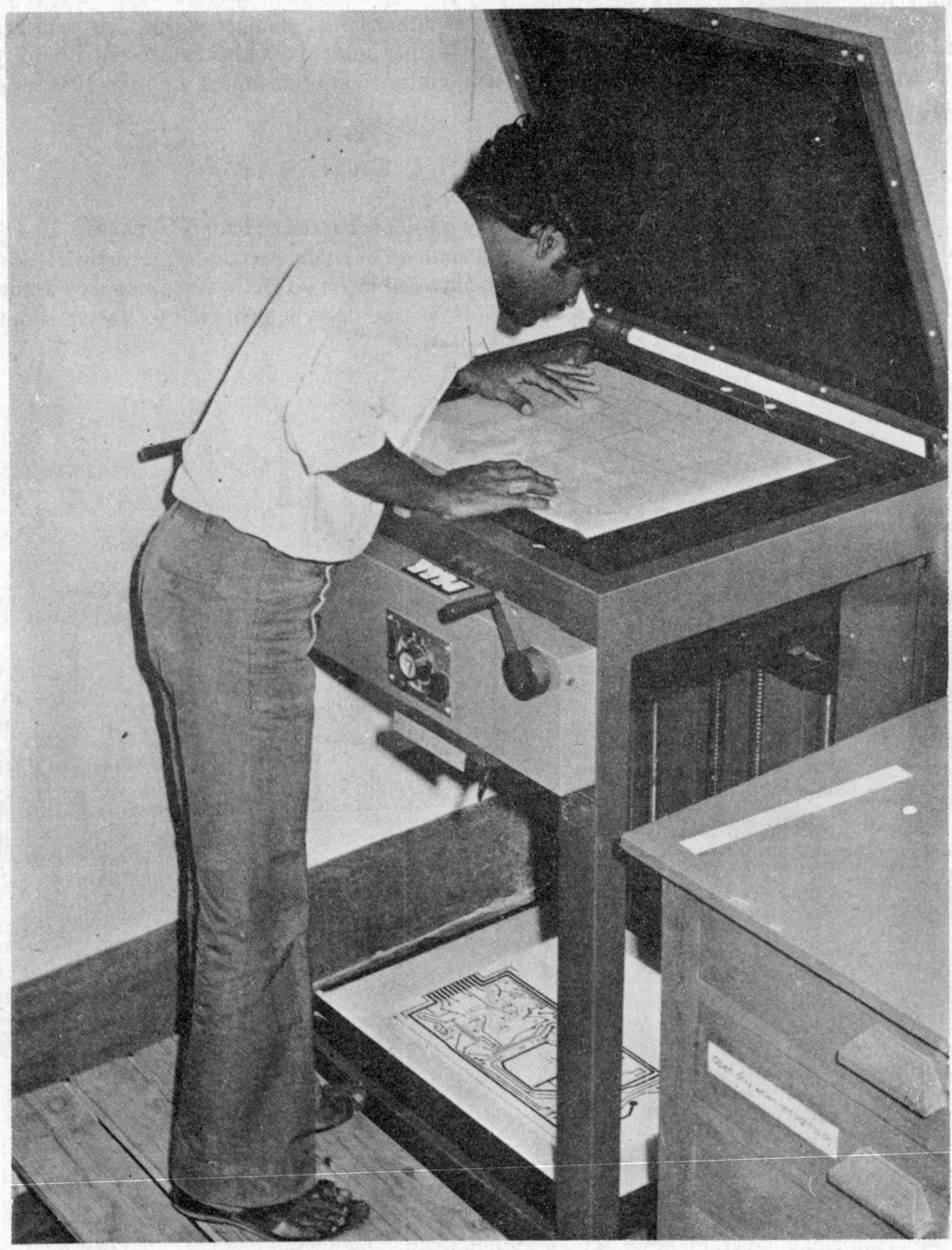

Fig. 10-12 Vertical-type reprographic camera with transilluminator: Helioprint Repromaster Mark III
(Photo CEDT)

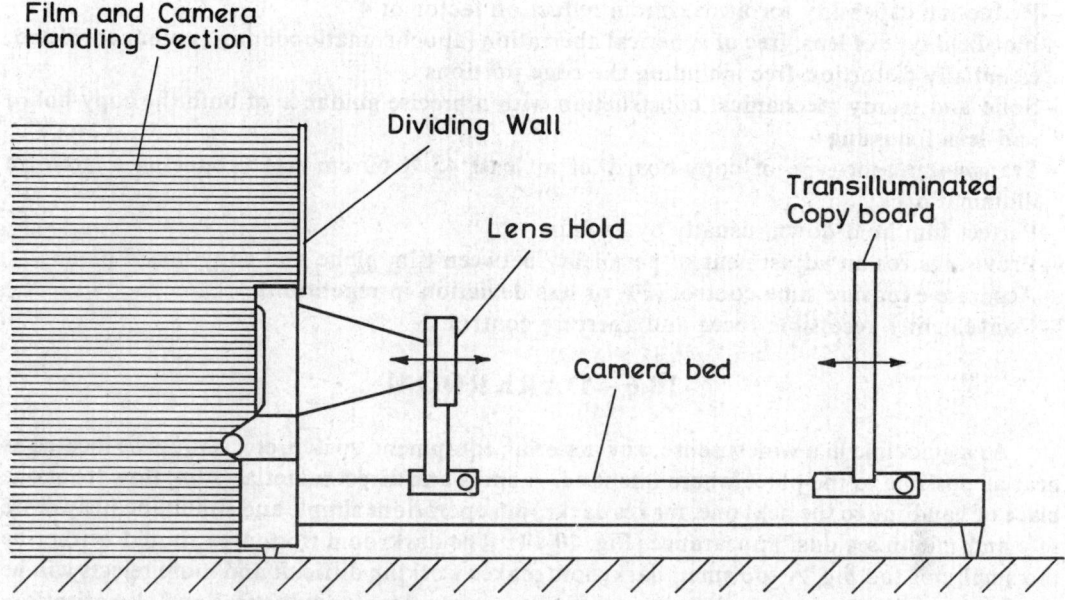

Fig. 10-13 Horizontal-type reprographic camera in divided room

focal length of the lens and the maximum size of the original (which is, in our case, the artwork). The relation between these factors shall be given here:

$$\text{Reduction ratio} = \frac{\text{Distance lens to artwork} - \text{Focal length of lens}}{\text{Focal length of lens}}$$

Usually, a maximum reduction ratio of 4 and a useful copyboard size of 45 × 60 cm is needed in PCB technology. If the focal length of the lens is 30 cm (typical value), the resulting distance from the lens to the artwork will be 150 cm.

The flat-field type of lenses used in high-precision reprographic cameras are computer designs and offer a low distortion and a high resolving power (resolution) over a reasonably large flat field. The lenses are made of glass with a high refraction index and low absorption of light. The focal length of the lens can not be made too small which otherwise would enable cameras of small dimensions: A lens with a small focal length has only a very limited resolving power and gives higher distortions which both are not acceptable in high-quality reprographic photography. Therefore, cameras are often of a very large size, sometimes up to 5 m in length.

The illumination of the copy-board needs careful attention: A transilluminator (through-illuminator) offers the best solution for artwork made on transparent foil. For taped artwork, it will provide typically a contrast ratio of better than 1000 : 1 compared to 50 : 1 which might result for the same artwork when illuminated with side lamps. Furthermore, energy consumption for the white fluorescent tubes in the transilluminator are a fraction of what side lamps would require. For a transilluminator of 60 × 75 cm useful area, a typical value is 650 W. The illumination of the copyboard must be uniform and is obtained by using milky glass as a diffusor between tube lights and artwork.

To sum up, the essential qualities of an accurate camera suitable for PCB technology are listed in brief:

—Reduction capability for a maximum reduction factor of 4
—Flat-field type of lens, free of spherical aberration (apochromatic coating), colour corrected, essentially distortion-free including the edge portions
—Solid and sturdy mechanical construction with a precise guidance of both the copy-board and lens focussing
—Transilluminator-type of copy-board of at least 45 × 60 cm size producing a uniform illumination
—Perfect film hold-down, usually by vacuum
—Provisions for an adjustment of parallelity between film plane and copy-board plane
—Accurate exposure time control (1% or less deviation in repetition)
—Conveniently accessible focus and aperture control.

10.6 DARKROOM

As a guideline in a widest sense, any material, equipment, switch, etc., should be located as near as possible to the place where it is used. A short and direct material (film) flow from one place of handling to the next one, makes darkroom operations simple and simultaneously more safe and minimises dust appearance (Fig. 10-14). The darkroom floor area should neither be too small nor too big. A too small darkroom makes working difficult and more rejects will be the result, while a too big darkroom is difficult to keep clean and dustfree and climatisation, where required, becomes very expensive.

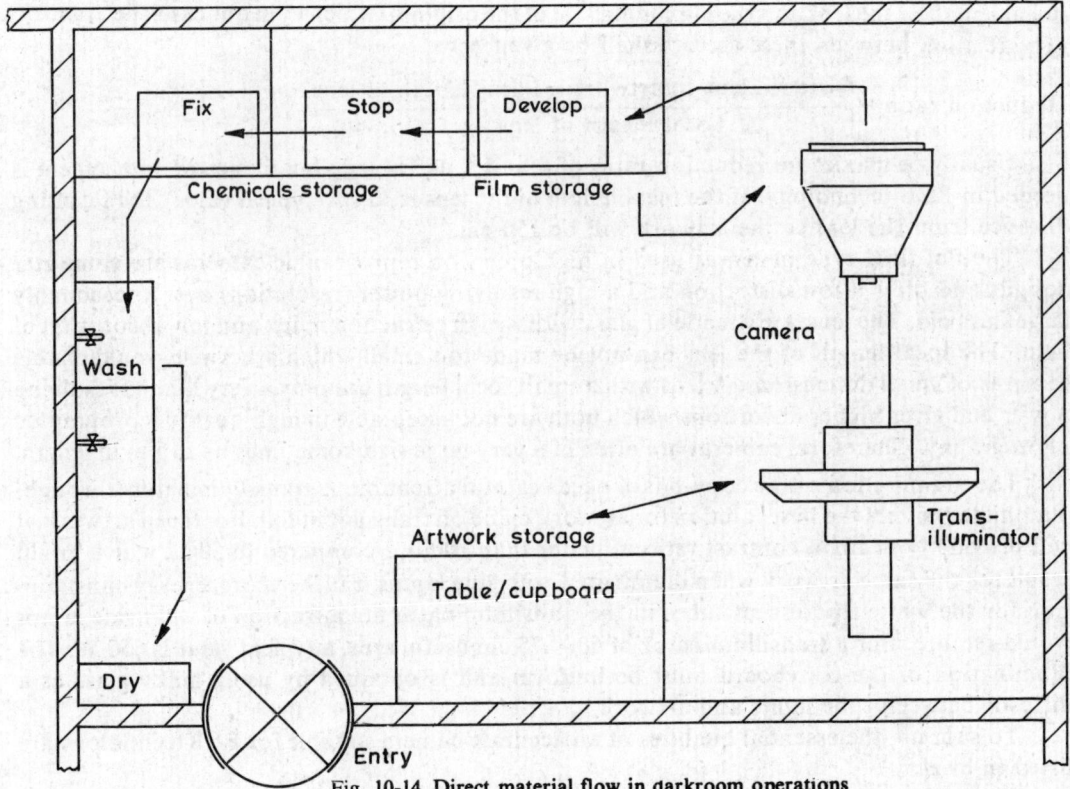

Fig. 10-14 Direct material flow in darkroom operations

The control of temperature and humidity has to match the dimensional stability requirements of the film masters to be produced. As already mentioned earlier in this chapter, where a dimensional film stability of 300 ppm is within acceptance, we can broadly tolerate temperature variations within $\pm 5°C$ and a relative humidity between 30 and 60%. For highest quality requirements, the temperature has to be kept constant at $21°C \pm 1°C$ and the relative humidity at $55\% \pm 2\%$. Such a strict control might really be necessary only in limited cases, although desirable everywhere. The costs for air conditioning are generally high and include the initial investment plus the enormous energy consumption once under operation. Therefore, the extensive use of such equipment must be minimised and avoided if possible. It will help to install the darkroom in a cool part of the building where there is minimum or no direct sunlight falling on the outside of the walls. All the heat producing equipment for film processing should as far as possible be kept out of the darkroom and placed in adjacent rooms (e.g., film drying cabinet).

The circulation of fresh and dustfree air should be to such an extent that the darkroom air gets completely exchanged, 3 to 6 times every hour. Where no air conditioner is used, a dry air filter is recommended. The use of fans inside the darkroom must be avoided because of dust turbulence caused. A slight atmospheric overpressure inside the darkroom prevents dust from entering through the doors and windows.

The prevention and removal of dust must be the prime concern of any person entering and working in the darkroom. The number of persons entering the darkroom should be restricted. The floor has to be covered with a material which is easy to clean (wet and dry). The walls should be painted with a washable paint of a matt finish. Windows and the doors should close hermetically so that neither light nor dust can enter. A typical darkroom door design is shown in Fig. 10-15. An arrangement with two ordinary doors or one door plus one light proof curtain with some light protected space in-between will also serve the purpose if it can be assured that both the sides are not opened at the same time. Locking of darkroom doors when working inside is not recommended since it will interrupt the photographer if somebody wants to enter.

The occurrence of vibrations in and around the darkroom should be minimum, and this must be considered in the planning stage: Exposure times can go beyond 100 sec and any vibration of the camera during these 100 sec will definitely reduce the quality of the work carried out. There can be external sources of vibration such as nearby railway lines, roads with heavy trucks passing, punching presses, etc. Internal vibrations can be produced by the air conditioner or by walking personnel if the floor is not well supported. The air conditioner (wherever required) has to be installed on shock-mounts on a cement foundation and as far as possible away from the camera.

It is good to have two sinks with running tap water in the darkroom; one sink will be used for rinsing of the films with water while the other is still available for hand-washing. If the mains voltage fluctuation is high, there will also be a need for a voltage stabiliser for the camera. The voltage stabilisers of magnetic type provide an output voltage which is constant within 2%.

The illumination inside the darkroom depends on the type of emulsions which are processed: Orthochromatic emulsions permit the use of the ordinary red darkroom lights fitted with a normal 15 W bulb. They are fixed at a distance of at least 1.5 m above the working area. A nearly absolute darkness is recommended when working with panchromatic film emulsions. At the maximum, darkroom lamps with a darkgreen filter can be tolerated. It is suggested to

Fig. 10-15 Light trap carousel as it can be fitted into the darkroom entry (Courtesy: Doschen India Pvt. Ltd., Post Vasai-401 201, Dist. Thane, India)

test the safety of the darkroom illumination by keeping a partially covered piece of film for 15 min in the working (processing) area of the darkroom. After regular development, no difference in transparency between the covered and the uncovered portion should be visible. Fig. 10-16 gives a suggestion for arranging the light switches for maximum safety and convenience.

List of accessories used in the darkroom:
—Darkroom alarm clock for control of processing timings
—Plastic or stainless steel trays of suitable size for developer, stop bath and fixing
—Film forceps to handle film during processing
—Bath thermometer

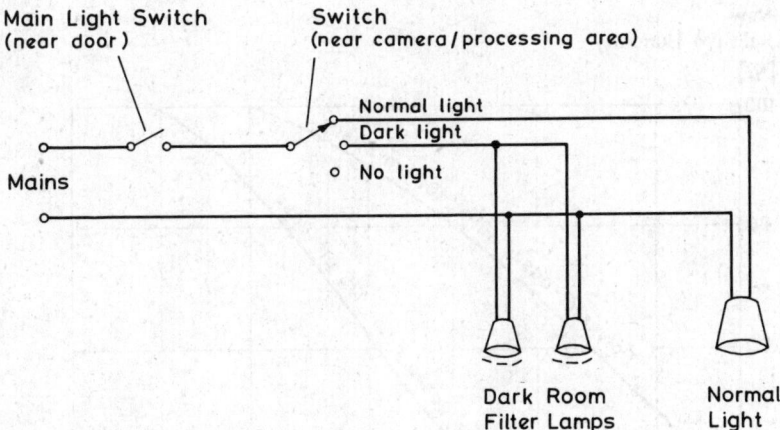

Fig. 10-16 Safe light switch arrangement in darkroom

—Trimmer and scissor
—Highly precise transparent scale for setting of the camera to the desired reduction scale
—Light-protected stock of drawers for storage of film materials
—Transilluminated inspection table
—Retouching utensils
—Stock of large drawers for artwork storage
—Refrigerator for storage of unopened film packages and processing chemicals.

10.7 FILM PROCESSING

Before coming to the actual exposure, film materials and also the artwork must have reached the dimensional equilibrium. If they have been kept in the darkroom during the past 12 hours, this would be the best preparation. The adaptation to a new temperature is quite fast and occurs within a few minutes. But it can take several hours for a film to get dimensionally stabilised after a change in relative humidity. The time to settle practically does not depend on the magnitude of the change but is linked with the new value of relative humidity and the base thickness. Typical figures for polyester based films are given in Fig. 10-17.

The storage of the hermetically sealed film packages is usally done in a refrigerator. This will maintain the film characteristics even beyond the expiry date which usually applies for storage under room conditions at 20°C. Once the package is opened, however, it cannot be put back into the refrigerator except if it has been sealed again. The storage of unsealed photographic materials in the refrigerator can cause irreversible changes in the emulsion because of the high relative humidity in the refrigerator.

Aperture Setting

To understand the optimum setting of the lens aperture, we have to look at the influence on resolving power and at the depth of field. After using ordinary amateur cameras, we know that it is, in most cases, desirable to stop down (close) the aperture to the maximum possible

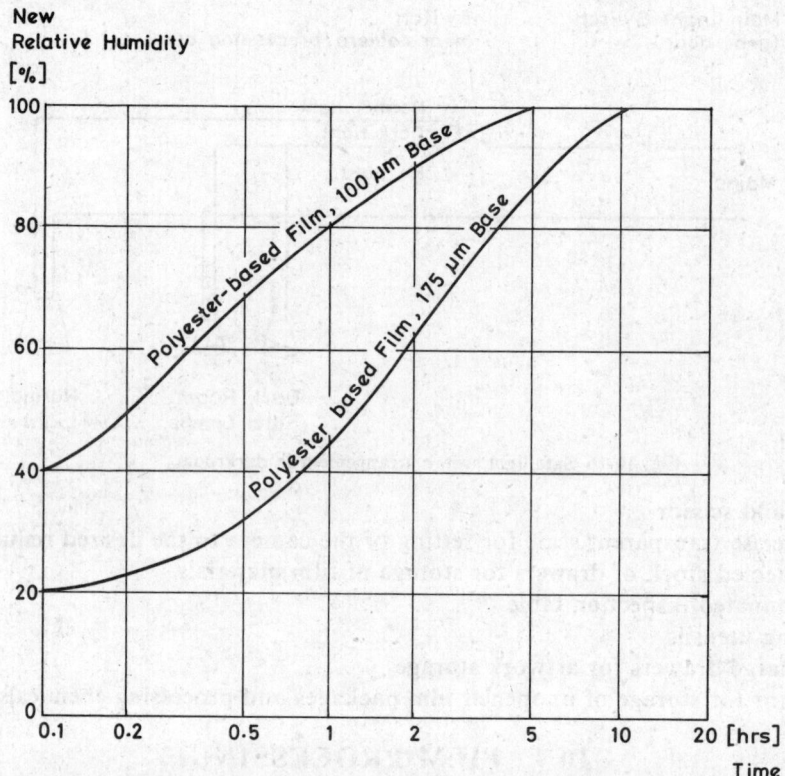

Fig. 10-17 Typical values of time required for dimensional adaptation to a new relative humidity

extent, in order to achieve a maximum depth of field (widest region of sharp focus as seen from the camera). In photographic pictures, this will be reflected by an excellent sharpness of the items right in front in the picture as well as in the sharply reproduced background. In reprographic photography, sharpness is relevant just in one plane at a fixed distance from the lens. But in these applications, too, a certain depth of field is desirable because of the difficulties in achieving a perfect focussing of the camera lens: The lens focussing is done at maximum brightness with the aperture fully opened which also gives minimum depth of field. Stepping down of the aperture thereafter increases the depth of field to a safe margin. On the other hand, the maximum resolving power of the lens is obtained with a fully opened aperture which means that stepping down of the aperture implies a quality reduction from this side, i.e., a reduction of resolving power. As a compromise, in practice, stepping down of the aperture is done for just 1-2 stops.

The total depth of field for near objects can be determined by the following formula [6]:

$$\text{Total depth of field} = \frac{2\,c\,f\,(m+1)}{m^2} \text{ [mm]}$$

c Diameter of the circle of confusion [mm]. This is a measure of the unsharpness produced in the projected image and means that a tiny dot in the original will be

reproduced in the projected image as a disc with the diameter c.
f Relative aperture, usually called *f number*
m Magnification

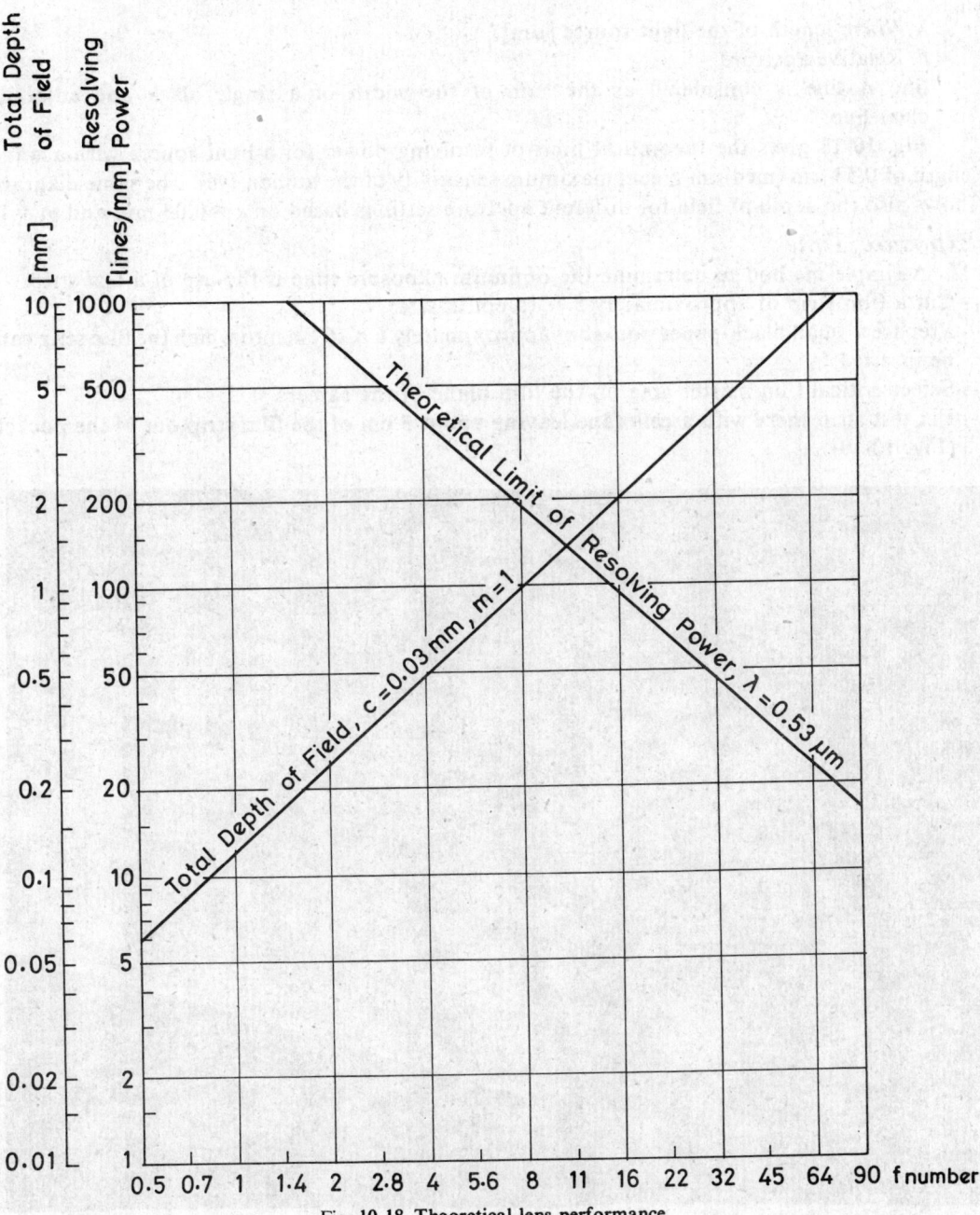

Fig. 10-18 Theoretical lens performance

And here you find the formula for the theoretically possible resolving power of a lens [6]:

$$\text{Theoretical limit of resolving power} = \frac{10^3}{1.22\,\lambda\,f} \quad [\text{lines/mm}]$$

λ Wave length of the light source [μm]
f Relative aperture
line A line is considered as the sum of the width of a single black and adjacent clear line.

Fig. 10-18 gives the theoretical limit of resolving power for a light source with a wave length of 0.53 μm (medium green, maximum sensitivity of the human eye). The same diagram shows also the depth of field for different aperture settings based on $c = 0.03$ mm and $m = 1$.

Exposure Time

A simple method to determine the optimum exposure time is the use of a test strip:
— Cut a film strip of approximately 5 × 15 cm in size
— Prepare a small black-paper pocket of approximately 6 × 16 cm into which the film strip can be inserted
— Select critical film master area on the film plane of the camera
— Fix test strip there with a cellotape leaving about 3 cm of the film strip out of the pocket (Fig. 10-19)

Fig. 10-19 Test strip ready for exposure (Photo CEDT)

— Expose for 5 sec
— Pull the pocket 3 cm further out and expose again for 5 sec
— Repeat the previous step until you have got about 5 exposures with exposure times of 5-10-15-20-25 sec.

After regular processing, the most suitable exposure time has to be selected on the base of high density in black areas, fully and unfogged transparency and sharp defined details (Fig. 10-20). The use of a magnifying glass and a transilluminated inspection table is necessary for the evaluation.

Fig. 10-20 Exposed test strip (Photo CEDT)

Development

The developer used has to be of a type recommended by the film manufacturer for that particular film. A major concern in developing is to keep all the parameters constant so that a negative with the same exposure will produce the same quality of negative at any time.

The temperature of the developer is easily checked with a bath thermometer. The recommended temperature of 20°C for ordinary developers will bring the best results if kept constant. If cooling is necessary, the developer tray should preferably be of stainless steel for a good thermal conductivity. The developer tray is then placed into another bigger tray filled with cold water. The importance of a *constant* developer temperature within 2°C has to be stressed; it means the elimination of one uncertainty which is possible without much extra efforts. In hot countries, it is often unavoidable to work with a higher developer temperature. This causes shorter developing times. The temperature of 28°C, however, is an upper limit for ordinary developers.

Each developer has a limited capacity to develop film materials which is usually given in the user's instructions. The developer is considered to be exhausted when the developing time for equal results has increased by 50% as compared to its fresh make-up.

An overworked or contaminated developer, or one that has been standing in an open tray for considerable time, will turn brown. At room temperature between 18 and 24°C, the working solution will maintain its properties in an open tray for just 4 hrs. The stock solution, kept in a stoppered bottle, will preserve its properties for 2 months if the bottle is half full and up to 6 months with a full bottle. These shelf life figures can be improved by keeping the solutions in the refrigerator and are reduced at higher temperatures.

Development technique: Use a developing tray large enough to enable film handling with ease. Put sufficient developer in the tray so that the film can get completely covered with it. All the handling of the film should be done by touching only the corners. In the processing trays, suitable photographic film forceps are used to move the film. Take strict care that no forcep is changed from one tray into another.

To start the development procedure, slide the exposed film sheet with the emulsion down through the solution, turn it quickly and place it into the solution and start immediately to lift the tray rotationally on each side by about 2 cm. This lifting of all the sides should occur in a rythm of about 5 sec for one full cycle and has to continue for the whole developing period.

Best results are obtained with fresh developer of 20°C. At the beginning, go for the recommended developing time and modify it only after sufficient experience. It is better to note down all the parameters which have given good results. Changing of the different developing parameters should be done only after having gained enough experience with the standard values. One will obtain interpretable results only if *one* parameter is changed at the time.

Stop Bath

After the development is over, the film is gripped with the forcep on one edge and lifted above the developing tray for maximum 3 sec which enables the excess developer to drop. Immediately thereafter, the film is immersed into the stop bath, always with emulsion side upward to avoid mechanical damages on the softened emulsion. The stop bath stops effectively the development action. The stop bath is acid-type and helps also to maintain the acidity of the fixing bath.

The film in the stop bath needs again moderate agitation. The type of stop bath and duration are recommended by the film manufacturers.

Fixing Bath

Also the fixing of the film occurs under frequent agitation of the bath. The temperature of stop- and fixing-bath are much less critical than the temperature of the developer bath. At higher temperatures, all the solutions will work faster while lower temperatures make them more passive. However, the variations should be within 18 and 26°C for reliable results. As a rule, the emulsion is properly fixed after immersion in the fixing solution for twice the time it takes for the milky appearance to clear completely. Normal light in the darkroom may be switched on after disappearance of the milky emulsion.

Film Washing

After all the undeveloped silver halides have been removed, the emulsion is still saturated with the chemicals of the fixing bath and some dissolved silver compounds. If they are not removed by washing, they will slowly decompose and attack the image, causing stained and faded film. The washing, therefore, is as important as the fixing of the film. The water is kept running to such an extent that a complete water replacement in the washing tank takes place once in 5 min. The duration of washing should be around 30 min or as recommended by the film manufacturer.

The effectiveness of the washing can be greatly improved if the film sheet, after fixing, gets first a brief water rinse under the tap to remove the excess fixing solution which otherwise would contaminate the washing tank.

The water from the tap usually contains solid matters like clay, silt, mud or silica. If mechanical damages through such solid impurities should occur, the use of water filter with a porosity of 50 μm is recommended as an effective remedy to trap the particles.

The water hardness, commonly measured in parts per million (ppm) of calcium carbonate ($CaCO_3$) will, if very soft (less than 40 ppm $CaCO_3$) tend to swell the emulsion. Excessive hardness (more than 200 ppm $CaCO_3$) migh cause troubles with residuals on the film after drying and affect the chemical mixed with it. This, luckily, occurs only in rare cases.

Drying

After washing, a dip in a wetting agent is a valuable aid for the drying process. For good results, drying should take place slowly at room temperatures. Most films dry sufficiently within one hour at room temperature and a relative humidity around 50%. If maximum dimensional accurary is required, you have to consider what has been mentioned under Section 10.4 before further usage of the film master.

10.8 INCREASING AND DECREASING LINE WIDTHS

In certain applications, it might be necessary to get a film master with widened or narrowed lines and patterns e.g., to compensate for underetching or plating overhang.

The basic requirements for changing of line width are:
—Point light source
—Vacuum printing frame or a spring-loaded pressure frame
—Film master positive if line widths have to be increased
—Film master negative if line widths have to be decreased
—Diffuser of at least film size to scatter the exposure light in a controlled manner (opalised glass or white translucent acrylic glass).

The principle of using diffused light can be seen in Fig. 10-21. The amount of spreading is primarly governed by the base thickness of the film and the exposure time. The longer the exposure time, the more the change in width. In case, still more spreading is required, an additional spacer can be inserted between the two films. Such spacers can be a piece of unexposed but fully processed film or clear acetate foil.

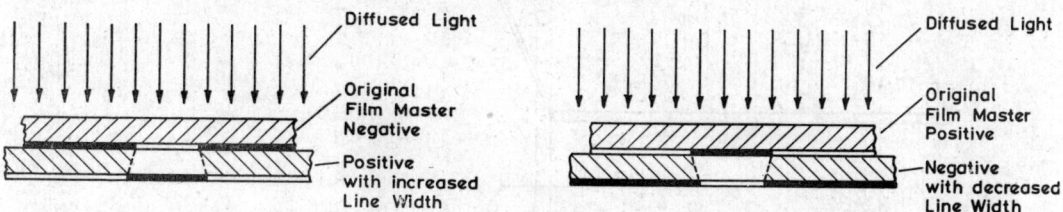

Fig. 10-21 Principles of increasing and decreasing of line widths

A contact copy of the positive with increased width will give the new film negative. For decreasing the line width, a contact copy has first to be made of the negative to provide the positive necessary for the process.

For maximum image sharpness with minimum rounding in the corners, it is essential that both the emulsion sides-face the light source. The diffuser is placed directly onto the glass of the printing frame (Fig. 10-22).

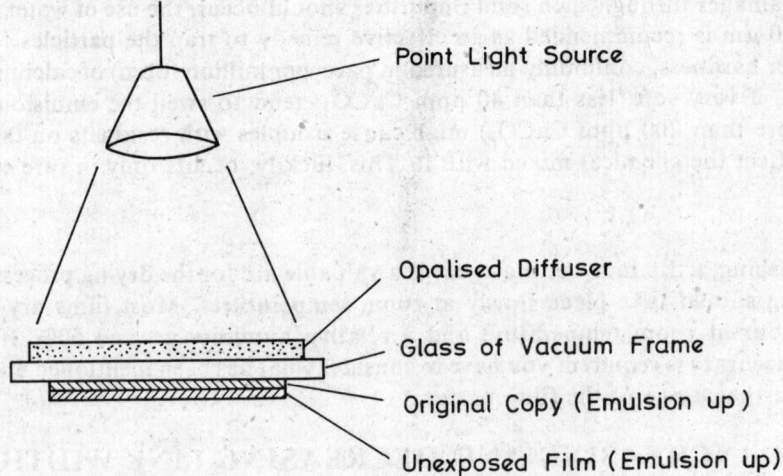

Fig. 10-22 Arrangement for changing line widths

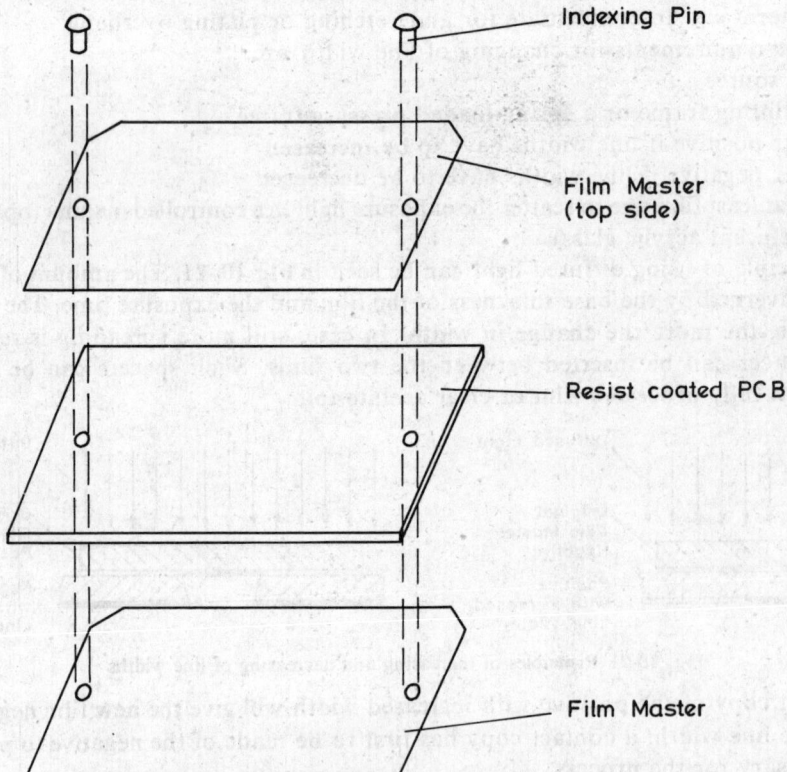

Fig. 10-23 Preparation for exposure of double-sided PCB with indexing pins

10.9 FILM REGISTRATION

The two film masters of a double-sided PCB, if used for photoprinting, have to be precisely aligned in a suitable manner for the exposure of the resist-coated PCB. Such an alignment can be done in various ways depending on precision and throughput requirements. Two possible solutions to this problem are discussed here which are widely used.

A typical method applied *for large PCB series* is the use of *indexing pins* (Fig. 10-23). It can provide a front-to-back registration of as accurate as 25 μm. All the PCBs have to be provided with two (sometimes also more) indexing holes which can be punched using a special punching tool, or drilled with a jig. The two film masters have also the same indexing holes at the corresponding position but punched with a film punch after they have been perfectly aligned under the microscope, emulsion facing emulsion. Sometimes the film masters are already punched before the exposure and indexing pins are provided with the camera both for the film as well as for the artwork. The indexing holes of the PCB serve also the positioning of the PCB for automatic drilling on NC machines (numerically-controlled machines) or for the drilling with a jig.

When assembling the PCB and film masters for exposure, the indexing pins are removed after the two film masters have been fixed in the corners on the PCB with adhesive tape. The assembly operation may also be done on a table with fixed indexing pins. It is also possible to fix indexing pins on the glass plate of the exposure unit and to do the assembly there. This eliminates the use of adhesive tape (Fig. 10-24).

Fig. 10-24 PCB exposure frame with indexing pins

196 PRINTED CIRCUIT BOARDS: DESIGN AND TECHNOLOGY

For *simpler PCB set-ups* and where no standardisation on an indexing system is possible, the use of a *sandwich* is suggested (Fig. 10–25). This method, if carefully carried out, can provide a front-to-back registration accuracy of approximately 150 μm.

The two film masters are fixed with adhesive tape on a spacer strip of the same thickness as the PCBs to be exposed the alignment is time on a transilluminated inspection table with a sheet of transparent acrylic glass between the two film masters. The alignment done by looking vertically through a piece of non-transparent plastic pipe helps to minimise alignment errors due to parallaxes (Fig. 10–26).

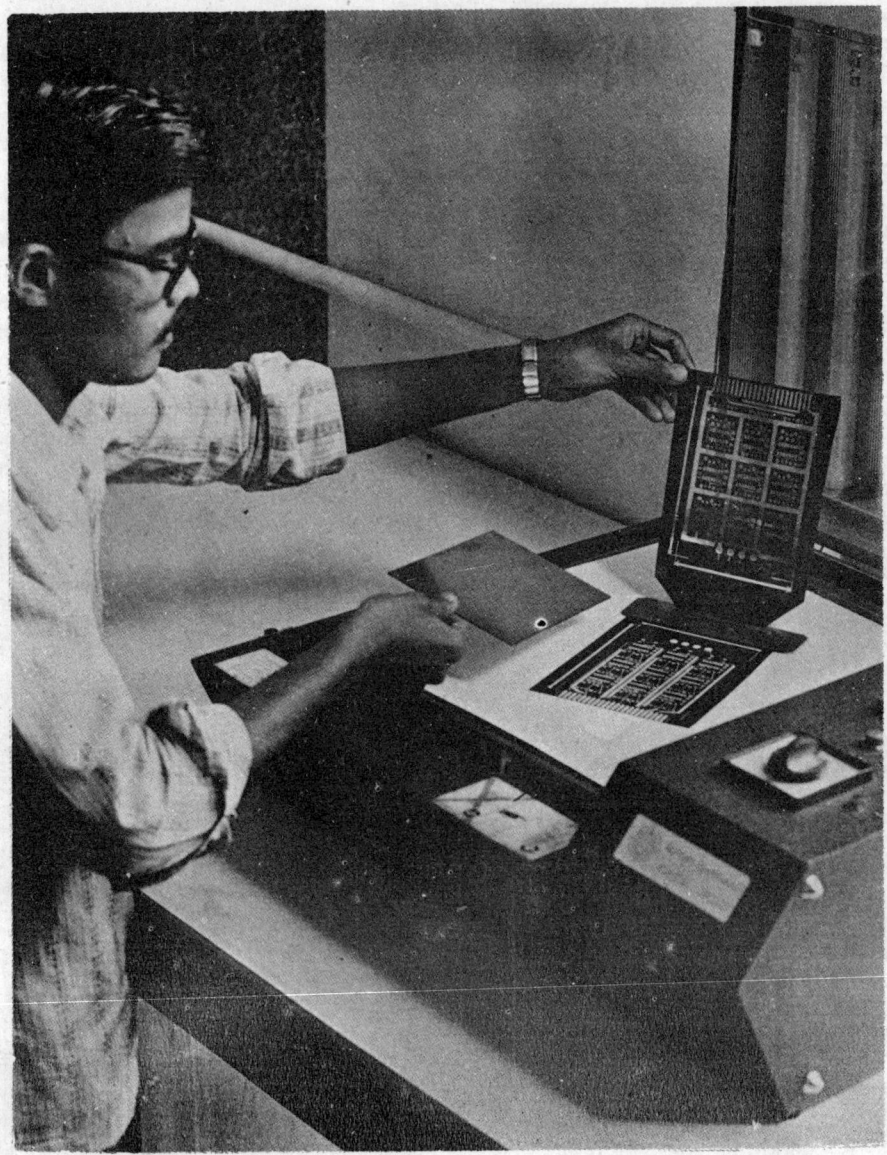

Fig. 10–25 Double-sided PCB exposure with 'sandwich' (Photo CEDT)

Fig. 10-26 Preparation of a 'sandwich' (Photo CEDT)

BIBLIOGRAPHY

1. COOMBS C.F.: *PRINTED CIRCUITS HANDBOOK,* McGraw-Hill Book Co., New York, 1967
2. *DIMENSIONAL STABILITY CHARACTERISTICS OF KODAK PRECISION LINE FILMS AND KODAGRAPH FILMS ON ESTAR BASE* Pubication No G-76, Eastman Kodak Co., Rochester, N.Y., 1977
3. FOLKERS I.: 'Neue Filme und Filmentwicklungsmaschinen fur Kameraarbeit. Plotter-Belichtung und Kontaktkopie', *EIPC PROCEEDINGS,* Zurich, June 1975, pg. 57-69.

4. d'HAENS J.: 'Dimensional Stability of Film Masters for Printed Circuit Production', *EIPC PROCEEDINGS*, Zurich, Febr. 1970, pg. 1-6.
5. HARPER C.A.: *HANDBOOK OF MATERIALS AND PROCESSES FOR ELECTRONICS*, McGraw-Hill Book Co., New York, 1970.
6. JACOBSON E.R.: *THE MANUAL OF PHOTOGRAPHY*, 7th edition, Focal Press Ltd., London, 1978.
7. HERRMANN G.: *LEITERPLATTEN*, Leuze Verlag, Saulgau/Wurtt. (Germany), 1978.
8. *KODAK PLATES AND FILMS FOR SCIENTIFIC PHOTGRAPHY*, Publication No. P-315, Eastman Kodak Co., Rochester, N.Y., 1973.
9. *KODAK PRECISION LINE FILMS, PRODUCTS FOR THE PRINTED CIRCUIT INDUSTRY*, Publication No. G-74, Eastman Kodak Co., Rochester, N.Y., 1978.
10. *LEITFADEN DER PHOTOFABRIKATION*, Publikation Nr. P-1, Kodak Aktiengesellschaft, Stuttgart (Germany), 1969.
11. MITTER V.: 'Photofabrication of Printed Circuit Boards', *ELECTRICAL & ELECTRONICS WORLD*, Vol. III, No. 2, Bombay, 1975, pg. 41-48

11
PROPERTIES OF COPPER-CLAD LAMINATES

11.1 INTRODUCTION

Although the number of different printed circuit base materials in common use is finite, the problems of material selection and quality control are almost limitless. The diversity in the applications of PCBs warrants an understanding of the copper-clad laminates themselves.

A laminate can be simply described as the product obtained by pressing layers of a filler material impregnated with resin under heat and pressure. The commonly used *fillers* are a variety of papers, or glass in various forms such as cloth and continuous filament mat. The *resins* could be phenolic, epoxy, polyester, PTFE (Polytetrafluoroethylene), etc. Each of these fillers and resins contributes intrinsically to the characteristic properties of the finished copper-clad laminates. It is further possible to manipulate the properties of copper-clad laminates by fine variations in the manufacturing process.

The large range of possible copper-clad laminates has been standardised in the national and international specifications. Thus, there are exactly laid down specifications for each copper-clad laminate grade, being defined by the resin/filler system and the minimum/maximum limits of the properties.

11.2 MANUFACTURE OF COPPER-CLAD LAMINATES

11.2.1 Materials

The basic ingredients of a copper-clad laminate are:
—Filler
—Resin
—Copper foil

Filler

Fillers are continuous webs of materials such as paper, glass cloth, etc., and are used as reinforcing agents. The papers used are kraft, alpha cellulose, rag or their combinations. The vast majority of printed circuits are made with paper based laminates as these are low priced and have easy machinability. Amongst various papers, the rag paper provides an electrically better laminate than the one made of alpha cellulose paper.

The glass filler is generally in the form of cloth woven of filaments. Glass cloth gives a laminate with a very high mechanical strength and a very low moisture absorption when epoxies are used as matrix.

Resin

The fillers described above are embedded in a matrix of a resin when laminated. Most widely used among all the matrix materials are the phenol-formaldehyde resins. Long experience with these has led to an almost perfect understanding of their behaviour. Epoxies, which are comparatively recent, are much costlier but they exhibit superior electrical and mechanical properties which are retained under hot and humid conditions. Polyesters, too, have good electrical and mechanical properties but they are restricted with respect to the type of filler applied. They also have a low chemical resistance.

Copper Foil

The copper foil which forms the surface of a copper-clad laminate is manufactured by the process of electro-deposition. A thin film of copper metal is deposited onto a slowly rotating corrosion resistant metal cylinder, whose lower portion is immersed in a copper rich electrolytic plating bath. As the cylinder slowly rotates in the bath, a thin copper deposit gradually builds up into an integral sheet of metal foil which can be gently peeled off from the cylinder surface at the point where the cylinder surface comes out of the plating bath.

The rolls of foil, as removed from the plating machine, are commonly called *raw foil*. Further process steps are applied to both surfaces of the raw foil in highly proprietary processes which increase the potential bond strength of that surface of the foil which is bonded. These processes also protect the corrosion-prone metal surface of the foil. The properties of commonly used foils are given in Table 11-1.

Table 11-1 Properties of copper foils. Extracted from [6b].

		Nominal thickness of foil	
		35 μm	70 μm
Nominal weight	[oz/sq. ft.]	1	2
	[g/m^2]	305	610
Weight tolerance	[%]	± 10	± 10
Thickness tolerance	[μm]	± 5	± 7.6
Purity of electrodeposited copper foil (silver counted as copper)	[%]	99.8	99.8
Resistivity at 20°C	[Ω g/m^2]	0.15940	0.15940

11.2.2 Manufacturing Process

The sequence of manufacturing steps involved in the making of a copper-clad laminate is shown, as a simplified diagram, in Fig. 11-1. At the impregnation stage, solutions of thermosetting resins are mostly used for treating the reinforcing web of a filler in an impregnating machine which is also called *treater* (Figs. 11-2 and 11-3). The web is dipped in the resin solution and it partially absorbs it. The amount of resin is further doctored by two precise squeeze rollers. In the treater oven, the solvents are removed and the resin is advanced in a semi-cured condition, called β-stage of cure. In this stage, the resin is tack-free but still capable of flowing under heat and pressure.

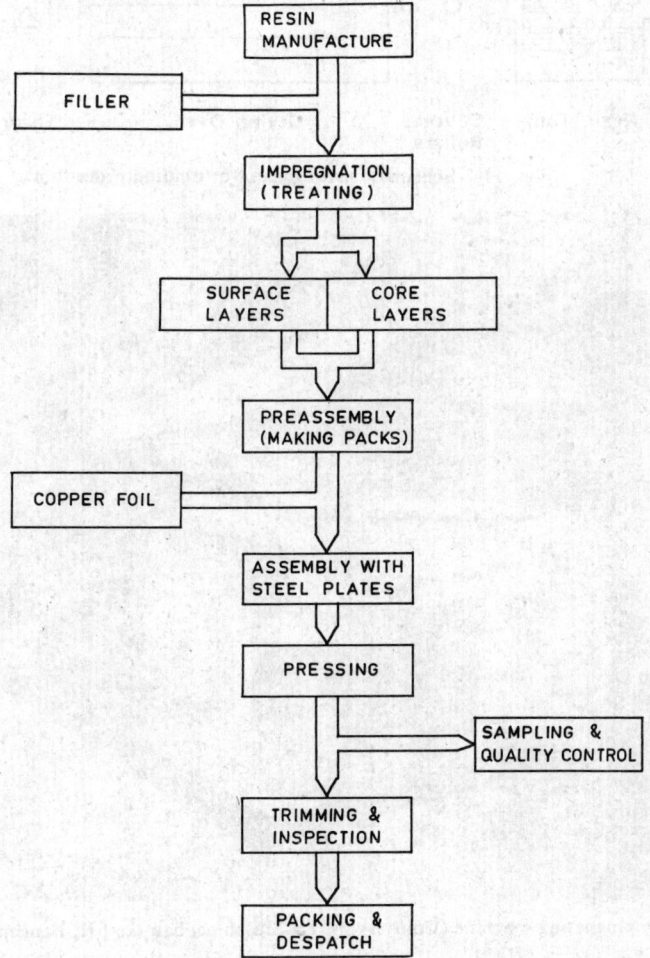

Fig. 11-1 Manufacture of copper-clad laminates

Such treated sheet layers are thereafter put into *packs*. The outer two sheets of the packs, called *surfaces,* have generally a different resin composition and the amount of resin in these is also higher than in the other layers. The surfaces serve the purpose of releasing the laminated sheets from the steel plates after moulding. They also impart an additional electrical resistance. The packs of treated material, in order to finally obtain the desired thickness, are controlled to close tolerances by weight or volume.

Each pack of treated layers is placed, along with an inspected copper foil, between two flat steel plates of a high surface finish. Several such assemblies are then charged into a multi-opening hydraulic press (Fig. 11-4).

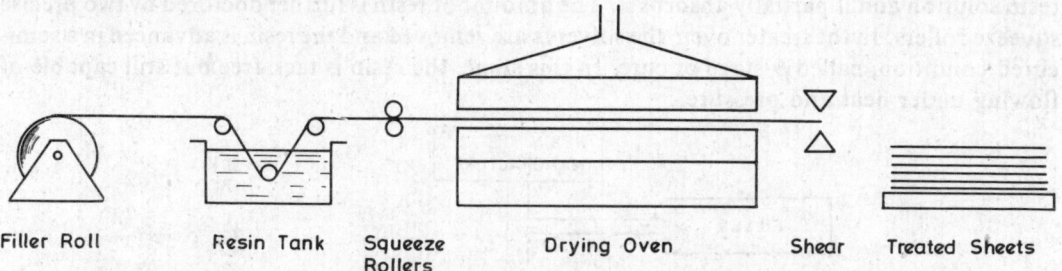

Fig. 11-2 Schematic arrangement of treating procedure

Fig. 11-3 Impregnating machine (Courtesy: VITS-Maschinenbau GmbH, Langenfeld, Germany)

The sheets are pressed at a high temperature (120°C to 170°C) and high pressure (20 to 110 kg/cm^2). In case of thermosetting resins, the resin first flows and then cures to a permanently infusible and insoluble state, forming a bond between the reinforcing filler and thereby resulting in a composite or a laminate. The pressed laminate is cooled inside the press under pressure and then removed, to be passed on to the finishing operations.

During the finishing operations, the edges of the pressed laminate are trimmed. The sheets are then individually inspected for surface appearance and thickness. From each batch of materials, samples are drawn on statistical quality control norms for the testing of other properties. The copper-clad laminates so tested and inspected are then cut into the required sheet-size and properly packed for despatch.

Fig. 11-4 Hydraulic press line for the production of copper-clad laminates (Courtesy: Siempelkamp GmbH & Co., Krefeld, Germany)

11.3 PROPERTIES OF LAMINATES

A copper-clad laminate, apart from its electrical and physical properties, must have a good copper-to-base laminate bond strength. The appearance of the copper side must be smooth and uniform. All these properties have to be retained during the production of the PCB and also under its working conditions. A designer of a PCB may need data on a wide spectrum of laminate properties such as dielectric strength, dielectric constant, dissipation factor, surface and volume resistance, foil-to-base bond strength, flexural strength, water absorption, flatness, flame resistance and appearance. During the production, properties like punchability, resistance to hot solder or peeling become important.

The nature of the resin used largely contributes to the thermal classification of the particular laminate. Phenolics yield a laminate of thermal class E (120°C). Some epoxies are specially formulated to obtain a higher thermal classification like class F (155° C) as compared to most other epoxies which are class B (130° C) laminate. The very high thermal classes H (180° C) and C (220° C) can be achieved by the use of silicone and polytetrafluorethylene (PTFE) resins.

All electrical and mechanical properties are affected by environmental conditions such as humidity, temperature, corrosive atmosphere, etc. Similarly, most of the electrical properties vary with changes in frequency. It is therefore necessary to understand the behaviour of insulating materials when subjected to various environmental conditions likely to be encountered in use. Almost all existing specifications have attempted to prescribe limiting values after exposure to simulated environmental and operational conditions.

11.3.1 Electrical Properties

Electrical properties of a laminate depend upon the electrical properties of the filler, the cured resin and the by-products of the curing reaction. The matrix acts as a waterproofing and therefore the amount of resin, the depth of penetration and the extent of the reaction affect the electrical properties. Each laminate absorbs *moisture* to some extent when exposed to high-humidity conditions. This absorbed moisture adversely affects the electrical properties. For 1.6 mm thick laminates, the approximate water absorption figures are as follows:

Paper phenolic (e.g., NEMA grades X, XX, XXX, etc.)	0.75-6%
Glass epoxy (e.g., NEMA grades G-10, G-11)	0.25%
Glass PTFE (e.g., NEMA grade GTE)	0-0.68%

Dielectric Strength

Dielectric strength is the ability of an insulating material to resist the passage of a disruptive discharge produced by an electric stress under specified conditions (Fig. 11-5A).

Most specifications use an AC voltage at 50 Hz on a piece of laminate whose copper surface is etched off, before placing it between two electrodes.

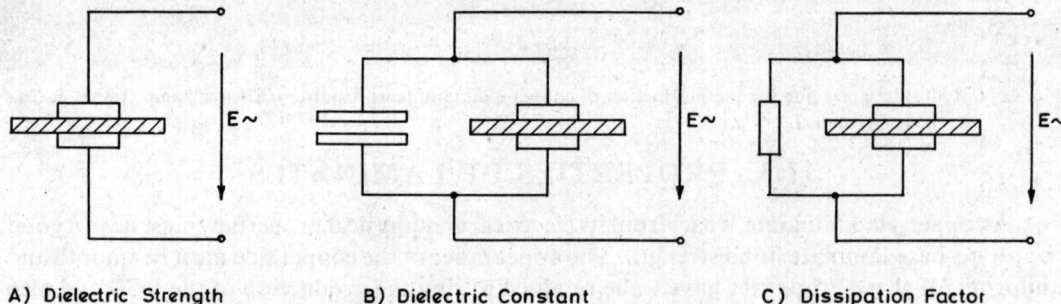

A) Dielectric Strength   B) Dielectric Constant   C) Dissipation Factor

Fig. 11-5 Schematic electrode arrangement for electrical tests

Dielectric strength depends on a large number of factors such as chemical composition, molecular structure, degree of moistening (which may cause sudden changes in the material structure and in its chemical composition due to washing out or dissolving of impurities), frequency and waveform of voltage applied, thickness of the specimen, degree of material ageing, cleanliness and roughness of surfaces, etc.

The effect of moisture on dielectric strength is a complex phenomenon and is related to the thermal behaviour of the dielectric. In moist and rather thick materials, thermal breakdown predominates: Since dielectric losses heat the material, the losses are increasingly growing. The moisture contained in a humid dielectric adds to the thermal conductivity which may be expected to raise the dielectric strength. But, as a matter of fact, the electric strength actually does not increase with moisture since a humid dielectric dries out due to its high electrical conductance and improved conditions of heat transfer. Especially in high voltage tests, the duration of voltage applied is in most cases far in excess compared with the fast moisture desorption from the material. Therefore, a careful interpretation of such test results is necessary.

Dielectric Constant

This is also referred to as permittivity. It is the ratio of the capacitance of the laminate under test and capacitance of the same electrode system with air replacing the laminate as dielectric medium (Fig. 11-5B).

Low values are best for high-frequency or power applications to minimise electrical power losses. Higher values are best for capacitance applications. The effects of temperature and frequency variations on the dielectric constant are unique for each material. Primarily, dielectric constant depends on the chemical composition of the material. In some laminates, moisture does not cause an increase in the dielectric constant or causes only a negligible increase.

Dissipation Factor

The dissipation factor is the ratio of parallel reactance to parallel resistance and is measured with the electrode arrangement as schematically shown in Fig. 11-5C. The dissipation factor is herewith the tangens value of the loss angle δ; hence $\tan \delta$ is identical with the dissipation factor. Fig. 11-6 gives the vector diagram of the equivalent parallel circuit.

The power factor ($\cos \theta$) and the loss factor are not identical with the dissipation factor although there exists a certain relation between all these factors:

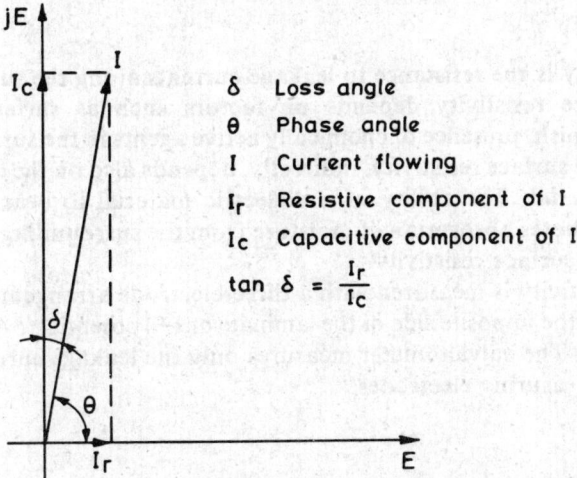

δ Loss angle
θ Phase angle
I Current flowing
I_r Resistive component of I
I_c Capacitive component of I

$$\tan \delta = \frac{I_r}{I_c}$$

Fig. 11-6 Vector diagram of equivalent parallel circuit

Power factor ($\cos \theta$) is the cosine of the phase angle θ. For low dissipation factor values, e.g., $\tan \delta < 0.10$, the corresponding power factor will have nearly the same value like $\tan \delta$; the power factor is therefore, erroneously, often mixed up with the dissipation factor.

Loss factor is, by definition, the product of the dissipation factor and the dielectric constant.

The dissipation factor is in direct relation with the resistive power losses in a laminate: The higher the dissipation factor, the higher the resistive power losses. It is therefore desirable, especially for electronic circuits operating at a high-power level, to use laminates with a low dissipation factor.

The ranges of dissipation factors that may be achieved with various combinations of fillers and resins, at 1 MHz, are as follows:

Paper phenolic laminates	0.02-0.08
Glass epoxy laminates	0.01-0.03
Glass PTFE laminates	0.0008-0.005

The dissipation factor of a laminate as such is not a constant factor: It varies with frequency, temperature and moisture absorbed in the laminate. An increasing frequency will usually show a decreasing dissipation factor while a higher temperature gives a higher dissipation factor. In most cases, the dissipation factor rises almost proportionally with the moisture content in the material; this may be attributed to the rise of current conduction through the moist dielectric. Because of such influences on the actual value of the dissipation factor, the values given in the data sheets have to be carefully interpreted: Particularly attention must be given to the conditions under which a dissipation factor has been determined.

Insulation Resistance

Insulation resistance of a base laminate is the ratio of voltage applied to the current flowing in the base laminate. Most test methods use a pre-determined DC voltage of 500 V. For practical use of this parameter, the measurements are carried out parallel to the surface of the base laminate as well as through the body of the base laminate. Such measurements are reported in terms of surface and volume resistivity.

Surface Resistivity

Surface resistivity is the resistance to leakage currents along the surface of an insulating material. The surface resistivity depends on factors such as surface humidity, surface cleanliness, surface finish, presence of chemically active agents in the surrounding atmosphere and temperature. The surface resistivity, indirectly, depends also on the chemical composition of the dielectric material: The ability of a dielectric material to bear electrostatic charges attracts dust and promotes absorption of moisture from the surrounding atmosphere resulting in a reduction of the surface resistivity.

The surface resistivity is measured with a three-electrode arrangement (Fig. 11-7A). The guard electrode keeps the opposite side of the laminate on (+) potential to avoid currents passing through the laminate. The galvanometer measures only the leakage current along the surface between the actual measuring electrodes.

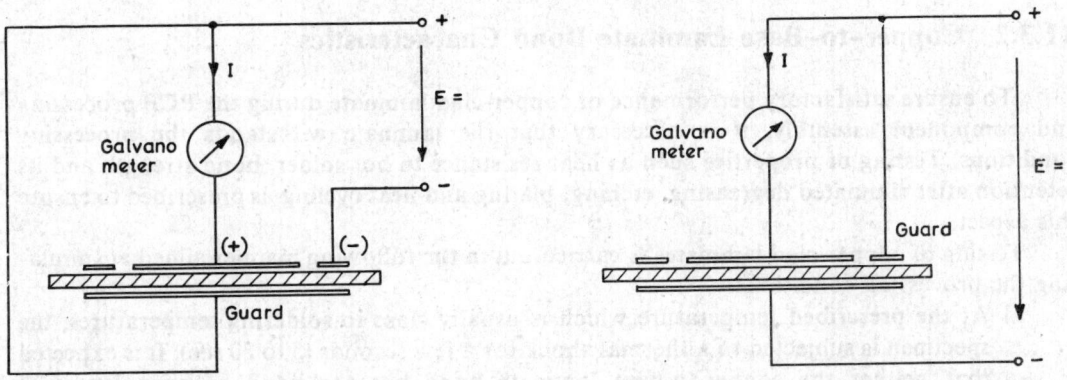

A) Surface Resistivity [Ω/square]   B) Volume Resistivity [Ω cm]

Fig. 11-7 Schematic electrode arrangement for measuring surface and volume resistivity

If a laminate is kept for some time in a humid environment, the surface resistivity will considerably decline. As a typical example [3], if paper phenolic laminate with an original surface resistivity of $10^{10}\,\Omega$ is subjected to an atmosphere of 96% RH and 23°C for a long duration, it shows $10^9\,\Omega$ after 10 days and just about $10^7\,\Omega$ at the end of 50 days. Such conditions may actually occur in practice, particularly in tropical countries.

Volume Resistivity

Volume resistivity is the measured resistance to leakage current through the body of an insulation material. Factors influencing the volume resistivity are the chemical nature of the material, temperature and moisture absorbed in the sample.

The measurement of volume resistivity is again carried out with a three-electrode arrangement (Fig. 11-7B). The guard electrode with (+) potential has to ensure that no surface leakage currents have an influence on the result. The galvanometer measures only the leakage current flowing between the measuring electrodes.

$$\text{Volume resistivity} = \frac{R \times A}{l}\,[\Omega\,\text{cm}]$$

R = resistance measured [Ω]
A = area of guarded electrode [cm^2]
l = thickness of sample [cm]

The volume resistivity, in general, falls rapidly with increasing temperature. As a typical example [3], a glass polyester laminate may show a decline in volume resistivity, if heated up from 100 to 150°C, to as low as 1.3% of the value at 100°C.

Volume resistivity must be considered as one of the most important electrical properties of a laminate and is much related to the moisture absorption property. Polymeric materials, e.g., PTFE, exhibit a very high resistance to moisture; their volume resistivity remains therefore almost unchanged even after exposure to highly humid conditions. But even with excellent characteristics of volume resistivity under humid conditions, the surface resistivity may drop to a low value resulting in a low total insulation resistance. Therefore, certain test methods check the insulation resistance of the sample *in total*.

11.3.2 Copper-to-Base Laminate Bond Characteristics

To ensure satisfactory performance of copper-clad laminate during the PCB processing and component assembly, it is necessary that the laminate withstands the processing conditions. Testing of properties such as heat resistance to hot solder, bond strength and its retention after simulated degreasing, etching, plating and heat cycling, is prescribed to ensure this aspect.

Testing of copper-clad laminates is carried out in the following manner aimed at simulating the processing conditions:
a) At the prescribed temperature which is usually close to soldering temperatures, the specimen is subjected to a thermal shock for a few seconds (5 to 20 sec). It is expected that neither the copper-to-base laminate bond nor the bond between layers of laminates give way.
b) By suitable arrangements, the copper foil is subjected to a mechanical pulling action. The force per unit width of the foil, which is required to peel it off, is then measured. This test is carried out under several conditions e.g., as received, at higher temperature, after exposure to solvent vapour, after the solder resistance test etc.

11.3.3 Physical Characteristics

Flexural Strength

This is a measure of the force per unit area, which a laminate strip will stand without fracture, when supported at ends and loaded in the centre. The values differ in two directions e.g., parallel to the length of filler (lengthwise) and perpendicular to it (crosswise). The crosswise values are in most cases lower.

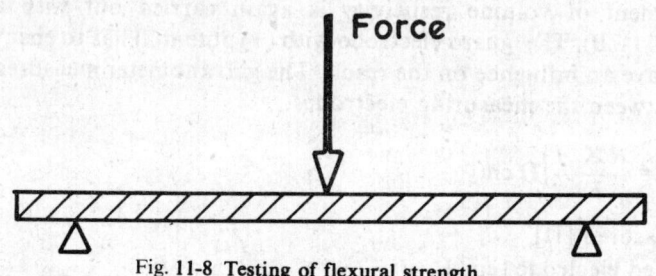

Fig. 11-8 Testing of flexural strength

Water Absorption

The amount of water absorbed by a sample of specified size immersed in distilled water for a specified period (generally 24 h) at a specified temperature (e.g., 20 to 25°C), is reported as water absorption in milligrams. The gain in weight can also be expressed as the percentage of increase over the initial weight.

This test, simple in nature, can only be used as comparator for various grades. It is difficult to draw an absolute and meaningful conclusion from the results. Certain electrical properties such as volume resistivity or dissipation factor after humidification are possibly better indicators for the moisture resistance of a composite material.

Thickness

The extent of manufacturing thickness *variation* and its control within limits is of particular significance when the PCBs are used with edge connectors.

Based on the experience in achieving nominal thicknesses in laminating, a maximum allowed departure is laid down in the form of *tolerances*, in all specifications. Very close tolerances required for multilayer circuits are obtained by making the laminates on precision presses of smaller sizes.

Warp and Twist

Warp or bow is the warpage along the edge of a sheet and twist is the warpage along the diagonal. Excessive warpage or twist presents problems at various stages of PCB making and assembly.

Copper-clad laminates are composites of two dissimilar materials, viz., copper foil and a synthetic resin-bonded base laminate with widely different thermal expansion/contraction characteristics as well as moisture resistance. The copper foil is impervious to moisture whereas the base materials absorb or loose moisture depending upon the environmental conditions. On account of these factors, copper-clad laminates exhibit a certain degree of departure from flatness in the form of warp and twist.

When the copper is etched away from a PCB surface, locked stresses are removed to a certain extent and dimensional changes occur in the card.

Flame Resistance

The electronics industry is becoming concerned with the problem of inflammability of materials used in the equipment. There is a great need to make laminates more flame-resistant.

When a laminate is in contact with a flame or a spark, it gets hot enough to cause a breakdown of the polymer structure. This evolves volatiles. The rapid oxidation of these free radicals produces flames. The reactions produce heat which further degrades the laminate and thus keeps the cycle going on and the laminate burning. Phenolic resins used in laminates are deliberately formulated to reduce the density of molecular cross-linking to give it enough flexibility and softness required for good punchability. This, unfortunately, increases the inherent flammability of the base laminate.

Flame retardant laminates are made by using glass cloth or special papers with resins to which special chemical flame-retardants are added. These additives decompose in the flame and react with the free radicals and this puts out the flame. Flame-retardants are carefully chosen and formulated. Extremely stable compounds which do not put out the flame or unstable ones which decompose at low temperature are avoided. Some flame-retardants reduce insulation resistance, some reduce moisture resistance, some reduce punchability. All the suitable retardants increase the laminate cost. These are the reasons why laminates are made to meet a particular flammability test or specification, rather than to have the maximum flame resistance possible.

Punchability

Among all the mechanical properties required for processing of copper-clad laminates to printed circuit boards, the most desirable may be punchability at room temperature. Punching is the simplest and quickest method of producing complex shapes from the laminate sheets.

Good punchability means no cracking, no lifting around the punched holes, smooth edges and smoothness inside holes, combined with a low shrinkage during and after punching especially when laminates are punched at elevated temperatures.

Over the years, improved resin systems have been developed, along with more suitable papers. They are combined under specific impregnating and pressing conditions to give maximum punchable laminates. Punchability is not achieved only by making laminates softer. Laminates must remain dimensionally stable and because of the more and more automated machinery, also free of warp and twist.

11.3.4 Copper Surface Standards

During the assembly of a laminate, when packs are introduced between steel plates, hard and minute particles (mostly the brittle resin flashes) get entrapped. At the time of pressing, such particles result in dents and pin-holes.

However, since almost 90% of the copper area gets etched out on a printed circuit board, the chances of such relatively rare defects occurring just on the critical (used) copper area are quite low. Also, the conductor widths are normally quite safe and actually over-dimensioned, with respect to the minimum width required as per the necessary current-carrying capacity. Therefore, a small hole, even in the used copper area usually does not matter.

Acknowledging that some incidence of pits, dents and pin-holes is inevitable, maximum tolerable limits for the defects have been specified by various sources. Table 11-2 gives such limits, when the copper-clad laminate is inspected in accordance with the test method prescribed by IEC 249-2A.

Table. 11-2 Type, size and permitted number of imperfections [IEC].

Type	Size (Length unless otherwise indicated)		Number of permitted imperfections	
			In any sheet of about 1 m^2 area	In any area of 300 × 300mm
Inclusions	> 0	$\leqslant 0.1$mm	any number	any number
	> 0.1mm	$\leqslant 0.25$mm	30	4
	> 0.25mm		0	0
Indentations	> 0	$\leqslant 0.25$mm	any number	any number
	> 0.25mm	$\leqslant 1.25$mm	13	3
	> 1.25mm	$\leqslant 3.0$mm or width = 1.0mm	3	1
	> 3.0mm	or width > 1.0mm	0	0
Bumps	> 0	$\leqslant 0.1$mm	any number	any number
	> 0.1mm	$\leqslant 4.0$mm or height = 0.1mm	10	2
	> 4.0mm	or height > 0.1mm	0	0

Notes:— Wrinkles or blisters of any size are not permitted.
— Total area of individual pinholes in any $0.5mm^2$ not to exceed the area of a circle with 0.125mm diameter.
— No scratches with greater than 0.01mm depth are permitted. Scratches with a depth of less than 0.01mm but more than 0.005mm should not be more than 1m in total length in an area of $1m^2$.

11.4 TYPES OF LAMINATES

The numerous grades of laminates obtained by the use of different fillers and resins are not merely random combinations. They make the best possible use of a particular filler and a matching resin to suit or fulfill definite requirements of the industry. Table 11-3 gives an indication of the range of properties for various types of laminates.

11.4.1 Phenolic Laminates

Phenolic resins usually consist of a solution of reaction product of phenol and formaldehyde in a solvent. The simplified reaction of phenol and formaldehyde to form a phenolic resin is shown in Fig. 11-9. As a partial or full alternative to the use of phenol in the resin, it is also possible to react cardphenol or cresol with formaldehyde. The phenolic resins are among the oldest plastics and are in wide use. They are inexpensive, non-hazardous and non-toxic.

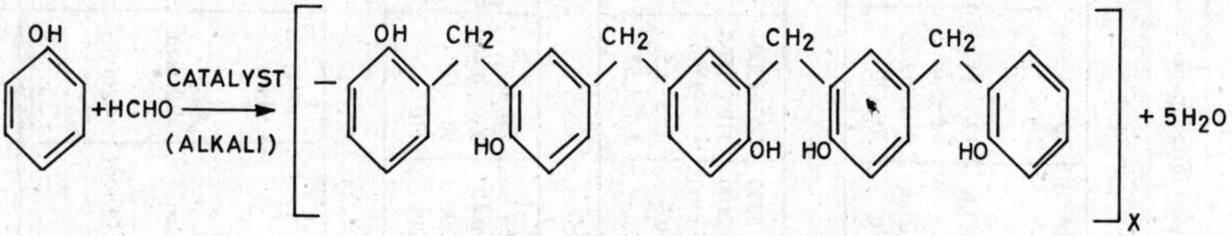

Fig. 11-9 Simplified reaction of phenolic resin

Phenolic resins are reinforced with paper fillers for copper-clad laminates. They are stable enough under a variety of conditions to suit a majority of applications. They definitely have an edge over other types of laminates as regards punchability and the ease of fabrication. These factors, coupled with their low cost and the long experience with them have made them the largest single class of copper-clad laminates in commercial use.

Since the phenolic resins are light or dark brown in colour, the use of a bleached variety of paper is necessary wherever some translucency is desired. Special additives are added to attain the extra properties of flame resistance and/or punchability at room temperature. Commercially available grades are hot punching variety, room temperature punching variety (cold punching) and flame-resistant variety, corresponding with the NEMA grades of XXXP, XXXPC and FR-2 respectively.

Paper phenolic copper-clad laminates are attacked by strong alkalis; acids have only a slight or no effect on them, depending on their concentration. Organic solvents do not affect the paper phenolic copper-clad laminates. One of the drawbacks of the paper phenolics is their poor arc resistance. Regarding moisture, phenolic laminates can be made with good resistance

Table 11-3 Chart indicating properties of various compositions of laminates. Extracted from [2]

PROPERTY	PHENOLIC		EPOXY		PTFE	MELAMINE		POLYESTER		SILICONE
	paper base	glass fabric	paper base	glass fabric	glass fabric	glass fabric	paper base	glass fabric	paper base	glass fabric

ELECTRICAL

Insulation resistance [MΩ]	35×10^6	$5 \times 10^4 - 10^6$	$10^5 - 5 \times 10^5$	$2 \times 10^4 - 10^5$	10^5	—	—	—	—	$2.5 \times 10^4 - 5 \times 10^6$
Volume resistivity [MΩ·cm]	$10^4 - 10^7$	10^9	4.5×10^4	—	—	—	—	—	10^7	—
Dielectric strength, step by step, thickness 3.2 mm [kV/mm]	10–32	11.8–23.6	12.7–21.2	9.8–27.5	5.9–17.7	8.6–23.6	15.7–23.6	5.9–13.7
Dielectric constant at 1 MHz	3.6–6.0	4.5–5.3	4.0–5.0	2.4–2.7	6.0–9.0	3.0–4.0	3.0–4.2	3.7–4.3
Dissipation factor at 1 MHz	0.02–0.08	0.01–0.03	0.01–0.025	$8 \times 10^4 - 5 \times 10$	0.011–0.025	0.007–0.03	0.02–0.03	0.005–0.01
Arc resistance [sec]	4–75	15–180	30–120	180	175–200	80–140	28–75	150–250

PHYSICAL

Tensile strength [kp/cm²]	550–1300	2300–5800	700–1300	810–1350	1700–5800	1220–4400	410–1000	680–2550
Flexural strength [kp/cm²]	700–2000	2700–7100	1100–1600	750–1150	2400–5800	850–6100	880–1900	680–2600
Shear strength [kp/cm²]	400–1000	1000–1700	610–830	750	1300–2400	810–1550	—	1120–1350
Water absorption [%]	0.2–6.0	0.04–0.3	0.15–0.5	0.02	0.2–2.5	0.15–2.5	0.1–5.0	0.07–0.65
Specific gravity [kg/dm³]	1.28–1.4	1.7–2.0	1.4–1.52	2.2	1.82–1.98	1.5–2.1	1.2–1.5	1.6–1.9

THERMAL

Thermal conductivity [10⁻⁴ cal/sec/cm²/°C/cm]	1.8–7.0	7.1–7.4	7.0	1.0–2.0	7.12	—	—	7.0
Specific heat [cal/°C/g]	0.38–0.41	0.35–0.40	0.35–0.40	0.2	0.23–0.40	0.25–0.26	—	0.25–0.27
Resistance to heat [°C] electr.	105–140	105–200	90–110	290	50–105	120–200	120–150	280
(UL temp. index rating) [°C] mech.	105–140	105–200	110	290	105–140	120	120	280

CHEMICAL

Effect of weak acids	none	none	none	none	none	none	none	none
Effect of strong acids	slight	slight	slight	attacked by HF only	—	some attack	some attack	some attack
Effect of weak alkalies	slight	slight	slight	none	none	slight to none	slight	very slight
Effect of strong alkalies	attacked	attacked	attacked	attacked	slight	some to severe	attacked	attacked
Effect of organic solvents	none	slight to none	slight	none	none on bleed-proof materials	some to severe	none	attacked by some

Notes: — Above values are based on test methods of ASTM
— Consult manufacturers before making a choice of the laminate

to water absorption but it is of much lower order when compared with the one of epoxy laminates. Effects of moisture on surface resistivity is shown in Fig. 11-10.

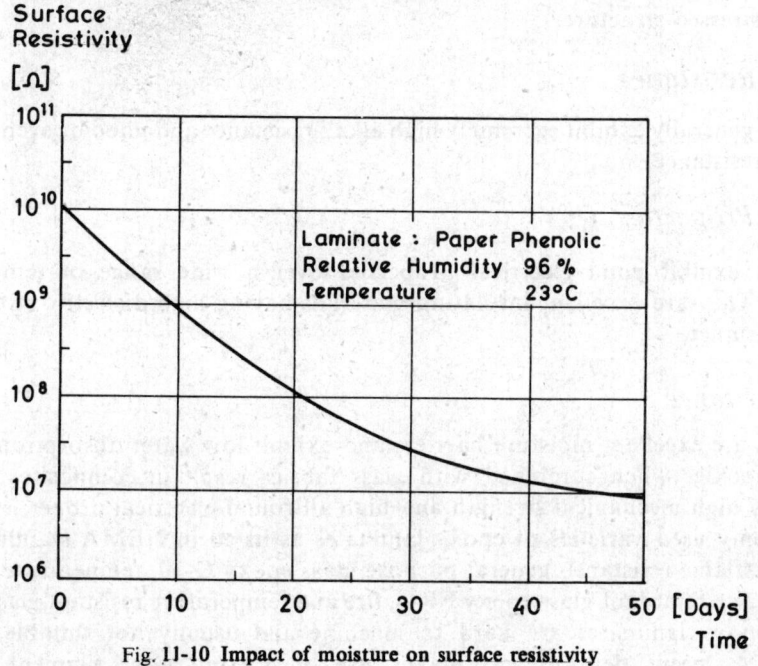

Fig. 11-10 Impact of moisture on surface resistivity

11.4.2 Epoxy Laminates

Epoxy resins became commercially popular in the 1950s. They are an important and rapidly growing class of resins. Generally, epichlorohydrin and a bivalent phenol is reacted to give the base resin. This again is reacted with suitable hardeners and accelerators to give a cross-linked product. None of the other resins has such a good combination of desirable attributes as available with epoxies:

Low Shrinkage

Epoxies react with their hardeners by direct addition, without evolution of volatile by-products and with very little chemical rearrangement. Unlike phenolics and polyesters, the epoxies exhibit very low shrinkage during cure.

Toughness

Cured epoxies are tough materials due to the long distance between cross-linking points and the presence of integral aliphatic chains. Phenolics and polyesters tend to be very brittle by comparison.

Adhesion

The chemical nature of epoxies, i.e., the polar hydroxyl and ether groups present, causes an outstanding adhesion. The low shrinkage further assists in the formation of a strong bond.

Mechanical Properties

Epoxies exhibit high mechanical properties, partly due to their low shrinkage and relatively unstressed structure.

Chemical Resistance

Epoxies generally exhibit extremely high alkali resistance and good or even excellent acid and solvent resistance.

Electrical Properties

Epoxies exhibit good electrical properties over a wide range of temperatures and frequencies. They are excellent insulating materials having high dielectric strength, arc and tracking resistance.

Water Resistance

Epoxies are excellent moisture barriers and exhibit low water absorption.

Thus, epoxies when combined with glass fabrics result in composites which exhibit exceptionally high mechanical strength and high all round electrical properties.

Commonly used varieties of epoxy laminates as listed in NEMA standards are paper epoxy FR-3 (flame resistant), general purpose glass epoxy G-10, temperature resistant glass epoxy G-11, fire retardant glass epoxy FR-4, fire and temperature resistant glass epoxy FR-5.

Glass epoxy laminates are hard to machine and usually not suitable for punching operations. A recent development makes use of a continuous filament glass mat as reinforcement. This is reported to give good punchability.

11.4.3 Polyester Laminates

Polyesters are solutions of unsaturated polyester resins in copolymerisable monomers such as styrene, etc. They are exclusively used with glass fibre reinforcement. The resins themselves possess good arc and track resistance which can be further enhanced by addition of fillers and additives. Their dimensional stability and water resistance are also good. They are not in extensive use in electronics industry since epoxies can often supply better dielectric properties.

Commercially available polyester glass laminates are claimed to comply with FR-6 grade of NEMA.

11.4.4 Diallyl Pthalate Laminates

They are glass reinforced and used almost exclusively as insulation in electronics. Their low-loss characteristics are attractive over a wide range of frequencies and up to their maximum temperature limit of 165°C. These types of laminates are specified in MIL-P8013 C.

11.4.5 PTFE Laminates

Polytetrafluorethylene is a thermoplastic which, when reinforced with glass, results in a laminate which has a low dielectric constant and a low dissipation factor under a wide temperature, humidity and frequency range. The electric strength is very high; PTFE laminates are therefore an excellent electrical insulation under severe environmental conditions. PTFE laminates are often used for very-high-frequency and microwave applications but also where high insulation resistance under humid conditions has to be maintained (e.g., input stages for high-impedance/low-current measuring instruments and amplifiers). However, the high costs for PTFE laminates rule out their widespread use.

11.4.6 Silicone Laminates

A small portion of copper-clad laminates are made of silicone resins with glass reinforcements. Silicone resins contain silicone, carbon, oxygen and hydrogen and are outstanding in their heat resistance. The glass-based laminates have excellent arc resistance as well as good electrical properties up to 250°C for an extended period and at higher temperatures for short periods of time. The typical comparative tracking index is 320 volts (IEC test methods). The silicone laminates are specified in MIL-P-997 type GS.

The silicones are relatively stable under heat. But they do not possess outstanding characteristics of strength at any temperature. Consequently, mechanical properties of a laminate containing silicone are not superior compared to the phenolics and epoxies.

It is also difficult to obtain a good copper foil-to-base material bond in the silicone-resin system.

11.4.7 Melamine Laminates

Melamine resins can be combined with a variety of reinforcing fillers but best properties result when glass fabric is used. The most significant property offered by these laminates is a high arc-resistance.

MIL-P-15037 type GM is a copper-clad glass melamine variety.

A major disadvantage of melamine glass laminate is the poor dimensional stability particularly when exposed to alternating cycles of high and low humidity. It is also difficult to obtain higher mechanical strengths with melamine-resin systems.

11.4.8 Polyimide Laminates

Polyimide is one of the most heat-resistant polymer known. The excellent retention of mechanical and electrical properties at elevated temperatures is due to the fused-ring nature of the aromatic constituents. Reinforced with woven glass fabric, the laminates find use in demanding military and aerospace applications and in special multilayer circuits. These laminates retain a higher copper bond strength at soldering temperatures than the general purpose epoxy materials. Epoxy laminates have very low peel strength at higher temperatures which can lead to lifted conductors after soldering.

11.5 SPECIFICATIONS AND TEST METHODS

Specifications covering all types of laminates are listed in Tables 11-4 and 11-5.
Important properties of the copper-clad grades, specified by IEC and IS, are listed in Table 11-6. Table 11-7 gives the properties of various grades by NEMA.

Table 11-4 List of major specifications applicable to copper-clad laminates

Number of specification	Content, remarks
BS 4584	These are specifications published in the U.K. titled 'Specifications for Metal-Clad Base Materials for Printed Circuits'. They consist of Part 1 giving details of methods of test and Part 2 to 8 with detailed specifications of various grades.
DIN 40802	German standard for copper-clad laminates.
IEC 249-1 Part 1	International Electrotechnical Commission—Test Methods
IEC 249-2 Part 2	International Electrotechnical Commission—Test Methods 5 detailed spec's covering different grades and their specified properties.
IS 5921 Part I	Indian Standards Institution. Test methods and general requirements of copper-clad laminates.
IS 5921 Part II	Indian Standards Institution. Detailed spec's for paper phenolic, economic grade.
IS 5921 Part III	Indian Standards Institution. Detailed spec's for paper phenolic, electrical grades.
IS 5921 Part IV	Indian Standards Institution. Proposed to be published. Spec's for glass epoxy grades.
JSS 51700	Joint Services Specifications, Ministry of Defence (India)— General requirements of copper-clad laminates.
JSS 51701	Joint Services Specifications, Ministry of Defence (India)— Detailed specifications of various grades.
MIL	Military Standard (USA). Some numbers and grades are listed in Table 11-5.
NEMA LI-1-1971 (R-1976) Part 10	National Electrical Manufacturers' Association (USA). Part 10 lists various grades of copper-clad laminates, their properties and test methods.

Table 11.5 List of some MIL standards (USA)

Number of specification	Description
MIL-P-13949D Type PH	Paper base, epoxy resin, hot punch, flame retardant
MIL-P-13949D Type PX	Paper base, epoxy resin, flame retardant
MIL-P-13949D Type GB	Glass fabric base, epoxy resin, temperature resistant
MIL-P-13949D Type GC	Glass fabric and non-woven fabric base, polyester resin, flame retardant
MIL-P-13949D Type GE	Glass fabric base, epoxy resin, general purpose
MIL-P-13949D Type GF	Glass fabric base, epoxy resin, flame retardant
MIL-P-13949D Type GH	Glass fabric base, epoxy resin, temperature resistant and flame retardant
MIL-P-13949D Type GP	Glass fabric base, polytetrafluroethylene resin
MIL-P-13949D Type GT	Glass fabric base, polytetrafluroethylene resin
MIL-P-22324A Type PEE	Paper base, epoxy resin, flame retardant
MIL-P-19161A Type GTE	Glass fabric base, polytetrafluroethylene resin
MIL-P-997 Type GS	Glass fabric base, silicone resin
MIL-P-15037 Type GM	Glass fabric base, melamine resin
MIL-P-8013C Types 1,2,3	Glass fabric base, polyester resin

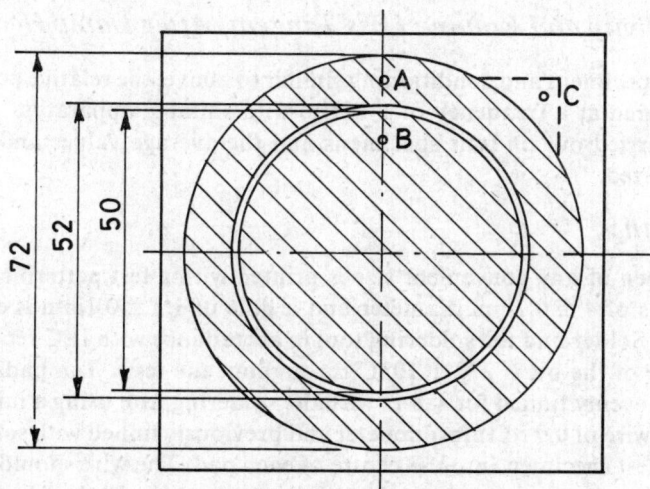

A,B,C = Points for Wire Attachment
Dimensions in mm

Fig. 11-11 Ring and disc pattern

11.5.1 Test Method Details

Descriptions of some test methods as specified by IEC and IS are given here.

Surface Resistivity After Damp Heat, Steady State

A ring and disc pattern as shown in Fig. 11-11 is used for the test. If the sample is of the type having copper foil on both sides, the copper foil on the other side is either completely removed or left intact.

Four 100 mm square test specimens are used and the *minimum* value of measurement on 4 specimens is reported as the value obtained.

Appropriate electrode attachments are carried out on the sample as illustrated in Fig. 11-7A. The specimen is subjected to the conditioning of 40°C at 90 to 95% RH for 4 days. Measurements are either carried out after recovery, by leaving the assembly recovering under prescribed standard atmospheric conditions for 1 to 2 hours or in the humidity chamber itself. The voltage applied for the test is 500 ± 50 V (DC) for 1 minute.

Volume Resistivity After Damp Heat, Steady State

A ring and disc pattern like the one used for surface resistivity is etched on four specimens. The test is carried out in a similar way to that of surface resistivity, however, with an appropriate arrangement of the electrodes as shown in Fig. 11-7B.

The minimum value of measurement on 4 specimens is reported as the value obtained on the sample.

Volume Permittivity and Volume Loss Tangent, After Damp Heat, Steady State

By using the specimens and conditioning similar to above, the relative permittivity and loss tangent are measured at a frequency of 1 MHz, with suitable apparatus.

The test is carried out on four specimens and the average values and maximum values obtained are reported.

Pull-off Strength

A test specimen of any convenient size is printed with a test pattern consisting of ten or more circular pads of 4 ± 0.1mm diameter and a hole of 1.3 ± 0.1mm is drilled through the centre of the pad. Solder and the soldering tool in accordance with IEC recommendations are used (temperature of the bit is $270 \pm 10°$C throughout the test). Ten pads are tested.

The pads are evenly tinned for 4 ± 1s with the soldering iron using a minimum amount of solder. A piece of wire of 0.9 to 1mm diameter and previously tinned with solder is soldered at a right angle to the test specimen into the centre of each pad. The wire should be attached to the foil in such a position that it passes through the foil into the drilled hole. The solder joint formed between the wire and the pad has to cover the entire area of the pad. The time taken for this soldering process shall be 4 ± 1s. During this soldering process and the subsequent cooling, the wire should not be moved. To ensure this, the wire and the test specimen may be held in a jig.

By means of a tensile testing machine, a load is applied by pulling the wire at a right angle to the test specimen. The load action is steadily increased at a rate of 5N/sec up to the maximum of 50 N until the pad separates from the base material.

The smallest of any of the loads required to detach the ten pads from the base material is reported as the pull-off strength. Breaking of a wire at or below the required values or wire pull-out is not to be considered as a failure but the test should be repeated. In no case should the same wire and the same pad be resoldered and pulled again.

Peel Strength

A test specimen of dimension $75 \pm 1mm \times 50 \pm 1mm$ is printed with the pattern as illustrated in Fig. 11-12. The number of specimens used must be sufficient for the peeling of a length of 25 mm for each of the four strips of foil.

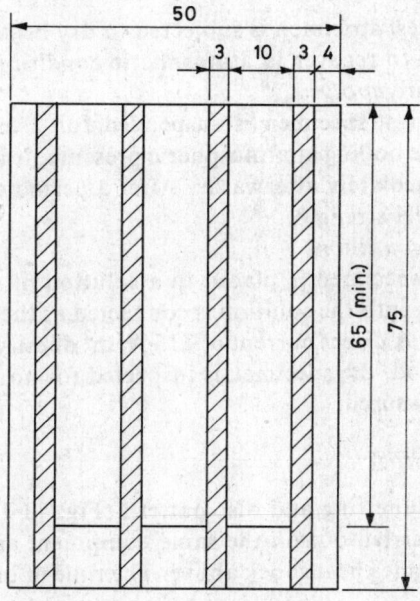

Dimensions in mm

Fig. 11-12 Specimen for peel strength measurement (IEC and IS)

The specimen is subjected to various conditions as described under a) to d) below. If it does not get blistered, the peel strength is measured as follows:

The copper foil is detached from the base material for a distance of about 10 mm from one end. The test specimen is supported in a suitable way, e.g., on rollers with the detached end of copper foil between them. The detached end of the foil is gripped over its entire width with a clamp.

Steadily increasing pull is applied in a direction perpendicular to the plane of the base material until the copper foil is peeled off at a steady rate of about 50 mm per minute. The force required to do this is measured. A deviation of maximum 5° from the vertical is tolerated. Foil of at least 25 mm length is peeled at this rate from each of the four strips.

The minimum load per unit width required to peel the foil during the test is taken as peel strength.

The conditions applied to the specimen prior to the peel strength measurement are given below:

a) *Heat Shock*

A bath of well stirred silicone or equivalent fluid, kept at $260 {}^{+5}_{-0}$ °C throughout the test is used. The temperature is measured at 25 mm below the surface.

The specimen is held in a horizontal position at a depth of 25 mm in a holder of a low heat capacity so that the temperature of the fluid is not brought below 260°C. The specimen is totally immersed in the fluid for the time specified. It is then allowed to cool between 15°C and 35°C and the peel strength is measured.

b) *Dry Heat*

The peel strength test specimen is subjected to dry heat in a drying chamber for 500 hours and allowed to recover in atmospheric conditions for 1 to 2 hours.

c) *Exposure to solvent vapour*

The peel strength test specimen is suspended for 2 minutes±5 sec. in the vapour of trichloroethylene boiling at atmospheric pressure. It is examined for delamination and blistering immediately afterwards. After a period of 24 hours, it is used for the measurement of peel strength.

d) *Simulated plating conditions*

The peel strength specimen is placed in a solution of sodium sulphate at 70°C. A carbon rod dipping into the solution is connected as the anode and the test specimen forms the cathode. A direct current of 215 A/m^2 density is passed for 20 minutes. At the end of this period, the specimen is inspected for nonadherence of the foil and its peel strength is measured.

Blistering After Heat Shock

One test specimen with the ring and disc pattern (Fig. 11-11) is used for this test.

The test is otherwise carried out on the same equipment and along similar lines as the conditions of peel strength under heat shock above. The time of immersion is the same as given in the relevant specification.

Immediately after removal, the specimen is inspected for blistering of the foil and delamination of the base material. The report states whether blistering or delamination has occurred on the specimen. A border of 1mm around the edge of the specimen is excluded from this testing.

Flammability

The material under test should first be heated to a sufficiently high temperature and for enough time to permit complete relaxation of stretching occurred during this processing. Four specimens of $125 \pm 5mm \times 13 \pm 1mm$ size are used after complete removal of the copper foil by etching. The edges have to be smooth. Each specimen is marked by scribing a line across it 25 ± 0.5 mm from one end.

The test is conducted in an atmosphere free from draught. The specimen is clamped in a rigid support at the end farther from the scribed line so that the longitudinal axis of the specimen is horizontal. Its transverse axis is inclined at $45 \pm 10°$ to the horizontal and the line on the specimen can be seen. A piece of clean wire gauze with 8 mesh/cm and 100 mm square is clamped in a horizontal position 10 ± 1mm below the specimen with 13 ± 1mm of the unsupported end of the specimen projecting beyond the edge of the gauze.

A bunsen burner of diameter about 10 mm and with a non-luminous flame about 25 mm high is placed under the free end of the specimen so that the tip of the flame just touches it. The burning time is measured in seconds from the instant of burner flame removal to that of the flame on the specimen going out.

Observation is made whether burning proceeds past the scribed line.

The test report states:
—The average of the four burning times.
—Whether the burning of any of the specimens proceeds past the scribed line.
—Whether the material melts or drips and if it drips also whether it burns.
—The thickness of the sheet under test.

Apart from the above IEC method, the Underwriters Laboratory method to measure the flame resistance is widely accepted and is given in Section 11.5.3. The NEMA method of rating the flammability is on similar lines to Underwriters Laboratory.

Flexural Strength and Water Absorption

The methods for these tests as such are not described in much detail here; they are simple and widely known.

The specimen for flexural strength measurement is 25 mm wide and at least 20 times thickness in length. The span of support is 16 times the thickness. Five specimens, each corresponding to A and B directions are used and the minimum average of flexural strength in two directions is reported.

In case of water absorption, three specimens of 50 mm × 50 mm size are used. They are subjected to immersion in distilled water for 24 hours at 20°C and the average value of water absorbed is reported in milligrams.

11.5.2 Solder Float Test by NEMA

This test is carried out on both etched and unetched specimens. The pattern to be etched in case of etched specimen is as in Fig. 11-13.

The specimen is floated in clear molten solder at the time and temperature specified in the specification for the grade being tested and with the copper side in contact with solder. The specimen is then removed, allowed to cool and examined for interlaminar blistering of the base material and blistering and delamination of the copper foil.

11.5.3 Underwriters Laboratories Tests for Flame Resistance

UL 94 is the standard for flammability testing of plastic materials. The tests are conducted either in horizontal or vertical positions. The materials are classified as 94 HB, 94 V-0, 94 V-1, 94 V-2.

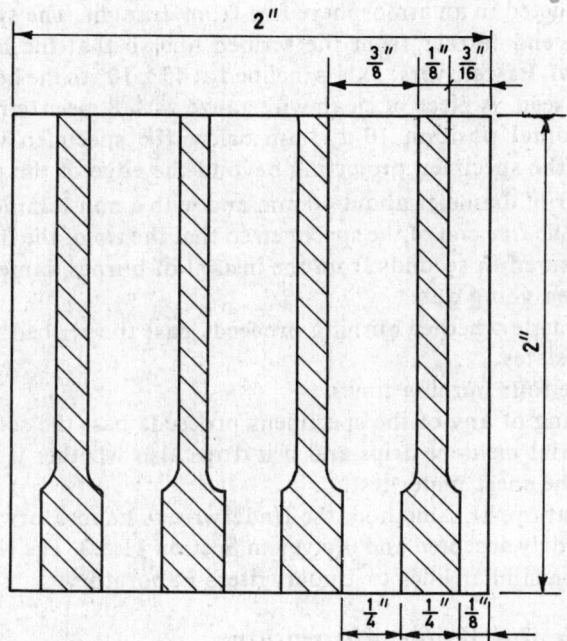

Fig 11-13 Pattern of etched specimen for solder float test (NEMA)

Horizontal Burning Test (94 HB)

Fig. 11-14 shows the general arrangement for this test. The equipment used and test set-up is described in detail in the standard.

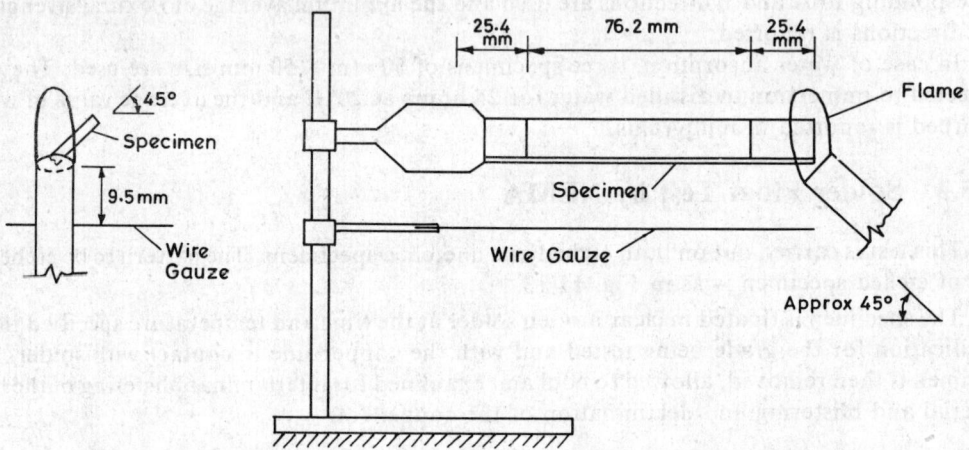

Fig. 11-14 Arrangement for UL horizontal burning test for '94 HB' classification

The test is conducted in a chamber free from draught. Three specimens of 127 mm length and 12.7 mm width are marked with two lines 25.4 mm and 102 mm from one end. They are conditioned for 48 hours at 23°C and 50% RH.

After obtaining a specified flame on a bunsen burner, it is applied at the free end at the lower edge of the specimen. The specimen, to an approximate depth of 6.4 mm, is subjected to the test flame for 30 seconds. If the specimen continues to burn after removal of the test flame, time for the flame front to travel from 25.4 mm mark to 102 mm mark is to be determined and the rate of burning has to be calculated.

The materials classed 94 HB should meet the following:
a) They should not have a burning rate exceeding 76.2 mm (3 inches) per minute over a 76.2 mm span (for specimens having thickness less than 3.05 mm).
b) They should cease to burn before the flame reaches the 102 mm reference mark.

Vertical Burning Test (94 V-O, 94 V-1, 94 V-2)

Five test specimens of 127 mm length and 12.7 mm width are used. They are conditioned for 48 hours at 23°C and 50% RH and for 168 hours at 70°C followed by cooling in a desiccator for 4 hours.

In a draught-free chamber, the specimen is supported from the upper 6.4 mm vertically by means of a clamp and ring stand. Its lower end is 305 mm above a horizontal layer of dry absorbent surgical cotton.

After obtaining a specified flame on the bunsen burner, the flame is centrally placed under the lower end of the specimen for 10 seconds. The test flame is then withdrawn and the duration of flaming of the specimen noted. When the flaming of the specimen ceases, the test flame is immediately placed again under the specimen and withdrawn after 10 seconds. Duration of flaming and glowing of the sample is noted. If the specimen drips molten or flaming material during either flame application, the burner may be tiled up to 45 degrees. The materials classified as 94 V-0, 94 V-1 or 94 V-2 have to meet the following requirements:

a) Specimens should not burn with flaming combustion after either application of the test flame for more than
10 seconds (94 V-0)
30 seconds (94 V-1, 94 V-2)
b) For each set of 5 specimens, herewith for totally 10 flame applications, total flaming combustion time should not exceed
50 seconds (94 V-0)
250 seconds (94 V-1, 94 V-2)
c) Specimens should not burn with flaming or glowing combustion up to the holding clamp.
d) The 94 V-0 and 94 V-1 material specimens should not drip flaming particles that ignite the dry absorbent surgical cotton below.
The 94 V-2 material specimens drip flaming particles which burn only briefly, some of which igniting the dry absorbent surgical cotton below.
e) After the second removal of the test flame, the specimens with glowing combustion should not persist beyond
30 seconds (94 V-0)
60 seconds (94 V-1, 94 V-2)

BIBLIOGRAPHY

1. COOMBS C.F.: *PRINTED CIRCUITS HANDBOOK,* McGraw-Hill Book Co., New York, 1967.
2. *GUIDE TO PLASTICS, 1978, PROPERTY AND SPECIFICATION CHARTS,* McGraw-Hill Book Co., New York, 1977.
3. HARPER C.A.: *HANDBOOK OF MATERIALS AND PROCESSES FOR ELECTRONICS,* McGraw-Hill Book Co., New York, 1970.
4. MASLOV V.: *MOISTURE AND WATER RESISTANCE OF ELECTRICAL INSULATION,* Mir Publishers, Moscow, 1975.
5. PERRY H.A.: *ADHESIVE BONDING OF REINFORCED PLASTICS,* McGraw-Hill Book Co., New York, 1959.
6. STANDARDS/SPECIFICATIONS
   a) INDIAN STANDARDS INSTITUTION, ISI.
   'Specification for metal-clad base materials for printed circuits for use in electronic and telecommunication equipment'.
   Part I: General requirements and tests, 1971.
   Part II: Paper phenolic copper clad laminated sheets, economic grade, 1974.
   b) THE INSTITUTE FOR INTERCONNECTING AND PACKAGING ELECTRONIC CIRCUITS, IPC.
   'Copper foil for printed wiring applications', Publication IPC-CF-150-B.
   c) INTERNATIONAL ELECTROTECHNICAL COMMISSION, IEC.
   'Metal-clad base materials for printed circuits'.
   No. 249-1: Test methods, 1968.
   No. 249-2: Specifications, 1970.
   No. 249-2 A: Specifications (Supplement): Surface finish, 1971.
   d) NATIONAL ELECTRICAL MANUFACTURERS ASSOCIATION (NEMA).
   Standard Publication No. LI-1-1971 (R 1976).
   e) UNDERWRITERS LABORATORIES INC., UL.
   UL-94, Standard for tests for flammability of plastic materials for parts in devices and appliances, 2nd edition, 1972.

12
BOARD CLEANING BEFORE PATTERN TRANSFER

The cleaning of the copper surface prior to resist application is an essential step for any type of PCB process using etch or plating resist. Insufficient cleaning is one of the reasons most often encountered for difficulties in PCB fabrication although it might not always be immediately recognised as this. But it is quite often the reason for poor resist-adhesion, uneven photoresist-films, pinholes, poor plating-adhesion, etc.

12.1 CATEGORIES OF SOILS

Organic Soils

In this category we find all kind of oils and greases. They seem to be omnipresent. Possible sources causing contamination could be the equipment used for shearing, drilling, punching, the air from the air compressor or contaminated cleaning solutions for such equipment. The classical remedy is degreasing with a suitable solution or a solvent. Also, the manual handling of the laminates will no doubt leave some marks behind like fingerprints and sweat marks. These marks are usually quite difficult to remove because of their corrosive nature. Only scrubbing or etching can help here.

A general characteristic of organic soils is their adherence to the metal surface, the tendency to form a film and to repel water. Their presence makes it even difficult to remove other soils such as oxides, hydroxides and salts.

Metallic Soils

Metallic soils are mainly oxides and corrosion products on the copper surface which are caused by atmospheric attack and residuals remaining after poor rinsing and drying. In particular, the sulphur compounds found in the air of an industrial environment can form metallic soils which are not easy to remove. Therefore, etching and scrubbing with abrasives has to be employed.

Particulate Matters

Under particulate matters, we group the particles from PCB machining and the remnants from the abrasive cleaning which get embedded into the copper surface. Water rinsing and the blowing with compressed air usually removes only the loose particles; hence for a thorough cleaning of particulate matters, wet *brushing* is recommended.

The list of soil categories is not complete; depending on the circumstances, still other contaminants could play a dominant role. But emphasis is given here on the ones appearing almost everywhere in PCB fabrication, and a proper treatment against these three categories will in most of the cases provide the required cleanliness. However, the details of the cleaning procedure applied might slightly change from place to place. This is because the composition of soils from these categories will not be the same at different places and the cleanliness actually required depends on the fabrication processes used.

Thus a PCB, meant for screen-printing, will be less sensitive to contamination than a PCB for photoprinting. And PCBs for photoprinting with wet-film resists are more sensitive to contamination than those for dry-film resists. For economic production, one has to choose the simplest cleaning procedure which fulfills the need for minimum rejects and the specified reliability.

In this chapter, a few guidelines and cleaning procedures are dealt with. The procedures are employed at various places and are some kind of a standard. But also other combinations of cleaning steps are possible and could be the optimum solution for that particular situation.

12.2 MANUAL CLEANING PROCESSES

Where cleaning has to be done with the simplest means or only for a limited quantity of PCBs, the cleaning process of Fig. 12-1 could provide the answer. We require just a sink with running water, pumice powder, scrubbing brushes and suitable tanks. Pneumatically or electrically operated scrubbing brushes will help to reduce the strenuous work and speed up the process.

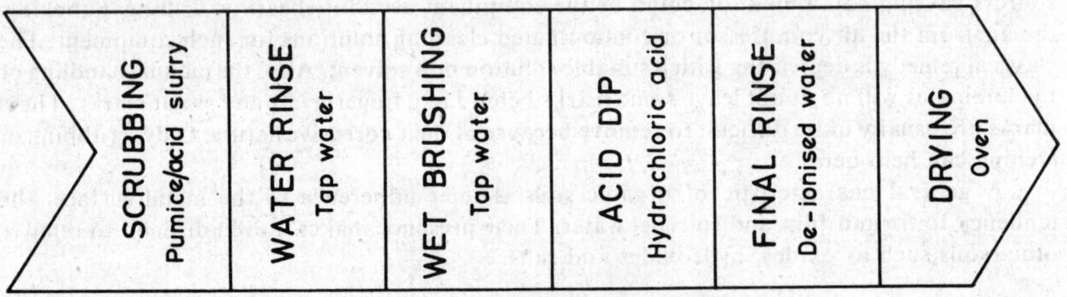

| SCRUBBING Pumice/acid slurry | WATER RINSE Tap water | WET BRUSHING Tap water | ACID DIP Hydrochloric acid | FINAL RINSE De-ionised water | DRYING Oven |

Fig. 12-1 Simple manual cleaning process

The first step, scrubbing with a pumice/salt solution aims at the removal of inorganic matters like particulates and oxides but it also performs degreasing up to a certain extent. Pumice powder is a cheap and well-suited abrasive for PCB cleaning and should always be used in a wet condition. Use a fine grade of pumice to minimise deep scratches on the copper surface. The unavoidable scratches should not be deeper than 0.5 μm for general applications.

A pumice slurry can be formed with water or, for a higher efficiency, with some acid or alkaline solution which will simultaneously improve the degreasing capability.

Acid-type solution to form a slurry with pumice:
  Citric acid   7–15 g/l
  Sodium chloride (salt)   30–45 g/l
  Nonionic wetting agent   0.7 g/l
  Water   to make 1 l

After scrubbing with the abrasive, a *water rinse* will remove most of the remaining slurry. But only a thorough wet brushing with a separate, clean brush removes also the pumice embedded into the copper surface, and not visible to the eye without a magnifying device. The formation of a completely wetted copper surface during rinsing indicates that the water repellant soils also have been removed. In further handling of the cleaned PCBs, all care must be taken not to touch the clean surface. Hold the cards only at the edges and use rubber gloves as far as possible.

The following *strong-acid dip* in hydrochloric acid (10 vol %) will remove residual alkali and metallic oxides and prepare the surface for maximum resist-adhesion.

The *final rinse,* for best results, has to be done by using de-ionised water. Only de-ionised water gives guarantee that no fresh contamination is brought onto the surface. If none or only limited quantity of de-ionised water is available, tap water rinsing is done under the risk of introducing troubles caused by water impurities. Especially tap water with a coloured appearance should be avoided.

Drying is usually done by blowing compressed air over the surface. The compressed air system must include a filter in the air pipe as close as possible to the blowing pistol in order to avoid contamination through oil from the compressor or condensated water from the pipe system. Sometimes it is desirable to keep the PCB for complete drying into an oven and keep it there for about 15 min at a temperature of 90° C.

Now, all dust has to be kept away from the cleaned boards. Any further handling of the cards is preferably done by wearing soil-free gloves. The time span until the next processing step which is resist coating or screen-printing, has to be made as short as possible to minimise the formation of fresh oxides.

If the degreasing capability has to be enhanced, a *solvent cleaning* step can be added at the beginning of the cleaning procedure (Fig. 12-2). With a clean, fuzzle-free piece of cloth, soaked in a solvent, the surface is gently rubbed regularly over the entire PCB area. Very efficient solvents for degreasing are of chlorinated type, such as trichloroethylene, perchloroethylene or trichloroethane. These solvents do not react chemically with the materials they dissolve. A thin layer of contaminants might be redeposited on the surface after the evaporation of the solvent but the following scrubbing and rinsing will remove it.

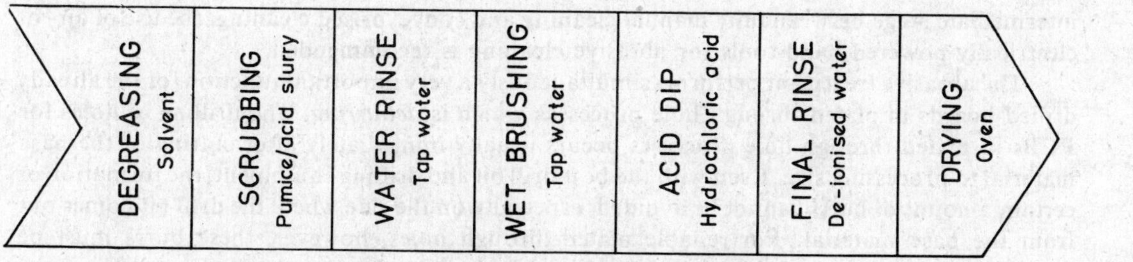

| DEGREASING Solvent | SCRUBBING Pumice/acid slurry | WATER RINSE Tap water | WET BRUSHING Tap water | ACID DIP Hydrochloric acid | FINAL RINSE De-ionised water | DRYING Oven |

Fig. 12-2 Manual cleaning process with solvent degreasing

One more manual cleaning procedure with a high degreasing capability is shown in Fig. 12-3. Instead of using a solvent, degreasing is efficiently carried out by a *soak cleaning* step prior to the acid dip. Soak cleaners react chemically with the organic soils. Soak cleaning is one of the earliest methods used for degreasing and can be of alkaline- or acid-type. Alkaline soak cleaners remove oils and greases quite efficiently nearly like solvents. But unlike solvents, soak cleaners have little tendency to redeposit soils on the surface of the PCBs and there is no severe air pollution threat.

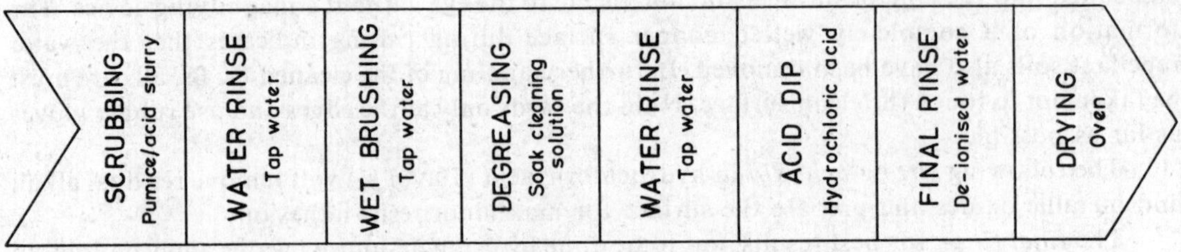

SCRUBBING Pumice/acid slurry	WATER RINSE Tap water	WET BRUSHING Tap water	DEGREASING Soak cleaning solution	WATER RINSE Tap water	ACID DIP Hydrochloric acid	FINAL RINSE De-ionised water	DRYING Oven

Fig. 12-3 Manual cleaning process with soak degreasing

Health Care

The vapours of the mentioned solvents are toxic to a certain extent. The typically first signs of intoxination by such vapours are headache, increased tiredness or diminishing appetite. Especially trichloroethylene is a known air pollutant and its use is therefore less recommended although it is the cheapest solvent in this group. A less harmful alternative is using trichloroethane.

When working with this kind of solvents, all care must be taken not to breathe in the vapours. Sufficient fresh air must circulate over the entire working area and directly into the free space. The maximum tolerable concentration of trichloroethylene vapours in working areas is 100 ppm which should be controlled with a gas detector.

Also, the strong chemicals used for soak cleaning and acid dip, call for suitable measures to prevent health hazards. Please see also Chapter 22 on these aspects.

12.3 MACHINE CLEANING PROCESSES

The question of when to go for mechanising and automising the cleaning process depends on local factors like availability, cost and reliability of manpower. In particular, the investment into automatic and fully conveyorised equipment is heavy and very often the machines remain underutilised after they have been acquired and put into operation. Therefore, as an intermediate stage between fully manual cleaning and conveyorised cleaning, the use of air- or electrically-powered hand-tools for abrasive cleaning is recommended.

The abrasive treatment performs simultaneously a very important function for the already drilled boards in plated-through hole processes which is *deburring*. The drilling of holes for PCBs in plated-through hole processes occurs usually immediately after cutting of the base material to processing size. Even with the best drill bit and drilling equipment, the formation of certain amount of burrs cannot be avoided, especially on the side where the drill bit comes out from the base material. For reliable plated-through holes, however, these burrs must be completely removed. And this should be done by abrasive cleaning.

There are three methods of mechanised abrasive cleaning which are commonly used: abrasive slurry scrubbing, sanding and rotary abrasive brushes. The first two methods are also suitably carried out with powered hand-tools.

Abrasive Slurry Scrubbing

Typically applied equipment for abrasive slurry scrubbing incorporates fast-rotating brushes made of fine plastic bristles. The brushes are continuously fed with abrasive slurry which is recirculated. The slurry may be prepared by adding pumice powder to water. Scrubbing machines require careful cleaning and maintenance because of the abrasive nature of the slurry. Such machines provide an excellent surface finish on the PCB surface and include also a final water brush/rinse cycle to ensure a complete removal of particulate matters and slurry rests.

For boards which are moderately soiled, the treatment in such machines completely cleanes them. If necessary, an acid-abrasive slurry capable of emulsifying light organic soils can be applied. A typical composition of such an acid-abrasive slurry is given here [2]:

Citric acid	7% (weight)
Trisodium phosphate	7% (weight)
Wetting agent, nonionic	0.1% (weight)
Pumice	85.9% (weight)
Use 100–300 g/l	
pH 3–4	

If this method is carried out with a powered hand-tool, the following guidelines have to be noted: The motor spindle shaft should be of light weight. The rotation speed should be kept low to hold the slurry in the fibre brush. None or only little down pressure should be used.

Sanding

Like abrasive scrubbing, sanding too removes heavy inorganic soils and particulate matters. In sanding, however, the abrasives have less tendency to get embedded into the copper surface. Sanding can be done wet or dry but does not remove much of the organic soils. Care must be taken not to remove too much copper.

Sanding as such is a good precleaner step and removes also the burrs around the drilled holes. It is carried out with conveyorised equipment and also with powered hand-vibrator sanders. The use of 280–600 grit water resistant abrasive paper is recommended for hand-vibrator sanders.

Sanding is usually followed by degreasing in a soak cleaner.

Rotary Abrasive Brushes

This type of machine is gradually taking over the field of abrasive cleaning. The rotary brushes are made of abrasive material (e.g., 3M Scotchbrite), available in different gradings such as used for grinding, polishing and finishing of the copper surface. In many designs of such equipment, the brushes also oscillate for a maximum uniformity of the process. The machines permit a clean operation and the investment is usually lower compared to other conveyorised abrasive cleaning machines because of the simple design. During operation, fresh water flows continuously over the boards to be cleaned, to remove residuals. Cleaning performance with respect to heavy oxides and other metallic soils as well as removal of particulate matters is excellent and the actual removal of metal is less compared to sanding operations. However, the removal of organic soils such as oils and greases is not complete and also some of these soils can be retained in the brushes. Therefore, the degreasing step with a solvent is recommended before brushing.

A complete cleaning process with rotary brushing, suitable for plated-through hole processes (already drilled holes) is schematically shown in Fig. 12-4 while Fig. 12-5 shows a compact cleaning machine incorporating rotary brushing.

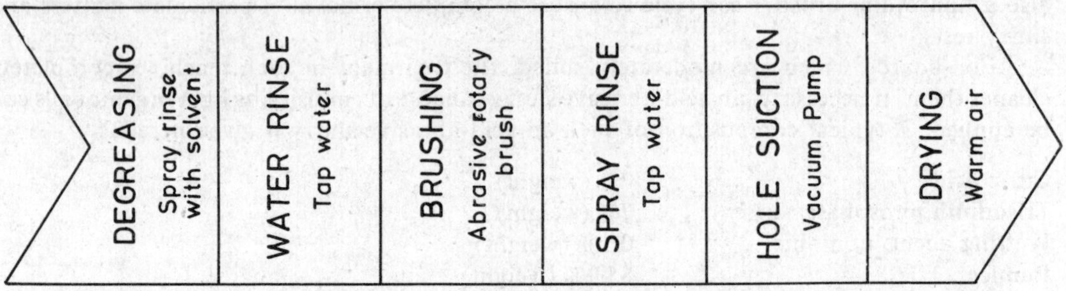

| DEGREASING Spray rinse with solvent | WATER RINSE Tap water | BRUSHING Abrasive rotary brush | SPRAY RINSE Tap water | HOLE SUCTION Vacuum Pump | DRYING Warm air |

Fig. 12-4 Scheme of a typical cleaning process with abrasive brushing for high-volume PCB production with plated-through holes

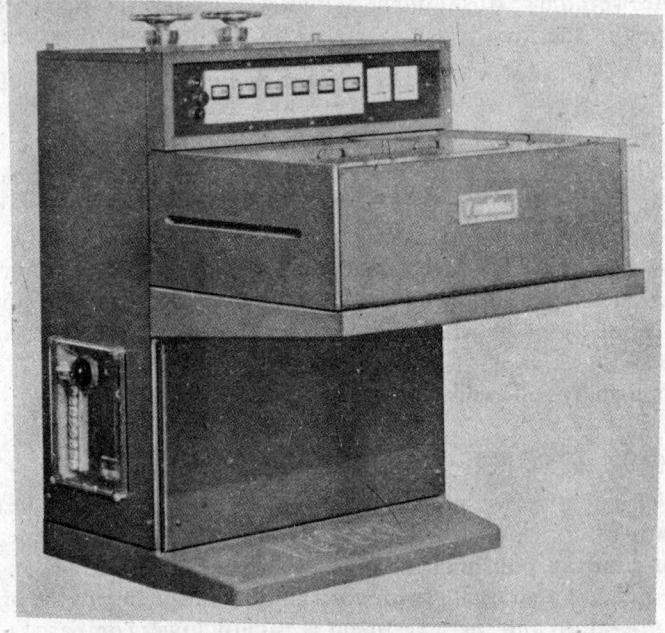

Fig. 12-5 Compact cleaning machine for printed circuit boards
(Courtesy: Wesero GmbH, Sprockhoevel, Germany)

12.4 EQUIPMENT TRENDS

In large-volume PCB production, the use of conveyorised equipment for all possible production steps is attempted. This is particularly in highly industrialised countries with high costs for labour. On the other hand developing countries usually do not have a PCB market of a big volume. And to satisfy the needs of a smaller market, however, several small- or medium-scale plants will probably function much better than one or two large-volume production plants. This must be kept in mind when planning in a developing country.

The latest trends for cleaning equipment in the highly-industrialised countries go for full conveyorisation. Since many of the PCB production steps are of a similar nature (cleaning-stripping) or even repeated (water rinse, drying, etc.), suitable equipment which fits into a *modular concept* has been developed. As long as such modules are from the same manufacturer, they can be added to production lines on an individual basis. The conveyor is driven by an electric motor in one of the modules. But each module, if the motor is provided, could as well work as a free-standing unit. If the process requirements keep changing, new modules can easily be added and others left out if not required.

As an example, the steps of a cleaning process are given schematically in Fig. 12-6. This process can be completely implemented by using modules.

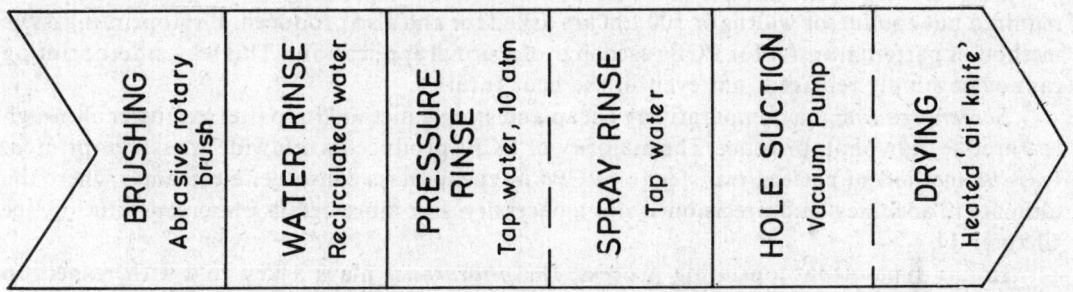

Fig. 12-6 Scheme of another cleaning process with abrasive brushing for high-volume PCB production with plated-through holes.

BIBLIOGRAPHY

1. COOMBS C.F.: *PRINTED CIRCUITS HANDBOOK*, McGraw-Hill Book Co., New York, 1967.
2. DEFOREST W.S.: *PHOTORESIST, MATERIALS AND PROCESSES*, McGraw-Hill Book Co., New York, 1975.

13
PHOTOPRINTING

The transfer of the conductor pattern, which is on the film master, onto the copper-clad base material, is done by two methods mainly: photoprinting and screen-printing.

Photoprinting is an extremely accurate process which is also applied to the fabrication of semiconductors and integrated circuits where the conductor widths are typically in the region of just a few microns (μm). However, in the production of PCBs, such extreme precision is not required but conductor widths of 100 μm are asked for and also produced. Photoprinting is *the* method of pattern transfer for PCBs used in professional applications. This way, photoprinting cannot be simply replaced, not even in the near future.

Screen-printing is a comparatively cheap and simple method for pattern transfer although less precise than photoprinting. The majority of PCBs produced worldwide are screen-printed. It is *the* method of pattern transfer for PCBs in cheap mass consumer electronics where the ultimate in accuracy and precision is not a necessity. For more details on screen-printing, see Chapter 14.

In any type of photoprinting process, the *photoresist* plays a key role with respect to economic production, quality, process reliability and other factors. But before going into more details, a glance has to be given to the very basic PCB fabrication processes currently employed.

13.1 BASIC PROCESSES FOR DOUBLE-SIDED PCB'S

Print-and-Etch Process

The print-and-etch process for single- as well as double-sided PCBs is the earliest process which has made PCB production possible at a large scale (Fig. 13-1). This process was the first answer to the steep increase in using semiconductors towards the end of the 1950s. The conductor pattern is transferred onto the copper foil surface with the help of a screen-stencil or by photoprinting with photoresist. The process is very simple and primarily does not require any plating. The investment into equipment can be kept quite low. In spite of the tremendous progress made in the development of new PCB processes, the number of PCBs made by the print-and-etch method is today still by far the largest and will remain so for some more time because of the low production costs. The PCBs made with this process are typically used in consumer-type electronics like radios, TVs, household appliances, and wherever the price of PCBs is a prime consideration.

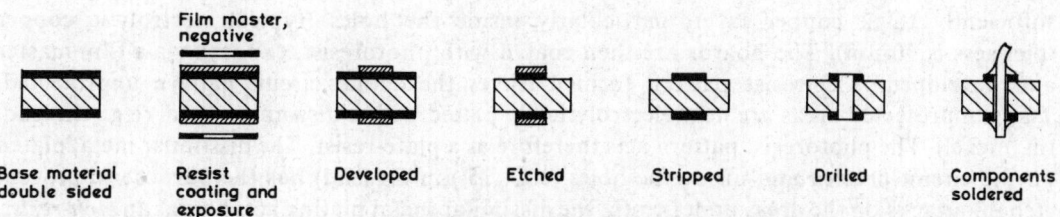

Fig. 13-1 Print-and-etch process

The limitations of the print-and-etch process, if used for double-sided PCBs, are basically in the missing pattern interconnections from one side to the other. If there are only a few interconnections required, they are often done with rivets or with wire pieces, soldered on both the sides of the board. It is also possible to solder the component leads, wherever required, on both the sides to establish contact between the two patterns. This method, however, makes it more difficult to replace or remove the components at a later stage.

Panel-Plating Process

Prior to the introduction of plated-through hole processes in PCB fabrication, the existing metallising baths had to be improved in reliability, stability and performance in order to enable an economic use of them in PCB fabrication. Also, the existing photoresists had to undergo further development to make them more resistant against electrolytic attacks which are more severe than etchant attacks.

To come to a profound understanding of the differences between the various processes for making plated-through hole boards, we have to know the difference between panel- and pattern-plating: In *panel-plating*, the complete copper-clad board surfaces get plated along with the inner surfaces of the drilled holes. The boards have neither been etched before nor do they have a resist pattern on the copper surface. In *pattern-plating*, a plated layer gets built up only on the unprotected circuit pattern areas and inside the holes; the other surface areas, where plating is not required, are protected by a resist image.

In the panel-plating process for plated-through hole boards (Fig. 13-2), the holes have first to be drilled. The boards (panels) go then for the electroless copper panel-plating, followed by the electrolytic copper panel-plating. The electroless copper layer provides a thin but fully conductive board surface which includes the walls of the already drilled holes. This layer enables thereafter to use the considerably faster electrolytic copper plating process to build up a

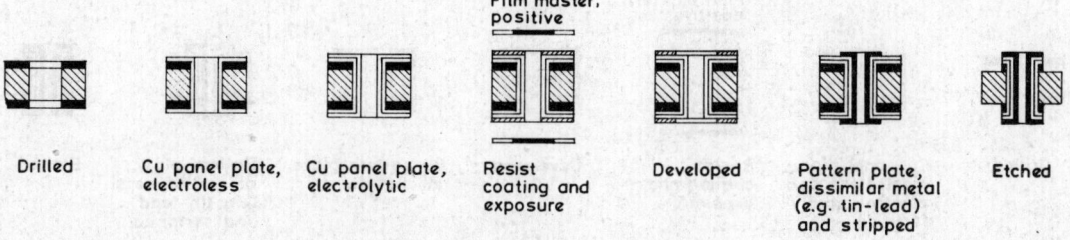

Fig. 13-2 Panel-plating process

sufficiently thick copper layer, particularly inside the holes (typical electrolytic copper thickness is 30 μm). The boards are then coated with photoresist, exposed via a film master and developed. The resist pattern formed leaves the actual circuit pattern unprotected: These unprotected areas are now electrolytically plated with a dissimilar metal (e.g., tin-lead, tin, nickel). The photoresist pattern acts therefore as a plate-resist. The dissimilar metal plated on the circuit pattern and inside the holes (e.g., 15 μm tin-lead) has the purpose to prevent etchant attacks on the areas underneath; the dissimilar metal plating acts here as an *etch-resist*. Etching in a suitable etchant is carried out after the plate-resist (photoresist) has been stripped.

Plate-resists: In Fig. 13-2, photoresist has been shown as a plate-resist. However, for mass production of uncritical plated-through hole boards, a screen-printed resist will probably be more economic. In any case, a very important requirement for plate-resists in such applications is that the resist does not go into the holes which would otherwise disable the plating of dissimilar metal in such areas. This requirement puts serious restrictions on the wet-film coating methods applicable and has helped dry-film resists to gain popularity.

The described panel-plating process is less often used because it has certain disadvantages if compared with the more common pattern-plating process. The main disadvantage is that most of the carefully built up copper plating has again to be etched off; the copper plating has to remain only in the circuit pattern areas and inside the holes. This fact is uneconomic with regard both to etching as well as to plating. Finally, it increases unnecessarily the overall copper consumption and the use of etchant. Since etchant saturated with copper is not too easy to get rid of because of its polluting nature, electrolytic copper panel-plating is only recommended for small production volumes and tenting processes. Another disadvantage is the increased underetching because etching has to remove both the initial copper foil as well as the additionally plated copper layer. Fine-line conductor PCBs are therefore very difficult to achieve, even if the initial copper foil has less thickness than the standard thickness of 35 μm.

Pattern-Plating Process

In the pattern-plating process (Fig. 13-3), we have basically the same process steps as in the panel-plating process: The main difference to the previously described panel-plating process steps is the printing of the plate-resist which is done before the main electrolytic copper plating is carried out. The comparatively thick, electrolytically plated copper layer is therefore only plated on the actual circuit pattern areas and inside the holes, from where it does not have to be removed by etching; hence the term pattern-plating.

The process starts with drilling of the holes. To make the holes electrically conducting, electroless-copper is panel-plated. This thin and rather sensitive copper layer has to be

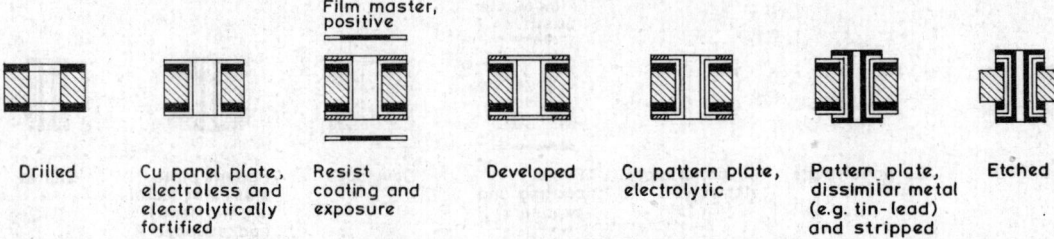

Fig. 13-3 **Pattern-plating process**

strengthened immediately by electrolytically plated copper of a few microns thickness. The panels are now ready to go for printing of the plate-resist. In the following (main) electrolytic copper plating step, copper is built up to the required thickness only in the circuit pattern areas and inside the holes. The process steps thereafter are electrolytic plating of dissimilar metal etch-resist, plate-resist stripping and etching.

The pattern-plating process is today the most often used subtractive PCB process for professional-type PCBs with plated-through holes.

Tenting Process

With the tenting process (Fig. 13-4), it is possible to produce double-sided PCBs with plated-through holes but without the use of a metal etch-resist. If the tenting process is compared with the print-and-etch process (Fig. 13-1), we can see that basically only two steps are additionally required for tenting which are the electroless and the electrolytic copper panel-plating. The positive resist pattern protects the actual circuit pattern from etchant attacks (etch-resist) and is printed onto the board after copper panel-plating. The resist covers also the plated holes like a tent in order to protect them during etching.

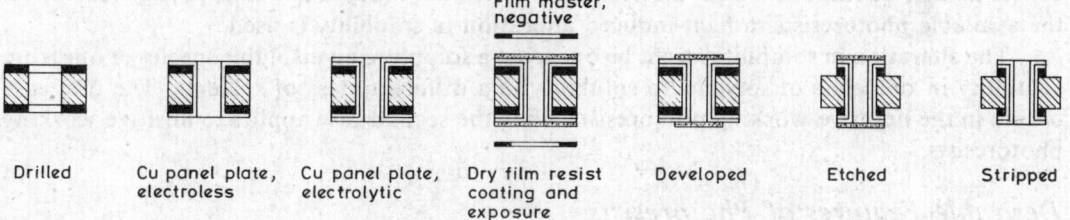

Fig.13-4 Tenting process

The tenting process came up with the availability of dry-film resists which are sufficiently thick (strong) to cover the holes. This process again applies electrolytic panel-plating with its inherent limitations. The use of thin-copper-foil laminates is recommended to reduce the total copper quantity to be etched. The tenting process is typically used in combination with hot-air levelling (see Chapter 15) which gives a protective tin-lead layer on the conductor surfaces and inside the holes. An alternative protection could be a final plating with electroless tin.

Any common etchant, in particular acid-type etchants, can be used in the tenting process. This freedom is possible because there is no metal etch-resist applied.

The tenting process is recommended for the production of professional-type PCBs with plated-through holes in smaller production volumes only, because electrolytic copper panel-plating is used. On the other hand, the tenting process is a comparatively simple process since the use of a metal etch-resist has been eliminated and any of the common etchants can therefore be used.

13.2 PHOTORESISTS IN GENERAL

Photoprinting of PCBs means essentially applying photosensitive material having the ability to form a continuous film which is sensitive to light or other radiation so that the

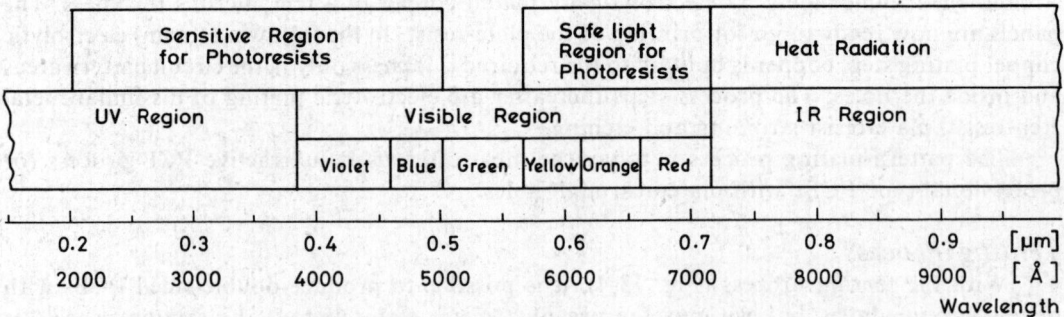

Fig. 13-5 Photospectrum and sensitive region for photoresists

unexposed (or exposed) areas of this film can be further processed without affecting the exposed (or unexposed) areas. This photosensitive material is called *photoresist* and the essential feature of it is that an exposure to proper radiation must produce a change in it to enable in later operations a clear discrimination between exposed and unexposed areas. In all the available photoresists, a light-induced alteration in solubility is used.

The alteration in solubility might be one from a soluble to an insoluble species or one from solubility in one class of solvents to solubility in a different class of solvents. The first case occurs in the negative-working photoresists while the second case applies to positive-working photoresists.

Desirable Features of Photoresists
—sufficiently light-sensitive to make its use economically attractive
—ability to produce a pin-hole free film of high uniformity with a short exposure time and a high resolution
—highly resistant against the chemicals used in further processing
—possibility to strip with satisfactory ease
—low price

In practice, it is hardly possible to have a photoresist incorporating all the desirable features simultaneously since they can oppose each other: A very thin resist film gives a high resolution but there will be more danger of pin holes and less chemical resistance. Or a resist with a high chemical resistance is probably more difficult to strip. All depends on the proper selection of a resist for the given purpose to provide the optimum results.

13.3 WET-FILM RESISTS

Wet-film or liquid-type resists were *the* photoresists exclusively used until the end of the 1960s. Only the appearance of dry-film resists has brought them serious competition, which are gradually replacing them more and more, especially for the more sophisticated process with plated-through holes. But still, there is a large field of applications today for wet-film resists where they continue to offer the economically best solution, such as in small-volume production of print-and-etch PCBs or as an etch resist in general. Wet-film resists remain therefore the *general* type of photoresist recommended for all non-specific applications.

The film thickness and uniformity of wet-films can be controlled within certain limits by the coating method applied and with the viscosity of the resist. The film thickness required, where applied as an etch resist, is typically 2.5 – 5 μm. The film thickness can, if available, be measured with a Betascope. However, a Betascope (also called β-ray backscatter) costs more than Rs. 50,000/- and is therefore only an essential asset where mainly plating thicknesses have regularly to be measured. Whether the film thickness and uniformity is sufficient or not, has to be judged by the etching results.

The type of coating process used is an important consideration since the film thickness and uniformity plays a major role in the subsequent process steps. If the film thickness is too little, a fog or a rough pattern image can be caused as well as a dull pattern finish and poor dye retention. Processing parameters thereafter can be adjusted for either one of the extremes in film thickness but not for both. A successful wet-film resist process must therefore have a coating method which produces reasonably uniform and constant film thickness.

If the wet-film resist has to serve as an etch-resist, a film thickness on the lower side usually works sufficiently well but if used as a plate-resist, maximum thickness must be achieved to withstand plating temperatures up to 70°C and plating current densities up to 8 amp/dm^2.

The *uniformity* of the film coating is expressed in % and is defined as

$$\text{Uniformity} = \frac{\text{minimum film thickness}}{\text{maximum film thickness}} \times 100 \ [\%]$$

To avoid critical processing, a uniformity of at least 60% is generally recommended.

The *viscosity* of the wet-film resist, once optimum results are obtained, should be maintained at the same value. There are various standardised methods for viscosity measurement used: A simple one is using a DIN-beaker (Fig. 13-6). Such a beaker is basically a cup with a hole of, say 3 mm or 4 mm in diameter in the bottom and the time [sec] it takes to empty the beaker is used as an index for the viscosity. The beaker is completely immersed in the resist; time measurement starts immediately after lifting of the beaker and ends with the first sharp break in the resist-stream draining from the beaker. The procedure has to be repeated about three times and the average is taken as the viscosity reference.

The adjustment of the viscosity, if required, is done by adding resist thinner. Suitable resist thinner is always specified and also supplied by the resist manufacturer.

The basic steps of wet-film resist process are shown in Fig. 13-7. The prebake and postbake steps may in many cases be omitted with modern wet-film resists. Baking is used only where hardness of the film needs further strengthening in order to achieve a high chemical resistance, e.g., as required for plate resists.

Negative-Working Resists

A negative-working resist is generally composed of a polymeric film-forming material along with a light-sensitive agent in a suitable solvent. The negative-working wet-film resists are known to the printing industry since more than a century. The early varieties were based on dichromates and on natural or later synthetic colloids. But only the introduction of sensitised polymers in the 1950s opened the field for widespread applications on an industrial scale for PCBs. The pioneering work was carried out mainly by Eastman Kodak Company, Rochester, N.Y., and led to the well known *KPR* resist which was for a long time a reference in the field.

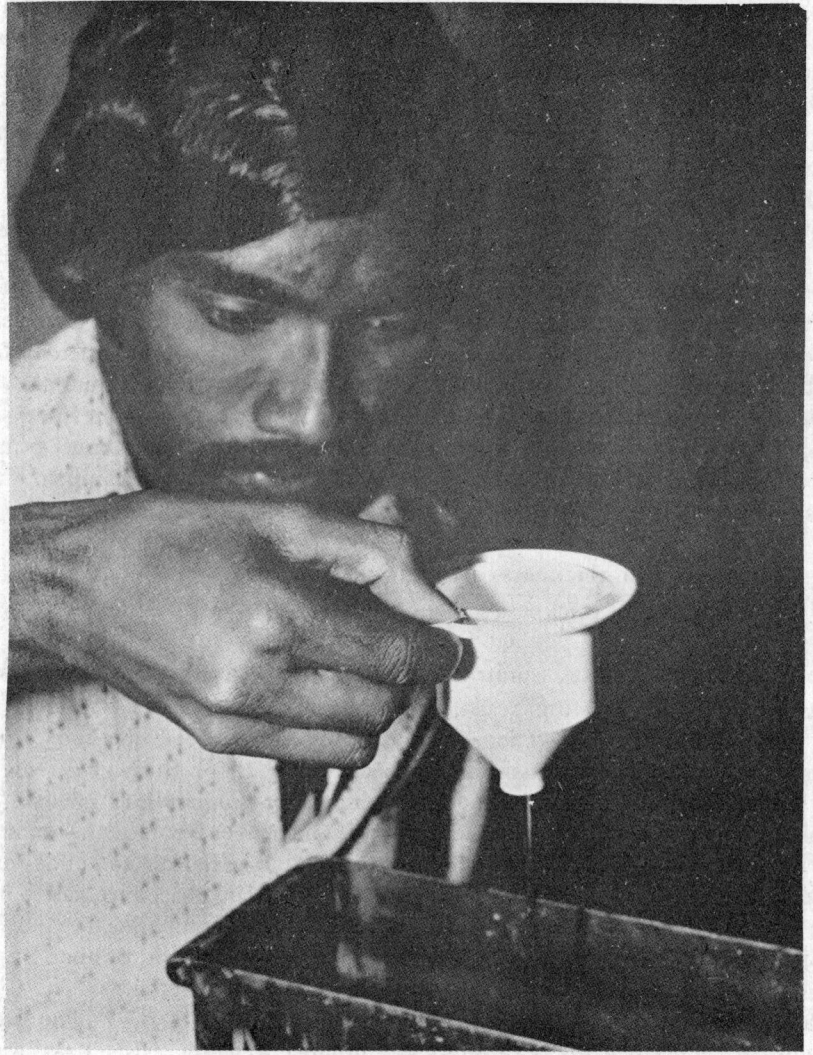

Fig. 13-6 Measurement of relative viscosity (Photo CEDT)

Modern wet-film resists of negative-type have characteristically a high chemical resistance, good image reproduction capabilities and are of low costs.

Positive-Working Resists

Positive-working resists, when exposed to proper light, suffer a change in chemical constitution like decomposition or isomerisation. The exposed product is still soluble but its solubility characteristic is sufficiently different from that of the starting material. Therefore with the proper choice of solvent, a selective removal of exposed material is possible, leaving the unexposed areas untouched.

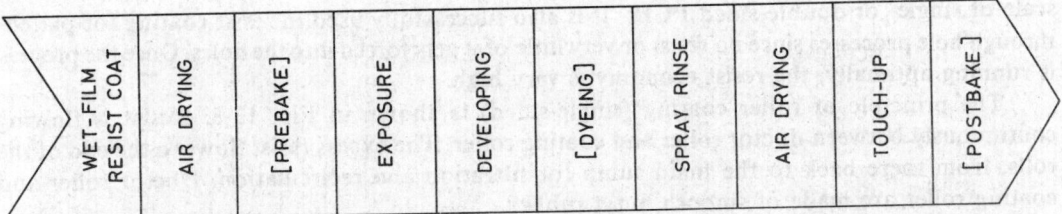

Fig. 13-7 Basic steps of a wet-film photoresist process. The steps in brackets [...] are carried out only if necessary, depending on photoresist and application.

Pioneering work and introduction of the first positive-working resist was done in the early 1960s by Shipley Company.

Positive-working resists are coloured and, exposed or unexposed, soluble in strongly alkaline solutions. They develop in mildly alkaline solutions. Compared to the negative-working resists, they offer less chemical resistance and they are more costly to procure. On the other side, they are extremely accurate and processing becomes simpler: There is no dyeing required and the developing and stripping in aqueous solutions permits the use of equipment made from readily available plastics. Stripping of positive-working resists can easily be done in conveyorised rinsing modules in an aqueous solution after the re-exposure to UV light.

The very high accuracy typically obtained with positive resists is a result of the development in an aqueous solution. Such solutions do not swell or distort the remaining pattern image. This characteristic is quite important and also compensates the resulting disadvantage where thicker films are required because of lower chemical resistance.

13.4 COATING PROCESSES FOR WET-FILM RESISTS

The choice of the proper coating method and equipment depends on factors like required film thickness and uniformity, output volume, single- or double-sided PCBs, capital investment possible, etc. In any case, photoresist coating is one of the more critical operations that determine decisively the overall quality and resolution of the finished PCBs. It should therefore be carefully studied when planning a new PCB set-up.

Flow Coating

Flow coating is the simplest and was the earliest method of coating with wet-film resist: A few drops of resist are poured onto the cleaned board. Then the board is tilted and rotated until the resist film covers the entire surface. The surface tension of the resist generally prevents the resist from spilling over the edges. The coating will also work with a glass rod placed parallel to the surface and touching the resist. Moving of the glass rod parallel to the surface from one side to the other will permit a quick distribution of the resist. The drying must occur in horizontal position.

Flow coating, however, does not provide a very uniform coating and its use is restricted to small PCBs in noncritical and short-run applications.

Roller Coating

Roller coating is a very efficient and satisfactory method in resist applications on a large-scale of single- or double-sided PCBs. It is also successfully used in resist coating for plated-through hole processes since no resist or very little of it gets forced into the holes. Once the process is running optimally, the resist economy is very high.

The principle of roller coating (single-sided) is shown in Fig. 13-8: Resist is flowing continuously between doctor roller and coating roller. The excess resist flows to the end of the rolls, from there back to the main sump for filtration and recirculation. Doctor roller and coating roller are made of smooth butyl rubber.

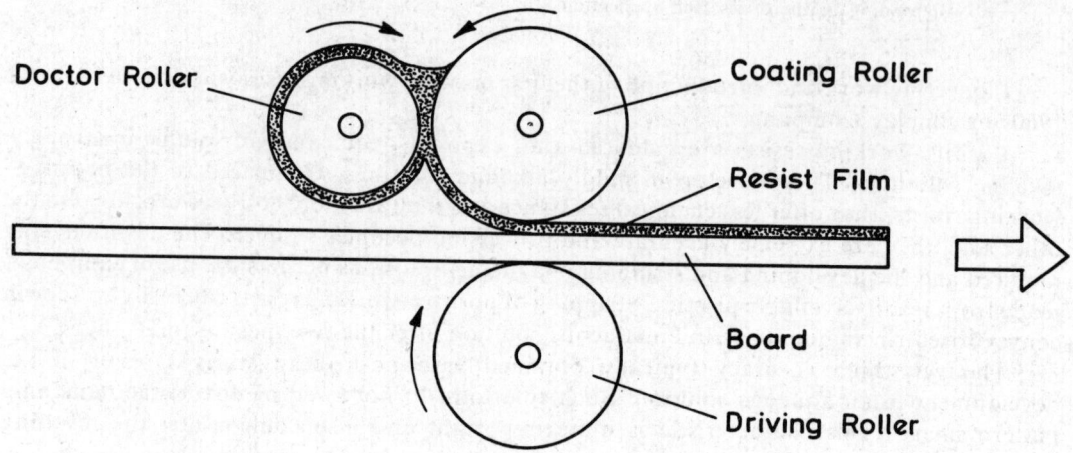

Fig. 13-8 Principle of roller coating

The coating uniformity is quite high and ranges within 65–95%. Once the optimum working conditions are established, the dependence on the operator is low. A good practice is the double coating: Boards which have passed through the rollers are immediately returned and a second coat is applied after the board has been turned by 90°. Double coating eliminates, to some extent, the defects from the first coating and the film thickness gets increased by 30–40%.

For double-sided PCBs, there are roller coaters available which simultaneously coat both the sides of the boards (Gyrex Corp., Santa Barbara, California). With ordinary, single-sided roller coaters, the second side of the board can be coated after drying of the first side.

The film thickness is affected by the conveyor speed, the resist viscosity and the pressure between doctor roller and coating roller. Increasing the conveyor speed will result in more film thickness, however, not in a proportional relation. Typical conveyor speeds are in the range of 2–5 m/min. A too low viscosity will give a very thin film and filled holes. High viscosity will give thick coatings and, if too high, bubbles in the dried film. Viscosity must be checked frequently since the resist is continuously exposed to air.

The roller coating process has also its specific disadvantages. It is a large-output process and will not work economically if the roller coater is used e.g., just for three times 10 min a day. Before starting coating operations, the roller coater should run for at least 20 min for resist temperature stabilisation and good resist uniformity. Extreme care must be taken to keep rollers and working parts clean when the coater is not in use.

Dip Coating

Dip coating is a very simple concept: The board is immersed vertically into a narrow tank (slit tank) and pulled out at a regular speed. Both the sides are coated simultaneously with a resist film of typically 2.5–4 μm thickness and a uniformity of 50–70%. For many applications dip coating is an *economic method*, especially if uniformity and thickness requirements are less stringent. The film thickness depends mainly on the withdrawal speed and the viscosity. A high withdrawal speed will give thicker films up to a certain limit (Fig. 13-9). The best withdrawal speed has to be experimentally determined for the particular resist used and is usually somewhere between 10 and 100 cm/min.

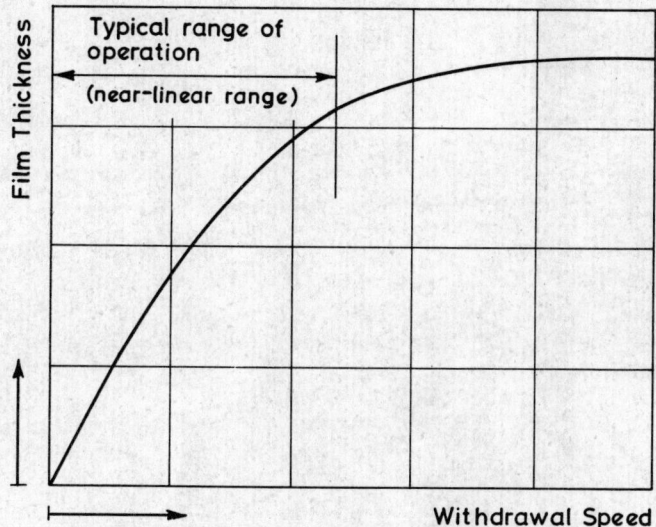

Fig.13-9 Film thickness in dip coating versus withdrawal speed

The most limiting factor in dip coating is the uniformity achievable. At the top of the board, film thickness is minimum while the maximum is at the bottom of the board. An immediate second coating does not add much to the thickness because the first coat gets again partly dissolved. Another disadvantage is the contamination of the resist which is brought in with the immersion of the boards. Such particles of board material and other solids must be frequently removed by filtration.

A dip coater realisation, designed, built and used at CEDT is shown in Fig. 13-10. Five boards of 200 × 250 mm size can be simultaneously coated. The lift in which the boards hang, is driven by a reversible and geared synchronous motor. The operation of the coater is controlled by the shield (visible in Fig. 13-10). If the shield is brought to the upper end position, the lift moves down and the cover of the resist tank turns vertical. After hanging in the boards, the shield is shifted down: This starts the withdrawing operation automatically. As soon as the lift reaches its upper end position, the tank is automatically covered and a warm air blower (tangential type) forces air from behind towards the shield and from there the air gets directed upwards for drying of the boards.

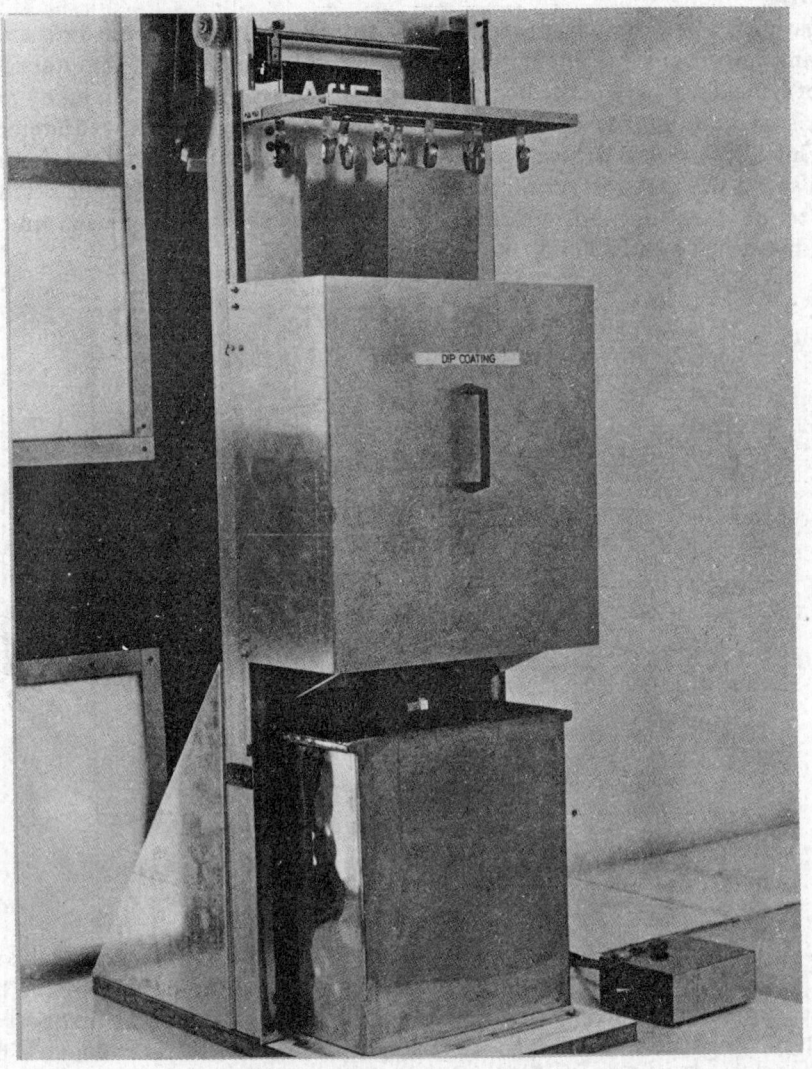

Fig. 13-10 Dip coater used at CEDT (Photo CEDT)

Spraying

Spraying offers a few advantages over the other coating methods:
— Rough and uneven boards can readily be coated.
— Wide range of control over film thickness through repeated coating/drying and adjustment of viscosity.
— Drilled holes are not or only little infected with resist.

On the other hand, the dependance on the operator is fairly high and resist lost because of overspray is significant. Much care must be taken to keep the spray gun always clean and the spray nozzle unclogged. Equipment-wise, the requirements are limited to an air compressor,

Fig. 13-11 Whirl coater used for resist coating of printed circuit boards
(Courtesy: Monotype India Ltd., Bangalore, India)

spray gun, air filter unit for the compressed air and a spray booth. For best results, the boards are kept horizontally or within 30° thereof. The coating is done by a series of evenly spaced vertical passes followed by a similar number of horizontal passes. For double-sided boards, the coating of the second side follows the coating and drying of the first side.

Whirl Coating

Whirl coating has long been used in graphic arts and printing industries before it became an important method for applications in microelectronics. In PCB applications, it is used where its low productivity rate and its considerable resist losses do not matter. However, it is not suited for the production of PCBs with plated-through holes; they would get infected by the resist.

Whirlers provide a film uniformity of at least 80% and better. The maximum thickness is found in the centre. Usually, the board is first kept rotating and then a few drops of resist are poured in the centre until a complete coverage of the entire surface occurs. Considerable losses of resist have to be accounted for, especially for boards rectangular, non-square shape. The spinning rate is typically in a range of 100–200 rpm. A higher viscosity and a lower spinning rate will give more film thickness. Multiple coatings are possible after drying of the previous coating. This includes coating of double-sided boards.

13.5 EXPOSURE AND FURTHER PROCESSING OF WET–FILM RESISTS

13.5.1 Exposure

The exposure is done in an exposure unit equipped with a light source which emits in the UV spectrum where the resists are most sensitive. The exposure units can be classified from the design side into fixed light-source type and moving light-source type.

In exposure units with a *fixed light-source* (Fig. 13-12), the board is kept in a vacuum frame which can be inserted like a drawer for exposure while the loading is done in its outer position. The control of the amount of exposure is achieved by controlling the time for which the board is exposed or by using integrating radiation controller. With instantly starting light-sources, the exposure is terminated by switching off the light-source; for light sources that are not instantly starting, the light-source is covered by a shutter and the energy reduced to a threshhold-level. There is usually no light collimation required because of sufficient distance between light-source and board to be exposed.

Fig. 13-12 Exposure unit with fixed light-source (Courtesy: Hansa Enterprises, Cooke Town, Bangalore-560 005, India)

Fig. 13-13 Exposure unit with moving light-source (Courtesy: Colight Photomechanical Products Inc., Minneapolis, USA)

In exposure units with a *moving light-source*, also called scanning printers, the board is also kept in a vacuum frame, but in a fixed position (Fig. 13-13). The amount of exposure is controlled by the moving speed of the tubular light-source. The power of the light-source in this is usually higher than with fixed light-source units. Light collimators are used since the boards are kept very close to the light-source. The moving light-source units are particularly useful for large production volumes.

Light Sources

The basic feature of a light-source suitable for the exposure of photoresists is a high radiation in the UV region. There are various types of light-sources in use (Fig. 13-14).

Mercury-vapour lamps: They have a long life and are available at reasonable cost. The emitted light-spectrum has a peak energy very close to the maximum sensitivity of most of the photoresists. Mercury-vapour lamps are also producing infrared radiation which could lead to undesired secondary effects through heating-up. About 5-30 min warm-up time is necessary until full intensity is reached.

Pulsed xenon lamps: These light-sources produce a rich radiation in the UV range and can be instantly started and switched off.

UV fluorescent tubes: Only little heat is produced. They are a very common type of light-source for fixed light-source units. They can be started like ordinary tube lights and have a long life. As a disadvantage, the broad spectrum of the light-source has to be mentioned.

Carbon arc lamps: An arc is produced between two carbon electrodes kept at a constant distance. They are very effective light-sources for any type of resist. They are, however, slowly disappearing because they produce gases and carbon particles which are harmful to health. In some countries, there are very strict limitations from official side to use them because of the mentioned health hazards. For these reasons we do not recommend their use.

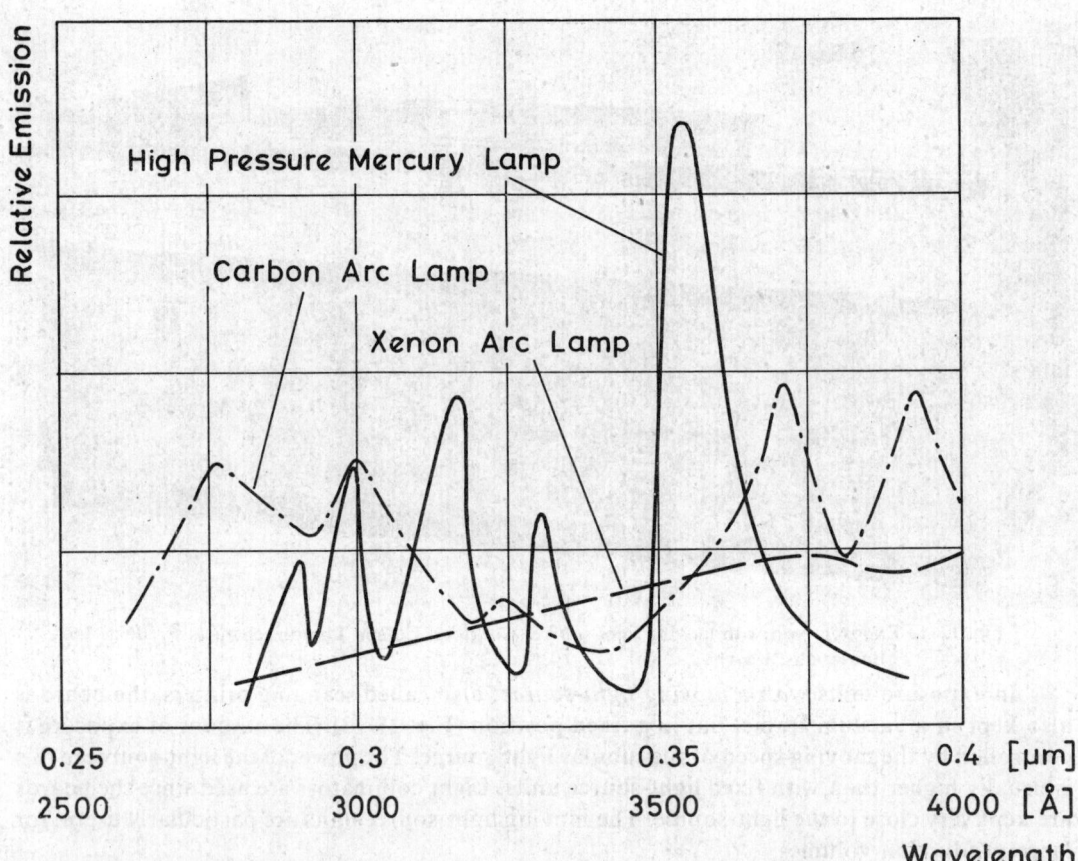

Fig. 13-14 UV light-emission spectrum of a few photoresist exposure light-sources

Safe Light Requirements

The illumination in the rooms where photoresists are handled, has to take care of the sensitivity spectrum of the resist. Safe light for the purpose is in the region of yellow-orange-red light (Fig. 13-5). Windows have to be covered with transparent plastic foil of one of these colours. Where yellow fluorescent tube-lights are not available, ordinary tube-lights can be covered with the same foil as used for the windows. Practical tests with coated boards kept lying for a certain time, will reveal whether the light in these rooms is safe or not.

When only negative-working resists are used, safe light is necessary till the developing is over. For the application of positive-working resists, it must be remembered that the resist, remaining after exposure/developing on the pattern image, is still light-sensitive and is chemically not changed by developing.

Determination and Control of Optimum Exposure

The photoresists offer a wide tolerance towards overexposure which makes their application much easier compared to other photographic materials. But in spite of this, it is necessary to check daily whether optimum exposure is maintained. Many factors have an influence such as an unstable mains voltage, changing resist film thickness, ageing of the light-source and others. For the determination and maintenance of the optimum exposure, it is essential to keep all such factors as constant as possible.

Exposure determination with a film master: We select a critical film master with a high interconnection density and fine conductor lines. The film master is fixed with adhesive tape at one edge onto the resist coated board. A series of exposures is made with a mask partially covering the film master. For each new exposure after shifting the mask, the exposure time is doubled: If the start is made with 10 sec, we get after 5 exposures, 6 steps which have been exposed in total for 310-300-280-240-160-0 sec. The inspection of the developed and dyed test piece should reveal at least one but usually more satisfactory sectors. The optimum exposure will be in the middle of these satisfactory sectors and the extent of such sectors will indicate the amount of exposure latitude. If the exposure series does not provide satisfactory sectors, it was out of limits and has to be repeated with a different start time or one more exposure step.

Exposure determination with a step tablet: Step tablet exposures are an excellent method to maintain exact control in resist exposure. A very often used step tablet is the *Kodak photographic step tablet No. 2* from Eastman Kodak Co., Rochester, N.Y. (similar to Fig. 13-15). It is a piece of film that has been exposed and developed to yield steps of increasing density from clear film base to complete opacity (21 steps, 0.15 density gradient). The test board is exposed via a step tablet and film master. If the optimum exposure is achieved in a range of 6 or 7 gray-steps (in midst the 21 steps) satisfactory results are obtained after regular processing of the test piece.

Fig. 13-15 Photographic step tablet for exposure determination (Photo CEDT)

13.5.2 Developing

Developing in a Slit Tank

It is a recommended practice to have two tanks for a cascaded development. The boards are kept under agitation in each tank for half of the total developing time. The second tank should always contain a fresh developer which is later used to replace the developer in the first tank, after exhaustion. The exhaustion of the developer can be observed by observing an incomplete removal of the resist portions to be solved and the resulting improper dye retention.

The slit tank is made as narrow as possible to reduce the evaporation losses. Suitable materials are stainless steel, glass, but also polypropylene. The tank must be closed when not in use. Proper ventilation is important to minimise health hazards.

Spray Developing

Spray developing is a very efficient method to develop the resist image. It provides a high degree of agitation at the board surface along with continuous rinsing. This is particularly useful for the development of thick resist films, especially for dry-film resists. The spray development is done in a specially designed spray tank or in conveyorised spray developing machines.

13.5.3 Dyeing

Dyeing may be required to render the developed resist image visible for inspection and touch-up. Correction of pinholes and other image defects are thereafter easily carried out in an easier way before the board goes for etching and plating. Dyeing is carried out immediately after the development of the pattern is finished. A dip in the dye tank for 5-20 sec, depending on the concentration, is usually sufficient.

Dyeing as such is a messy operation and the loss of some chemical resistance in the resist has to be taken into account. It is therefore avoided, wherever possible.

The dye solutions contain the same solvents as used in the developers. Therefore tanks for the dye are made of the same materials as those used for the developing tanks.

13.5.4 Touch-up

Defects in the resist film image are touched-up under a magnifying glass with a resin which is resistant to the processing solutions. Inks for screen-printing of PCBs are often used for this operation.

The amount of touch-ups required is an excellent indicator of the quality of the previous processes. The best case is when no touch-up is required by taking adequate care of all the previous steps. Especially helpful can be a daily inspection of the exposure vacuum frame and the film masters used.

13.5.5 Postbaking

Postbaking of the processed image is applied where a high durability is essentially needed, e.g. if the resist-image serves as a plate-resist. Postbaking might also be necessary for thicker resist films which tend to retain some solvent.

Postbaking is done at temperatures around 100°C for about 10 min. However, the optimum must be found experimentally. It must be remembered that postbaking should be done just to the minimum extent required. Too much postbaking will introduce more difficulties in stripping of the resist.

13.5.6 Stripping

After the resist has performed its duty, it has to be completely removed from the board. On a small scale, it is usually done by scrubbing and with the help of a solvent or a commercially available stripping solution. On a large scale, conveyorised spray units are employed which work with strong stripping solutions. The contact with such stripping solutions must be kept minimum because of possible attacks on the base materials. The stripping procedure ends usually with a solvent rinse followed by a water rinse. Very tough resist images can sometimes

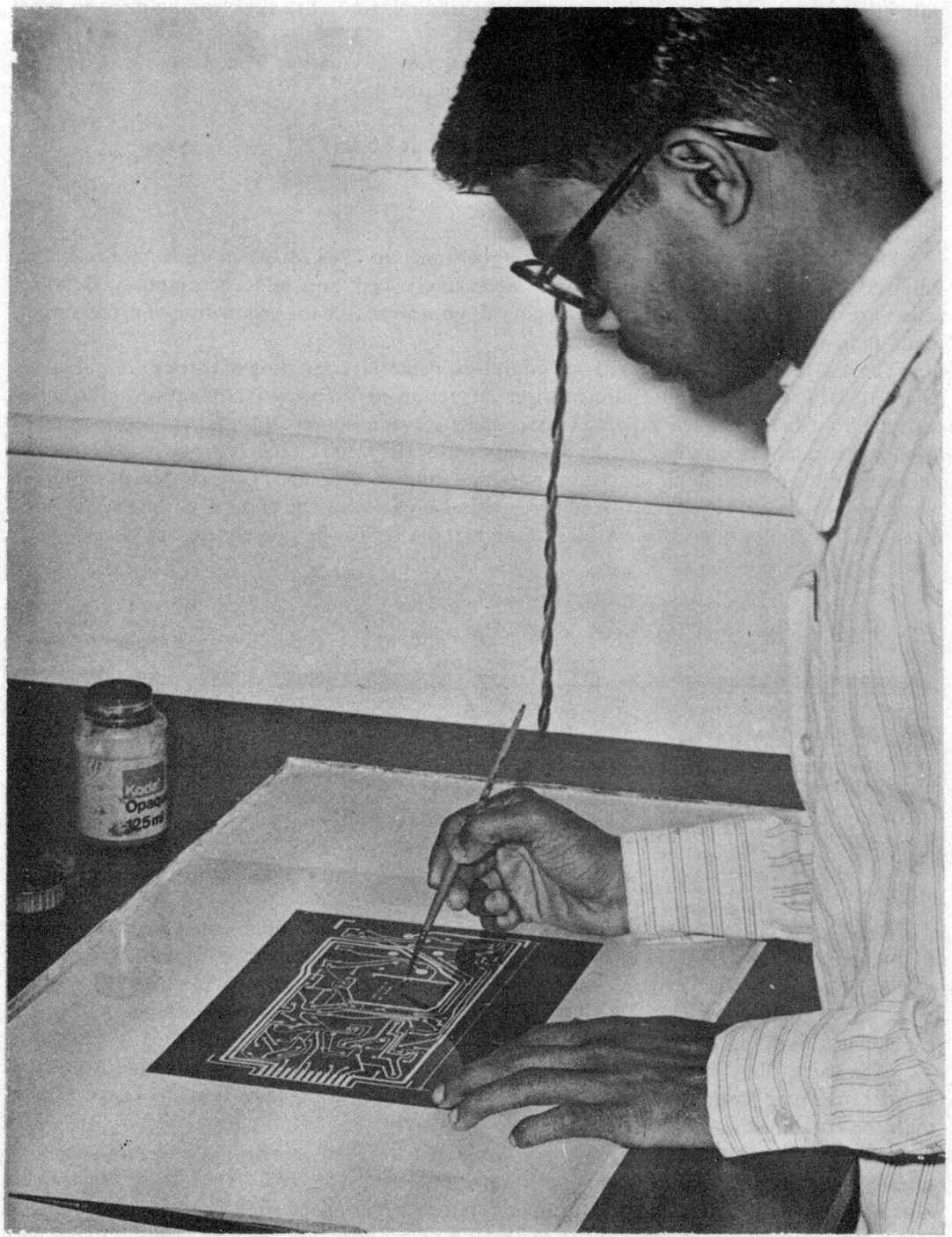

Fig. 13-16 Touch-up operation on the transilluminator (Photo CEDT)

only be stripped satisfactorily by a cleaning procedure, similar to the cleaning prior to resist application which includes rotary brushing.

The stripping of positive-working resists is basically easier. The stripping is done in alkaline solutions after the board has been re-exposed.

13.6 DRY-FILM RESISTS

13.6.1 Features of Dry-Film Resists

Dry-film resists have, since their first appearance in 1968 (du Pont de Nemours & Co., followed by Dynachem Corp.), found increasingly fast and wide acceptance. Dry-film resists are today the most important category of photoresists in the production of professional-grade PCBs with plated-through holes.

The dry-film resist is delivered as a composite material consisting of three different layers (Fig. 13-17). The polyolefine separator sheet gets removed just prior to lamination of the board with the dry-film resist. The polyester cover sheet serves as a protection for the sensitive resist layer and is removed only before the development of the resist image. It protects therefore the photoresist layer through all the critical handling until developing. This includes the exposure which is done via the cover sheet. And here we find a big advantage of the dry-film resists which is their *high immunity against physical damages*, thanks to the polyester cover sheet.

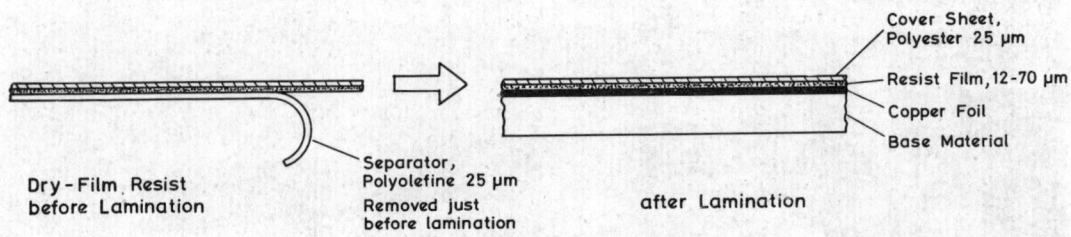

Fig. 13-17 Composite structure of dry-film resist

The photoresist layer is available in different thicknesses in the range of 12.5–70 μm. The PCB manufacturer has therefore a wide choice to select the optimum resist film thickness for his particular applications. Typical resist film thicknesses in applications as an etch- or plate-resist are ranging between 12.5–40 μm. The higher-thickness films, above 37μm, are particularly useful in the *tenting process* (Fig. 13-4).

The thick film layers available with dry-film resists also offer an effective *protection against overhang formation* in plating processes (Fig. 13-18), since a film thickness of more than the combined plating thickness can be chosen.

The coating of the boards with dry-film resist is called *lamination* and this process is usually less critical than the coating with wet-film resists. The *uniformity* of dry-films is very high and is therefore not a critical parameter in the further processing. However, care must be taken for the cleaning of the boards: Adhesion of dry-films on surfaces contaminated with

PHOTOPRINTING

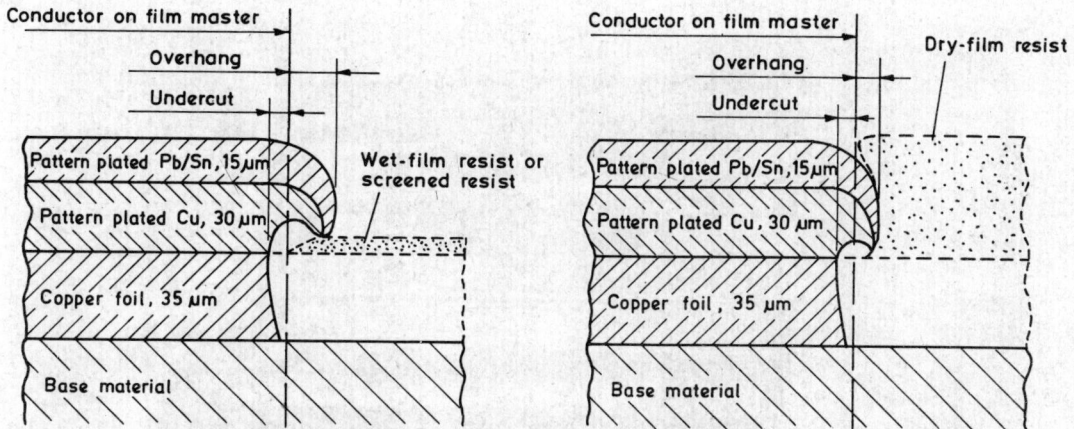

Fig. 13-18 Reduced overhang with dry-film resists (typical example)

traces of oil and grease is even more critical than with wet-film resists. This also because such defects may here not be immediately detected. Surface defects like deep scratches also harm because they are not completely filled with dry-film resist so that etchant can enter between the resist film and the copper surface, from the resist pattern flanks.

The lamination is carried out in a laminator (Fig. 13-19). The resist film is heated up for a specific time and pressed onto the copper surface. The separator sheet is automatically removed just before heating occurs. Coating speeds are typically around 1.5 m/min and the temperature to which the resist is heated up is around 110°C. After lamination, the board is separated from the continuously coming dry-film by a knife-cut. Since the dry-film resist supplied from the rolls is always of same width, a certain wastage of material is unavoidable.

Both the board sides are laminated simultaneously which is otherwise only possible with roller or dip coating. However, in the case of dry-film resists, the drilled holes will remain absolutely clean. Dry-film lamination of single-sided boards can be done by putting two boards back to back or by removing one of the dry-film rolls.

Another aspect of the dry-films is a *simpler processing procedure:* After lamination, no drying of the resist film is necessary and a dye is incorporated in the pre-dyed resist film layer. The higher costs for dry-film resists are partially compensated for by the simplified processing and the lower reject rate when used on an industrial scale for highly sophisticated boards with plated-through holes.

13.6.2 Categories of Dry-Film Resist

There are two different types of dry-film resist in broad use; they are categorised by the manner they are developed.

Solvent-Developing Dry-Film Resists

They form the most widely used category of dry-film resists and are negative-working. For the development, 1,1,1-trichloroethane (methyl chloroform) at room temperature is used. The development equipment must be made of stainless steel to withstand this strong solvent.

Fig. 13-19 Dry-film laminator for simultaneous lamination of both sides of the PCBs (Courtesy: E.I. du pont de Nemours & Co. Inc. RISTON Product Division, Wilmington, DE 19898, USA)

Aqueous-Developing Dry-Film Resists

They are also negative-working but are developed in aqueous- or semi-acqueous solutions, depending on the variety used. The processing with aqueous solutions is gaining more and more importance because neutralisation and disposal is much easier as compared with solvents, and because less health-risks for the personnel are involved. In addition to this, aqueous solutions permit the use of equipment made of PVC or other plastics which brings equipment costs down. One major drawback of this category of dry-films is their limited resistance against alkaline solutions with a pH-value of more than 8.5. This fact sets some restrictions to the choice of processes subsequently applicable.

13.6.3 Exposure and Further Processing of Dry-Film Resists

This section will deal only with the aspects which are different from the processing of wet-film resists.

Exposure: The peak sensitivity to UV radiation of all the currently available dry-films is around a wavelength of 0.35 μm. The most recommended light-source for the dry-film exposure is the mercury-vapour lamp. Satisfactory results, however, are also obtained with carbon-arc lamps and UV fluorescent tubes. Optimum exposure is best controlled with the step-tablet method.

Because of the thick film layer, the exposure of dry-film resists is longer than that required for wet-film resists. The production of heat during exposure has therefore to be considered, especially when exposing with mercury-arc lamps. Heat energy can continue the polymerising process initiated by light. A warmed up exposure frame could therefore lead to overexposure in spite of an otherwise correct exposure time.

Development is carried out in spray developing equipment. They provide the necessary effectiveness for the development of thicker film-layers, as typically used with dry-film resists.

The *dyeing* step has been completely eliminated. All dry-film resists are supplied in a pre-dyed condition.

The *touch-up* operations are drastically reduced with dry-film resist thanks to the cover sheet which is removed just before developing. But some touch-up may still be required in most of the cases.

Postbaking should not normally be necessary. If it is applied, then it should be done only to the minimum extent, to minimise difficulties in stripping.

Stripping: Scrubbing action will be mostly required for room temperature solvent-strippers while the heatable solvent-strippers are sufficiently active to strip without scrubbing.

BIBLIOGRAPHY

1. BAUMANN G.: 'Neue Entwicklungen auf dem Gebiet der photopolymeren Filmresiste', *EIPC PROCEEDINGS*, Zurich, Nov. 1971, pg. 6–18.
2. COOMBS C.F.: *PRINTED CIRCUITS HANDBOOK*, McGraw-Hill Book Co., New York, 1967.

3. DEFOREST W.S.: *PHOTORESIST, MATERIALS AND PROCESSES*, McGraw-Hill Book Co., New York, 1975.
4. HARPER C.A.: *HANDBOOK OF MATERIALS AND PROCESSES FOR ELECTRONICS*, McGraw-Hill Book Co., New York, 1970.
5. KOLF G.: 'Untersuchungen uber die Galvanoresistenz von COPYREX-Photolacken', *EIPC PROCEEDINGS*, Zurich, Febr. 1970, pg. 56–76.
6. *PHOTOFABRICATION METHODS WITH KODAK PHOTOSENSITIVE RESISTS*, Publication No. P-246, Eastman Kodak Co., Rochester, N.Y., 1971.
7. STOCKLEY F.: 'Positive-working photo resists, applications in printed circuits and microelectronics', *EIPC PROCEEDINGS*, Zurich, Febr. 1970, pg. 93–107.

14

SCREEN-PRINTING

14.1 SCOPE OF SCREEN-PRINTING

The process of screen-printing, since long, is well known to the printing industry. Its inherent capabilities of printing a wide variety of inks on almost any kind of surface including glass, metal, plastics, fabrics, wood, etc., found their way into an extremely broad field of applications.

But only the development of *dimensionally highly accurate and stable fine-mesh fabrics* of monofile materials (mainly polyester and stainless steel) made the screen-printing process applicable to the fabrication of printed circuit boards.

Screen-printing offers the advantage of *wide control on the ink deposition thickness* through the selection of suitable mesh density/thickness and ink composition. In the production of PCBs, it is successfully employed in printing of

—etch resists
—plate resists
—solder stop lacquers
—notation printing

The screen-printing process is in particular suitable *for large production volumes*. However, the preparation of a screen can also be economically attractive for series of 100 PCBs or below. While photoprinting is basically the more accurate method to transfer a pattern onto a board surface, screen-printing is a considerably cheaper way to do a sufficiently accurate job for large series. With the screen-printing process, one can produce (provided of course, one has enough experience with it), PCBs with a conductor width of as low as 0.5 ± 0.1 mm and a registration error of just 0.1 mm on an industrial scale with a high reliability. Therefore, screen-printing is the method by which worldwide the largest number of PCB patterns are printed. The screen-printed boards are typically processed in the print-and-etch process but in many cases also in plating processes for PCBs. Screen-printing is therefore suited in application where the ultimate in precision and accuracy is not needed.

The actual screen-printing can easily be carried out with a simple frame arrangement on a work bench. Up to 100 prints per hour can be expected this way but the dependence between production quality and operator's skill and care is high.

Semi-automatic and fully-automatic screen-printing machines are mainly used in countries where the relative costs for labour are very high. Such machines are basically *not* more accurate than hand screen-printing, but they offer less dependence on the personnel, higher output (up to 1,000 prints per hour) and better cleanliness.

In its basic form, the screen-printing process is very simple (Fig. 14-1). A screen fabric with uniform meshes and openings is stretched and fixed on a solid frame of metal or wood. The circuit pattern is photographically transferred onto the screen, leaving the meshes in the pattern area open, while the meshes in the rest of the area are closed. In the actual printing step, ink is forced by the moving squeegee through the open meshes onto the surface of the material to be printed. The ink deposition, in a magnified cross-section, shows the shape of a trapezoid (Fig. 14-2).

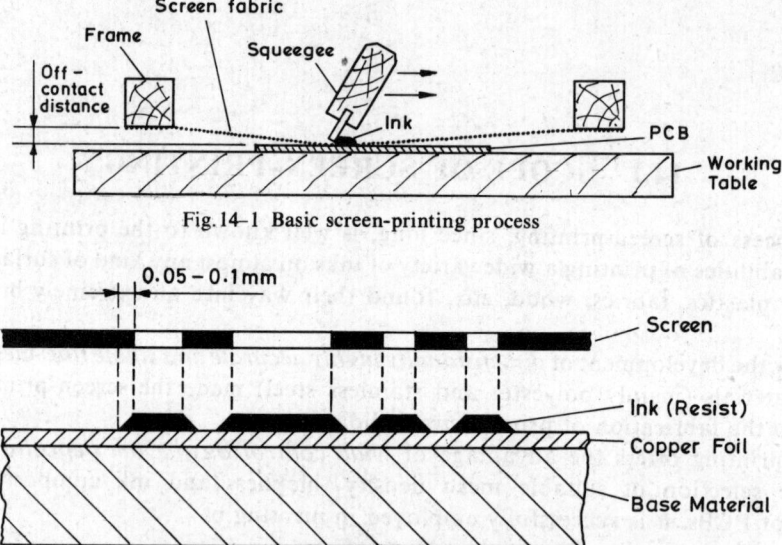

Fig. 14-1 Basic screen-printing process

Fig. 14-2 Screen-printing ink deposition in the shape of a trapezoid

14.2 SCREEN FABRICS

The selection of the best suited fabric for the given purpose is a step that should not be neglected. There are many factors depending on the type of fabrics chosen, such as
— thickness of the ink (resist) deposited
— contour sharpness: finer fabric will give finer details
— durability of the screen
— chemical resistivity, e.g., when printing with aggressive inks
— mechanical sensitivity to handling
— resilience
— and many other interdependent factors.

The thickness of the ink deposited is primarily proportional to the thickness of the screen fabric. But also other fabric parameters like opened screen area and mesh number (mesh/cm) contribute, in a roughly proportional way, to the ink deposition. We can therefore say that for a thick ink deposition also, a thick screen fabric has to be used while a thin ink deposition will need a thinner, finer screen fabric.

The screen fabrics used for the production of PCBs are almost exclusively made from two

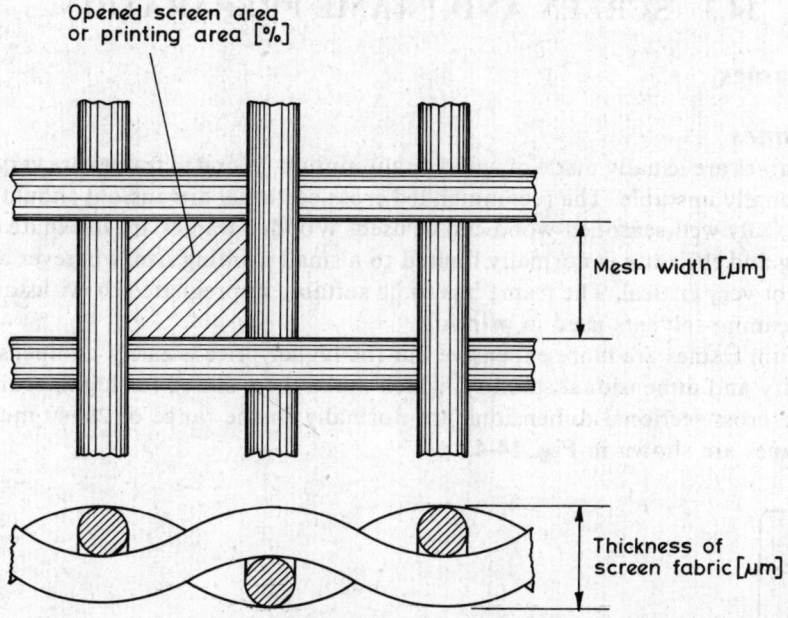

Fig.14-3 Geometry of a screen fabric

materials: Monofile polyester fabric and stainless steel fabric. Other materials, like silk or nylon fabrics, have too many inherent disadvantages and are used only in exceptional cases.

Monofile Polyester Fabrics

Monofile polyester fabrics are most widely used in the fabrication of PCBs and in particular for long-run jobs. They are very durable, have a high resilience and, compared to stainless steel fabrics, they are cheaper and less sensitive to mechanical damages. In all those applications, however, where maximum precision is asked for (for instance, if the registration error should be less than 0.1 mm) one has to use the more precise stainless steel fabric.

The polyester fabrics typically used for etch and plate resists have a mesh number in the range of 100–140 mesh/cm. For solder stop lacquers, where sufficient lacquer thickness is important, a mesh number of 50–90 mesh/cm will give the desired results. If a UV-curable solder mask is printed, one can go up to 120 mesh/cm.

Stainless Steel Fabrics

They are used in screen-printing, wherever maximum registration accuracy is first priority. Stainless steel fabrics are easy to clean but they are more sensitive to mechanical damages in handling and do not have the long life of polyester fabrics.

The price of stainless steel fabrics is also higher. Therefore, they are recommended only where their high dimensional stability is a must or, sometimes also, where their high chemical resistance is needed.

Stainless steel fabrics provide excellent results in hand-printing. For best results in printing of etch and plate resists, the mesh number is normally chosen within the range of 120–140 mesh/cm.

14.3 SCREEN AND FRAME PREPARATION

14.3.1 Frames

Simple Frames

These frames are usually made of wood or aluminium. Wooden frames are very cheap, but also dimensionally unstable. The recommended cross-sectional dimensions should be roughly 40 × 40 mm. Only well-seasoned wood can be used. Wooden frames are still quite common in hand printing and their use is normally limited to a small printing size, wherever dimensional accuracy is not very critical. The frame has to be suitably impregnated to withstand the ink, water and cleaning-solvents used in printing.

Aluminium frames are more expensive but the higher price is easily compensated for by their durability and dimensional stability, which make them suited for high-accuracy screen-printing. The cross-sectional dimensions are normally in the range of 20–40 mm and some preferred shapes are shown in Fig. 14-4.

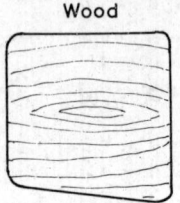

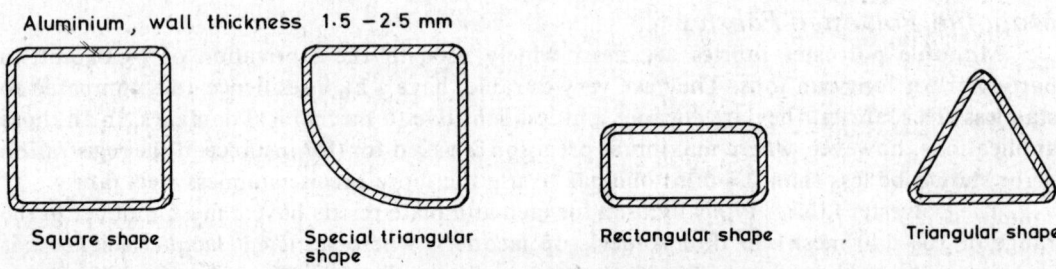

Fig. 14-4 Cross-sections of simple screen-printing frames

Self-Spanning Frames

They offer the big advantage that the tension for the screen fabric can be set with its inner frame (Fig. 14-5). Therefore no special equipment is needed to fix the screen fabric under tension. They are well suited for smaller printing units (no investment in special spanning equipment) and where maximum printing accuracy is not needed (special spanning equipment provides a still more perfect spanning).

Self-spanning frames can be made of wood or of special aluminium profiles.

Frame Size

The frames should have a size of not less than 400 × 400 mm for convenient handling. With respect to the pattern to be printed, only 10–25% of the screen area should be utilised for the pattern. This will provide sufficient space between the squeegee and the frame; thus printing distortions due to too high tension in the outer areas will be minimim (Fig. 14-6).

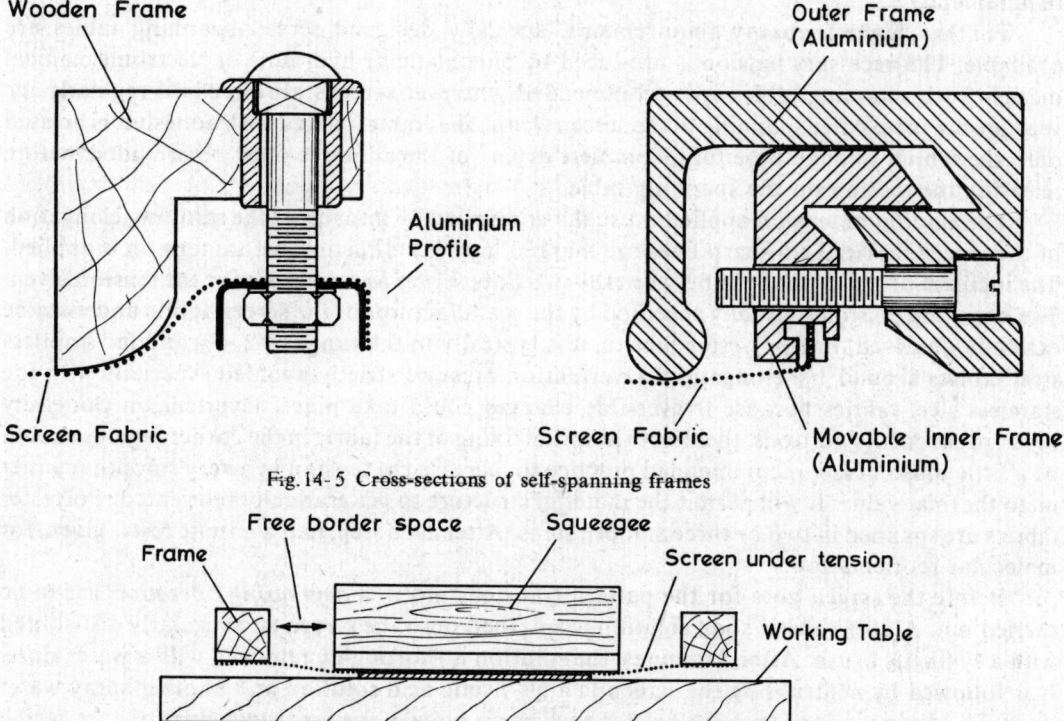

Fig. 14-5 Cross-sections of self-spanning frames

Fig. 14-6 Sufficient free border space minimises printing distortions

14.3.2 Fixing the Screen onto the Frame

The manner in which this operation is carried out, contributes to a great extent to the quality of the final printing results. An equally spanned screen at the optimum tension will give fine and undistorted prints and the screen will have a long life. The first task always is to clean the frame where the screen fabric has to be fixed on it. All marks of inks or adhesives have to be carefully removed and the edges must be rounded so as not to damage the delicate and sensitive fabrics. For wooden frames, sand paper can be used while metallic frames are cleaned with solvents.

The screen fabric is usually aligned in the frame in such a manner as to obtain parallelity between the threads and frame. For applications without much accuracy requirements, wooden frames are used and the screen fabric is fixed with staples and adhesives, under maximum tension uniformity applied by hand. It is the simplest method and no special equipment is used. However, not too much of printing quality and reliability can be expected with this simple approach.

A better control on the tension can be given with the self-spanning frames. The fabric is fixed with a special adhesive on the moving frame-part and spanned after drying of the adhesive. The adhesive used must have the following characteristics: Fast action, elasticity, stability under temperature variations and resistance towards water, solvents and cleaners used in printing. There are single- and two-component adhesives available which fulfill most of these requirements.

For the highest quality requirements, specially designed screen spanning tables are available. The necessary tension is produced by pneumatic or hydraulic or electromechanical means. Such spanning tables are recommended wherever screens are prepared regularly for very precise jobs. After spanning of the screen fabric, the frame, coated with adhesive, is pressed onto the fabric. Enough time for a complete drying of the adhesive must be provided before releasing the tension on the spanning table.

The amount of tension applied is usually controlled by measuring the resulting elongation of the screen fabric. Two sharp lines are marked in either direction before tension is applied. The increase of the distance in between the two lines serves as a measure for the tension given. The optimum tension is usually specified by the manufacturer of the screen fabric and must be carefully observed. For polyester fabrics, it is typically in the range of 2–4% and for stainless steel fabrics around 1% elongation. Overtension must be strictly avoided especially with the stainless steel fabrics because irreversible changes could take place. Overtension can easily occur in the corner portions, therefore the initial fixing of the fabric in the corner regions should be a little loose. It is a recommended practice to increase the tension in a very smooth manner up to the final value. It will permit the material structure to get gradually reoriented. Polyester fabrics are spanned in two or three smooth steps. After each step, half a minute rest is given, for molecular reorientation.

Before the screen goes for the pattern transfer onto it, a *degreasing operation* has to be carried out: A 20% caustic soda solution is splashed onto the screen and regularly distributed with a synthetic brush. After 5 minutes, the solution is thoroughly removed with a water rinse. It is followed by neutralising the screen in a 5% acetic acid solution and another spray water rinse. The drying is done with the help of a warm air blower or a fan. But beware of dust, which could be deposited on the screen. Touching of the screen with fingers is now forbidden.

14.4 PATTERN TRANSFER ONTO THE SCREEN

There are two different methods in use, and each method has its own advantages and disadvantages. With the *direct method*, the screen is prepared, by coating a photographic emulsion directly onto the screen fabric and exposing it in direct contact with the film master. The developing opens the meshes in the pattern area. The *indirect method* makes use of a separate screen process film, supported on a backing sheet. The film master is exposed onto this film followed by film developing which dissolves the unexposed areas of the film and leaves the rest on the backing sheet. The film on its backing sheet is thereafter pressed onto the screen fabric and sticks there. Finally the backing sheet is peeled off, opening all those screen meshes, which are not covered by the film pattern.

The direct method provides very durable screen stencils with a higher dimensional accuracy but the finest details are not reproduced.

The indirect method is more suitable for smaller series and where the finest details have to be reproduced. The indirect method is faster but dimensionally less accurate and the screen stencils are less durable, more sensitive to mechanical damages and interruptions in printing.

14.4.1 Direct Method

Coating

Screen-process photographic emulsions are usually based on polyvinyl alcohol, polyvinyl acetate or polyvinyl chloride and contain some dye to make it visible during processing. Certain brands are available on the market with a separate sensitiser (potassium dichromate, ammonium dichromate or sodium dichromate) which has to be added and mixed to the emulsion before use. Other brands are available in pre-sensitised state. When using these emulsions, one has to follow strictly the instructions given by the manufacturer with respect to mixing and resting time before use.

The use of screen-process photographic emulsions has certain similarities with wet-film photoresists. The light-sensitive region is roughly the same for both. This means that working with screen-process photographic emulsions must be done in rooms illuminated with yellow, orange or red light of low intensity. Exposure is done with the same type of light sources as they are used for the photoresist exposure. The exposure similarly brings about a polymerisation in the exposed areas which makes these portions insoluble in the developer.

The actual coating of the screen is done by regular distribution of a sufficient quantity of photographic emulsion with a squeegee all over the screen surface. More uniform layers are obtained by using a channel-type of coating device (Fig. 14-7), passing from bottom to top over the slightly tilted screen.

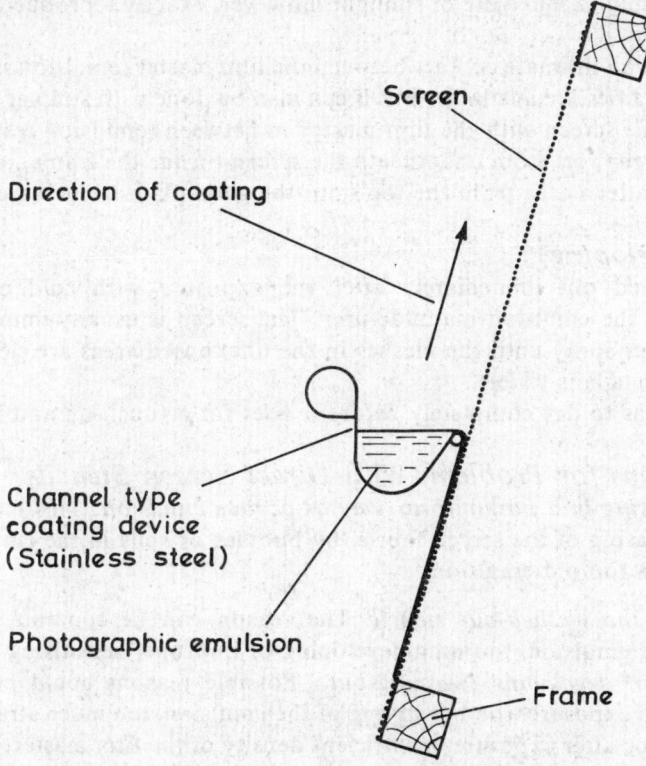

Fig. 14-7 Coating with a channel-type coating device

To provide sufficient film thickness, several coatings are necessary, alternatively applied on both the sides. Complete drying in horizontal position (for certain emulsions upto 20 min) is required before the application of the next coat. It needs some experience to find out the optimum number of coatings (Fig. 14-8). A thicker film basically needs more exposure, is more durable but reproduces less details, while a thin film gives very fine details but is less durable.

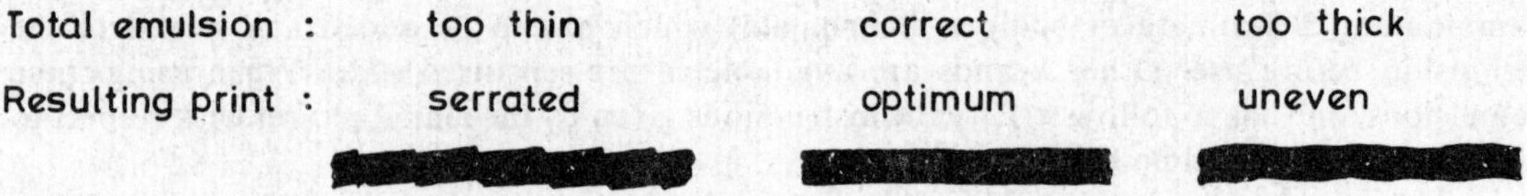

Total emulsion :	too thin	correct	too thick
Resulting print :	serrated	optimum	uneven

Fig. 14-8 Emulsion thickness affects printing results

Exposure

The exposure time depends on many factors like sensitivity of the emulsion, emulsion thickness, spectral sensitivity of the emulsion, light-spectrum and power of the light-source, relative humidity, distance from light source to screen, etc. The exact exposure time, usually in the range of several minutes, must therefore be experimentally determined. The same test methods as used for the photoresist exposure determination are applicable (see Chapter 13).

Also, the light-sources themselves are similar to those used for photoresist exposure. It is even possible, simply to make use of sunlight; however, exactly reproducible results will then not be possible.

To guarantee an intimate contact between the film master (emulsion side) and the screen, special vacuum-frames are available. But it can also be done with simpler means by putting a glass plate over the screen with the film master in between (emulsion resting on the screen). With the suitable support from underneath the screen (inside the frame), more weight can be put outside the pattern area from the top onto the glass plate to increase contact pressure.

Washing (Developing)

This is carried out immediately after the exposure with cold or warm water, as recommended by the emulsion manufacturer. The screen is usually immersed in water and rinsed with a water spray until the meshes in the unexposed areas are cleared and the sharp pattern with all details is visible.

The screen has to dry completely before it goes for a touch-up and for printing trials.

Possible Reasons for Problems with Direct Screen Stencils

Emulsion drying in a nonuniform way or porous emulsion: This may be caused by an insufficient degreasing of the screen fabric, by bubbles or soils in the emulsion, by irregular coating or using a too old emulsion.

Porosity in the washed-out stencil: The reason can be too thin coating, too short exposure, too old emulsion, too long developing or improper sensitising.

Difficulties in developing (washing-out): Possible reasons could be too long waiting after coating until exposure, too hot drying of the emulsion, too much stray light reaching the emulsion before or after exposure, insufficient density of the film master, insufficient contact between film master and screen during exposure.

14.4.2 Indirect Method

Photographic Screen Process Films

These films consist usually of a film emulsion on a backing sheet. The backing sheet is finally removed after sticking the emulsion onto the screen.

The films are commercially available either in a pre-sensitised or in an unsensitised state. Unsensitised films must be sensitised before exposure and processing. It can be done by immersion of the film in a tray containing the prescribed sensitising agent, for the recommended time. Another method is to brush the sensitiser onto the emulsion and to then place it, emulsion side down, on blotting paper which has been thoroughly wetted by brushing sensitiser onto both the sides.

It is clear that irregular sensitising will lead to irregular screen stencil results. Therefore, take all the care to keep processing parameters, solution concentrations, timings, etc. at a constant value.

Exposure

The exposure of screen process films is best done in a vacuum frame in direct contact with the film master. But watch carefully: The exposure must be done through the backing sheet. The film master emulsion therefore rests on the backing sheet and not on the screen process film emulsion. This will cause some diffusion and unsharpness in the pattern which, however, can be minimised by a short exposure time (powerful light source). As an exposure light source, point light sources with a high UV radiation give, again, best results.

If we use unsensitised film, the exposure should occur soon after sensitising, when the film is still wet. To protect the vacuum frame, thin transparent plastic sheets can be put in-between.

The proper amount of exposure is here still more important than in the direct method: The emulsion should be hardened by exposure to about just two-thirds of its thickness and the remaining third serves as an adhesive layer on the screen fabric (Fig. 14-9). In the direct method, a thorough hardening of the emulsion has to be achieved: This makes exposing less sensitive towards over-exposure.

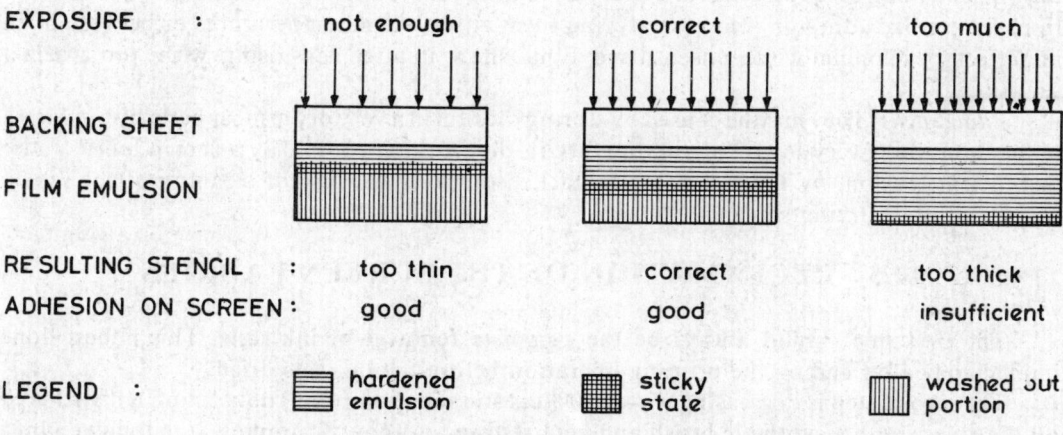

Fig. 14-9 Indirect method: Exposure of screen process film

Washing (Developing/Hardening)

Pre-sensitised films: Before washing, they are developed (hardened) in a solution as prescribed by the film manufacturer. This operation determines not only the hardness, but to some extent also the thickness of the emulsion.

Films sensitised directly before exposure: We can immediately start with the washing procedure. The film is immersed in a tray with warm water (45–50°C), emulsion side down. With a mild spray water rinse from below, the film is treated until the pattern appears with all details. The film is afterwards rinsed in cold water for 1/2 min.

Transferring the Film onto the Screen

The wet film, after the water rinse, is placed on a thin glass plate of slightly bigger size than the film. The emulsion side has to look upwards while the backing sheet is in contact with the glass support. Then the perfectly clean screen is lowered and brought into contact with the emulsion. Weight might be put on the frame to maintain a perfect and uniform contact between emulsion and screen fabric for drying. Make sure that you do not have too high a contact pressure: This would cause the threads to cut the emulsion. The excess water on the screen can be sucked off with a sheet of paper, but never enforce the drying with *hot* air. Fans, however, are suitable to effect a complete drying. After drying, the backing sheet is peeled off and the screen area surrounding the film pattern is coated with a suitable filler lacquer. Touch-up is done, wherever there are pinholes and other defects in the pattern.

Possible Reasons for Problems with Indirect Screen Stencils

Emulsion gets separated from backing sheet while washing: This might occur after the film has been stored at high relative humidity or if it has been exposed from the emulsion side and not, as recommended, through the backing sheet. Pre-sensitised films show this separation effect also after too long a developing time or with an imperfect developer; with unsensitised films, the effect might come from an unsuitable sensitiser.

Too little adhesion to the screen fabric: Adhesion can be insufficient because of a wrong exposure, too high a temperature in the working room, too warm or too fast film drying, insufficiently degreased screen fabric or a very high or very low relative humidity. Pre-sensitised films have a bad adhesion also after storing them at too warm temperatures or because of an imperfect developing while unsensitised films show it after sensitising with too warm a sensitiser.

Film cracks: They may be caused by storing the film in a very dry atmosphere, by too high a temperature or a very low relative humidity in the working rooms, by a screen fabric under insufficient tension, by too much off-contact distance or because of cleaning of the screen stencil with fast-drying solvents.

14.5 RECLAMATION OF THE SCREEN FABRICS

The first step should always be the complete removal of ink rests. This is best done immediately after ending the printing operation before the ink gets dried.

The second step is degreasing with a 20% caustic soda solution. The solution is distributed on the screen with a synthetic brush and kept resting for about 5 minutes. It is followed by a water rinse.

The next step depends on whether a direct or indirect method screen stencil has been used.

Direct Method Screen Stencil

Generally, the decoating chemicals recommended by the emulsion manufacturer should be used. For emulsions based on polyvinyl alcohol, a solution of sodium hypochlorite is often applied with a content of active chlorine of maximum 4%. The solution can be applied onto the screen for maximum 15 minutes. After that, the screen is washed with a high-pressure water rinse.

If there are still residuals remaining on the screen, never try with higher concentrated decoating chemicals or abrasives. This might damage the fabric. The residuals can easily be removed by rubbing with a piece of cloth which has been soaked in butylacetate or xylol or toluol.

Before applying a new coating, the screen has in any case, to be treated with a 20% caustic soda solution, rinsed, neutralised with a 5% acetic acid solution and again rinsed.

Indirect Method Screen Stencils

The decoating operation of these stencils is usually much easier. In certain cases, an immersion in hot water with brushing only gives the result. Film manufacturers recommend normally their own chemicals. In difficult cases, the same treatment as for directly-coated screen stencils will lead to the desired result.

Here, also, a treatment with 20% caustic soda solution, rinsing, 5% acetic acid solution and rinsing again, makes the screen finally ready for further use.

Caution

Most of the chemicals used for decoating are of a polluting nature and may also cause health hazards to the personnel. Such chemicals should be used only in smallest useful quantities. They have to be properly neutralised after use for disposal.

In many countries, laws restrict the use of these chemicals in order to care for a healthy natural environment. But even in countries where such laws are not yet framed or enforced, it is a matter of industrial ethics and of common sense to follow the same rules whose only scope is to avoid catastrophic pollution situations.

See also Chapter 22 on such problems.

14.6 PRINTING

14.6.1 Inks (Resists)

The ideal screen-printing ink should have many features which cannot be combined: It should dry rapidly on the PCB but dry slowly on the screen; it should be highly resistant against all the chemicals but easy to be stripped; it should be easily visible but not contaminate the cleaners, etc. The resists actually used offer, therefore, always a compromise among the desirable qualities.

The inks used in screen-printing of etch or plate resists are mostly vinyl based and can be broadly classified into two groups which are the resists soluble in solvents and the resists soluble in alkaline solutions.

Solvent-soluble resists: They are highly resistant against chemical attacks in both the acidic as well as the alkaline range. Stripping has to be done with solvents like trichloroethylene or trichloroethane e.g., in stripping machines.

Alkaline-soluble resists: The use of these resists has become very economic: No solvents are applied and minimum pollution is caused. The stripping can simply be done by spray-rinsing with a caustic soda solution (2.5%) at a temperature of 30–40°C. The limitation of these resists is that they withstand only etchants or plating electrolytes in the acidic range. But this does not pose serious problems with etchant and plating electrolytes available nowadays. The trend is definitely towards the much cleaner alkaline-soluble resists because of the more and more severe environmental problems (and resulting laws) coming up in most of the countries.

Another advantage is that the alkaline-soluble resists, if used as an etch resist, do not have to dry completely before etching. This means a simplified and shortened drying process.

For contour sharpness and thickness of the screen printed resist pattern, proper viscosity is the main factor to be concerned about. But also the thixotropic properties (characteristic flow properties under pressure) of the resist are of great importance besides the viscosity. Resists with the proper viscosity but not the right thixotropy may flow into the holes or widen the contours of the printed pattern.

Fig. 14-10 Adjustable had-screen-printing set-up (Courtesy: The Agron Service Ltd., Milano, Italy)

14.6.2 Hand Screen–Printing

Hand screen-printing is carried out on a sturdy work bench. The prepared screen is fixed in a frame holder which is held by a precise hinge-mechanism, preferably adjustable in the two horizontal axis. A squeegee is chosen with a width of 2–4 cm more than the pattern width.

Sufficient resist of adjusted viscosity is brought onto the screen. The first few prints are made on clean newspaper until the pattern gets perfectly printed. The resist, for the printing pass, must form a wave in front of the squeegee along its entire length. The squeegee is pulled and pressed firmly across the pattern on the screen. After the printing pass, excess resist and squeegee are returned to the upper screen end. The screen frame is raised and preferably kept by a counterweight balance.

The boards to be printed are kept in printing position either by a frame formed by spacers of similar thickness or with indexing pins (recommended for double-sided PCBs).

The off-contact distance (distance between the surface to be printed and the screen) is a compromise between pattern registration and contour sharpness. A typical value is around 2 mm. If it is more, contour sharpness gets further improved but more registration error is introduced because the moving squeegee presses the screen down by the off-contact distance, thereby causing an elongation of the screen fabric and simultaneously of the pattern stencil on it. Especially for screen stencils made by the indirect method, the off-contact distance must be kept lowest possible, because the film can withstand only little stress due to elongation.

The most important variables and their impact on resist thickness, contour sharpness and registration accuracy are given in Table 14-1 as a guide in practice: The table shows clearly that screen-printing means choosing a compromise between the various desirable qualities.

Table 14-1 Screen-printing variables and their impact on the printing result. Symbols in brackets indicate possible impacts of a secondary nature

		Resist thickness	Contour sharpness	Registration accuracy
Screen fabric,	Polyester	−	+	−
	Steel	+	−	+
Screen coating thickness,	5–15 μm	−	+	
	15–30 μm	+	−	
Squeegee hardness,	soft 60–70 Shore	+	−	+ (−)
	hard 80–90 Shore	−	+	− (+)
Squeegee pressure,	less	+	−	+
	more	−	+	−
Off-contact distance,	less		−	+
	more		+	−
Printing speed,	less	+ (−)	+ (−)	+ (−)
	more	−(+)	−(+)	−(+)
Resist viscosity,	less	− (+)	− (+)	+ (−)
	more	+ (−)	+ (−)	− (+)

14.6.3 Screen–Printing with Machines

The screen-printing with semi- or fully-automatic equipment gives very regular results and probably less dependence on the personnel. However, most of what has been said under hand screen-printing is equally applicable to screen-printing with machines.

There is a broad variety of designs of screen-printing machines available. To deal with all of them is not possible in this book. But a few details are mentioned here:

The squeegee can be driven by different means such as pneumatically, electro-mechanically or hydro-pneumatically. It is very important that the squeegee speed should be steplessly regulated, down to the minimum speed.

The machines are mostly supplied with an adjustable printing table. A big help in alignment and adjustment of the printing table (PCB position) is the provision of micrometer screws and dial gauges.

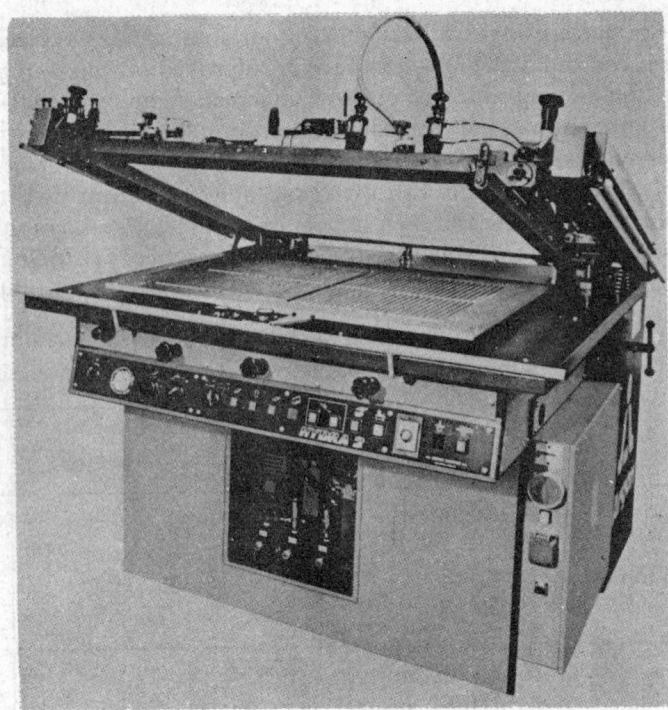

Fig. 14-11 Automatic screen-printing machine, pneumatically driven
(Courtesy: The Agron Service Ltd., Milano, Italy)

14.6.4 Screen Cleaning

Cleaning of screen stencils becomes necessary for interruptions of more than half an hour (or, sometimes even less, depending on how fast the ink dries). Screen-cleaning must also be done carefully after finishing a series of prints. Dry ink is usually quite difficult to remove.

The cleaning is, in any case, done from the squeegee side. The excess ink is first removed with a spatula and put into a separate tin because it cannot be used again for high-quality prints. With a soft piece of fuzzle-free cloth, wetted with cleaning solution, the flat-laying screen is carefully rubbed while sucking paper (newspaper) is put from below. The cleaning solution used (e.g., thinner or solvent) should not attack the screen fabric or stencil pattern, within a reasonable time.

As a general rule, screen contact with the cleaning solution should be minimum and of short duration especially for indirect stencils. The cleaning operation is only finished when all the meshes are freed from ink remainings.

14.7 TROUBLE-SHOOTING

If the screen stencils do not give the durability that is expected, various reasons could be responsible for this:

Direct Stencils
—insufficient degreasing before coating
—too little tension of the screen fabric
—insufficient coating thickness
—emulsion not hard enough (underexposed)'
—ink contaminated
—ink contains aqueous substances.

Indirect Stencils
—insufficient degreasing
—too little tension of the screen fabric
—screen fabric too coarse
—wrong exposure of the film
—off-contact distance too much
—working room temperature too much
—relative humidity too low
—screen stencil has been cleaned with solvents evaporating too rapidly
—screen stencil has been cleaned from the emulsion side and not, as recommended, from the squeegee side.

BIBLIOGRAPHY

1. COOMBS C.F.: *PRINTED CIRCUITS HANDBOOK*, McGraw-Hill Book Co., New York, 1967.
2. HARIDAS V. K.: 'Screen printing techniques for the electronics industry', *ELECTRONICS FOR YOU*, Vol. XI, No. 1, New Delhi, 1979, pg. 27–30.
3. HERRMANN G.: *LETTERPLATTEN*, Leuze Verlag, Saulgau/Wurtt. (Germany), 1978.
4. SCHALT E.A.: 'Der Siebdruck in der Elektronik', *FAG INFORMATION*, FAG SA, Lausanne (Switzerland), 1972.
5. SCHEER H.G.: *SIEBDRUCK, MONYL INFORMATION*, Zuricher Beuteltuchfabrik AG, Ruschlikon (Switzerland), 1967.

15
PLATING

15.1 INTRODUCTION

From a practical stand point, printed circuit boards may have to be stocked before being taken for assembly of components. It is expected that the circuit board retains its solderability for long periods of several months so that reliable solder joints can be produced during assembly.

Limitation of Bare Copper

Copper being the conducting material on the printed circuit board and forming the conducting track patterns, it is liable to tarnish, if exposed to atmosphere over long periods of time, endangering solderability.

Protection of Copper Tracks

It has therefore become customary to protect the copper tracks. However, protection by means of lacquer leaves much to be desired in the long-term storage properties of the lacquer, variations in thickness, composition and curing cycles, thus causing unpredictable deviations in solderability.

Plating a solderable metal over the copper tracks has hence become a standard practice to afford solderable protection to the copper tracks.

With the advent of the double-sided, plated-through hole boards, plating copper possessing high ductility and thermal-shock resistant properties has assumed a new significance.

Edge Connectors

Modular construction of electronic equipments led to the use of edge-board connectors with spring contacts mating with suitably designed connector tabs on printed circuit boards. Such contacts are required to possess a high degree of wear resistance and a very low contact resistance. This triggered off intensive research in precious metal plating, the most commonly used material being gold.

Plating Techniques

Plating of a metal can be accomplished on a copper pattern by three methods. They are:

a) Immersion plating
b) Electroless plating
c) Electroplating.

15.2 IMMERSION PLATING

Immersion plating is the deposition of a metallic coating on a substrate, by chemical replacement, from a solution of a salt of the coating metal. The substrate metal reduces the atoms of the metal in solution on itself by reducing the atoms from their ionic state in solution.

Advantages of immersion plating are simplicity, minor capital expense and the ability to deposit in recesses.

On the other hand, immersion plating thickness obtainable is usually extremely limited, because as soon as the substrate metal is completely or almost completely covered by the deposit, the reaction stops or slows down to the rate at which substrate metal is available through pores or discontinuities in the coating. Such limited thickness gives only short-term protection to printed circuit patterns and hence is used only in cases where immediate assembly of components is envisaged.

For practical purposes, this category is concerned with only two coating metals: Tin and its alloys and gold.

15.2.1 Tin Immersion Plating

Tin is immersion plated on copper conductors in printed circuitry; the typical thickness being about 3μm or less in an hour.

Copper is normally more noble (less readily reactive) than tin, and thus it would not be expected that the reaction

$$Cu + Sn^{++} \longrightarrow Sn + Cu^{++}$$

would take place simultaneously. Sn^{++} and Cu^{++} signify respective metal ions positively charged due to lack of two electrons. However, the electrode potential of copper can be made much more negative by the incorporation of a complexing agent for copper in the tinning solution.
Electrode potential: The copper on which tin is to deposit forms an electrolytic cell and the potential at the electrode which exists on account of ionisation is called the electrode potential.

As a complexing agent, cyanide ions are used in some formulations; thiourea also forms stable complexes with copper and is an ingredient of some other solutions.

Tin Bath Composition
A typical composition for an immersion tin bath for printed circuits is given below:

Stannous chloride, $SnCl_2 \cdot 2H_2O$	9.5–19 g/l
Thiourea, $CS(NH_2)_2$	80–90 g/l
Hydrochloric acid, HCl (conc.)	10–20 ml/l
Bath working temperature	60°C
Immersion time	5 min

15.2.2 Immersion Plating of Gold

A typical formulation for immersion gold which can deposit about 0.025 μm in 30 min is given below:

Gold(I) cyanide, AuCN	2.4 g/l
Potassium cyanide, KCN	12 g/l
Batch working temperature	80°C

The gold coatings obtained thus are relatively pore-free in spite of thinness and in view of the fact that the volume of gold deposited is about twice that of copper displaced.

15.2.3 Rinsing

It is very important to rinse thoroughly after immersion plating to eliminate all residual chemicals to ensure an acceptable shelf-life of a few weeks for optimum soldering. It is preferable to have a double rinse, the final rinse being made with de-ionised water.

15.3 ELECTROLESS PLATING

15.3.1 Electroless Plating Processes

A unique combination of characteristics is given with electroless plating processes which renders them very useful:

a) The throwing power is essentially perfect; deposits can be made on any surface to which the solution has free access, with no excessive build-up on projections or edges. The throwing power is the ability of the electrolyte to cover areas recessed from the anode as related to straight cathode areas.
b) Deposits are often less porous than electrodeposits.
c) Power supplies, electric contacts, bus bars and electrical measuring instruments are not required.
d) Deposits can be produced on non-conductors (with proper pre-treatment).
e) Some deposits have unique mechanical, chemical or magnetic properties.

15.3.2 Electroless Copper Plating

With the advent of the plated-through hole technology, it has become necessary to provide a copper layer on a drilled hole on its insulating area to obtain electrical conductivity for subsequent electrolytic copper deposition. This is most commonly achieved by what is known as electroless plating of copper, wherein copper is reduced from its ionic state in solution, by means of a chemical *reducing agent* rather than by an electric current. The reducing agent reduces the metal ions in solution to neutral metal atoms for deposition. The process provides a continuous build-up of metal coating on a suitable substrate by simple immersion in an appropriate aqueous solution. A chemical reducing agent in the solution supplies electrons for the reaction, $M n^+ + ne \longrightarrow M^\circ$, but the reaction takes place only on a *catalytic surface*. A catalytic surface is a surface on which a noble metal like palladium is seeded to trigger the chemical reduction process.

15.3.3 Chemical Reduction

Most commercial electroless copper baths employ formaldehyde as the chemical reducing agent. The reducing power of formaldehyde increases with the pH.

The pH of a solution is equal to the numerical value of the exponent in the denominator of the fraction that expresses the hydrogen ion concentration. (In a more mathematical language: The pH is the negative logarithm of the concentration of hydrogen ions, i.e., $pH = -\log CH^+$.) For example, in a neutral aqueous solution in which the hydrogen-ion concentration is 10^{-7} N or $1/10^7$ N, the pH is 7. Solutions with a pH < 7 are called acidic and pH > 7 are called alkaline.

The alkalinity in the baths is provided by sodium hydroxide. Other ingredients include a copper (II) salt, usually copper sulfate; a complexing or chelating agent to hold the copper in solution; and various stabilisers and other additives. The most common chelating agent is tartrate ion, but several others are used, including EDTA, amines and amine derivatives and glycolic acid.

The drilled printed circuit board is *activated* by coating with a thin film of palladium by an activator and then immersed in the copper bath. The following equation illustrates the copper deposition process using formaldehyde as the reducing agent.

$$CuSO_4 + 2HCHO + 4NaOH \xrightarrow{Pd} Cu + 2HCO_2Na + H_2 + 2H_2O + Na_2SO_4$$

15.3.4 A Typical Electroless Copper Bath

A typical electroless copper plating sequence is as follows:
1. Vapour degrease, pure trichloroethylene, 1 min
2. Rack boards, off-vertical
3. Alkali soak cleaning
4. Rinsing, cold water
5. Copper etching, ammonium persulfate 240 g/l, room temperature, 2 min
6. Spray rinse, cold water
7. Acid dip, sulphuric acid H_2SO_4 (25% vol.), 1 min
8. Spray rinse, de-ionised water
9. Acid dip, hyrdrochloric acid HCl (25% vol.), 2-5 min
10. Catalyst, 5 min

Formulation:		
	$PdCl_2$	700 mg/l
	$SnCl_2$	35 g/l
	$SnCl_4$	4.5 g/l
	HCl (conc.)	500 g/l

11. Double rinse, de-ionised water, 2 min
12. Accelerator, 10 min, H_2SO_4 110 g/l
13. Double rinse, de-ionised water, 2 min
14. Copper reduction, 30 min

Formulation:		
	$CuSO_4 \cdot 5H_2O$	13 g/l
	$KNaC_4H_4O_6 \cdot 4H_2O$	66 g/l
	NaOH	20 g/l
	Mercaptobenzothiazole	0.013 g/l
	HCHO (37%) with 12.5% methanol as preservative	38 ml/l

15. Spray rinse, cold water
16. Neutralisation dip, sulphuric acid H_2SO_4 (25% vol.), 1 min
17. Spray rinse, cold water
18. Proceed to flash plate copper in acid copper sulfate bath.

15.4 ELECTROPLATING

15.4.1 Principle

The passage of a unidirectional current through a solution is associated with the movement through it of charged particles called ions. The terminals leading the current into the solution are electrodes: The pole at which the chemical reaction of reduction takes place is the cathode and the pole at which the chemical reaction of oxidation takes place, is the anode. In an electrolytic cell, the total process of decomposition due to the passage of a current is called electrolysis.

15.4.2 Faraday's Laws

The fundamental principle of electrolysis is understood from Faraday's laws which are as follows:
1. The quantity of any element (or radical i.e. group of elements) liberated at either the anode or cathode during electrolysis is proportional to the quantity of electricity that passes through the solution.
2. The quantities of different elements or radicals liberated by the same quantity of electricity are proportional to their equivalent weights.

If 1 coulomb (1 A sec) deposits 1.118 mg of silver, then 10 coulombs will deposit $10 \times 1.118 = 11.18$ mg of silver. It is immaterial whether a current of 1 A flows for 10 sec or of 2 A for 5 sec, or any other current and time that yield 10 coulombs.

The equivalent weight of an element is its atomic weight divided by its valence in the compound present.

Since 1 coulomb deposits 0.0011182 g of silver, it requires $107.88 \div 0.0011182 = 96,488$ coulombs to deposit equivalent weights in grams of silver or any other element. This quantity of electricity (often rounded off to 96,500 coulombs) is known as one faraday. It is equivalent to about 26.8 Ah or about 1.1 A-days. It is sometimes convenient to remember that the equivalent weight in grams of an element is deposited in about 1.1 A-days.

For a more exhaustive treatment of the mechanism of electro-deposition, the reader is advised to refer to the references in the bibliography, in particular [6].

15.4.3 Preplate Treatment

For all electroplating processes employed in printed circuit manufacture, it is essential to operate preferably in the pH range which is acidic as most of the plating resist materials for the purpose withstand admirably acid-type electrolytes. In the following paragraphs, a typical sequence of preplate treatment is suggested which will by and large suffice for all plating requirements. It cannot be overemphasised that cleanliness required for printed circuit plating is of a very high order and mostly more stringent than what can be tolerated in the decorative plating field.

15.4.4 Purity of Water

The water used for preparing and topping up electrolytes as well as for final and critical rinses must invariably be deionised to maximum 5 ppm dissolved solids. The dissolved solids are usually measured by recording the conductivity of the solution using a conductivity meter. 5 ppm roughly corresponds to 1 μ mho cm. Ordinary cold water rinses also use water which is softened and contains dissolved solids of not more than 500 ppm. Such high-purity water is obtained by employing ion exchange columns which are available from reputed manufacturers like Ion Exchange (India) Limited. All preplate treatment solutions are also prepared using deionised water to prevent any drag-in into the electrolytes.

15.4.5 Typical Preplate Treatment

A typical treatment sequence is described below:

1. Vapour degrease, use of an organic solvent such as trichloroethylene. A mixture of trichloroethylene and perchloroethylene in the ratio of 60 : 40 gives best results. This mixture, however, cannot be used when the panels are imaged since most resists are attacked by chlorinated hydrocarbons.
2. Alkali cleaning, 1 min

Formulation:	Sodium hydroxide, $NaOH$	25% (vol.)
	Sodium metasilicate, Na_2SiO_3	40%
	Sodium tripolyphosphate, $Na_5P_3O_{10}$	10%
	Sodium carbonate, Na_2CO_3	23%
	Sodium linear alkyl sulfonate (40%)	2%
	Concentration	30–60 g/l
	Current density	200–500 A/m^2
	Bath temperature	50–60°C

3. Spray rinse
4. Copper etching, ammonium persulfate 240 g/l, room temperature, 2 min.

   Use proprietary *microetches* if treatment is on electroless copper of little thickness. Proprietary microetches are dilute etchant solutions incorporating suitable wetting agents and inhibitors in order to mildly etch the thin electroless copper surface without sizable metal removal to render the surface more receptive to subsequent electroplating for better adhesion.
5. Spray rinse
6. Sulphuric acid H_2SO_4 (25% vol.), room temperature, 1 min
7. Spray rinse
8. Acid dip, in an appropriate acid to suit the type of electrolyte used, to prevent drag-in:
   —sulphuric acid for acid sulfate
   —hydrochloric acid for acid chloride
   —fluoboric acid for acid fluoborate, etc.

15.4.6 Plating Installation Considerations

The plating installation is usually on acid-proof construction using *vitrified ceramic* or refractory bricks. (Vitrified ceramic is ceramic fired at glass blowing temperatures to render it resistant to most acids and alkalies). The installation is made of polypropylene tanks mounted on epoxy-powder coated steel racks and polyvinyl chloride enclosures and covers. All electrolytes need individual electrical control with timer and buzzer. Continuous filtration and temperature control will have to be provided. Most metals plated require anode and cathode hooks to be of stainless steel (Monel metal is required for fluoborate electrolytes). Fig. 15-1 shows a typical plating installation.

Fig. 15-1 Manual plating installation (Courtesy: HBS Equipment Division, Los Angeles, Ca. 90040, USA)

15.4.7 Copper Plating

Copper plating is done to build the plated-through holes in double-sided or multilayer PCBs.

As the boards are subject to soldering during assembly and subsequently to widely fluctuating environments in service, the copper deposited is required to be very highly ductile, fine-grained, with excellent adhesion and exceptionally good thermal expansion properties. Several formulations have been marketed and the most widely used is the acid sulfate high acid, low metal, high throwing-power bath. As the drilled holes are perpendicular to the board, a high throwing-power bath enables adequate deposition inside the hole without excessive plating on the flat board surface.

Acid Copper Electrolyte

Copper (in the form of copper sulfate)   15–20 g/l

Sulphuric acid	150–225 g/l
Chloride (in the form of HCl)	0.02–0.08 g/l
Addition agents	as required
Bath working temperature	24–32°C
Anode current density	150–300 A/m^2
Cathode current density	300–600 A/m^2

The plating rate will be about 40 minutes for 25 μm Cu at 300 A/m^2. The most common addition agents employed are gelatin, thiourea and sodium potassium tartrate.

Analysis

Copper (weekly)

1. Pipette a 5 ml sample into a 250 ml Erlenmeyer flask.
2. Add ammonium hydroxide (conc.) until the solution becomes dark blue. Boil to vaporise excess ammonia.
3. Add 2 g of ammonium bifluoride and 5 ml of glacial acetic acid.
4. Cool and add 30 ml of potassium iodide solution (10%).
5. Titrate with standard 0.1N sodium thiosulfate solution until the solution becomes light in colour.
6. Add 2 g of ammonium thiocyanate and 2-3 ml of starch indicator (1%) solution. Continue titrating with 0.1N sodium thiosulfate until the colour changes from dark blue to grayish white, and remains light for at least one minute.
7. Copper in g/l = ml of Na$_2$S$_2$O$_3$ × N$_{Na_2S_2O_3}$ × 12.7.
8. Correct bath concentration to restore operating conditions by necessary addition or depletion, as the case may be.

Sulfuric acid (weekly)

1. Pipette a 10 ml sample into a 250 ml Erlenmeyer flask.
2. Add 150 ml of distilled water and 5 drops of methyl-orange indicator solution.
3. Titrate with standard 0.1N NaOH until the colour changes to a pale green (greenish yellow).
4. Sulfuric acid in g/l = ml of NaOH × N$_{NaOH}$ × 4.90

Chloride (weekly)

1. Add 5 ml of nitric acid (conc.) and a 5 ml sample (carbon-treated and filtered) of copper bath to a 25 ml volumetric flask.
2. Add 10 ml of ethylene glycol and dilute to under 24 ml. Mix thoroughly.
3. Add 1 ml of 0.1N silver nitrate solution. Quickly bring to 25 ml and mix immediately and thoroughly.
4. Prepare a reference solution by carrying out steps 1-3 above, then dilute to 25 ml and mix thoroughly.
5. Allow to stand for 30 ± 2 min, then measure the absorbance versus reference sample at 440 mU in 1 cm cells.
6. Chloride in mg/l = 152 × absorbance at 440 mU.

Addition agents:

Follow supplier's recommendation.

Copper-Plating Trouble-Shooting

Problem	Cause	Remedy
Streaky, banded, dull	Lead contamination Oil contamination	Low-current electrolysis Activated carbon treatment
Stressed, cracked deposits	Excessive additives, or breakdown products	Hydrogen peroxide and activated carbon treatment
Poor throwing power	Impurities build-up	Low-current electrolysis and activated carbon treatment
Skip plating	Degraded brighteners Incorrect current Bad contact	Carbon-treat Correct current Clean contacts
Nodules in holes	Poor drilling or suspended particles in solution	Filter solution

15.4.8 Tin Plating

Pure tin is the most commonly used plated finish for printed circuits. Its advantages are good solderability, and easy plating control.

Tin is plated from acid sulfate type electrolytes similar to the following:

Typical Formulation

Stannous sulfate, $SnSO_4$	54-90 g/l
Free sulfuric acid, H_2SO_4	40-70 g/l
Phenol sulfonic acid, $HOC_6H_4SO_3H$	30-60 g/l
Gelatin	2 g/l
β-naphthol, $C_{10}H_7OH$	1 g/l
Bath working temperature	20-30 °C
Cathode current density	200-500 A/m^2
Anode current density	200 A/m^2

The plating rate will be about 10 min for 10 μm Sn at 2 A/dm^2.

Analysis
Tin (biweekly)
1. Pipette a 5 ml sample into a 250 ml Erlenmeyer flask containing 100 ml of water and 35 ml of HCl (conc.).

2. Add 1g of $NaHCO_3$ and, after foaming ceases, 2.0 g of iron powder. Boil until the iron dissolves.
3. Add another gram of $NaHCO_3$ and cool rapidly under running water.
4. Add 5 ml of starch indicator and titrate rapidly with standard 0.25 N standard iodine solution until the solution turns blue and the blue colour persists for half a minute.
5. Total tin in g/l = ml of iodine $\times$ ($N_{12} \times 11.87$)

Free mixed acids (monthly)
1. Pipette a 5 ml sample into a 250 ml Erlenmeyer flask, and add 50 ml of ammonium oxalate solution (4%).
2. Add 10 drops of methyl-red indicator and titrate with standard 1.0 N NaOH solution until the colour changes from red to yellow.
3. Mixed acid to be added

$$g/l = (8 - \text{ml titration with 1.0 N NaOH}) \times \frac{130}{8}$$

A second flask containing 50 ml of ammonium oxalate and 10 drops of the indicator can be used as a standard for end-point comparison.

Tin-Plating Trouble-Shooting

Problem	Cause	Remedy
Rough, grainy deposit	Suspended dust or stannic tin	Filter or decant and correct tin content
Streaked and striated deposit	Organic contamination	Active carbon treatment
Low current density, dullness	Additive inadequacy	Correct by addition
High current density, burn	Low metal content	Analyse and correct
Skip plating	High metal, low acid	Analyse and correct

15.4.9 Nickel Plating

Nickel is most often used as an undercoat for gold to act as a barrier layer to copper migration as well as to increase the wear resistance of the coating thereby reducing the minimum gold deposit needed for a given application. As nickel passivates very rapidly in normal ambients, an activation is necessary before plating gold to ensure proper adhesion. The normal coating thickness range is 4-6 μm.

The deposits are usually semibright and plated from Watts type bath with appropriate stress-reducers.

Typical Nickel Plating Bath Formulation

Nickel sulfate, $NiSO_4 \cdot 6H_2O$	330 g/l
Nickel chloride, $NiCl_2 \cdot 6H_2O$	45 g/l
Boric acid, H_3BO_3	38 g/l
Stress reducers and addition agents	as required

pH 3-4
Bath working temperature 45-65°C
Cathode current density 250-1000 A/m^2

Sodium lauryl sulfate and sodium allyl sulfonate are ingredients in many proprietary addition agents for nickel.

Analysis

Nickel (biweekly)

1. Pipette a 2 ml sample into a 250 ml Erlenmeyer flask.
2. Add 90 ml of distilled water and 10 ml of ammonium hydroxide (conc.).
3. Add 'Eriochrome Black T' indicator, 1/2 g, solid.
4. Titrate into standard 0.06 M EDTA solution (Na$_2$) until the colour changes from greenish to a deep bluish violet colour.
5. Nickel in g/l = ml of Na$_2$ EDTA × (M Na$_2$ EDTA × 29.35).

Chloride (biweekly)

1. Pipette a 5 ml sample into a 250 ml Erlenmeyer flask.
2. Dilute to 100 ml with distilled water, adding a little powdered calcium carbonate (1 g) if the pH is below 4.0.
3. Add 1 ml of potassium chromate indicator (5%).
4. Titrate with standard silver nitrate solution (0.1N) until a drop gives reddish-orange colour to the silver chloride precipitate.
5. (Nickel chloride · 6H$_2$O) in g/l = ml of AgNO$_3$ × N$_{AgNO_3}$ × 23.8.

Boric acid (biweekly)

1. Pipette a 2 ml sample into 125 ml Erlenmeyer flask.
2. Add 25 ml distilled water, 10 ml of saturated potassium ferrocyanide and 5 drops of bromocresol purple indicator (9% in ethyl alcohol).
3. Titrate with standard 0.1N NaOH until the greenish precipitate present in the flask becomes a distinct blue. Record reading.
4. Add 5 g of mannitol, shake and then add 5 drops of phenolpthalein indicator. The sample becomes green again upon addition of mannitol.
5. Continue titrating with the standard NaOH until the colour of two solutions changes from green to blue to violet-blue. Record the burette reading.
6. Substract step 3 from step 5 = ml of NaOH. Boric acid in g/l = ml of NaOH × (N$_{NaOH}$ × 30.9).

Nickel-Plating Trouble-Shooting

Problem	Cause	Remedy
Pitting	Low metal contact	Analyse and correct
	Low pH	Correct pH
	Contamination	Carbon filtering
Stressed deposits	Wrong operating conditions	Correct pH, temperature, if contamination is not suspected
Blackish deposit	Low pH	Correct pH (nickel carbonate slurry)

15.4.10 Gold Plating

Gold, being by nature soft, is always plated alloyed with minute quantities of indium, cobalt or nickel to provide the desired hardness. Gold is plated on edge connector tabs to provide low contact resistance as well as high wear resistance, to permit a few hundred insertions, as all modular construction and easy serviceability of plug-in type modules depends on these characteristics.

It is customary to plate 5 μm gold over copper or 2-3 μm over a nickel undercoat of 4 μm. Both are equivalent specifications giving nearly equivalent performance.

Typical Formulation

Gold (metal) as $KAuCN_2$	6–10 g/l
Cobalt	0.15–0.25 g/l
Complexing salts (as density), primarily potassium citrate	10–25° Be
pH	3.8–4.5
Bath working temperature	30–40°C
Current density	0.3–1.5 A/dm^2

The deposition rate will be about 13 min for 5 μm at a current density of 1.0 A/dm^2.

Analysis
Gold (weekly)
1. Pipette a 25 ml sample of gold plating solution into a 500 ml Erlenmeyer flask.
2. Add 25 ml sulfuric acid (conc.) *under a hood* and heat to dense white fumes (take utmost care not to breath in the extremely poisonous cyanide fumes).
3. Continue the fuming until the solution becomes transparent. Add 2 ml of nitric acid.
4. Allow to cool, then carefully dilute to 125 ml with distilled water.
5. Filter out the gold on an ashless paper, and wash thoroughly with hot water.
6. Ignite the filter paper containing the gold in a weighed crucible. Cool and reweigh.
7. Gold in g/l = gain in crucible weight [g] $\times$ 40.

Gold Plating Trouble-Shooting

Problem	Cause	Remedy
Gassing and pitting	Low metal content	Correct metal content
Peeling over nickel	Passivated nickel	Active nickel with 50% HCl or proprietary activators
Resist breakdown	Conductivity low	Correct conductivity or density
Yellow precipitate	pH is low	Correct pH (Potassium hydroxide)
Blackish deposit	pH is high	Correct pH (Citric acid)

15.4.11 Tin-Lead Plating

Tin-lead alloyed in the region of 60 : 40 has been found to have the lowest melting point of all tin alloys and hence has been adopted as a printed circuit coating. It is more commonly plated from fluoborate-type electrolytes. On account of its lower melting point, it is possible to reflow it and thereby achieve edge coverage of conductor tracks as also avoid residual stresses resulting in whisker growth as in the case of tin in outer atmospheres. However, reflowing plated tin-lead entails danger of thermal shock, discolouration and *warp* which are not easy to control with varying circuit patterns unless the design stage itself visualises these problems and takes care of them by intentional symmetry and control (The patterns are laid out symmetrically so as to ensure a uniform expansion and contraction in all directions).

*Typical Tin-Lead Electrolyte**

Stannous fluoborate, $Sn(BF_4)_2$	38 g/l
(Tin)	15 g/l
Lead fluoborate, $Pb(BF_4)_2$	19 g/l
(Lead)	10 g/l
Free fluoboric acid, HBF_4	400 g/l
Peptone	5 g/l
Bath working temperature	24 °C
Current density	150 A/m^2

Analysis

Lead (weekly)
1. Pipette a 10 ml sample into a 400 ml beaker and dilute to 200 ml.
2. Heat almost to boiling and add dropwise an excess 10 ml of H_2SO_4 (20%).
3. Cool in running water and let stand for 1 hour.
4. Filter through a weighed Gooch crucible, washing four times with cold sulphuric acid (4%), twice with water, and finally with alcohol (50%).
5. Dry in an oven at 110°C for half an hour, cool and weigh.
6. Lead in g/l = weight of lead sulfate [g] $\times$ 68.3.

Free fluoboric acid (weekly)
1. Pipette a 10 ml sample into a 100 ml Erlenmeyer flask.
2. Titrate directly against a black background to a turbid end-point with standard 1.0 N NaOH solution.
3. Fluoboric acid (free) in g/l = ml of NaOH $\times$ ($N_{NaOH} \times 8.78$).

Tin (as stannous)
1. Pipette a 5 ml sample into a 500 ml Erlenmeyer flask containing 50 ml HCl (37%). Swirl.
2. Add 5 ml of 0.1% starch solution and 1 g sodium bicarbonate.
3. Titrate rapidly with 0.1N iodine solution to a purplish blue end-point having at least a 10 second duration
4. Stannous tin in g/l = ml 0.1 N I_2

US Patent 3554878

Tin-Lead Plating Trouble-Shooting

Problem	Cause	Remedy
Rough, grainy deposit	Suspended dust or stannic tin	Filter or decant and correct tin content
Streaked or striated deposit or gritty finish on reflow	Organic contamination	Active carbon treatment
Low current density, dullness	Additive inadequacy	Correct by addition
High current density, burn	Low metal content	Analyse and correct
Skip-plating	High metal, low acid	Analyse and correct
Dull finish on reflow	Excess lead	Check lead content and correct
Prominent freezelines on reflow	Excessive plating thickness	Check current
Dewetting on reflow	Bad surface	Check treatment

15.5 ALTERNATIVE FINISHES

15.5.1 Limitations of Tin-Lead Plating

The plate-and-etch method of manufacture of printed circuit boards uses tin, tin-lead or gold as an etch-resist to withstand the etchant and as such leaves an overhang of plated metal on the conductor edges after etching, which is inimical to the ultimate reliability of the PCB. The process of tin-lead plating and fusing involves difficult control of the alloy ratio over the varying patterns as also problems with thermal shock, discolouration and *warp*. Moreover, it is realised that tin or tin-lead could be plated over relatively unclean surfaces resulting in poor solderability later.

Some researchers therefore devised means of coating a uniform 60 : 40 solder to print-and-etch type of boards by contacting with molten solder rather than by plating.

Three methods are in use for this purpose and will be discussed here.

15.5.2 Roller Tinning

The bare copper boards are suitably fluxed and then coated with solder from a roller, dipping in a bath of molten solder. This is mostly used for mass-produced single-sided boards as it cannot be used for plated-through holes. It gives a relatively thin coat of solder which is mostly adequate for a few weeks storage.

15.5.3 Centrifugal Tinning

In this case, the board is dipped in a bath of molten solder and then swung in a centrifuge to throw off excess solder. This will not give uniformity in thickness from the centre to the edge of the board due to the different centrifugal force from the centre to the edge of the board.

284   PRINTED CIRCUIT BOARDS: DESIGN AND TECHNOLOGY

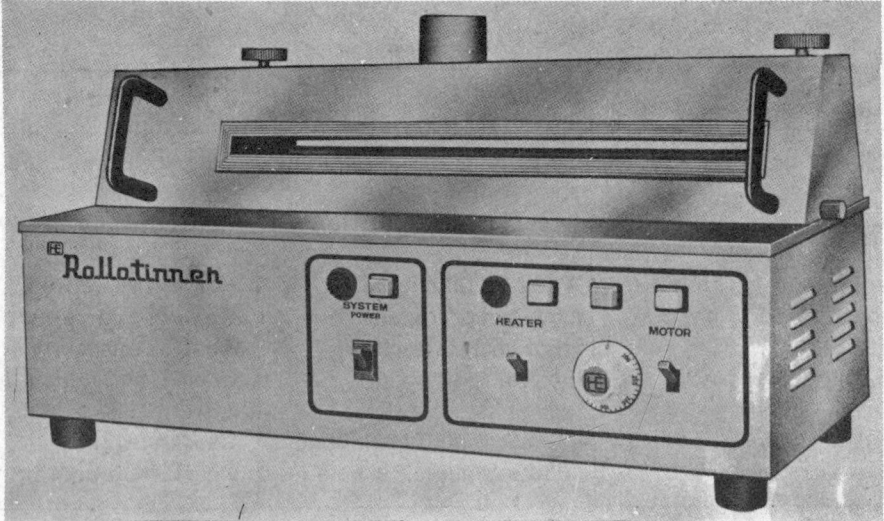

Fig. 15-2 Equipment for roller coating of solder on PCBs (Courtesy: Hansa Enterprises, Cooke Town Bangalore-560005, India)

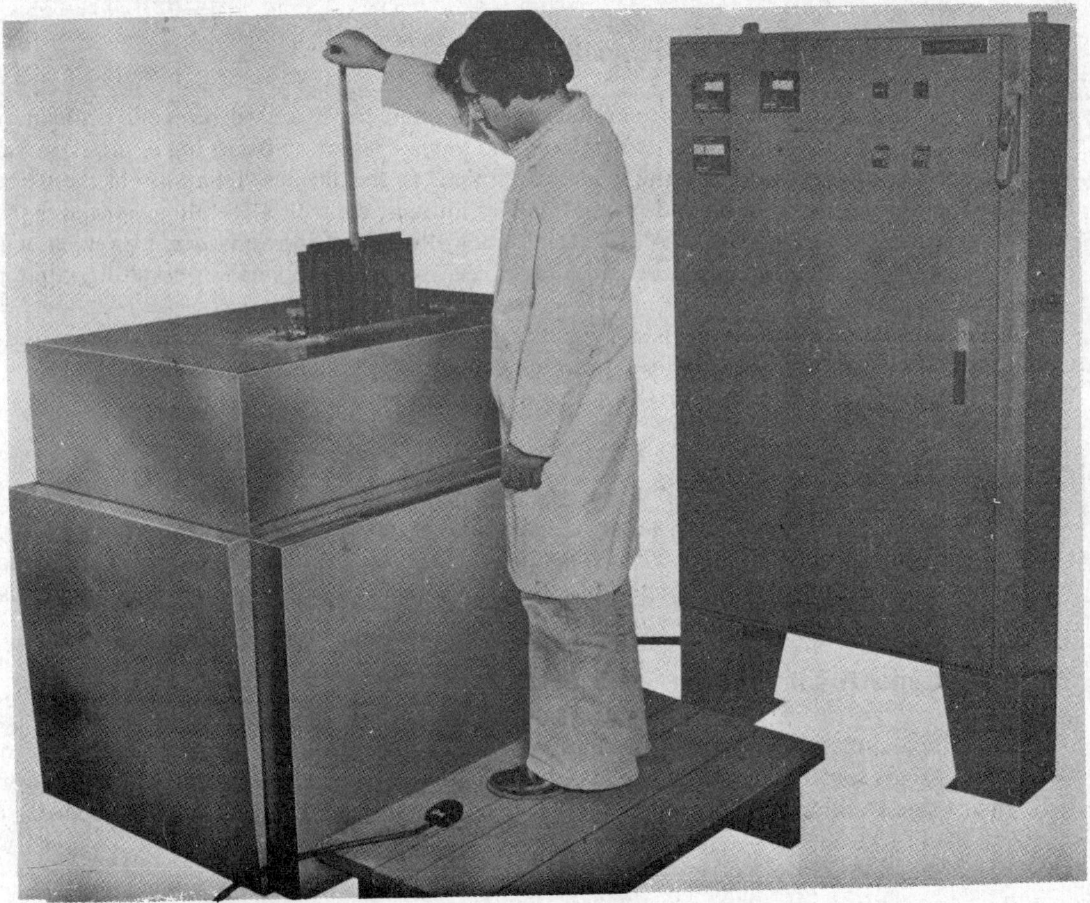

Fig. 15-3 Manual equipment for hot-air levelling (Courtesy: Electrovert, Montreal, Canada)

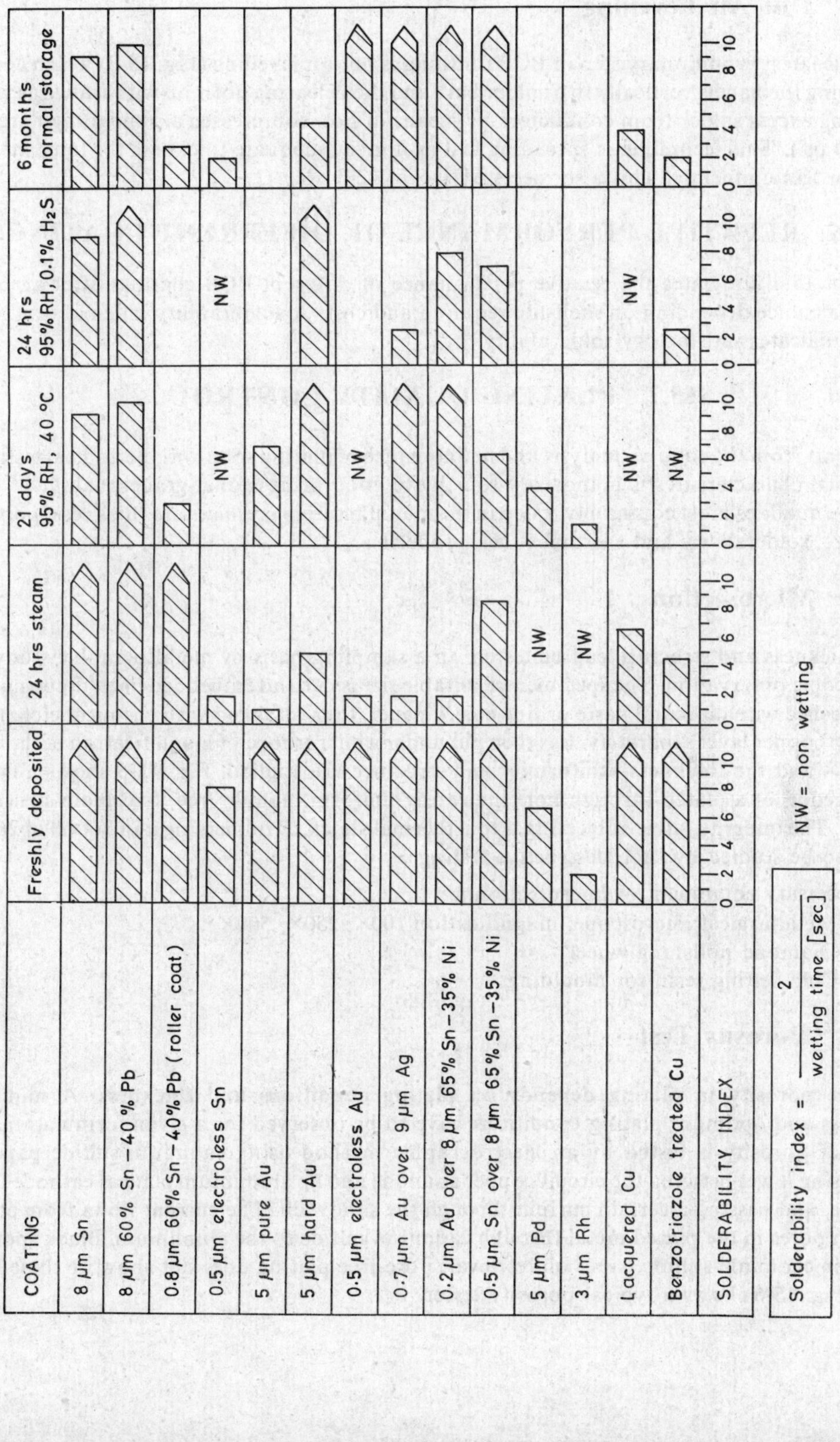

Fig. 15-4 Coating type versus solderability chart

$$\text{Solderability index} = \frac{2}{\text{wetting time [sec]}}$$

NW = non wetting

15.5.4 Hot-Air Levelling

The latest evolutionary step in PCB finishing is hot-air levelling (Fig. 15-3) which consists of dipping the panels vertically in a solder bath and then clearing holes on withdrawal as well as shedding excess solder from conductors by means of hot, compressed air under high pressure (85-120 psi). This technique is spreading fast as it gives adequate thickness for long shelf-life more or less uniformly. This also meets MIL-STD-275 Rev C.

15.6 RELATIVE PERFORMANCE OF DIFFERENT COATINGS

Fig. 15-4 illustrates the relative performance of different PCB coatings after ageing to enable a choice depending on shelf-life required and cost. A solderability index above 2 after ageing indicates satisfactory solderability.

15.7 PLATING QUALITY CONTROL

Apart from the routine analysis and maintenance of plating solutions, periodical testing of functional characteristics is mandatory to achieve good professional-grade quality.

The qualities most commonly tested in the manufacturers premises are thickness, porosity, structure, solderability and thermal shock properties.

15.7.1 Microsection

Thickness and structure can be tested on a sampling basis by metallographic study and microscopic observation. The specimen of suitable size is cut and embedded in cold-curing resin and polished with diamond paste or fine emery paper, then attacked with chemical etchants to show up copper layer separately. It is then put under a microscope of magnification of the order of $100 \times$ and the thickness, uniformity and structure are studied. Fig. 15-5 shows a typical microsection of a plated-through hole indicating uniform, nodule free, fine grain structured deposit. The integrity, after subjecting it to a thermal shock by immersing it in hot oil of $260°C$, may also be studied by such destructive testing.

Necessary equipment of microsectioning:
—Metallurgical microscope; magnification $100\times$, $250\times$, $500\times$
—Diamond polishing wheel
—Cold-setting resin for moulding.

15.7.2 Porosity Test

The porosity in plating depends on plating conditions and thickness. A minimum thickness and optimum plating conditions have to be observed for a given formulation. The extent of porosity is tested by an electrographic method using cadmium sulfide paper by interposing it wet between the circuit copper as anode and an aluminium plate as cathode under pressure, and passing a certain current through the sandwich. The current flows from copper through pores in the plated metal through cadmium sulfide to the aluminium. Black spots are found on cadmium sulfide paper on retrieval. Pore-free plating does not show up these black spots. Fig. 15-6 shows a typical porosity tester.

PLATING

Fig. 15-5 Microsection of a plated-through hole

Fig. 15-6 Electrographic porosity tester (Courtesy: Fischer Instrumentation (GB) Ltd., Newbury Bersk. RG 14 5RU, England)

15.7.3 Solderability Test

The solderability of any given coating can be tested by subjecting the fluxed specimen of usually one inch square to a contact with dross-free solder at the recommended temperature (typically 240°C/3 sec) and then the cleaned specimen is observed through a magnifier with a magnification greater than 10×. The specimen must exhibit 95% complete wetting. The parameter may be set depending on actual customer's soldering equipment. Fig. 15-7 shows a typical solderability tester.

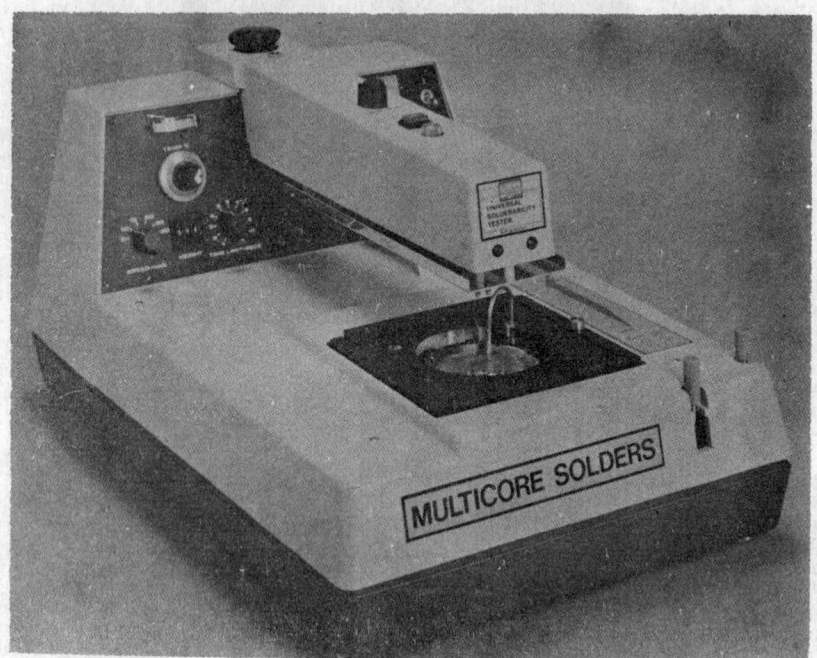

Fig. 15–7 Solderability tester (Courtesy: Multicore Solders Ltd., Hertfordshire, England)

15.7.4 Thickness Test

Non-destructive methods of thickness measurement have come into vogue for the past few years enabling the manufacturer to control his process more closely. The most commonly used method for PCB finishes is the beta-ray-backscatter method. This uses a radio-isotope emitting beta-rays, mounted on a suitable probe. The β-rays impinge on the surface whose thickness is to be measured and the quantity reflected depends on the atomic number of the coating and its variance with respect to the base metal. By pre-calibration with a standard, an electronic counting equipment is made to interpret the reflection in terms of thickness in microns. By this method, gold, tin, tin-lead on copper and copper on epoxy and photoresist on copper can be measured to great advantage in process control. Fig. 15-8 shows a typical instrument for this purpose. Fig. 15-9 shows a probe for plating thickness measurement inside plated-through holes.

PLATING

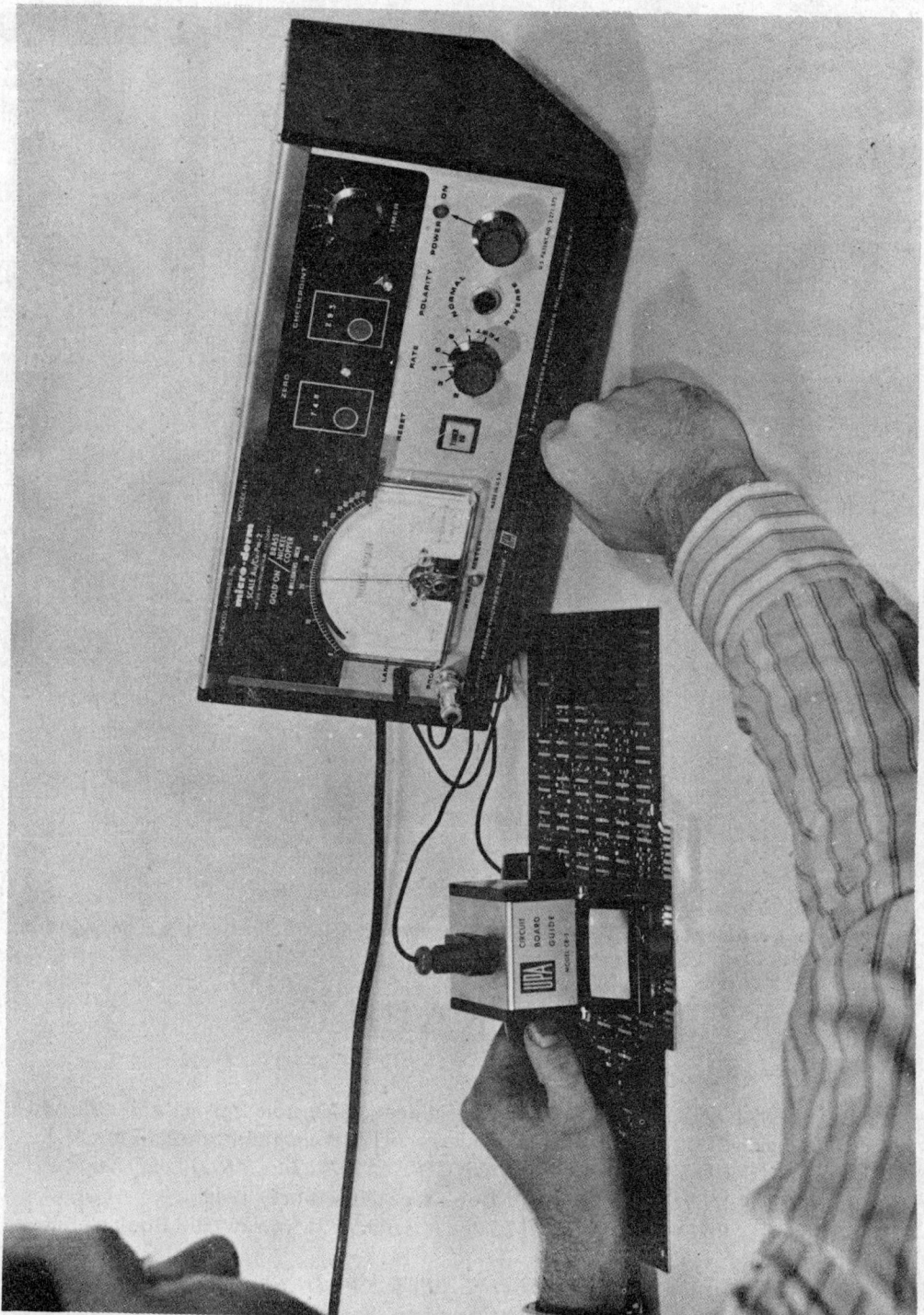

Fig. 15-8 β-ray backscatter with probe for plating thickness measurement (Courtesy: UPA Technology Inc., Syosset, New York, USA)

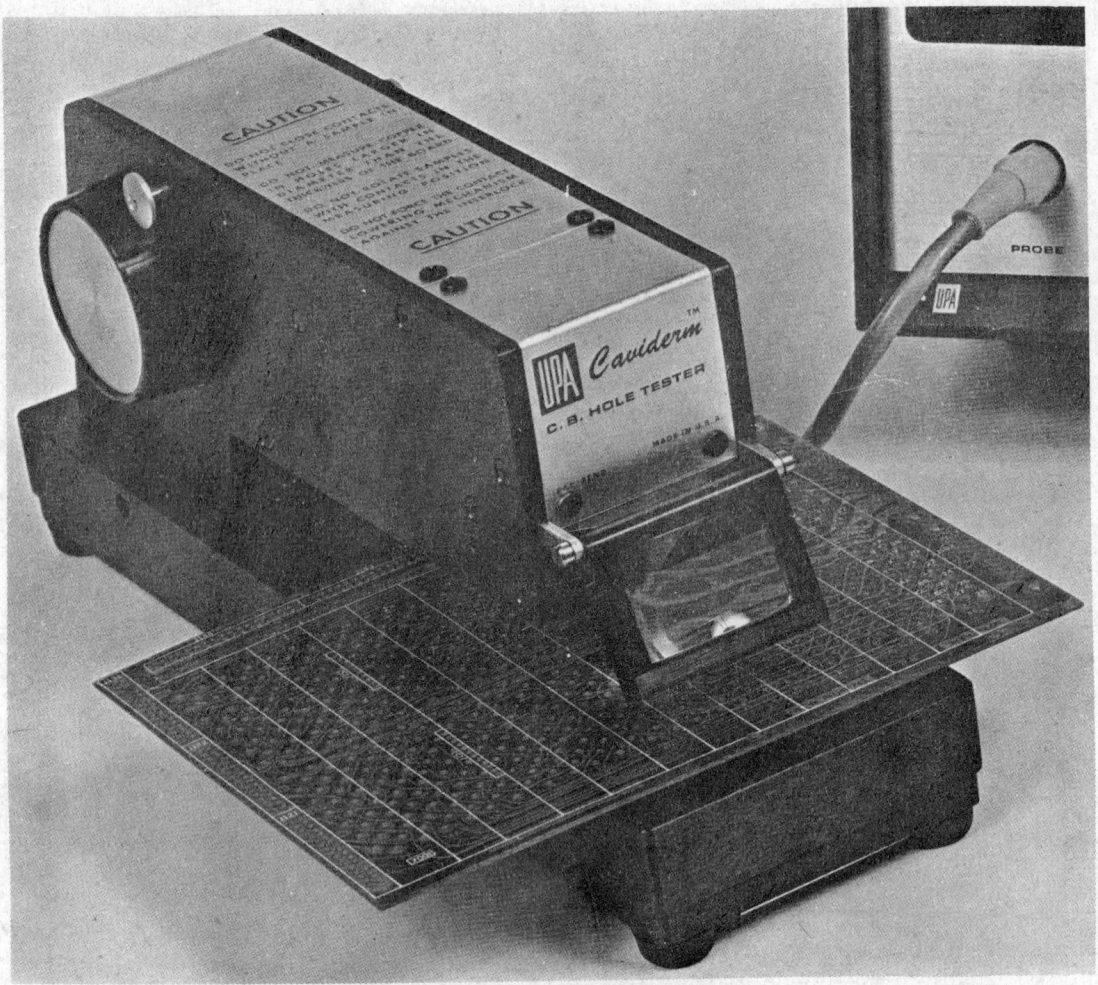

Fig. 15-9 Probe for measuring plating thickness inside plated-through holes (Courtesy: UPA Technology Inc., Syosset, New York, USA).

BIBLIOGRAPHY

1. ACKROYD M.L. AND MACKAY C.A.: 'Solders, solderable finishes and reflowed solder coatings', *CIRCUIT WORLD,* January 1977, Wela Publications Ltd., U.K.
2. BLUM W. AND HOGABOOM G.B.: *PRINCIPLES OF ELECTROPLATING AND ELECTROFORMING,* McGraw-Hill Book Co., New York, 1949.
3. COOMBS C.F.: *PRINTED CIRCUITS HANDBOOK,* McGraw-Hill Book Co., New York, 1967.
4. GARDNER FOULKE D.: *ELECTROPLATER'S PROCESS CONTROL HANDBOOK,* Robert E. Kreiger Publishing Company, Huntington, New York, 1975.

5. LANGFORD K.E.: *ANALYSIS OF ELECTROPLATING AND RELATED SOLUTIONS*, Electroplating & Metal Finishing, Teddington, Middx., England, 1951.
6. LOWENHEIM F.A.: *ELECTROPLATING*, McGraw-Hill Co., New York, 1978.
7. MISSEL L.: 'Gold plating of printed circuit boards and wires', *CIRCUIT WORLD*, October 1976, Wela Publications Ltd., U.K.
8. RANTELL A and HOLTZMANN A.: 'Metallisation of circuit boards using $SnCl_2$/$PdCl_2$ catalysts', *CIRCUIT WORLD*, January 1976, Wela Publications Ltd., U.K.

16
ETCHING

16.1 INTRODUCTION

In all subtractive PCB processes, etching is one of the most important steps: The final copper pattern is formed by selective removal of all the unwanted copper, which is not protected by an etch resist. This looks very simple at first glance but in practice there are factors like underetching and overhang which complicate the matter especially in the production of fine and highly precise PCBs.

Underetching

During the etching process, it is expected that the etching progresses vertically. However, in practice there is also an etching action sideways which attacks the pattern below the etch resist. If the etching action is not stopped immediately after all the unwanted copper has been removed, underetching will continue and can lead to a considerable reduction of conductor line widths.

Underetching can be minimised by keeping the etching time as short as possible (fast working etchant and exact control on etching time) and by a pressurised, perpendicular discharge of the etchant towards the surface to be etched (spray etching). A common term to express underetching, especially when screen- or photoresists are used, is the *etch factor*, defined as the ratio of etching depth (copper foil thickness) to the width of the side attack (Fig. 16–1).

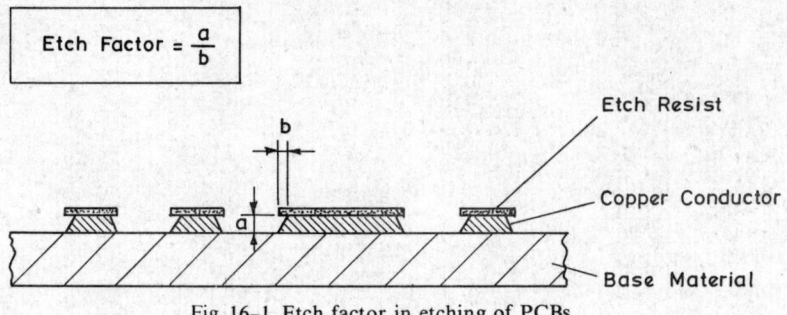

Fig. 16-1 Etch factor in etching of PCBs

Overhang

The exact control over conductor widths is further complicated where metal etch resists are used i.e., in pattern plating processes: The metal plating built up shows growth sideways also, resulting in overhang (Fig. 16-2). The difficulties with overhang are much reduced with the use of dry-film resists. Dry-film resists are available in thicknesses of as much as 70 μm and can therefore act as an effective barrier against sideway growth of plating layers. Another effective remedy against overhang, if solder-metal etch resist is applied, is the fusing operation, in which the solder plating is melted after etching and covers thereafter the side flanks of the copper conductors.

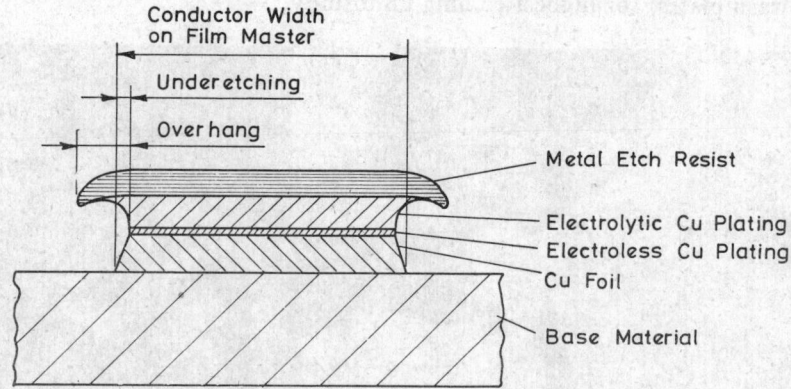

Fig. 16-2 Overhang in pattern-plated PCBs

In critical cases, the influences of overhang and underetching are compensated for by modifying the conductor widths already on the film master. See also what has been said under Section 10.6 of this book on increasing or decreasing of conductor line widths of the film master.

16.2 ETCHING MACHINES

Etching of PCBs, as required in modern electronic equipment production, is usually done in *spray-type etching* machines.

Tank or bubble etching, in which the boards, kept in a rack, were lowered and fully immersed into the agitated etchant, has almost disappeared. Tank etching could not provide the fast, precise and uniform etching which is desired in today's electronic equipment production. Another disappearing method is the etching with *splash-type etching* machines. In these machines, the etchant is thrown by centrifugal forces via a rotating device in the centre of the etching machine onto the surface of the surrounding boards.

Spray Etching

In spray etching, the etchant is pumped under pressure from the sump via a pipe network to the nozzles and from there gets splashed onto the boards. Spray etching machines can offer very high etching uniformity and a fast etching rate. The high etching uniformity is achieved by spraying through a full number of equally distributed nozzles in ring supply. A high etching rate is the result of the continuous flow of fresh etchant over the boards and the air oxygen absorbed in the etchant through spraying.

From the design point of view, spray etching machines can be classified into the laboratory type and the conveyorised etching machines.

Laboratory-type spray etchers: They are preferably used in a batch-type production set-up. The boards are kept in a frame for the complete etching process and the following water rinse. The number of boards going for etching with each loading operation depends on the size of the frame and the size of the boards. This is the limiting factor in the productivity rate of laboratory-type etchers. Such etchers are available for single-sided etching and for double-sided etching which also permits back-to-back etching of twice the quantity of single-sided boards. There are designs available with a fixed frame to keep the boards but with rotating frames also (swash plates) for highest etching uniformity.

Fig. 16-3 Swash-plate loading into a laboratory-type spray-etching machine (Photo CEDT)

Conveyorised etching machines: Machines of this type are employed today in all major PCB industries because of their high productivity (up to several square meters of laminates per hour), their excellent etching uniformity (designs with adjustable nozzles for regularity down to $\pm 1\%$) and the short etching time required (faster conveyor speed). There are designs with fixed nozzles but for high etching uniformity, wiping or oscillating nozzles are used. In PCB production transfer-lines, where the etching machine is the speed limiting factor, there are often two or even three etching machines (modules) cascaded to utilise the full capacity of all the other equipment in the line.

The typical material used for spray-type etching machines is PVC, a plastic which withstands almost all the different etchants currently used. PVC can easily be machined, bent, welded and adhesively joined. PVC is available in transparent or in different colours and is reasonably priced. It can withstand continuous etchant temperatures up to 60°C until it starts softening. However, for safety, heating elements are usually connected via a safety relay which switches off at 57–58°C.

The etching chambers are covered with transparent PVC. This permits control of the etching process. Furthermore, removable top covers are secured with an automatic pump-off switch to disable an accidental removal of the cover while spraying is on.

Metal parts in contact with the etchant have to be made in titanium. Heaters are encapsulated either in titanium or in quartz glass. Sophisticated designs also include etchant cooling to have a perfect control on the etchant temperature. Etchant pump inlets are usually provided with a filter to protect the pump from crystallised salts or undissolved pieces.

If wiping or oscillating nozzles are used, the movement should be synchronised with the conveyor speed to provide also the high etching uniformity when thin copper layers are etched at a faster conveyor speed.

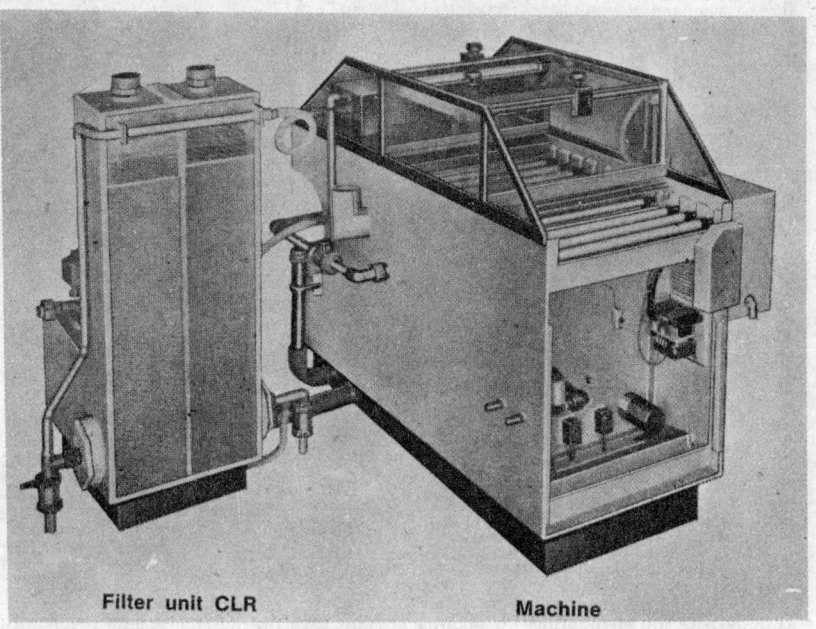

Fig. 16-4 Compact, conveyorised spray-etching machine (Courtesy: Resco, Milano, Italy)

Trends in Etching Machine Designs

With the introduction of conveyorised etching machines, making PCBs in transfer lines has become possible. While the early conveyorised etching machines included also rinsing in a separate chamber, the transfer line concept progressed further and became a module concept, in which each operation is done by an individual module. The different modules can be combined together with a minimum amount of interconnecting work. If a new process is introduced, fresh modules can be added to the existing transfer line and others removed.

The concept of modules incorporates a high degree of flexibility in combining the modules to suit almost any PCB process. This flexibility again gives in the long run a better utilisation of the fairly high investment into a conveyorised transfer line.

There are different levels of automation possible. The simplest etching line consists of input module with drive (boards are put on this), etch module, rinsing module and output module (boards are taken out from this). This combination has exactly the same performance as a conventional conveyorised etching machine with built-in water rinse. But the modular system offers flexibility: If for instance air drying is additionally desired, an air drying module is added without the need of a new conveyor-drive or much extra space. Other modules available are auxiliary etch module (for cascading of etchers); rinse module with or without recirculation, heating or filtering; inspection module; anti-pollution rinse module; etchant regeneration module; drying module with features like squeezing, vacuum suction, hot air, and other modules.

Fig. 16-5 Modular-concept etching-line for alkaline etchant with rinsing and etchant regeneration facilities
(Courtesy: Hollmuller Maschinenbau, Herrenberg, Germany)

16.3 ETCHANT SYSTEMS

16.3.1 Comparison of Etchants

When it comes to the choice of the most suitable etchant system for a PCB production process, there are many factors to be considered. First it has to be matched with the etch resist used. Screen- and photoresist can be either solvent or aqueous soluble type. The resists soluble in aqueous solutions are not suitable for alkali etchants, but they offer sufficient other advantages with respect to little environmental pollution and easy removal. With metal etch resists, only gold plating can withstand all the currently used etchants. Solder plating cannot be used as etch resist in ferric chloride or cupric chloride etchants. The suitability of various etch resists with respect to the most important etchants is shown in Table 16-1.

Table 16-1 Matching of etch resist and etchant.

ETCH RESIST ETCHANT	Screen resist	Photo resist	Solder metal resist	Tin metal resist	Gold metal resist
Ferric chloride	+	+	−	−	+
Cupric chloride	+	+	−	(+)	+
Chromic acid	+	+	+	+	+
Alkaline ammonia	+)*	+)*	+	+	+

Legend: +   recommended
    (+)   possible but critical
    −   impossible
)*   except aqueous soluble resists

Not only the etch resist should be compatible with the etchant but also the etching speed (related to corrosiveness), copper solving capacity, etchant price and pollution character come into the picture in the overall economy of the etchant (Table 16-2).

Table 16-2 Operational characteristics of different etchants.

FACTOR ETCHANT	Corrosiveness	Neutralisation, disposal problems	Toxicity	Required ventilation	Operational costs incl. disposal
Ferric chloride	High	Medium	Low	Low	Medium
Cupric chloride	High	Low	Medium	Medium	Low
Chromic acid	High	High	High	High	High
Alkaline ammonia	High	Medium	Medium	High	High

Among the etchants listed in Tables 16-1 and 16-2, *ferric chloride* was the earliest one used on a massive scale. With the availability of etchants which can be regenerated and which are compatible with the common metal etch-resists, ferric chloride has lost its importance. Because of its unproblematic handling, however the author recommends ferric chloride for small PCB facilities where etching is only occasionally carried out for a small number of boards.

Cupric chloride has replaced ferric chloride in large-scale etching of print-and-etch type boards. Cupric chloride can be regenerated and the saturated etchant can be disposed by supplying it to chemical industries for further use.

For the etching of boards with metal etch-resists, *chromic acid* was once widely used because of its high corrosiveness which works well with all kinds of metal resists. A high corrosiveness gives generally a fast etching rate and little underetching. However, because chromic acid has an *extremely high toxicity*, it is hardly used now, while *alkaline ammonia* based etchants are the *recommended choice today* for etching of metal-resist boards.

In handling of any etching chemicals, suitable safety precautions have to be observed and the necessary safety devices like acid-proof gloves, aprons, face shields, filter masks etc., have to be within an easy reach.

After this more general introduction, the individual etchant systems are now dealt with in more detail.

16.3.2 Ferric Chloride

Ferric chloride shall be explained here first because it is an etchant very simple to use especially in small-scale PCB production. In high-volume production, ferric chloride is not of much importance because it can hardly be regenerated and it attacks the common metal etch-resists.

Chemistry

Free acid to attack the copper is formed by the hydrolysis reaction.

$$FeCl_3 + 3H_2O \rightarrow Fe(OH)_3 + 3HCl \qquad (1)$$

The copper is oxidised by the ferric ions, forming cuprous chloride (CuCl) and ferrous chloride ($FeCl_2$).

$$FeCl_3 + Cu \rightarrow FeCl_2 + CuCl \qquad (2)$$

Cuprous chloride (CuCl) oxidises further in the etching solution to cupric chloride ($CuCl_2$).

$$FeCl_3 + CuCl \rightarrow FeCl_2 + CuCl_2 \qquad (3)$$

The built-up cupric chloride ($CuCl_2$) itself reacts also with copper and forms cuprous chloride (CuCl).

$$CuCl_2 + Cu \rightarrow 2CuCl \qquad (4)$$

Advantages

The high corrosive power of ferric chloride leads to short etching times and little underetching. Ferric chloride matches well to photo- and screen-printed resists, both for solvent and alkali soluble ones. Also, shelf-life is very long, a feature important for small set-ups where only occasional etching is done. Ferric chloride can dissolve copper up to approximately, 120 g/l (practical limit) if the prolonged etching time is acceptable.

Proprietary ferric chloride formulations include anti-foaming agents and other additives. They give the advantage of less odour and fuming, more even etching due to better surface wetting, longer life and less slime formation.

Disadvantages

A major drawback of ferric chloride etching is the impossibility to regenerate it economically. This practically rules out its use in bigger set-ups where a constant etching speed is a must. The dependence of etching time versus copper content in the etchant is shown in Fig. 16-6. Metallic etch resists are attacked by ferric chloride which rules it out as an etchant for pattern plating processes. The copper reclamation of the spent etchant is not possible economically.

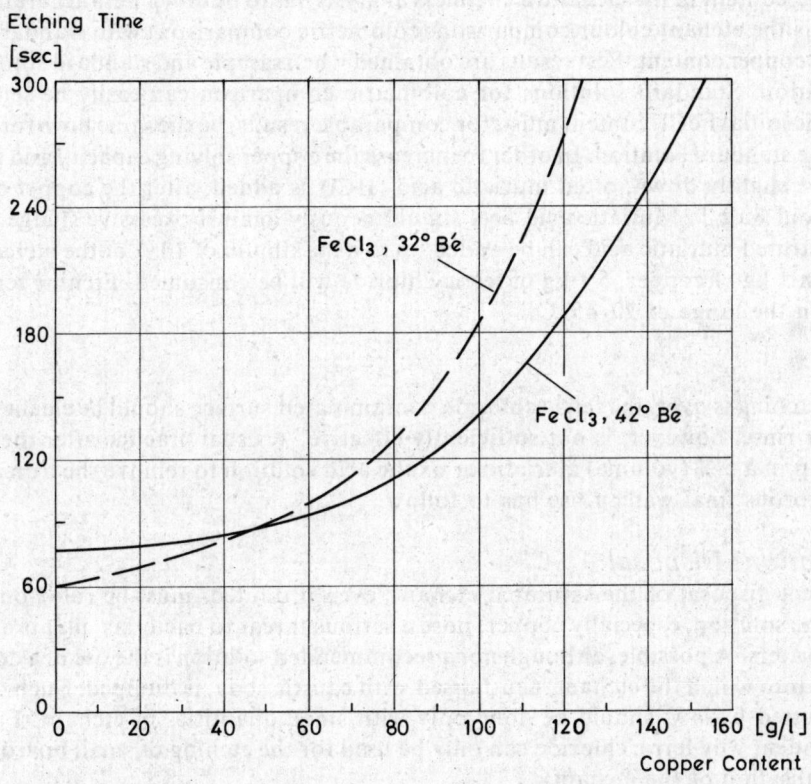

Fig. 16-6 Etching time with ferric chloride at 43°C, spray etching, 35 μm Cu [3]

Operation

Ferric chloride is available as lump $FeCl_3 \cdot 6H_2O$ and as aqueous solution with or without additives. The working concentration can vary within certain limits [1]:

		Low etching power	Optimum		High etching power
$FeCl_3$	[g/l]	365	452	530	608
$FeCl_3$	[% weight]	28	34	38	42
Specific gravity	[kg/dm³]	1.275	1.353	1.402	1.450

All the values are related to room temperature (20-25°C)

In practical etching with ferric chloride, it is essential to know about the conditions and state of the etchant. Control is easiest done by observing the etching time while other parameters like etchant temperature, copper foil thickness and etching pressure are kept constant. Absolute values for the etching time range cannot be given since this depends on the equipment and solution strength used for etching. But practical saturation is somewhere around, when etching time is two to three times what it was with fresh mix-up. To know the exact copper content in the etchant, a chemical analysis has to be done. Less accurate but useful in practice is the etchant colour comparison (colometric comparison) with standard solutions of a known copper content. Best results are obtained when sample and standard solutions are in a 2 : 1 dilution. Standard solutions for colometric comparison can easily be self-prepared; however, the initial $FeCl_3$ content must, for comparable results, be the same both for the sample as well as the standard solution. In order to increase the copper solving capacity and to bring the etching time slightly down, often muriatic acid (HCl) is added, after the copper content has reached about 80 g/l. Muriatic acid acts simultaneously against excessive sludge formation. The concentrated muriatic acid can be added up to a maximum of 10 % of the etchant volume.

To etch 1 kg of copper, 5.1 kg of ferric chloride will be consumed. Etching temperatures should be in the range of 20–45° C.

Rinsing

After etching is over, the ferric chloride contaminated surface should be cleaned. A simple spray water rinse, however, is not sufficiently effective. A usual practice after the first water rinse is a dip in a 5 % (volume) muriatic or oxalic acid solution to remove the iron and copper salts. A vigorous final water rinse has to follow.

Neutralisation/Disposal

A direct disposal of the saturated etchant, even if diluted, must be ruled out, since the metals in the solution, especially copper, pose a serious threat to bacterias, plants and animals in natural waters. A possible, although *not* a recommended solution is the use of a deep sump in the ground into which the etchant, neutralised with caustic soda, is dumped. Such dumping, if at all permitted by law, should be done only with small quantities of etchant. From this, it becomes evident why ferric chloride can only be used for the etching of small board quantities.

Neutralisation of the etchant:

1 kg $FeCl_3$ — 0.74 kg caustic soda
1 kg $FeCl_2$ — 0.63 kg caustic soda
1 kg Cu   — 1.26 kg caustic soda

The disposal method actually suggested here is the dumping *after* the heavy metals (Cu, Fe) have been precipitated as described under Section 22.5.1

16.3.3 Cupric Chloride

Performance-wise, cupric chloride can be compared with ferric chloride: It provides rapid etching and causes thereby little underetching. In addition to this, cupric chloride offers the advantage of easy regeneratability with low quantities of cheap and commonly available liquid chemicals plus the possibility of a relatively easy disposal and a still higher copper solving capacity of practically up to 150 g/l. It does not produce sludge formation as ferric chloride does.

Cupric chloride basically cannot be used with metallic etch-resists. Only gold and, under certain conditions, tin can also withstand the metallic etch-resists. Compared to ferric chloride, bath maintenance needs more care.

Cupric chloride is today the most economic solution in etching of print-and-etch type of boards at a larger scale. It is also from the pollution point of view, considerably cleaner than ferric chloride: Saturated etchant is readily accepted by many chemical industries for further use.

Chemistry

The copper surface gets attacked by the cupric chloride ($CuCl_2$) while cuprous chloride ($CuCl$) is formed. Cuprous chloride as such is not effective in further etching.

$$Cu + CuCl_2 \rightarrow 2CuCl \qquad (1)$$

The hydrogen peroxide (H_2O_2) and muriatic acid (HCl) available in the etchant, regenerate cuprous chloride back to the etch-active cupric chloride. This regeneration capability of cupric chloride etchant is made use of in regular regenerating by adding of the mentioned two chemicals.

$$2CuCl + 2HCl + H_2O_2 \rightarrow 2CuCl_2 + 2H_2O \qquad (2)$$

Copper is mostly attacked by the process of equation (1) but the availability of muriatic acid and hydrogen peroxide directly supports also the copper etching as long as the volatile hydrogen peroxide is contained in the etchant in its basic form. As copper solving continues, hydrogen peroxide will then get absorbed in the formation of cupric chloride out of cuprous chloride (equation (2)).

Operation

To start etching with cupric chloride, there are two methods to obtain the initial etchant concentration. The quantity of etchant prepared should be only two-thirds of the maximum etchant volume of the machine in order to make the adding of regenerating chemicals possible.

- A) Diluting saturated cupric chloride: If saturated $CuCl_2$ etchant is available, it has to be suitably diluted in order to get a copper concentration of approximately 40 g/l. This makes it necessary to know the copper content of the used etchant within 10 % limits. This can be found out by colometric comparison or, if the chemical balance of the etchant has been disturbed, by a chemical analysis of the copper content. The dilution occurs by adding water and muriatic acid (HCl) in a ratio of 5 : 1.
- B) Preparing fresh cupric chloride: This process requires a few hours time for dissolving the copper which is added in pure form. The ingredients are added in the following order and quantities:
    77% Water
    20% Muriatic acid (HCl) in 32% concentration
    3% Hydrogen peroxide (H_2O_2) in 35% concentration

Then copper has to be added *immediately* in the ratio of 40 g per liter etchant. To enable it to dissolve quickly, the copper should be in the form of chips or wires or any other form providing a large surface. The etchant temperature for this is best kept around 40 °C. This operation of dissolving copper should not take too much time, in order to avoid evaporation of the volatile hydrogen peroxide until it is absorbed in the bath.

Etching

The etchant temperature is preferably around 45 °C. This temperature is optimum for a good etching speed and moderate fuming. Exhaust facilities for the corrosive fumes are recommended. Otherwise care for sufficient free air flow must be taken.

Control of the copper content: The rising copper content in the use of cupric chloride is reflected in the gradual lowering of the etching speed. With spray-type etching machines, an etching speed between 10–30 μm/min can be expected. The copper content can be determined exactly with chemical analysis or by measuring the oxidation-reduction potential with a suitable device. Less but still sufficiently accurate, for most practical purposes, is the check of the specific gravity or the colometric method by comparing with the colour of standard solutions that have a known copper content.

After using the etchant for some time, less etch performance might be observed. The reason is probably due to water evaporation and a lower content of free oxygen. In such a case, the missing water volume is compensated for by adding water and muriatic acid in a 5 : 1 ratio.

Manual Regeneration

As the etching time and copper content increase more and more, the etchant has to be regenerated by adding the chemicals which promote the formation from cuprous chloride to cupric chloride as shown in equation (2). The need for regeneration can also be seen in a change of the etchant colour from light green to the darker olive colour.

>Regeneration chemicals:
>Muriatic acid (HCl) in 32 % concentration
>Hydrogen peroxide (H_2O_2) in 35 % concentration

HCl and H_2O_2 are added in the ratio of 2.4 : 1 in a total quantity of approximately 2 % but a still high cuprous chloride content makes it necessary up to 4% of the total etchant volume. To regenerate 1 kg of dissolved copper, 2.4 l HCl and 1 l H_2O_2 will be sufficient.

Important for safety: The two regenerating chemicals must be added one after the other or simultaneously poured into the tank at two separate places. The two highly concentrated regenerating chemicals should never come directly together, not even in small drops. It would cause a strong exothermic reaction.

How to recognise wrong dosage of regenerating chemicals: Too much H_2O_2 brings excess oxygen into the etchant. The result will be a foaming etchant at a dangerously increasing temperature because of the exothermic reaction. Too much HCl brings the etching speed down because of insufficient oxygen in the etchant. At the same time, the etchant colour will turn brownish. In both cases of wrong dosage, chemicals are wasted and an unbalanced etchant composition is left behind.

Manual regeneration can be carried on until the maximum etchant level in the machine is reached. Then the saturated etchant has to be removed and again fresh etchant with a copper content of 40 g/l is prepared as described earlier under operation.

To maintain cupric chloride etchant at optimum performance level needs more detailed observations and experience than the control of ferric chloride. Manual regeneration is therefore recommended mainly for smaller set-ups where certain irregularities do not start a chain reaction in the whole production process. For more stringent requirements, the use of automatic regeneration equipment is suggested which is available, both working on a continuous or a discontinuous base.

Rinsing
After the first spray water rinse, a neutralising dip in a 5% (volume) HCl solution is recommended. This has to be followed by another rigorous spray water rinse.

Neutralisation / Disposal
Copper-saturated cupric chloride etchant is a byproduct which is readily accepted by various chemical industries for their use. The price they pay for saturated etchant may even cover most of the expenses incurred for buying the etching chemicals. Should a neutralisation be required, the following guidelines can be helpful:

Neutralisation : 1 kg Cu — 1.26 kg Caustic soda
1 l HCL — 0.41 kg Caustic soda

16.3.4 Chromic Acid

Chromic acid etching is usually carried out with the addition of sulfuric acid, hence it is also called chromic-sulfuric acid etching.

Although this etchant is *highly toxic*, it is unfortunately still often used in India, probably based on difficulties in the availability of alkaline etchants; also because of a lack of sufficient knowledge and experience with the alkaline etchants.

Due to the high toxicity of chromic acid etchants, their use is *absolutely not recommended*. This is not only because of the heavy pollution caused but also because of the health hazards for the personnel. When starting new etching facilities for boards with metal-etch-resist, only alkaline etchants are recommended today.

A common feature of any etchant for metal-resist boards, is the higher complexity and costlier operation of the etchant, compared to etching with ferric or cupric chloride. Daily maintenance of the etchant is a must. Chromic acid etching requires additionally very efficient exhaust facilities for the etching machine and rigorous protection devices for the personnel (strict use of acid-proof rubber gloves, aprons, face shields, filter masks).

Chromic acid etchants show high corrosiveness but solder metal-resist is not attacked by it because of the formation of insoluble lead sulfate. It is possible to solve up to 60 g Cu per liter etchant. The high corrosiveness will even attack ordinary plastics, rubbers and electrical wire insulation.

Chemistry
Chromates are formed of chromic acid and water by the following reactions:

$$2CrO_3 + H_2O \rightarrow Cr_2O_7^{-2} + 2H^+ \tag{1}$$
$$Cr_2O_7^{-2} + H_2O \rightarrow 2CrO_4^{-2} + 2H^+ \tag{2}$$
$$CrO_4^{-2} + H^+ \rightarrow HCrO_4^- \tag{3}$$

The actual reaction between copper and the chromates is expressed in the next formula:

$$3Cu + 2HCrO_4^- + 14H^+ \rightarrow 3Cu^{+2} + 2Cr^{+3} + 8H_2O \tag{4}$$

The hydrogen ions (H^+) are usually supplied to the process by adding sulfuric acid

(H_2SO_4). The overall reaction can then be shown as follows:

$$3Cu + 2CrO_3 + 6H_2SO_4 \rightarrow Cr_2(SO_4)_3 + 3CuSO_4 + 6H_2O \qquad (5)$$

Chromic acid crystals (CrO_3) and sulfuric acid (H_2SO_4) are commerically available; however, if compared with ferric or cupric chloride etchants, etching costs will be considerably higher. The etching rate for copper is not very high with chromic-sulfuric acid formulation but it can be increased by the addition of N_2SO_4, iodine compounds, NaCl, $AgNO_3$ or mercuric chloride. *Note* that muriatic acid (HCl) should never be added which otherwise would result in the formation of the highly poisonous chlorine gas.

Operation

Two formulations shall be suggested [2]. The first one uses Na_2SO_4 as an accelerator, while iodine compound must be added to the second one for the same purpose.

	1	2
CrO_3	240 g/l	480 g/l
Na_2SO_4	40.5 g/l	—
H_2SO_4 (96%)	98 ml/l	31 ml/l
Cu	—	4.9 g/l

In proprietary etchants, usually wetting, antifoaming, gelating agents and catalysts are also added.

The copper content is controlled by the colometric method (comparing colour of standard solutions with a known copper content). The density should be maintained around a value of 1.26 kg/dm^3 at 20°C or 1.24 kg/dm^3 at 27°C. If necessary, water is used to adjust the density. The density is measured with a hydrometer. If the density scale is given in Baume' gravity, the following formula can be used to calculate the specific gravity (USA definition):

$$\sigma = \frac{145}{145 - °Be'}$$

σ = specific gravity [kg/dm^3]

°Be' = degree Baume' gravity

The etchant temperature for optimum results will be around 30°C while the pH value is at about 1.1.

Safety

The first safety rule is *not to use* chromic acid at all because of its high toxicity. If at all it has to be used, the following points must be strictly remembered:
— Chromic acid causes heavy burns and etchings on the skin, eyes and clothes: The handling of chromic acid etchant should only be done with acid-proof gloves, aprons and a face shield.

— There is danger of fire if chromic acid comes in contact with inflammable materials: Do not smoke where chromic acid is stored and keep the acid away from inflammable materials.
— Chromic acid fumes also attack the breathing path: Exhaust facilities for the etching machine and efficient air flow are essentially required.
— Keep all the eatables away from the etching and chromic acid storing room.

Neutralisation/Disposal

A regeneration technique has been developed which reduces the chromic acid consumption to one third. The investment of minimum 200,000 $, however, rules out its use for a commerical production of PCBs. The spent etchant, therefore, is usually disposed.

The disposal of chromic acid etchant *poses a very serious problem*. Both the chromic as well as the copper ions are harmful to biological processes. The commonly used disposal methods are burial after neutralisation (where at all permitted!) or the supply to a recognised disposal agency with *proper* facilities to extract the poisonous content of heavy metals. The reader is advised to refer also to the related Section 22.5.1.

16.3.5 Alkaline Ammonia

Alkaline etching processes have been industrially introduced in the early '70s in order to overcome the shortcomings of the other etchants for metal-resist boards (chromic acid, persulfates, etc.). The early alkaline etchants were known for their low corrosiveness and low copper solving capacity. Due to further development efforts, today's alkaline etchants are able to provide continuous etching rates of approximately 30–60 μm Cu/min at a solved copper contents of 150 g/l in the etchant: Alkaline etchants have become greatly competitve in etching performance with all other types of etchants.

Alkaline etchants match well with the common metal-resists and in particular with solder metal. Efficient exhaust facilities are essentially required to remove the unpleasant and unhealthy ammonia fumes (maximum acceptable concentration 100 ppm NH_3).

Chemistry

The chemical reactions involved in alkaline etching are rather complex. Most of the etchants are *proprietary formulations* and details are hardly available.

The copper etching occurs basically by oxidation, solubilising and complexing. Cupric ammonium complex ions $Cu(NH_3)_4^{+2}$ are formed by ammonium hydroxide and ammonium salts in combination with copper ions. Up to 225 g Cu/l can be solved if a prolonged etching time can be accepted which results with the etchant reaching its limits of copper solving capacity.

Operation

Alkaline etchants used are mostly proprietary formulations: Complete process handling details have therefore to be obtained from the supplier.

Nevertheless, three formulations shall be given here as typical representatives of the alkaline etchant family [2]:

	1	2	3
NH_4OH	3.0 Mol/l	6.0 Mol/l	2-6 Mol/l
NH_4Cl	1-5.0	5.0	1-4.0
Cu (as metal)	—	2.0[1]	0.1–0.6
$NaClO_2$	10.4	—	—
NH_4HCO_3	0-1.5	—	—
$(NH_4)_3PO_4$	—	0.01	0.05-0.5
NH_4NO_3	0-1.5	—	—

[1] starter solution only.

The functions of the different ingredients are:

NH_4OH	complexing agent, holds Cu in solution
NH_4Cl	improves etching speed, copper holding capacity and solution stability
Cu^{+2}	copper ions as oxidising agent to react with and to dissolve metallic copper
$NaClO_2$	similar function as copper ions above
NH_4HCO_3	buffer to preserve solder metal surfaces
$(NH_4)_3PO_4$	preservative for solder metal surfaces
NH_4NO_3	increases etching speed and preserves solder metal surfaces

The stability of alkaline etchants is critical. In certain formulations, two solutions are mixed together directly before actual use. Oxidisers and complexing agents are added separately.

Typical etchant operating conditions are:

Temperature	50° C
pH	7.8–8.2
Specific gravity	1.17–1.20 kg/dm³
Copper concentration	135–180 g/l
Etching speed	35–50 μm Cu/min

Regeneration

Alkaline etchants are usually operated in the open-loop or closed-loop regeneration mode (Figs. 16-11 and 16-12).

Closed-loop regeneration needs substantial investment for special equipment and its economic use is restricted to large-scale production units. The open-loop regeneration can be manually carried out, but automatic equipment is also available. The specific gravity is controlled in the open-loop mode: If the gravity exceeds a certain value, replenisher chemicals are added after removal of some quantity of saturated etchants. Replenisher chemicals have to match the type of process used and are different from case to case.

Neutralisation/Disposal

Also, for alkaline etchants, direct disposal is *strictly ruled out*. Since in most cases

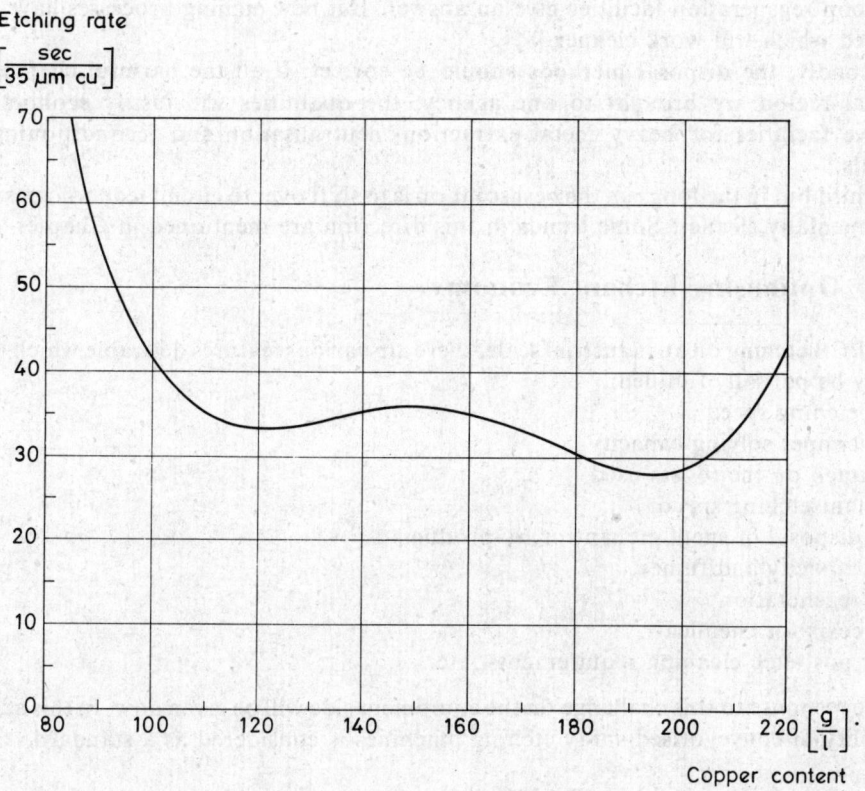

Fig.16-7 Typical etching rates with alkaline etchant versus copper content in etchant [2]

proprietary etchants are applied, it is a usual procedure to return the saturated etchant to the supplier who normally has both the facilities to extract the heavy metals as well as to recondition the etchant for reuse. Only bulk consumers of alkaline etchants go in for their own closed-loop regeneration facilities which finally get the highly concentrated copper salts separated. Such highly concentrated copper salts can then easily be sold to chemical industries because of the high copper content.

16.4 MINIMISING POLLUTION

The widespread use of chemicals as required for PCB etching and plating as well as water for rinsing purposes, has everywhere taken on critical dimensions, with respect to our biological environment. A point has been reached in industrial pollution where only strict measures, enforced by laws, can help not to disturb further the biological equilibrium of our planet. This need of the hour has received global meaning and no dispensation can be given to such a responsibility.

Coming to the practical implications: Any disposal of etching or plating chemicals, especially of larger quantities, must be carried out in a proper manner which does not pollute the environment. This is an expensive way but the only accepted solution today.

Firstly, the consumption of harmful chemicals should be to the minimum. Open-loop and closed-loop regeneration facilities give an answer. But new etching processes have still to be developed which will work cleaner.

Secondly, the disposal methods should be correct: If all the harmful chemicals of an industrial region are brought to one agency, the quantities will justify economically the expensive facilities for heavy metal extraction, neutralisation and reconditioning of such chemicals.

A third but in the long run the best solution is to shift over to circuit technologies which are environmentally cleaner: Some trends in this direction are mentioned in Chapter 18.

16.4.1 Optimising Etchant Economy

In PCB etching on an industrial scale, there are various features desirable which in practice can only be partially fulfilled:
— High etching speed
— High copper solving capacity
— No attack on the resists used
— Constant etching speed
— Easy disposal of spent etchant or by-products
— Little toxicity and fumes
— Easy regeneration
— Low costs for chemicals
— Little post-etch cleaning requirements, etc.

The response to this challenge on the equipment side will be given now. In this account, the availability of conveyorised spray etching machines is considered as a standard.

Simple Batch-Production Etching

This is the simplest arrangement possible for etching of PCBs. The etchant is used until saturation or until the etching speed becomes too slow. The etchant is then disposed and fresh solution is filled into the etching machine.

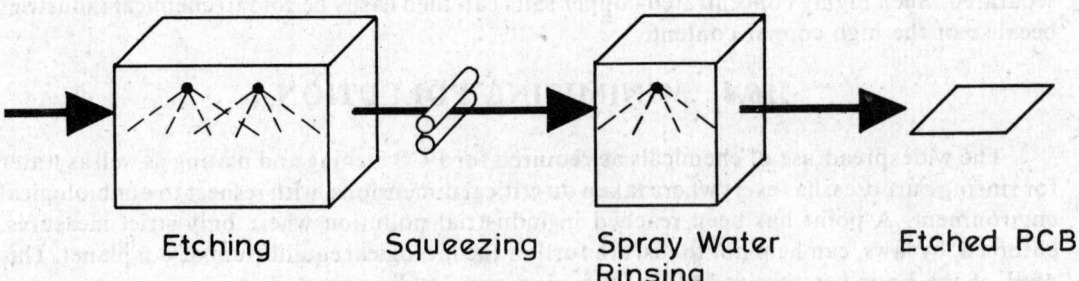

Fig.16-8 Simple batch-production etching

Batch etching is applied in laboratories and small industries where small series of PCBs have to be etched occasionally. When etching of a new batch of PCBs is started, the optimum etching time (conveyor speed) has to be determined first. A typical etchant used in this method is ferric chloride.

Continuous-Feed Etching

A small steady stream of fresh etchant is either continuously or periodically flows into the etchant sump while an equal quantity of partially saturated etchant is removed.

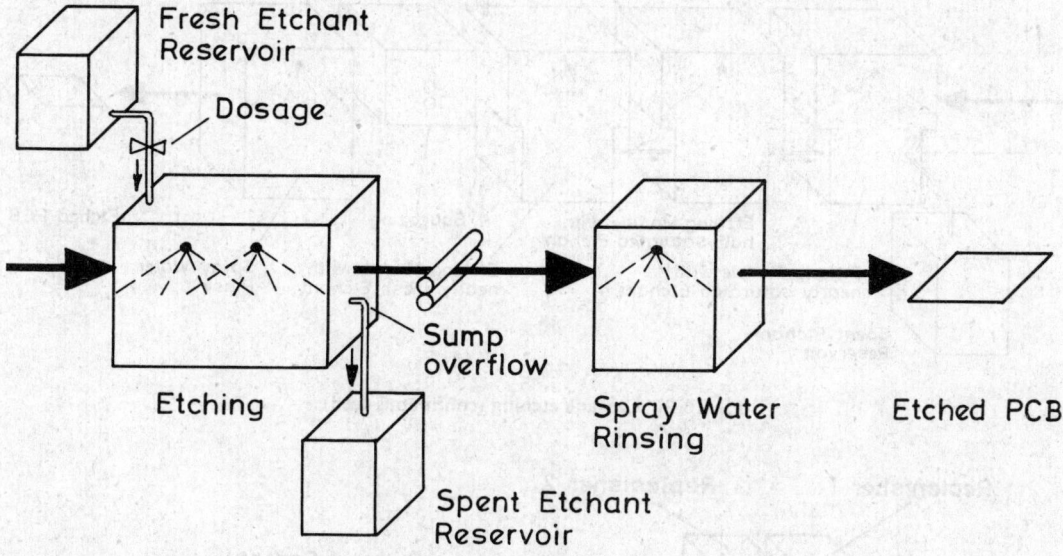

Fig. 16-9 Continuous-feed etching

To utilise the full copper solving capacity of the etchant would mean etching with a very slow etching speed (conveyor speed). This however would give only a low productivity rate. The system is therefore operated in a mode with only partially saturated etchant which gives more or less constant results at a reasonable etching speed. However, copper solving capacity of the etchant is not fully utilised. The most typical etchant in continuous-feed etching is ferric chloride.

To improve the etchant economy in continuous-feed etching, there are often several etching modules cascaded. While the first module contains almost saturated etchant, the following inter-connected modules have decreasing copper content in the etchant and the last module operates with nearly fresh etchant. There are usually three or four modules cascaded and the etchant flow between the modules goes via sump overflows. With this arrangement, reasonable etching speed is obtained in combination with a practically full utilisation of the etchant copper-solving capacity. A typical problem with cascaded etching systems is to maintain the copper content within certain limits in each one of the modules.

Open-Loop Regeneration

While the previously discussed methods utilised the addition of premixed full-strength etchant, the open-loop regeneration employs the adding of chemical additives (replenishers, regenerating chemicals) in order to maintain the etching performance at a constant level.

In the automatic mode, the composition of the etchant is monitored via a sensor for the pH

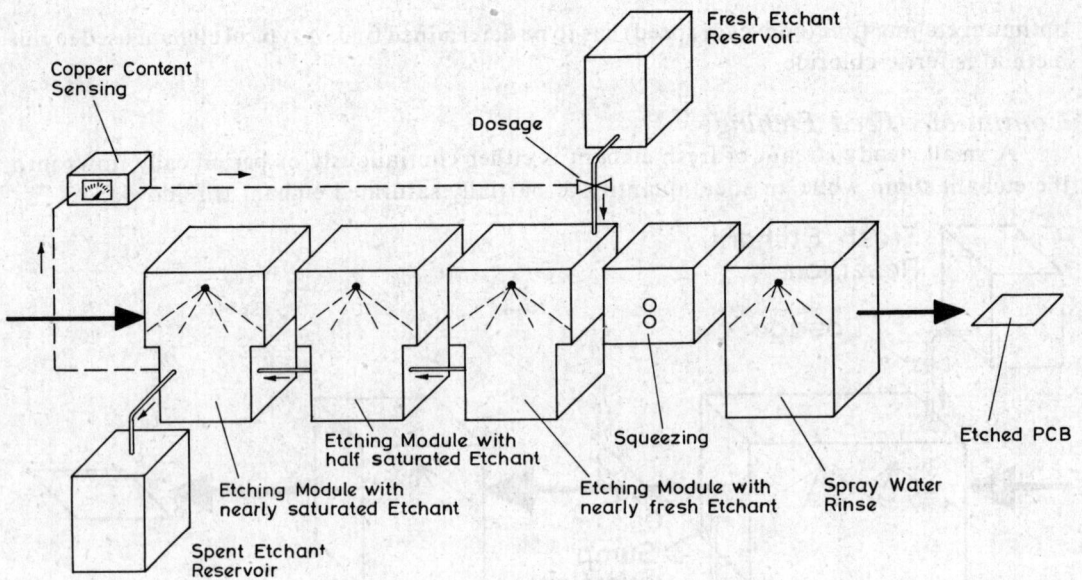

Fig. 16-10 Cascade etching (continuous-feed)

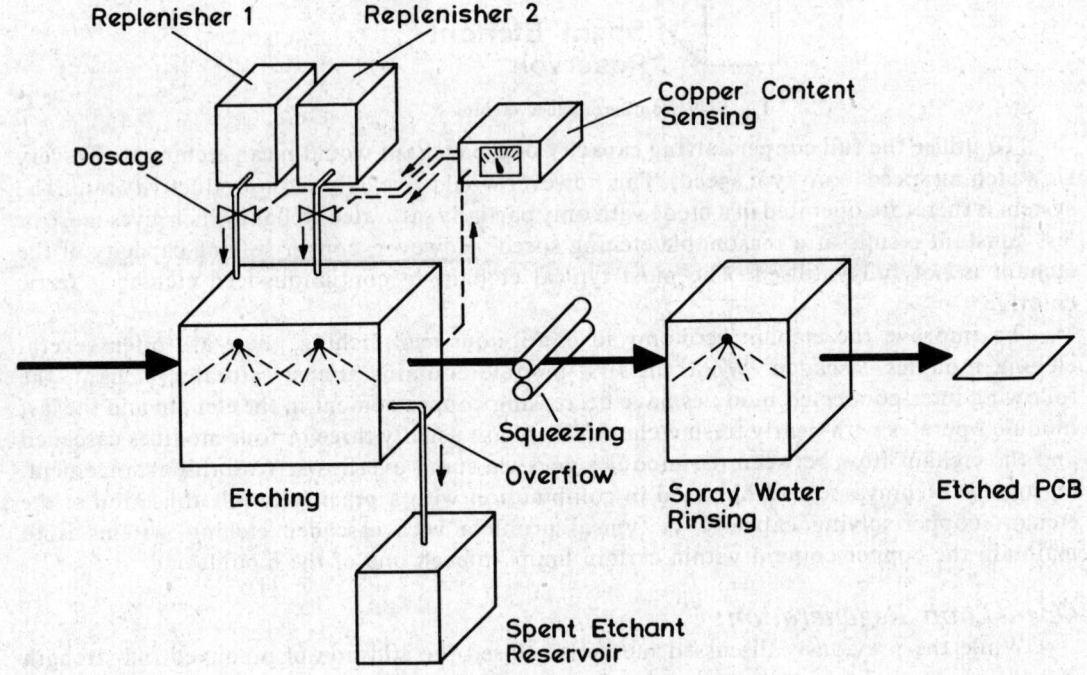

Fig. 16-11 Etching with open-loop regeneration

value, the oxidation-reduction potential (redox potential), specific gravity or colour. The most typical etchant used in open-loop regeneration is cupric chloride which permits this way to solve typically 130 g Cu per liter etchant spent. Open-loop regeneration can as well be carried out *manually*, especially for smaller production volumes. But it needs careful monitoring of the etchant composition.

Closed-Loop Regeneration

The principal feature of closed-loop regeneration is the removal of copper containing by-product from the etchant main stream while the Cu-purified etchant is returned to the etchant sump. The investment into ancillary equipment is very high but with respect to the economy of etching chemicals, constant etching performance and environmental pollution, it is an efficient solution.

The high complexity of the system needs meticulous maintenance since a constant etching quality has to be provided while there is a wide variability in chemical constraints and production cycles.

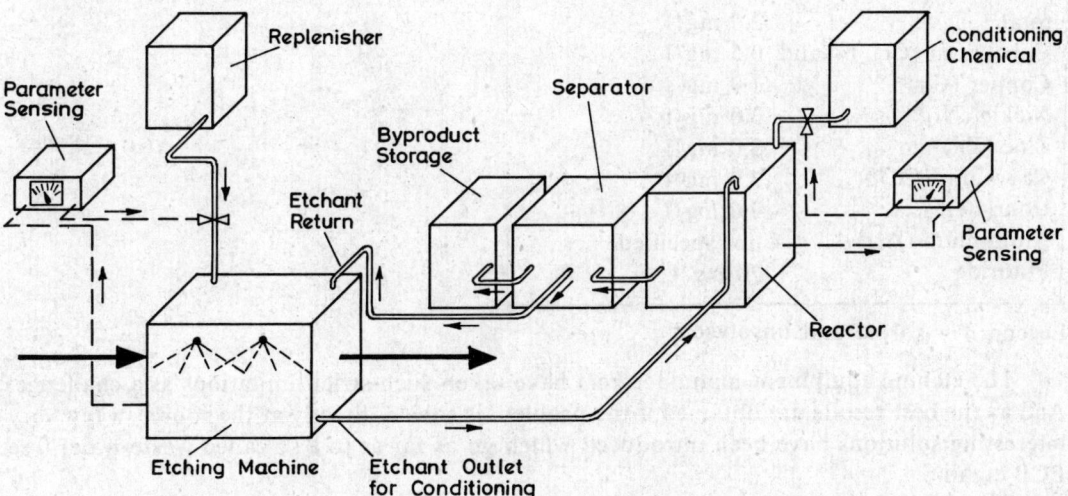

Fig. 16-12 Etching with closed-loop regeneration

Among the etchants economically suited for closed-loop regeneration, we find cupric chloride, ammonium persulfate and alkali etchants. For each individual one of these etchants, the reactor/separator system will be completely different to suit the particular chemistry. The conditioning chemicals promote the formation of copper salts in the reactor which is in certain cases further supported by a chilling of the etchant. The copper salts are then filtered out in the separator and stored as a by-product in a special tank. This by-product with its high copper content can usually be sold without problems to chemical industries for their further use.

16.4.2 Waste-Water

The pollution limits of waste-water are governed today in most of the countries by special rules and laws which have to be strictly followed.

As an example, the requirements for water disposed into surface waters in the Federal Republic of Germany are given here (status as in 1970):

Temperature	30°C
pH	6.5–9.0
Sediments	0.3 ml/l after 2 hrs.
Oils and greases, extractable by petroleum ether	10 mg/l
Sulfates (SO_4)	not specified
Cyanide (CN) destroyable by chlorium	0.1 mg/l
Free chlorium	0.5 mg/l
Chromium (Cr)*, total	2.0 mg/l
Chromium (Cr), 6-valid	0.5 mg/l
Copper (Cu)*	1.0 mg/l
Nickle (Ni)*	3.0 mg/l
Zinc (Zn)*	3.0 mg/l
Cadmium (Cd)*	3.0 mg/l
Iron (Fe)*	2.0 mg/l
Aluminium (Al)*	not specified
Fluoride	20 mg/l

Legend: * = solved and unsolved.

The etching equipment manufacturers have taken such strict limitations as a challenge. And as the best results are obtained if the problem is solved directly at the source, a few very interesting solutions have been introduced which go as far as to a so-called waste-water free PCB etching.

The most important factor is to minimise the drag-out of the etching machines. This is done by the use of squeezing rollers after the etching machine but also after each rinsing step. Do not forget that even the best squeezing device does not help if not properly adjusted and maintained.

The next step is the applying of recirculation and high-pressure spray rinsing modules. Cascading of such modules brings a thinning factor with respect to the copper content in the rinse water in the range of 20–60 times per module. Fig. 16-13 shows a typical arrangement with open-loop regeneration as used for high-volume etching with cupric chloride. The set-up shown has an etching capacity of 40 m^2 of double-sided boards per hour with copper foil of 35 μm thickness. The manufacturer of this high-performance set-up (Hollmuller, 7033 Herrenberg, Federal Republic of Germany) claims a fresh-water consumption of just 2m^3/h with a copper concentration in the outcoming water of approximately 0.06 ppm which is about 0.5 mg/l. This low copper concentration makes it possible to discharge such water directly into the sewer system. The system is called 'waste-water free'; a better term would probably be 'not-harming waste-water system'. The low content of copper residues after the last rinsing permits the

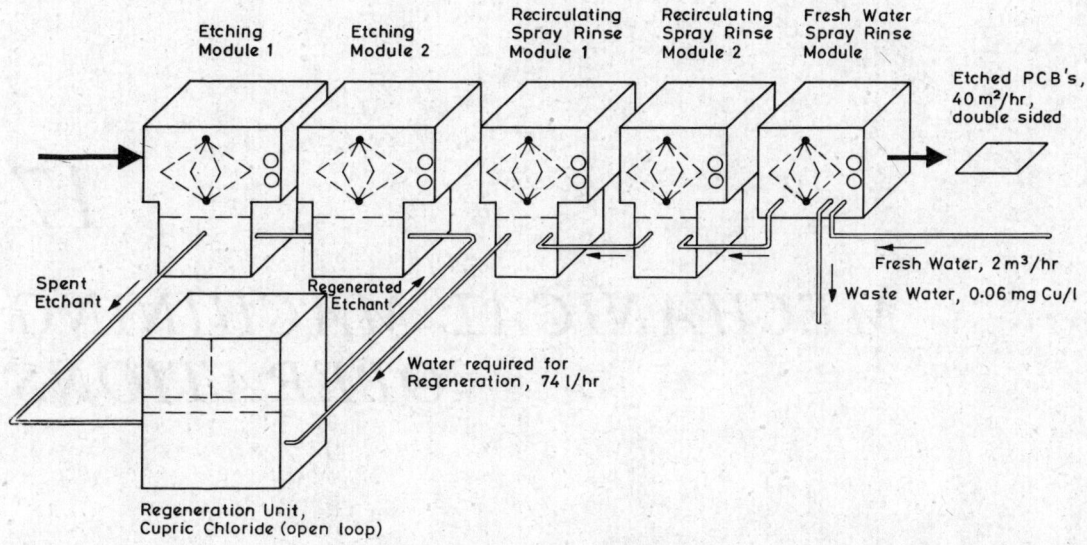

Fig. 16-13 Minimising waste-water production

advancing of boards directly to the stripping module or drying and storing them in this condition.

If it is not possible to minimise water contamination directly at the source to a level acceptable for the sewer system, the waste-water has to be collected in a tank and chemically treated with methods incorporating sedimentation, ion exchange, precipitation or flocculation. More on this can be found in Chapter 22.

BIBLIOGRAPHY

1. COOMBS C.F.: *PRINTED CIRCUITS HANDBOOK*, 1st edition, McGraw-Hill Book Co., New York, 1967.
2. COOMBS C.F.: *PRINTED CIRCUITS HANDBOOK*, 2nd edition, McGraw-Hill Book Co., New York, 1979.
3. DECKER G.: 'Aetzmaschinen und Aetzverfahren zur Herstellung Gedruckter Schaltungen', *EIPC PROCEEDINGS*, Zurich, Febr. 1970, pg. 27-55.
4. GURIAN M.I.: 'Steady-state regeneration etching: A technology review', *ELECTRONIC PACKAGING AND PRODUCTION*, USA, July 1978, pg. 32-36.
5. HERRMANN G.: *LEITERPLATTEN*, Leuze Verlag, Saulgau/Wurtt. (Germany), 1978
6. 'Leitfaden fur die Herstellung ein- und zweiseitig Gedruckter Schaltungen' (Leaflet), Laif Electronic GmbH, 5202 Hannef/Sieg (Germany), 1974.
7. 'Regeneration Unit for Cupric Chloride, Type ARC 75' (Leaflet), Hollmuller, 7033 Herrenberg (Germany).
8. SCHALT E.A.: 'Das Aetzen mit $CuCl_2$ (Kupferchlorid)', *FAG INFORMATION*, FAG SA, Lausanne (Switzerland), 1974.
9. SCHALT E.A.: 'Die $CuCl_2$–Regenerierung mit der RDCH 72', *FAG INFORMATION*, FAG SA, Lausanne (Switzerland), 1974.

17
MECHANICAL MACHINING OPERATIONS

17.1 INTRODUCTION

Under mechanical machining operations, we understand PCB production steps such as hole drilling, laminate cutting to processing size or final size and connector polarisation slot cutting.

Unlike other PCB processing steps, mechanical machining operations cannot be carried out on transfer-line equipment. They account therefore for a substantial part of manual handling of the boards. This also gets reflected in the costing of these operations. Tolerances, wherever required, should be made as narrow as they are functionally really needed. A good knowledge of the laminates, technology and equipment available will help in specifying tolerances which make sense.

The standardisation on a few basic PCB formats can help to bring the costs for contour cutting substantially down: Tools and jigs can be applied which otherwise would not be justified because of high costs. Of course, standardisation is not applicable in all cases but could be done especially in the field of professional PCBs.

It is in the nature of these operations that only a few guidelines can be given here. There are too many factors which vary from case to case such as base material composition, equipment, tools, operator's habits, etc. The ultimate solution and data is based on one's own experience, as many will agree. Comparing it with the machining of wood or metals, one can note some basic differences for the machining of PCB laminates:

a) Machining forces must be kept low. Excessive machining forces could cause a partial *delamination* because of the inherent laminate structure. The interlaminar bond strength as such is weaker than the strength of the reinforcement.
b) *Tool sharpness* is very important in order to get an acceptable machining finish. Blunt and dull tools can cause chipping because of the resin brittleness.
c) PCB laminates can show a fairly high amount of *resilience*. This tends the laminate to move away from the cutting edge but springing back occurs when the deforming forces are removed.

To consider this behaviour, sharp cutting edges and oversize-dimensioned tools are used.

Most of the base materials used can be classified into one of the two categories: Those with paper reinforcement and others with glass reinforcement. Base materials with glass reinforcement show typically a 10 times higher abrasive effect on the cutting tool used than those with paper reinforcement. Therefore, tools made of tungsten-carbide materials are preferred for glass reinforcements.

To maintain satisfactory working conditions, dust collectors are recommoded wherever dust is produced. It must alse be noted that certain people show allergic reaction to the dust of glass epoxy base. Possible remedies are the use of a barrier cream on the unprotected skin or wearing of clothes and gloves covering such areas. However, clothing should not be tight especially around wrists and neck.

17.2 SHEARING

Shearing is a suitable cutting method for almost any kind of base material of less than 2 mm thickness. If boards thicker than 2mm are cut, the finish of the edges will become coarse and unclean and therefore unacceptable.

The shearing of glass reinforced laminates is less difficult in obtaining a perfect edge finish. Paper reinforced laminates require often a heating up to a temperature in the range of 60–100°C so as to obtain a clean edge. Whether there is a need to heat up the laminate depends also on the precision and the conditions of the shearing machine.

Shearing Machines

Laminate shearing is preferably done on a motor-driven plate shear with a cutting length of at least 1,000 mm. Plate shears used for laminate cutting should not be used for sheet-metal cutting because perfect blades are a must for clean laminate shearing. When shearing starts, the board to be cut is first firmly pressed down with a spring loaded hold-down device before shearing is executed. This hold-down is an absolute must to prevent the otherwise unavoidable shifting of the board while shearing it through. The hold-down device should be at a very close distance to the cutting blades.

The treadle-type shearing equipment is used for the maximum laminate thickness of 1.6 mm or less.

Shearing Blades

Both blades can be of rectangular shape but the lower one is often provided with a free angle of up to 7° (Fig. 17-1). For shearing of different kind of laminates, a length-wise angle between the blades of 1–1.5° will give best results. For glass epoxy materials, up to 4° can be tolerated. For a clean edge finish, the gap between the cutting edges of the two blades has to be less than 0.25 mm.

Shearing is widely used in cutting the laminates to board-processing size. There are machines available with very sophisticated stopper designs to permit an accurate dimensional reproducibility. Larger machines can cut several 100 kgs of base material per hour. This figure of course will be much smaller if the board format has to be changed often.

For the cutting to final size (which is also called *clean cutting*) the board to be cut is aligned in such a way that corner marks are flush with the fixed shearing blade. Because of unavoidable parallax errors, board dimensions will come within 0.3–0.5 mm tolerance: Such a large

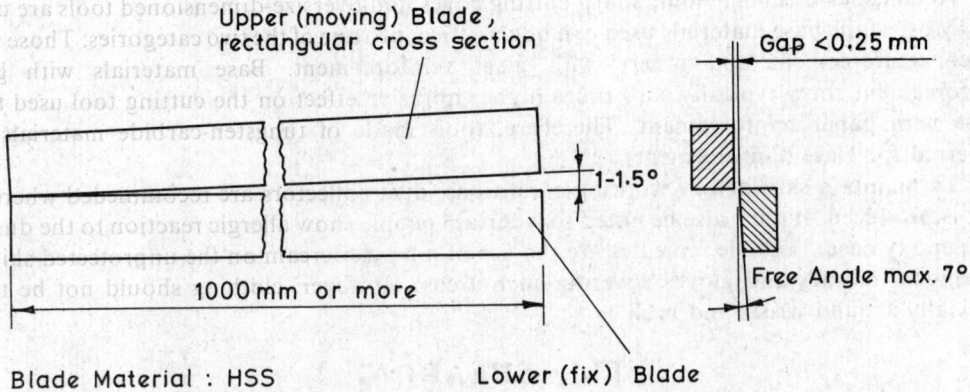

Fig. 17-1 Blades for shearing of PC laminates

tolerance is unacceptable for many applications. In such cases and also wherever the edge finish is found insufficient, other methods are applied for clean cutting. Hand-operated plate shears with precision stoppers are quite often usefully applied for clean cutting but one has to look for a good hold-down mechanism.

17.3 SAWING

The sawing of laminates by means of a circular saw is the preferred method of many laminate manufacturers. PCB manufacturers, too, often use these sawing machines, which are specially designed for this purpose. The edge finish is smooth and acceptable for clean cutting. The dimensional tolerances are of the same order as in shearing (0.3–0.5 mm), unless a jig with registration pins is used, which provides a ± 0.2 mm tolerance field.

Circular Sawing Machines

Sawing machines of the moving-table type give the best results. The design should be such that no noticeable drop in cutting speed occurs during operation. This is achieved by heavy pulleys with more than one V-belt. The optimum saw blade speed can vary in the range of 2,000–6,000 rpm which calls for suitable provisions for speed control. The precision of the bearings has a direct impact on the edge finish: No play should be felt when inspected by hand. The gap between saw blade and bench should be minimum to give a good support to the board near the cutting edge.

Circular Saw Blades

For paper phenolic materials, HSS blades with a diameter of approximately 300 mm are used at a speed of 2,000–3,000 rpm. Toothing should be around 1.2–1.5 tooth per cm circumference. The time between resharpening is multiplied by a factor of 3–5 if tungsten-carbide tipped saw blades are used at a speed of 2,500–6,000 rpm.

Glass-epoxy materials are cut with tungsten-carbide tipped blades or, for still better performance, with diamond wheels. The higher initial costs of such blades are offset by the much longer life and improved edge finish.

MECHANICAL MACHINING OPERATIONS

Fig. 17-2 Circular sawing machine for PCBs, moving table type, with dust collector (Courtesy: FAG SA, Lausanne, Switzerland)

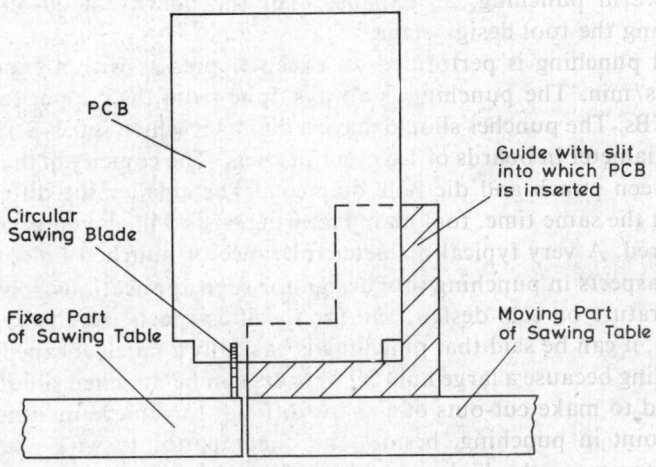

Fig. 17-3 Sawing of polarisation slots

Operation

For safety purposes, the blade should always be covered by a suitable guard device. The circular saw is adjusted in such a manner that the free height of the blade above the boards to be cut is in the range of 10–15 mm. A bad edge finish can be caused by a blunt or badly sharpened blade, by too much play in the bearings, by a too coarse toothing or by an incorrect feed rate. Thick materials need a lower feed rate while thin materials can be cut faster.

Sawing of Polarisation Slots

Polarisation slots are used in the connector area of the PCBs to prevent an incorrect insertion of the board into the matching contact receptacle. Sawing of polarisation slots is carried out after clean cutting because an accurate reference edge is needed. The reference edge, for sawing the slot, is firmly pressed towards a guide. Since the polarisation slot position is related to the pattern rather than to the board contours, the contour tolerances have to be accordingly adjusted.

17.4 PUNCHING

Punching of holes into PCBs (also called piercing) is a widely used method in high-volume production of consumer-type boards made of paper phenolic and paper epoxy laminates. The punching of glass epoxy boards is not common because of cracks formed inside the hole.

A cross-section of a punched hole will show a conic shape and partially a rough surface. Punching of holes is therefore not compatible with professional PCB plated-through hole processes. For plated-through holes, only drilling provides an exactly cylindrical shape and a high surface smoothness. Another restriction in hole punching are the difficulties arising with small diameter punches (damage, fracture): A thumb rule for minimum diameter is half the board thickness.

Where paper phenolic material is punched, the cold-punching varieties are usually preferred. This is not only because heating up of the boards is adverse to a rationalised production but also because of dimensional accuracy which can be better controlled in cold punching. In warm punching, an expansion of the boards of 20–40 ppm/°C has to be considered during the tool design stage.

The actual punching is performed on excenter presses with a capacity of 10–30 t and 100–200 strokes/min. The punching is always done from the copper foil side in the case of single-sided PCBs. The punches should have a diameter which is 0.1–0.12 mm bigger than the desired whole diameter in boards of 1.6 mm thickness. The conicity of the holes depends on the difference between punch and die hole diameter. The smaller the difference, the lesser the conicity, but, at the same time, tool manufacturing costs will rise enormously with the higher precision required. A very typical diameter tolerance of punched holes is ± 0.1 mm.

For other aspects in punching tool design for such applications, it is suggested to refer to the general literature on tool design, and for specific aspects, also to [1] and [2].

To sum up, it can be said that punching is basically a much cheaper way to get holes into PCBs than drilling because a large number of holes can be punched simultaneously. Punching can also be used to make cut-outs of a shape difficult to achieve by other methods. A rather heavy minus point in punching, besides the uncompatibility with plated-through holes in subtractive processes, is the high cost factor for tooling. Punching will therefore become economically attractive only if a PCB quantity of at least 2,000 numbers is required.

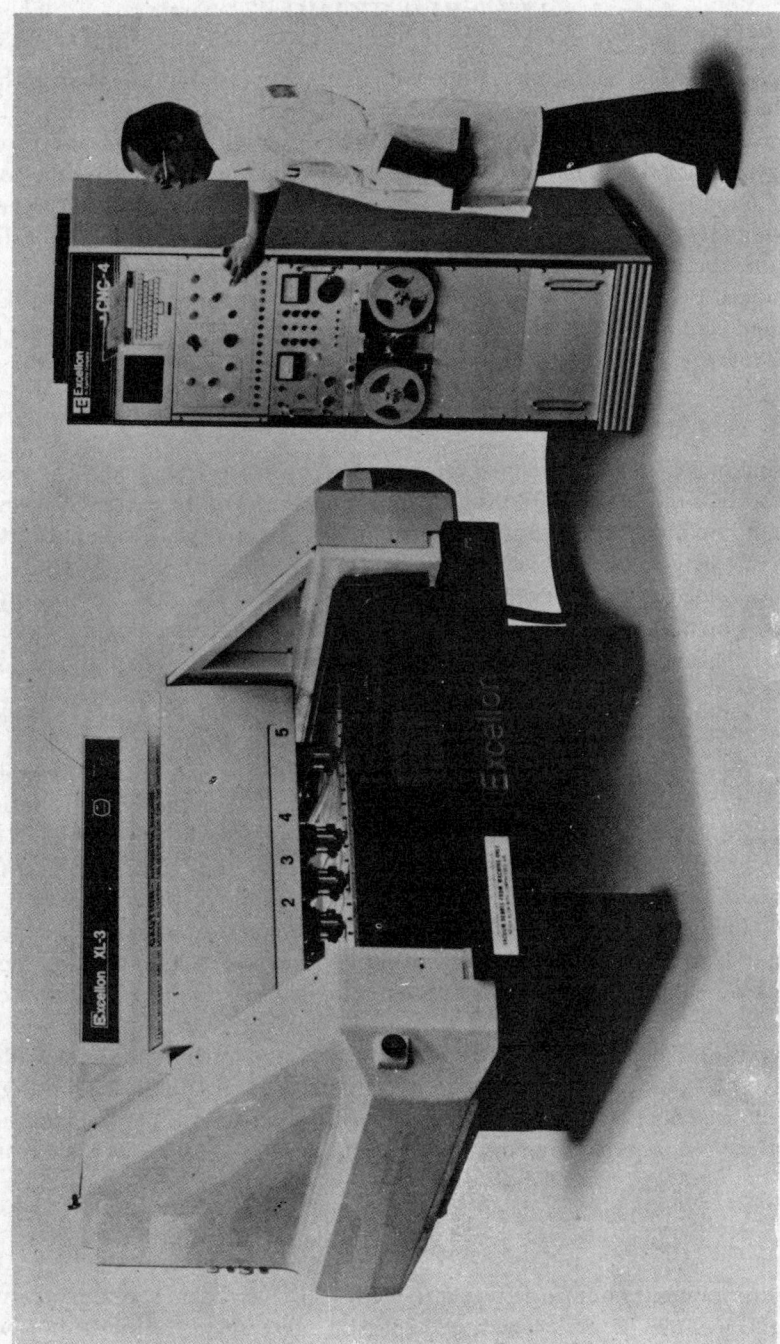

Fig. 17-4 CNC routing machine with 5 spindles (Courtesy: Excellon Europe GmbH, Dreieich, Germany)

17.5 BLANKING

When the clean-cutting operation is done with a punching tool (rather than with a shear or a saw), it is called *blanking*.

In some cases, hole punching and blanking are done with the same tool in the same operation. Therefore, most of what has been said under punching applies equally to blanking.

Blanking is sometimes also used for glass epoxy materials but the edge finish does not look very clean although it is functionally acceptable. With blanking tools, PCB dimensions can be kept within a tolerance field of $\pm (0.1$–$0.2)$ mm.

Where standard PCB formats with a pin registration system are prevalent, blanking can be a very efficient and economic solution for clean-cutting of PCBs. This also includes PCBs with plated-through holes.

17.6 MILLING

Clean-cutting of PCBs by milling is preferred where a good edge finish and exact overall dimensions are a required feature. Milling of PCB contours can be carried out on wood- or metal-working milling machines, either of horizontal or vertical type. The usual cutting speeds are in the range of 1,000–3,000 rpm with straight or spiral tooth HSS milling cutters. Here, too, use of tungsten-carbide tools will result in a much longer life of the cutter, especially for glass-epoxy materials. For milling, a strong support to the PCBs must be given, if necessary with backing plates. This is to avoid possible delamination.

17.7 ROUTING

In routing, the clean-cutting (contouring) is done with the help of a pin router machine (copier router). The boards are moved past a vertical side mill with the aid of a routing jig. The routing jig is guided in relation to the mill by holding it against a bushing which is concentric with the mill. Internal registration holes are used to position the PCBs on the routing jig.

Routing provides an excellent edge finish similar to milling. The dimensional tolerances are within $\pm (0.1$–$0.2)$ mm. Compared with blanking, it is much slower; its speed depends on the quantity of material to be removed. But a routing jig is much cheaper than a blanking tool. Sometimes blanking with slight oversize is applied followed by routing to get a smooth surface finish.

The most recent developments in routing include the introduction of CNC routing machines (CNC = Computer Numerical Control) with several spindles. Routing jig is not required with the machine, therefore, there are no tooling costs. The information on the contour is fed, either via a paper tape or directly to the computer and is recalled from there as required.

17.8 DRILLING

Drilling of component mounting holes into PCBs is by far the most important mechanical machining operation in PCB production processes. Holes are made by drilling wherever a superior hole finish for plated-through hole processes is required and where the tooling costs for a punching tool cannot be justified. Therefore drilling is applied by all the professional-grade PCB manufacturers and generally in all smaller PCB production plants and laboratories.

The importance of hole drilling into PCBs has further gone up with electronic component miniaturisation and its need for smaller hole diameters (diameter less than half the board thickness) and higher package density where hole punching is practically ruled out.

17.8.1 Hole Diameter Tolerances

To compensate for the laminate resilience the drill bit diameter is chosen 0.05 mm bigger than the hole diameter expected. The following hole diameter tolerances have been generally accepted wherever no other specifications are mentioned:

$D \leqslant 0.8$ mm $\rightarrow \pm 0.05$ mm
$D > 0.8$ mm $\rightarrow \pm 0.1$ mm
$D =$ hole diameter

The production of holes with diameter tolerances as specified above should not need any special attention from the manufacturer: A suitable drilling machine with a correctly sharpened drill bit will provide these results. Where smaller diameter tolerances are specified, the additional efforts required will be clearly reflected in the higher costs for drilling.

In plated-through hole processes, the usual practice is to drill holes about 0.15–0.25mm bigger in diameter than the component lead to be inserted. This is necessary to compensate for the hole diameter reduction caused by plating inside the holes. For ordinary holes without plating provision, the diameter allowance can be within 0.10–0.20mm.

17.8.2 Drilling Machines

Drilling machines for PCB applications are available in a wide range of designs and principles. But one common feature of all PCB drilling machines is the high-speed range which

Fig. 17–5 Low-cost drilling machine for drilling by direct sight (Courtesy: Punjab Recorders Ltd., S.A.S. Nagar, Mohali, India)

is required for an economic and efficient drilling of the various base materials. The speeds mostly used are around 20,000–50,000 rpm. The minimum spindle speed available should not be less than 15,000 rpm for glass reinforced laminates (tungsten-carbide drill-bits) or 8,000 rpm for paper reinforced laminates and HSS drill-bits. Very sophisticated drilling machines have spindles operating up to 100,000 rpm. While very simple machines use belt drives or fast running DC-motor spindles, there are electronically controlled high-frequency spindles employed in almost all the more sophisticated drilling machines. The spindle speed can thus be controlled over a wide range and is kept constant. In some cases, the spindles are water-cooled with the help of a water circulation and cooling system built into the machine.

Drilling by Direct Sight

This is the simplest method: The board is positioned manually under the point of the drill bit. Such equipment is very simple and cheap but the accuracy of the hole positioned cannot be expected to be much better than ± 0.25 mm.

The accuracy of hole positions will be improved by using artwork pads with an inside hole of about 0.4 mm diameter. An optical device projecting a cross-hair target showing the centre position onto the board is very helpful for the operator's convenience. However, the accuracy limits mentioned above can hardly be improved, not even with a cross-hair projector.

Drilling by Optical Sight

The drilling spindle is mounted underneath the table while an optical magnifying system above the table projects the target onto a viewing screen. The magnification of the optical system is 10 times. Subdued illumination is required for convenient observation of the screen. For board centering, sight marks consisting of concentric circles and a cross-hair are utilised on the viewing screen. The drilling operation is actuated by means of foot switch that raises the drilling spindle. The position accuracy can, in this way, be controlled to within ± 0.1mm.

Drilling by direct or by optical sight needs the full concentration of the operator. The resulting fatigue, especially of the eyes, can be adverse to the precision expected from the final product. A possible solution to this problem is the part-time engagement of the operator in drilling while the remaining time is spent in doing some other work on an exchange base.

In plated-through hole processes, drilling is usually done directly after cutting the laminate to the board processing size, before a pattern is printed or formed on the board surface. The question is how to find the hole positions for drilling, by direct or by optical sight. In smaller production quantities and where drilling occurs from below, a film master can be fixed with adhesive tape onto the top of the board to enable centering. Also, coating of the board with cheap wet-film resist, followed by photoprinting, development and dyeing, is a solution for this problem and is sometimes used. However, wherever a large output is requried, jig and NC-drilling are preferred.

Jig Drilling

With jig drilling, also called semi-automatic drilling, the productivity goes up to 30–50 strokes per minute. The resulting accuracy is comparable to drilling by optical sight (±0.10 mm) but more uniform results are obtained.

A jig can be made out of 3 mm acrylic glass or brass. The holes in the jig are drilled by optical sight using a film master of the respective pattern on the top of the plate. An alternative method is screen- or photo-printing of the pattern onto the top side of the plate, followed by

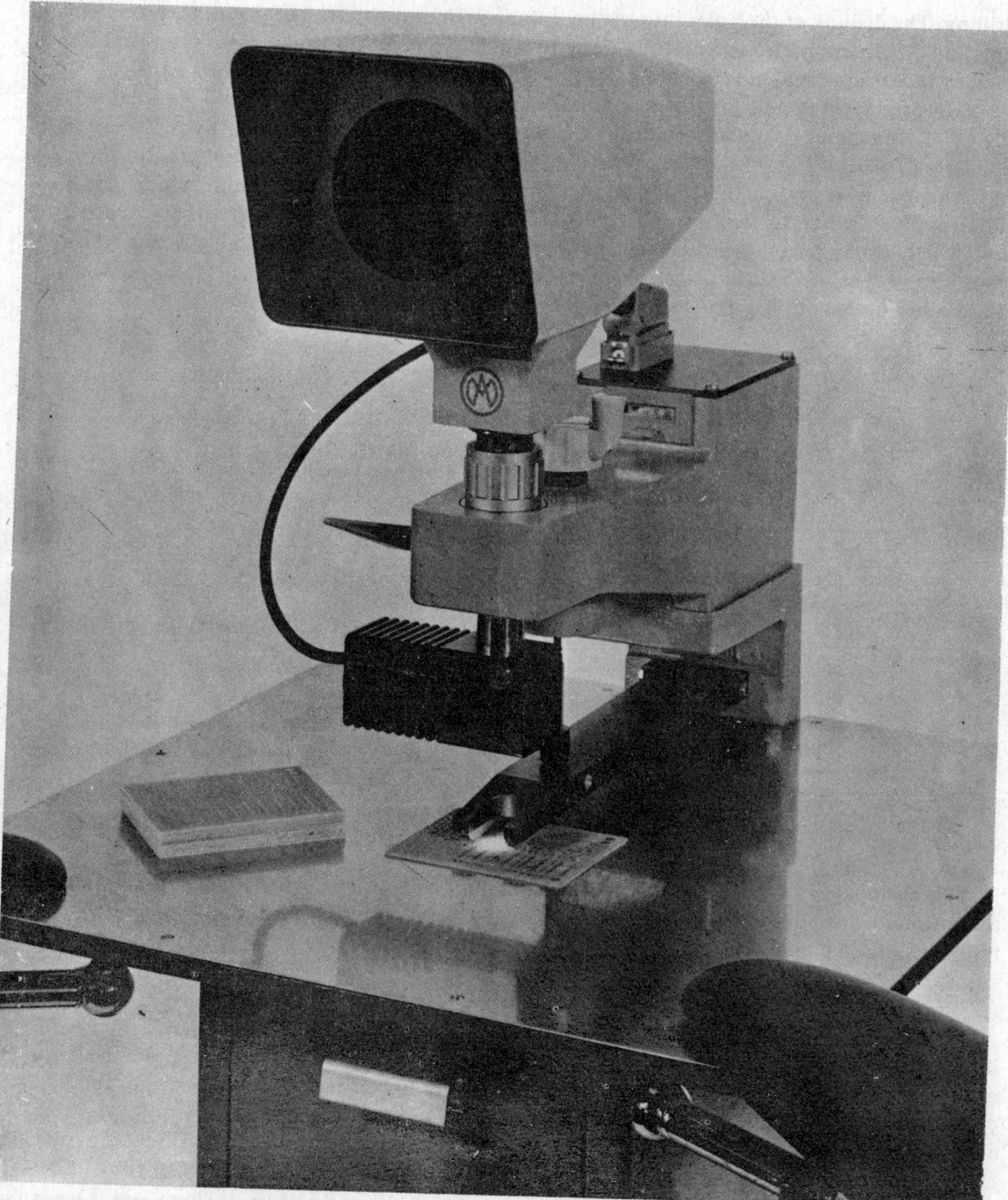

Fig. 17–6 Typical machine for PCB drilling by optical sight (Courtesy: Contact-Systems Ltd., Egg/Zurich, Switzerland)

drilling. The drilling of the holes into the jig is done with utmost care because of their reference function. Counter-sinking is carefully done to suit the type of position sensor (spring tracer) used in the drilling machine.

The principle of jig drilling in machines designed for drilling by direct sight is explained in Fig. 17-7. Machines designed for drilling by optical sight can be modified by replacing the optical part with a suitable sensing device as suggested by the machine manufacturer.

The machines and methods discussed so far make use of only one spindle which permits simultaneous drilling of maximum five boards in a stack. To increase productivity, specially designed jig drilling machines are available, operating with up to 8 spindles, all governed by the same jig and sensor. In certain designs, the movement of the table is controlled via the jig sensor while other designs utilise the control of the spindle position via the sensor.

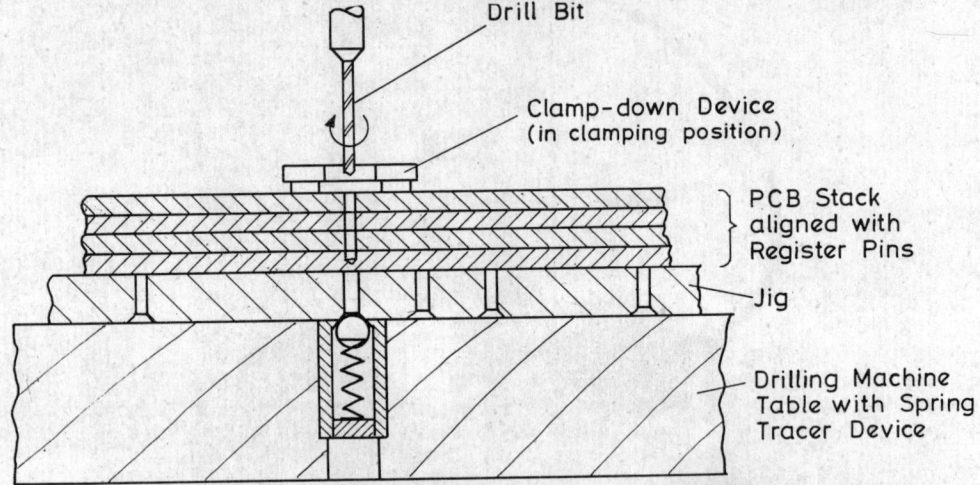

Fig. 17-7 Jig drilling on a machine for drilling by direct sight

To ensure that all the holes are drilled in jig drilling operations, the routes corresponding to the various drill bit diameters can be indicated on the jig by different colours and controlled by a resetable stroke counter.

NC Drilling

NC drilling (Numerically Controlled drilling) is an automatic drilling process where the information on the hole position, sequence and diameter is stored on a paper tape and the table or spindle arrangement moves automatically according to this information. Some of the NC drilling machines on the market can operate up to 240 strokes per minute with a position accuracy of ± (0.025–0.050) mm for single spindle, or ± (0.050–0.075) mm for multi-spindle machines. An NC drilling machine with 8 spindles at 240 strokes per minute and stacks of 4 PCBs can therefore produce theoretically 460,800 PCB holes per hour. In practice, however, it is not possible to reach the theoretical limits because time must be provided for drill-bit changing board loading and unloading.

NC drilling is a very precise and extremely efficient method for the high-volume

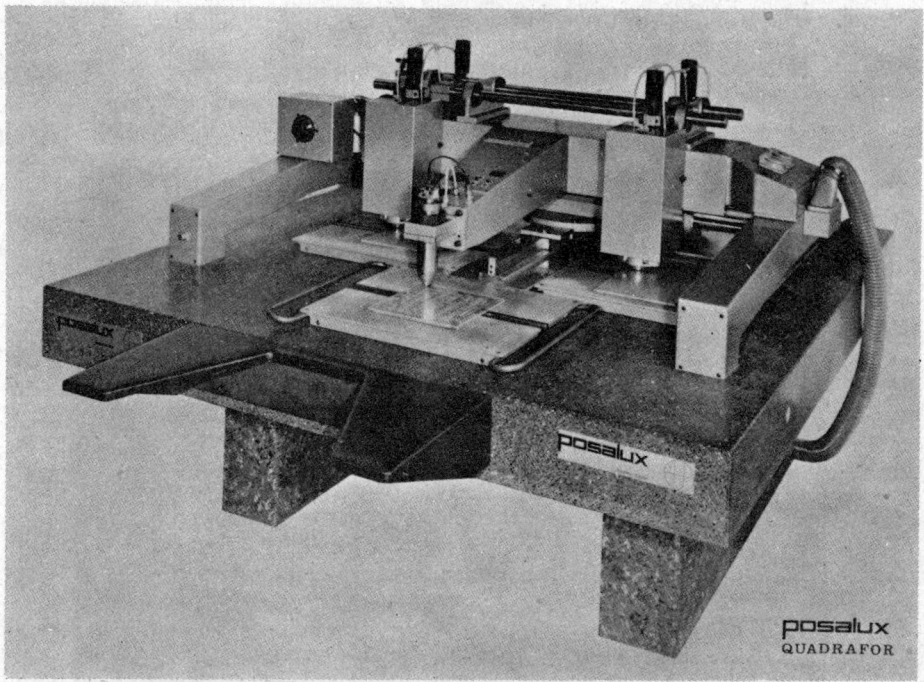

Fig. 17-8 Jig drilling machine with four spindles and air cushion supported working table
(Courtesy: Posalux SA. Bienne, Switzerland)

professional PCB manufacturer. The result dependence on the operator has been largely eliminated and the output corresponds to the quantities produced in automatic PCB transfer production lines.

The high degree on positional accuracy achieved, matches well with fine-line-conductor PCBs where the small solder pad size (e.g., width of annular ring typically 0.25 mm) permits only small hole position deviations.

The punched tape can be derived from a film master on a digitising machine. A digitising machine looks quite similar to a machine for drilling by optical sight. However, instead of the drilling spindle, position sensors are provided along with the moving working table so as to give digital information about the hole position. The working table with the film master is moved from pad to pad, controlled on the viewing screen and the position data are automatically punched into the paper tape. Where computer-aided design has been employed in designing the PCBs, the paper tape is directly available from there and no digitising machine is required.

The latest developments in automatic drilling include automatic drill-bit changing, fast positioning to enable up to 400 strokes/min and CNC-type of control. CNC drilling machines can directly be programmed so as to eliminate the need of a digitising machine. They are supplied with an optical device to drill the first board by optical sight while the position information of the holes gets simultaneously stored.

What has not been considered in the given account of drilling machines, is the investment aspect. The costs are naturally directly linked to the sophistication. A simple machine for direct-sight drilling may cost 250 dollars but a multi-spindle NC drilling machine can easily be

Fig. 17-9 CNC multiple spindle drilling machine (Courtesy: Klingelnberg GmbH, Remscheid, Germany)

several hundred times more. At this point, the initial fascination about the technical possibilities ends and economical thinking has to begin. For economic drilling, cost factors for manpower, equipment-investment, number of PCBs per series, turn-around time and others have to be kept in healthy relations with each other: They will also be different from place to place.

General Requirements for Drilling Machines

Bearings: The life of the drilling spindle depends fully on the life of the bearing system. The bearings have to withstand a continuous high-speed operation without giving vibrations; they should not produce more noise or sound with continued use. There are only a few, highly-specialised PCB drilling-spindle manufacturers. Their spindles can be found in most of the PCB drilling machines in the market. To evaluate the performance of a drilling machine is therefore easier if the qualities of the spindles are already known.

The bearings should be adequately protected from abrasive dust which is caused in glass-epoxy drilling.

Speed range: While 15,000 rpm is considered the minimum useful speed, it is desirable to have different speeds available, covering at least the range of 20,000–50,000 rpm which will be used mostly.

Clamp-down device: It should hold down the boards being drilled with sufficient pressure to overcome the occasional warping of the boards. Clamp-down devices are a must especially in stack-drilling operations Clamping-down devices are usually pressed down by electro-magnets as soon as the drilling stroke starts.

Feed rate: Best results are obtained with feed rates adjustable without steps in a range of 0.01–0.05 mm per drill bit revolution. The feed rate can be at the upper limit for drilling of just one board. The more the combined drilling depth, the lower the feed rate to be chosen.

Bushes: Bushes are made of tungsten-carbide materials and are preferred where the ultimate in position accuracy must be achieved. The drill-bit is guided by the bush directly where the drilling operation starts. The bush also suppresses burr formation at the drill-bit entry point

Bushes must be considered as highly precise tools and they are manufactured with tolerances in the μm-range to give maximum precision and a long life of 120,000–150,000 strokes.

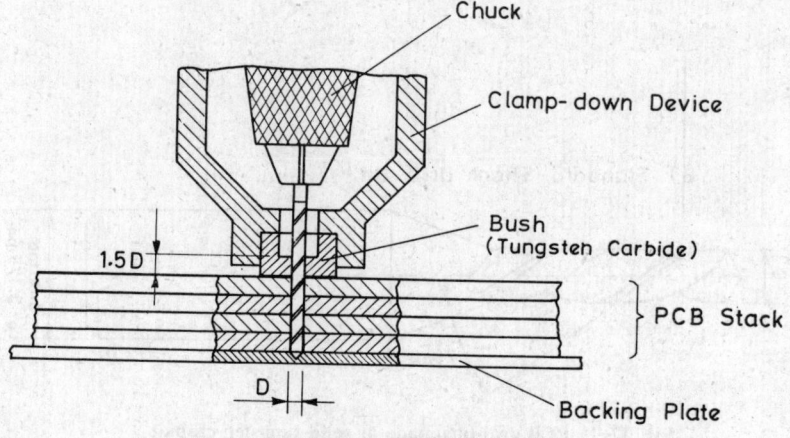

Fig. 17-10 PCB drilling with a bush

The optimum length, where the drill bit is guided by the bush is 1.5 times the drill-bit diameter. The drill-bits suitable for drilling with bushes are of the straight-shank type with the shank diameter equal to the working (nominal) diameter. Straight-shank drill-bits are cheaper than the drill-bit with standard shanks. But for each drill-bit diameter, a separate bush and collet has to be provided which again adds to the operational costs.

Changing of drill-bits: In drilling practice, the time required for drill-bit changing is expressed as unproductive time. It should therefore be carried out as quickly as possible; this is achieved by an easy accessibility of the chuck and an easy drill-bit depth adjustment.

17.8.3 Drill Bits

The drilling of holes into paper-phenolic laminate with drill-bits made of high-speed steel (HSS) does not cause much difficulties. However, if the same drill-bits are used for glass-epoxy materials, the cutting edges will become blunt after a short time and resharpening has to be done. Therefore drill-bits made of tungsten-carbide are widely used in PCB drilling operations, even for paper-phenolic laminates.

Drill-bits made of tungsten-carbide are also preferred because of a very high operation time between resharpening and also excellent quality of holes. Based on this fact, the present section will deal exclusively with drill-bits made of tungsten-carbide materials.

The tungsten-carbide drill-bits employed are usually of the solid types as shown in Fig. 17-11. Straight-shank drill-bits are used in drilling with bushes which give perfect guidance very close to the point where the bit enters the boards and thus enable very accurate hole positions. Standard-shank drill-bits are used without bushes. They are costlier because of more tungsten-carbide material involved. On the other side, the standardisation on one particular shank diameter will be cheaper on the drilling equipment side (only one collet type required).

A) Straight-Shank Drill-Bit

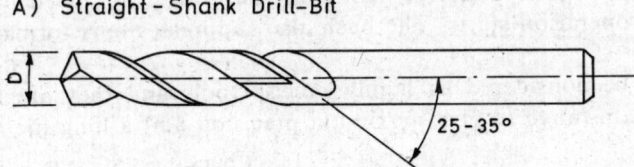

B) Standard-Shank Drill-Bit

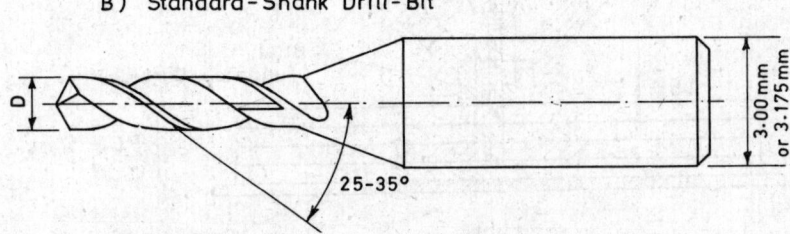

Fig. 17-11 PCB drill-bits made of solid tungsten carbide

Tungsten-carbide is an extremely hard-wearing sinter material. Due to its hardness, it is very brittle: Simple dropping of the drill-bit on the drilling machine table might be sufficient to cause damages to the cutting edge, visible under a magnifying glass. The personnel handling these drill-bits must be properly instructed so that they are fully aware of the special care needed.

17.8.4 Drilling Practice

Cutting Speed

For tungsten-carbide drill-bits, the cutting speeds recommended vary in the range of 70–200 m/min. The cutting speed depends firstly on the laminate to be drilled but also on the number of boards (stack drilling), feed rate, drill-bit type and the hole quality expected. A few general guidelines can be given here as a help in finding the optimum performance in practice:

NEMA grade	Cutting speed
G–10	150–200 m/min
G–11	70 m/min
FR–4	110–150 m/min

The conversion from cutting speed to drilling spindle speed for a particular diameter can be done with the help of Fig. 17–12. In general, too high a speed will shorten the period between resharpening of the drill-bit while a slow speed means a lower production yield. Cutting speed and feed-rate must herewith be optimised for high production, long tool life and acceptable hole quality.

Stack Drilling

A common method to increase the efficiency in drilling is stack drilling: Here several boards, stacked together with registration pins, are drilled simultaneously. The yield will be multifold according to the number of boards in the stack. But there are practical reasons limiting this number, like swarf clearing and hole position deviation. The deviation from the actual hole position centre in a stack of 5 PCBs can easily be 0.02–0.05 mm for the last drilled board in the stack. The deviation will be minimum for the first board and gradually increases with maximum for the last board. It is therefore recommended to limit the combined drilling depth to 8 times the smallest drill-bit diameter in the case of standard-shank drill-bits and for straight-shank type the corresponding factor is 10. In straight-shank drilling operations with bushings, the guiding bush length must be considered as additional drilling depth because of swarf clearing requirements.

Backing Plate

In NC- and jig-drilling machines, special provisions must be made to enable the drill-bit to drill completely through all the boards but without drilling into the drilling machine table. A common method is the use of a backing plate out of the same material and size as the boards drilled. The backing plate can be thin, but one must ensure that the last PCB, too, is completely drilled through but not the backing plate.

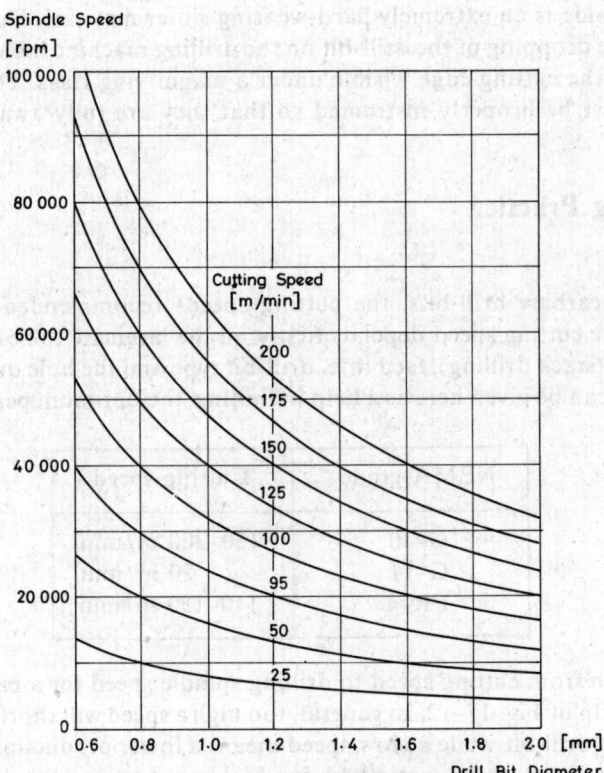

Fig. 17-12 Selection chart for spindle speed

Backing plates cannot be used again after the drilling operations; they should be as cheap as possible. In glass-epoxy board drilling, the cost for glass-epoxy backing plates would be too high. A cheap variety of X- or XP-grade of paper-phenolic is used for backing purposes.

It is also possible to perform drilling without backing plates by limiting the drilling depth in such a manner that the last PCB in the stack gets only halfway drilled. This PCB is thereafter put at the top of the next stack for drilling. This method, however, reduces the production volume to some extent and also hole position deviations will be more. It has already been mentioned that the last PCB in the stack shows maximum deviations from the ideal hole centres. Still, it is an accepted method to avoid costs for backing plates.

Resharpening

The number of holes which can be drilled between the resharpenings of the drill-bits is linked with the material drilled, board thickness, spindle speed and feed-rate. As a rule, we can expect a yield with glass reinforced laminates around 5,000 – 10,000 holes. A high cutting speed will give a shorter life for the cutting edge. This will be observed with materials to be drilled at a high cutting speed such as FR-5 and G-11 grade.

The need for resharpening is recognised by burr formation at the place where the drill bit enters the board. Drill bits should never be used when they are completely blunt. This not only produces burrs but also the probability for breaking grows rapidly. In multispindle machines, *all* the drill-bits are changed as soon as one of them produces burrs.

As a general practice, resharpening can be carried out 4 times with the same bit. After that, drill-bits are no longer fit for a good hole quality because the side flanks too begin to suffer of the abrasive nature of glass reinforced materials. This can also be recognised by decreasing diameter.

When drilling *multilayer* boards, only new drill-bits have to be used and not resharpened ones. This is because even a very small amount of smearing in the hole cannot be tolerated for making reliable through-contact between the different conducting layers.

The resharpening of tungsten-carbide drill-bits is done on special grinding machines (Fig. 17-13). Only such special machines provide the high degree of centricity needed for resharpening of these small drill-bits. Resharpening by a manual method is completely ruled

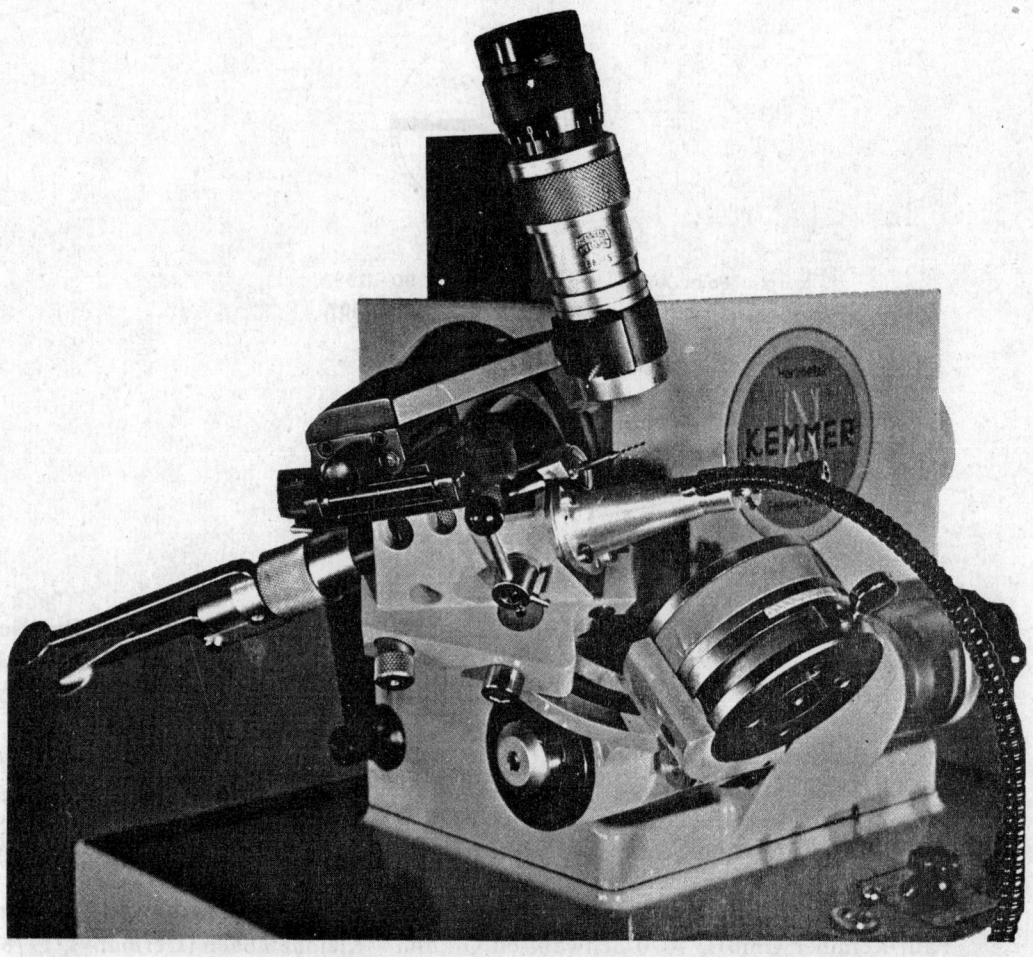

Fig. 17-13 Special grinding machine for resharpening of tungsten-carbide drill-bits
(Courtesy: Paul **Kemmer** GmbH, Schwabisch Gmund Kleindeinbach, Germany)

out. The special grinding machines are preferably equipped with a diamond wheel for this purpose and resharpening is done dry. Point and primary clearance angles are given in Fig. 17-14.

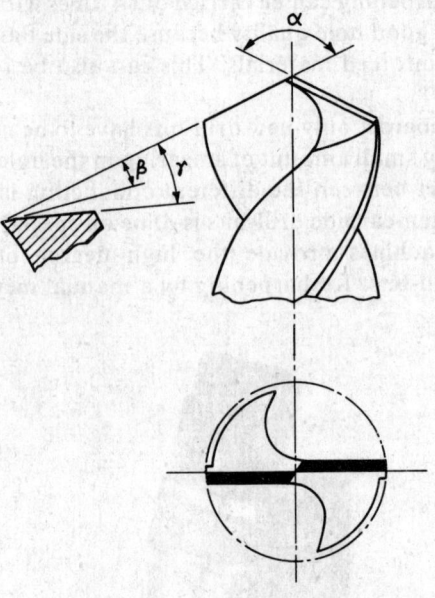

α	Point Angle, Paper-Phenolic	90 – 115°
	Glass-Epoxy	118 – 130°
β	1st Primary Clearance Angle	15°
γ	2nd Primary Clearance Angle	30°

Fig. 17-14 Angles for drill-bit resharpening

BIBLIOGRAPHY

1. COOMBS C. F.: *PRINTED CIRCUITS HANDBOOK*, McGraw-Hill Book Co., New York, 1967.
2. HERRMANN G.: *LEITERPLATTEN*, Leuze Verlag, Saulgau/Wurtt. (Germany), 1978.
3. LUND P.: *GENERATION OF PRECISION ARTWORK FOR PRINTED CIRCUIT BOARDS*, John Wiley & Sons, Chichester, 1978.
4. 'Recommended methods for machining FORMICA industrial laminates', Formica Technical Information Leaflet No. 1, Formica India Limited, Poona–411 001 (India), 1975.
5. 'Solid tungsten carbide tools for the machining of printed circuit boards' (Leaflet), Paul Kemmer GmbH, 7070 Schwabisch Gmund—Kleindeinbach (Germany), 1978.
6. WENG G.: 'Bohren von Gedruckten Schaltungen', *EIPC PROCEEDINGS*, Zurich, Febr. 1970. pg. 7–26.

18
PCB TECHNOLOGY TRENDS

Although the technology of PCB production has grown well and matured during the past few decades, its development has not come to a standstill. A permanent pressure arises out of the continuous miniaturising tendency prevalent in electronics and the falling prices for ICs and other components. Because PCB costs could only partly follow this downward trend, the relative price share of PCBs in the overall cost of electronic equipment went up.

There is a hard struggle going on in the field of PCB technology to find new solutions. This chapter will give an account of the major new streams which have mostly been developed in the USA. Since the accents are, therefore, naturally given by the material and market conditions there, we must mention here that these solutions are not universally applicable with the same success. With particular reference to PCB technology, most readers will agree that any PCB production set-up forms an individual solution which should be best adapted to the local conditions and demands. There are many more processes which are not mentioned in this book. But there are definitely still more possible solutions not yet tried out. This chapter should therefore also encourage in the search for new, genuine ways to electronic interconnecting techniques with less and less of the old handicaps.

18.1 FINE-LINE CONDUCTORS WITH ULTRA-THIN COPPER FOIL

18.1.1 Ultra-Thin Copper Foils

The manufacture of thin copper foils is a rather difficult process. Even today, there are only a few major companies supplying copper foils to the large number of PCB base material manufacturers. For a long time, the foil thickness of 17.5 μm appeared to be the practical limit for an economic foil production. During the past few years, a good deal of effort has gone into the development of manufacturing processes for still thinner copper foils with the result that ultra-thin foils are now available to the laminate and PCB manufacturers. Ultra-thin foils are defined here as foils with a thickness in the range of 5–10 μm. These foils are supplied on G-10 and FR-4 laminates which ensure sufficient bond strength between copper foil and laminate surface.

The price for ultra-thin copper-clad laminates is considerably higher due to the high costs of the foil. The price difference is around 20–30% (US market). On the other side, there are distinct advantages in PCB performance and processing which easily compensate for the higher price of the copper-clad base material. And as more and more PCB industries and customers are switching over to PCBs based on ultra-thin copper foil (presently 5-10% of the total PCB market [6]), there is an excellent scope for a substantial replacement of the standard 35 μm copper foil.

The ultra-thin copper foils are supplied either with a protecting aluminium foil or as bare copper foil. The aluminium protection itself is available as chemically or mechanically strippable (hand peeling) depending on the base material suppliers.

18.1.2 Pattern-Plate Processing with Ultra-Thin Foils

With the exception of the aluminium foil removal, all the processing steps are the same as the conventional ones in the pattern-plating process (Fig. 18-1) and basically the same equipment is used.

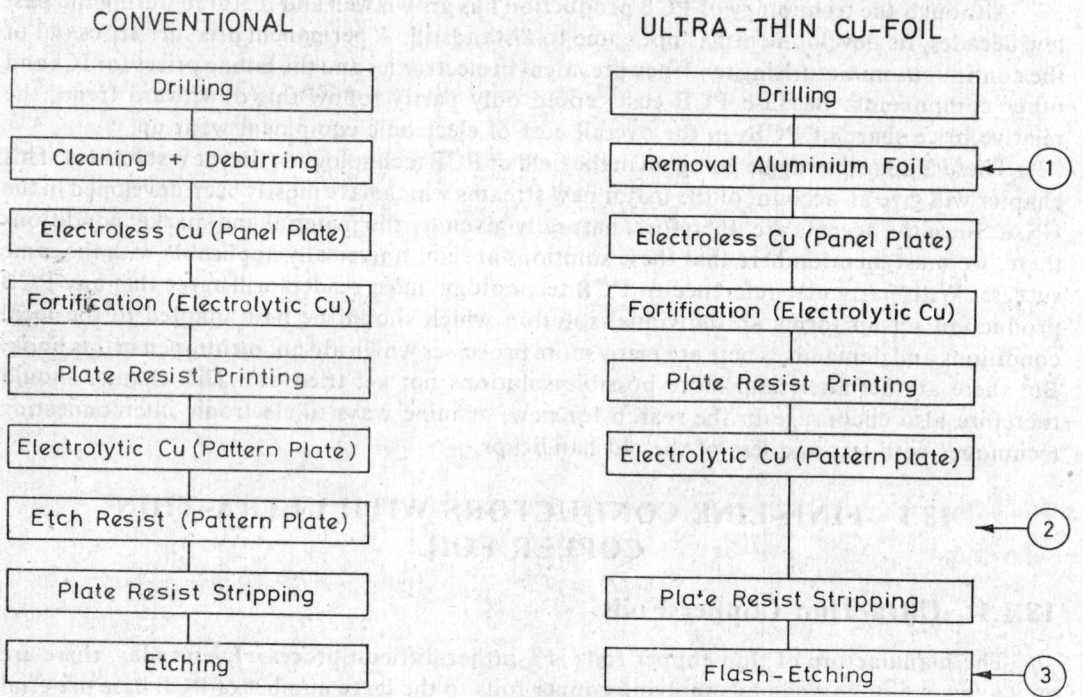

Fig. 18-1 Simplified PCB processing with ultra-thin foil

Let us have a look at the differences compared to the conventional pattern plating process:
1) The aluminium foil not only protects the copper surface but it widely suppresses the burr formation on the copper layer; drilling is done through the aluminium foil and burr is mainly formed where the drill-bit enters which is here the aluminium foil. Where necessary the minor burrs on the copper layer can easily be removed by a brief etch-dip after the aluminium protection has been removed.

The essential cleaning/deburring step in conventional processing is therefore replaced by the much easier to perform aluminium foil removal which uncovers the clean copper surface as it can directly go to the electroless Cu plating without further surface treatment. The chemical removal of the aluminium is simply done by etching in a tank with sodium hydroxide. A good air circulation is a must because of the fumes formed. The exhausted etchant, sodium aluminate in sodium hydroxide, is a solution extensively used as flocculating agent in waste water treatment and does not give problems if supplied to users of such processes.

2) The pattern-plating of an etch resist has become superfluous which means a considerable processing cost reduction. After flash etching, the copper surface can be protected with a solderable lacquer or solder hot-air leveled using a solder mask. Where tin-lead pattern plating is desired, it can be added like in conventional processing.

3) The final flash-etching removes the ultra-thin copper layer still between the pattern. The simultaneous thickness reduction of the copper conductor pattern does not cause problems, as it has been reported in [2].

Since only 5 μm Cu has to be removed, the etching chemicals required can be reduced to as low as 10% of conventional processes and the throughput in etching goes up by a factor 5 for the same equipment. These are really desirable attributes, also from the environmental point of view.

18.1.3 Advantages and Disadvantages with Ultra-Thin Copper Foils

Processing

Economic advantages result through the elimination of the cleaning/deburring step which is replaced by a much simpler aluminium foil removing step. Furthermore, there is no need for a pattern plated etch resist. Costs involved for etching of ultra-thin foil are about 25% compared to the etching of 35 μm foil.

PCB Performance

As the flash-etching has to remove just 5 μm of copper, the resulting under-etching will be very minimum. This provides a much better control on conductor widths and permits the production of fine-line PCBs with a remarkably high yield (Fig. 18-2). It even permits fine-line conductors of less than 0.1 mm width at good economical rate. It is therefore possible to run safely two or even three conductors between component pads in a 0.1" grid. Such features help to increase the package density enormously and densities are achieved where earlier only multilayer boards could provide the answer.

Where the PCB designer plans to make full use of the fine-line capabilities of PCBs made with the ultra-thin foil process, some cautionary remarks are nevertheless necessary: It is very clear that the smaller the conductor width/spacing and the smaller component solder pad width chosen, the higher is the need for a perfect registration of the patterns printed on the two sides. Also pinholing caused by dust becomes far more serious.

But where fine-line PCB production is changed over from conventional to ultra-thin foil processing and the same conductor width/spacing is maintained, there will definitely be a decrease in rejects plus better processing economy.

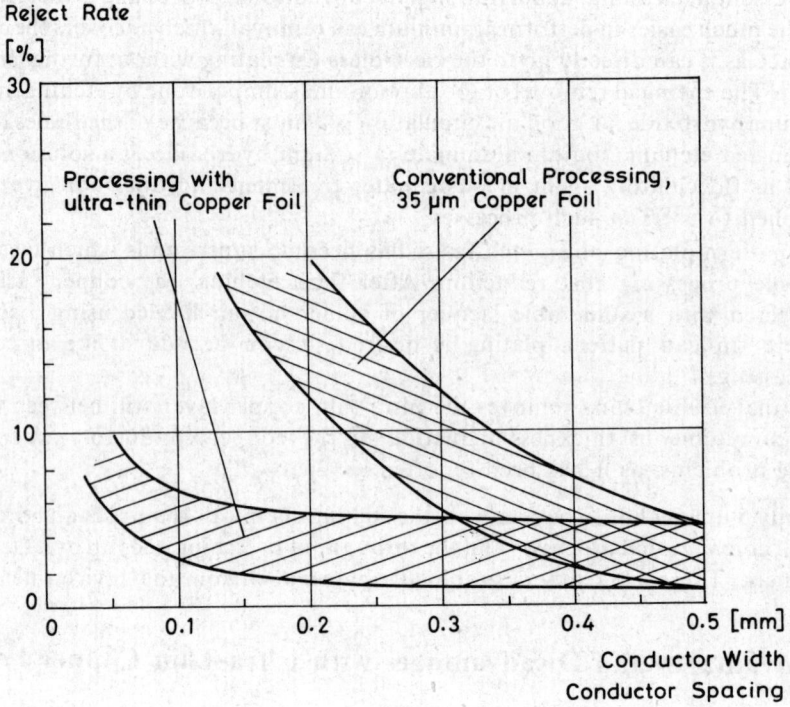

Fig. 18-2 Fine-line conductors: Improved yield with ultra-thin copper foil [2]

Ultra-Thin Copper-Clad Base Material—The Disadvantage

The production of ultra-thin copper foils, as it has already been pointed out, is a rather difficult process which is successfully handled by only a small number of companies. Because of this, the costs for ultra-thin copper-clad laminates are higher than for ordinary laminates. Also, more dependence on the material suppliers has to be accounted for because there are only a few of them. But it can be hoped that these handicaps will be overcome with more industries going in for processing PCBs with ultra-thin foil technology.

18.2 MULTILAYER BOARDS

The purpose of this section on multilayer boards is to give a brief scope of multilayer boards in the context of PCB technology trends. For a detailed study of the aspects of multilayer boards, the reader is advised to refer to Chapter 19 in which multilayer boards are dealt with in depth.

What is a Multilayer Board

Multilayer boards consist of a certain number of thin PCBs stacked together and adhesively joined with insulating prepregs to form one rigid board. Electrical connection between the different conductive layers is done with plated-through holes (Fig. 18-3). Where the conductors are connected with a plated-through hole, the conductor width is increased to slightly more than the hole diameter. The hole drilling thereafter will expose the bare copper of the conductor around the hole for the plating-through process.

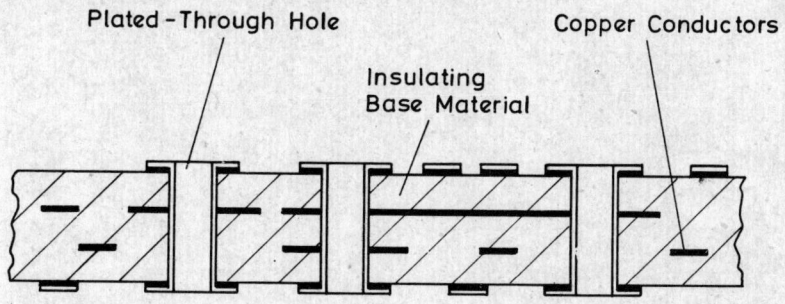

Fig. 18-3 Cross-section of a multilayer board with 4 layers

Processing

Processing corresponds by and large to the production of double-sided PCBs with the essential addition of the laminate bonding step. But considerably higher dimensional precision must be achieved, increasing with the number of layers accommodated. In particular, care should be taken under the bonding press to get maximum registration accuracy.

Where are Multilayer Boards Used

They are preferably used
— where overall PCB dimensions, number and arrangement of interconnections are not feasible on a double-sided PCB,
— where reduction in electronic equipment weight and volume is the prime concern,
— where decoupling and shielding of the interconnections is important for the functioning of the electronic circuit,
— where signal frequencies require a certain wave impedance of the interconnections to minimise signal deformation or signal delays and where these parameters have to be kept at a constant value for each board produced.

The production volume of multilayer boards has steadily increased and is now about 30% of all the PCBs produced on the PCB market in the USA [13]. However, it seems that this figure represents roughly a saturation value which is contrary to the expectations evoked only a few years ago. Reasons are the availability of alternative high-interconnection-density boards and improved fine-line PCBs with conductor widths of as low as 0.05 mm.

18.3 MULTIWIRE BOARDS

18.3.1 Concept

With the introduction of multiwired boards along with the related production equipment by PCK Technology Division of Kollmorgen Corporation, Glen Cove, USA, a new alternative to the established interconnecting techniques has been created and has proved its usefulness particularly in high-interconnection-density applications.

338   PRINTED CIRCUIT BOARDS: DESIGN AND TECHNOLOGY

Fig. 18-4 Finished multiwire board as it goes for component assembly (Courtesy: PCK Technology Division of Kollmorgen Corporation Glen Cove. USA)

Instead of printed and etched conductors, multiwiring makes use of insulated wires for interconnections. The mechanical stability for the multiwire board in its simplest form is obtained from an insulating laminate board as used for ordinary PCBs. After an adhesive preparation of the surface, wires are placed with a special NC wiring machine (Fig. 18-5) to form the desired interconnection pattern. Thereby, a *high-density* wire network is formed with cross-overs which are here absolutely permitted. If there is a need for still more interconnections, the other side of the board can as well be utilised. After wiring, a top layer of insulating material is added under the laminating press. In the drilling operation, the wires are cut, leaving their cross-sections exposed. Plating-through of the holes provides the means of interconnection both internally as well as for the external components.

18.3.2 Techniques in Multiwire Board Production

The information on the component hole positions has initially to be drawn as circles in a grid sheet which forms a system of well-defined coordinates. The interconnection details are put down in a list, giving the coordinates of all the holes to be electrically linked by wires. The

PCB TECHNOLOGY TRENDS

Fig. 18-5 Multiwire machine with 4 wiring heads (Courtesy: PCK Technology Division of Kollmorgen Corporation, Glen Cove. USA).

information on the coordinates prepared in this way is then punched on cards in such a way as to be digestible for the computer system employed.

The task of the computer is to transform the coordinate informations into a code which will be understood by the NC wiring machine and the NC drilling machine. A typical computer used with the multiwiring system from Kollmorgen Corp., is the IBM 1130 with 16K word of storage. Compared with the conductor routing in the conventional computer-aided artwork design, the result here is a 100% routing because of the cross-overs permitted. Only a few minutes of computer processing time is required for this. The output is in the form of punched paper tape suitable for the NC machines.

The base material plate for the multiwire board can be a single- or a double-sided PCB as desired. In most of the cases, this solution is preferred because broad ground and supply lines can be realised. The wires used in multiwiring have a diameter of only 0.15 mm. This corresponds to a printed conductor of 0.6 mm width on a 35 μm foil which may be too little for ground- and supply-lines. At the same time, the base PCB provides accommodation for gold-plated connector parts as well as for test points.

The base board is now coated with an adhesive layer, fixed on the NC wiring machine and wiring is carried out. The wire used is of high ductility and has a polyimide insulation of 18 μm thickness. In a cross-over point, the combined insulation thickness of both the wires will result a breakthrough voltage of more than 1,500 V which is sufficient for all practical purposes.

The wiring head (Fig. 18-6) contains a pressure foot, dispensing stylus and cutter. To make a wire connection, the pressure foot is lifted by 1 mm above the board and the wire is fed through the stylus. The pressure foot is then pneumatically lowered and activated with ultrasonic energy. The heat generated partially melts the adhesive layer on the board with the result that the wire gets embedded. The movement of the machine table pulls the wire from the stylus which gets continuously embedded. The table movement is such as to form a wire network or grid with 1.27 mm (0.05") equidistance. At the termination point of the wire, which must always be a hole position, the pressure foot is lifted and the knife actuated to cut the wire.

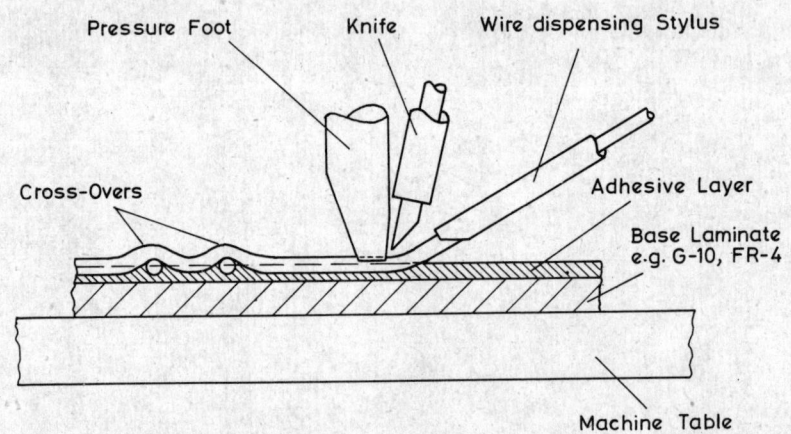

Fig. 18-6 Multiwiring principle

Where there is a need, several layers of wires can be laid on either side of the board. The different layers on the same side have to be electrically separated by a prepreg in between. Each additional layer of wires will add 0.5 mm to the board thickness. If the board has to be fitted into a card guide, an appropriate border area must be kept free of wires. Such information has also to be fed to the computer at the beginning. After the end of wiring, the top insulation layer is added and cured under the lamination press for a specified time and temperature.

The drilling of holes cuts the embedded wires and reveals their blank cross-section. In a further step, the polyimide wire insulation in the holes is chemically stripped back by about 0.05–0.1 mm in order to provide a bigger surface for plating.

The through-hole plating finally establishes contact between the wire ends and provides interface to the electronic components. After contour cutting and assembly, the multiwire boards can be soldered like any other kind of PCB either by hand, dip or wave soldering.

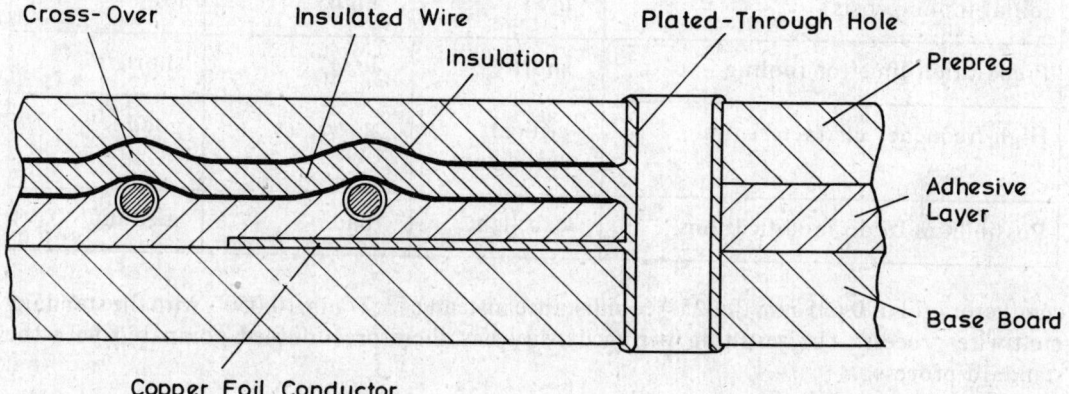

Fig. 18-7 Multiwire board with plated-through hole and single-sided PCB base

18.3.3 Where to Apply Multiwire Boards

The advantage of multiwire boards is basically in the quick implementation of high-density electronic circuits. On the other hand, a fairly high initial investment is required which could in certain cases also be utilised to improve or update existing PCB facilities for a faster turn-around time.

Multiwiring is economically attractive in the production of smaller series of high-interconnection-density boards (more than 300 component holes per dm^2). A comparison with alternative methods is given in Table 18–1.

18.3.4 High Conductivity Multiwiring

Multiwiring was introduced in the early '70s. Today, there are a number of companies in the USA, Europe and Japan who have got the licence from PCK Technology Division of Kollmorgen Corporation to produce multiwire boards. Multiwiring machines working on the same principle as used by the inventor, are now also built in Japan by Hitachi Seiko Ltd.

For applications where a still higher interconnection density is desirable (e.g., for the increasing use of chip carrier circuits), the *High Conductivity Multiwire Process* has been introduced in 1979. It is basically an improvement of the multiwiring principle and uses still the same type of wire with a diameter of 0.15 mm. But the improved wiring head design enables a

Table 18-1 Comparison of different methods to make interconnections of a high density.

	Wirewrap techniques	Multilayer techniques	Multiwire techniques
Rel. costs per board < 50 boards 50-2500 boards > 2500 boards	high high high	very high high/medium medium/low	medium medium/high high
Initial tooling costs	low	high	low
Preparation time for tooling	short	long	short
High-frequency characteristics	critical/sufficient	good	good
Possibilities to do modifications	excellent	bad	medium

conductor grid of 0.635 mm (0.025") equidistance instead of 1.27 mm (0.050") with the standard multiwire process. The resulting wiring density has therefore doubled compared with the standard process.

18.4 SUBTRACTIVE-ADDITIVE PROCESS

The subtractive-additive process, as it is described here, is typically applied in PCB plants specialised in the bulk production of paper-based laminates for consumer electronics like TVs (where there is sometimes a need for plated-through holes but where a professional type plated-through hole board would not be economical).

18.4.1 Principle of the Subtractive-Additive Process

The production process (Fig. 18-8) consists of first manufacturing a double-sided PCB with the ordinary print-and-etch method. It is followed by a mask printing step to cover all the conductors and pads where no copper deposition is wanted. The electroless copper plating finally adds a copper layer of approximately 15 μm thickness inside the holes and on the uncovered pads. The printed mask serves also as a solder stop in the component soldering process.

18.4.2 Process Details

Laminates
Since the process aims at high-volume consumer boards, paper-based laminate of grade FR-2 is mostly used. The laminate has to come in a pre-catalised condition so that no further activation for the electroless copper plating is needed after hole drilling. The catalyst seeding in

PCB TECHNOLOGY TRENDS

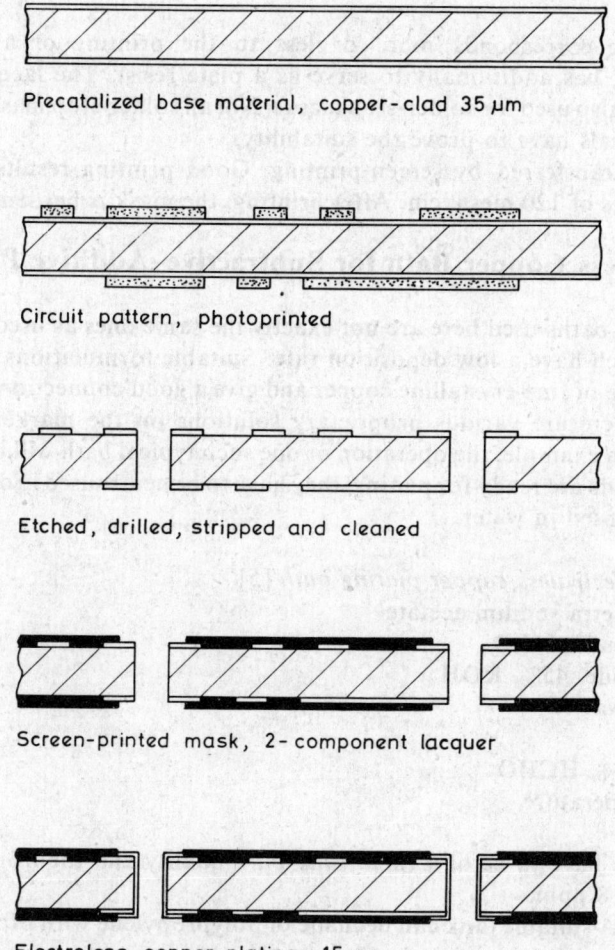

Fig. 18-8 Principle steps of subtractive-additive process

the laminate is done by adding around 10% palladium into the phenolic resin which should not change the insulation qualities too much. The copper foil can be of ordinary type, such as 35 µm thick, on both sides.

Print-and-Etch Processing

Pattern transfer is done by the cheap screen-printing process where conductor line widths and spacings are usually more than 0.2 mm. For finer details, e.g., for printed IF inductor coils, photoresist processes are applied using wet or dry-film resists. The etchant can be the same as used in mass production of print-and-etch PCBs (e.g., ferric chloride or cupric chloride).

The drilling step, thereafter, is followed by resist stripping and cleaning/deburring. Rotary brushing machines are recommended for an economic mass production. The cleaning/deburring has to be succeeded by a high-pressure rinsing which also perfectly cleans the holes and exposes fully the palladium seeds inside the holes for the electroless copper plating process.

Mask Printing

Mask printing corresponds more or less to the printing of a solder-stop mask. However, the mask has additionally to serve as a plate resist. The lacquers used are of 2-component type as also used for solder-stop masks. Not all solder-stop mask lacquers are suited for this purpose; trials have to prove the suitability.

The mask is transferred by screen-printing. Good printing results are obtained with stainless steel fabrics of 120 mesh/cm. After printing, the mask is hardened in a drying oven.

18.4.3 Electroless Copper Bath for Subtractive–Additive Process

The coppering baths used here are not exactly the same ones as used in other electroless copper plating which have a low deposition rate. Suitable formulations must give relatively high deposition rate of fine-crystalline copper and give a good connection with the copper foil of the pattern. There are various proprietary solutions on the market which fulfil these requirements. As an example, the operation of one such typical bath will be explained further. But before the boards are ready for plating, they have to be neutralised in diluted sulphuric acid (10%) and again rinsed in water.

Formulation for electroless copper plating bath [5]:

Ethylene diamine tetra sodium acetate	26g/l
Cupric chloride, $CuCl_2 \cdot 2H_2O$	10g/l
Potassium hydroxide, 45%, KOH	30ml/l
Sodium cyanide, NaCN	50mg/l
Wetting agent	1ml/l
Formaldehyde, 37%, HCHO	6ml/l
Bath working temperature	60°C

The chemicals have to be of a *chemically pure* quality and the iron content in the bath should not exceed 8 ppm.

Plating tank: A suitable tank can be made of polypropylene with provision for rocking of the boards parallel to the axis of the holes to be plated. A separate compartment of the tank is provided for adding the chemicals (mixing compartment) and is fitted with a stirring device to provide a regular motion of the bath via the mixing compartment.

Bath preparation: After the tank has been filled to two-thirds with deionized water, the heater and the stirrer are switched on. The chemicals are added in the above order; but before adding a new chemical, the previous one has to be completely dissolved. Formaldehyde is added just prior to hanging-in of the boards.

The deposition rate is roughly 1.5 μm/h. For a copper deposition thickness of 15 μm, the boards are kept for 10 hrs in the bath. More thickness can be obtained with a still longer plating duration.

The theoretical consumption of bath chemicals per hour for a plating area of $1m^2$ is as follows:

Cupric chloride, $CuCl_2 \cdot 2H_2O$	36 g
Formaldehyde, 37%, HCHO	51 ml
Potassium hydroxide, 45%, KOH	78 ml

However, parallel-running side reactions are not included in the above specifications. Also not considered is the deposition of chemicals on the tank walls and bottom.

For all practical purposes, it is suggested to add the above chemicals once every hour after an analysis of the copper content in the bath. Formaldehyde and potassium hydroxide are added in stoichiometric relation to the cupric chloride. A complete analysis on the content of all the chemicals in the bath should be done once every day. Detailed procedures for this are given in Ref. [5].

It is typical for electroless copper plating baths that after some time copper gets deposited on the tank walls and bottom, resulting in an increasing and uncontrolled consumption of chemicals. A remedy is, continuous filtering with a filter of 1 μm pore width or complete tank cleaning once a month.

After finishing plating, the following process steps have still to be executed: Rinsing, neutralisation in diluted sulphuric acid (10%), again rinsing and 2 hrs tempering in a warm air oven at 120°C to improve ductility of the deposited copper. The freshly formed oxide layer is removed by putting the board into a sulphuric acid solution of 5% to which sodium persulfate (10%) has been added. Soldering is established and maintained by a final dip in commercial solder lacquer.

18.5 SEMI-ADDITIVE PROCESSES

In semi-additive processes as well as in additive processes, the PCB production starts without any copper foil on the base material. The difference between semi-additive and fully additive processes is that for the latter the complete circuit conductor pattern is grown in an additive and electroless manner on the base while semi-additive processes need first an electroless panel plating to enable an electrolytic pattern plating followed by differential etching. Fig. 18-9 shows such a semi-additive process using adhesively coated and seeded laminate.

The semi-additive process looks almost similar to the processing with ultra-thin foil, but the thin foil has in the semi-additive process first to be established by electroless coppering which gives also a deposition inside the holes.

18.5.1 Base Materials for Semi-Additive and Additive Processes

There are basically three different types of laminates available for additive applications: swell-and-etch type (sometimes also called *buttercoat* type), sacrificial foil type and adhesively coated type. While the classification concerns the laminate surface, there one can furthermore distinguish on the base material side the *seeded* and *nonseeded* types. The resin/filler systems can be of paper-phenolic, paper-epoxy, glass-epoxy or other types.

Swell-and-Etch Laminate

These laminates are supplied bare with resin surfaces. The surfaces have to be treated with an organic solvent causing softening and swelling of the epoxy resin (buttercoat). This is followed by a chromic-sulphuric acid treatment (etching) which provides finally adequate adhesion for electroless copper plating.

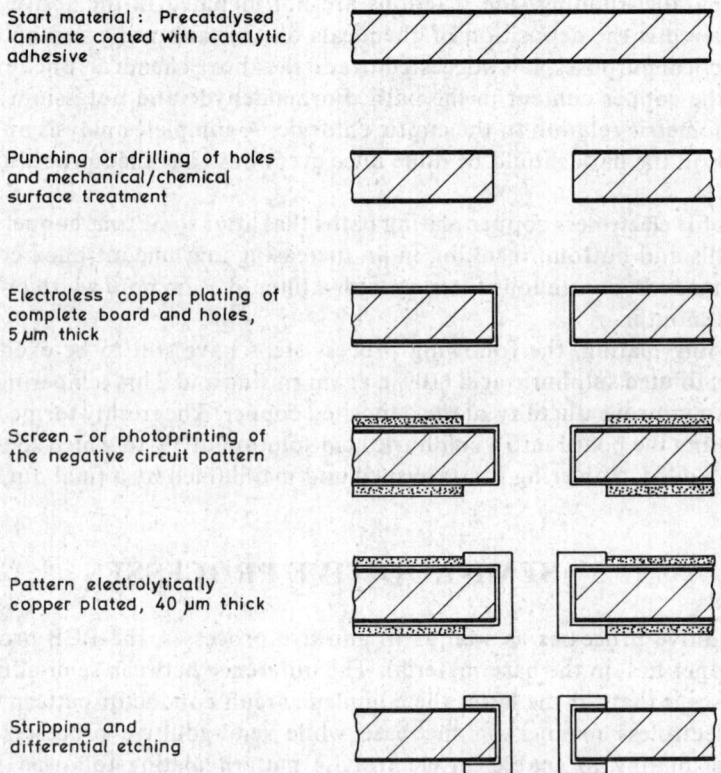

Fig. 18–9 Principle steps of semi-additive process

Sacrifical Foil Laminates

They are available with an anodised aluminium foil bonded to the epoxy-resin surface of the laminate. The foil is etched off after drilling, just prior to electroless coppering. The foil removal leaves the anodised surface texture back on the clean laminate surface which provides the bonding sites for electroless copper deposition.

Adhesively Coated Laminates

Adhesive top layers can be added in a lamination process. The adhesive layer consists of thin proprietary elastomer of 20–35 μm thickness. If the laminate is of seeded-type, the adhesive layer must also contain seeds. To ensure sufficient bond strength for the pattern, brushing and chromic-sulphuric acid treatment followed by neutralisation of the surface are required.

Seeded and Nonseeded Laminates

The nonseeded laminates are typically used in subtractive PCB processes where a catalytic sensitation is done before the electroless copper deposition.

The three laminate types for additive processes are basically available either in seeded or unseeded condition. Seedling consists of the dispersion of an electroless copper catalyst into the polymeric resin prior to lamination. Mostly seeding is done by using tin-palladium. Seeded laminates, primarily catalysed for additive processes, result in a superior electroless copper deposition with a better adhesion to the base material.

18.5.2 Semi-Additive Processing

As an example, a semi-additive PCB process will be described here in more details. The process uses adhesively coated but unseeded laminates.

Base material: Semi-additive processing can be successfully carried out with paper-phenolic, paper-epoxy, glass-epoxy and other laminates. After cutting the laminate to board processing size, drilling or punching of the holes is carried out.

Preparing the adhesively coated surface: The mechanical surface treatment consists of abrasive brushing with water and pumice powder. It cleans the surface but also results in a rough surface finish which is desirable for the chemical surface treatment later on. The brushing must be followed by a high-pressure water rinse to remove all the pumice residues, also those inside the holes.

The chemical surface roughening is done with chromic-sulphuric acid. The boards are kept for 10 minutes in such a solution at a temperature of 40°C. The extremely aggressive solution is prepared with sulphuric acid (96%) into which sodium bichromate has been added at a rate of 300 g/l. This treatment is followed by a 1 min water rinse (recirculated). Because of the high toxicity of chromic-sulphuric acid, the boards are treated for another 10 minutes in a sodium hydrogen sulfite solution (20 g of sodium hydrogen sulfite per liter water). The boards are ready for electroless coppering after again rinsing them for 10 minutes in water.

Electroless copper deposition: A first step is pickling (removal of corrosion products with acid) of the boards for 1 min in diluted muriatic acid. The solution is prepared by adding one part of muriatic acid (37%) to two parts of water.

The surface has to be activated with a catalyst. The activation is done for 15 minutes in a commercially available Palladium activator. This step is exactly similar to the ordinary plated-through hole processes. A water rinse for 10 minutes has to follow. Where seeded laminates and seeded adhesive coatings are used, the activation step is eliminated.

The activation-acceleration step is done in hydrofluoboric acid for 8 min. Strength of the acid: 3% in aqueous dilution. It is followed by a water rinse (1 min) and rinsing in deionised water (1 min). The boards are now brought into the tank with the electroless copper bath. The plating thickness required for paper based laminates with punched holes is 5 μm while 2 μm is sufficient for glass-epoxy boards with drilled holes. The electroless coppering forms a quite critical part of the whole processing: the deposition should be fast to provide a smooth and stress-free copper layer of a high ductility. The electroless copper baths in semi- and fully additive processes are therefore different from the ones used in ordinary plated-through hole processes. It is a typical practice to go for a commercially available bath which is already working satisfactorily at other places.

As an example, a possible electroless copper bath formulation is shown here [5]:

Copper sulfate, $CuSO_4 \cdot 5H_2O$	10 g/l
Ethylene diamine tetra sodium acetate	60 g/l
Sodium cyanide, NaCN	50 mg/l

Formaldehyde, HCHO (35% aqueous solution)　　8 ml/l
pH (adjusted with caustic soda)　　12.8
Bath temperature　　40°C

The tank should be made of polypropylene with a separate compartment for the adding of chemicals and with a stirring agitator which effects a continuous circulation of the bath. Rocking of the boards is a must for regular deposition thickness.

The boards are held in a suitable rack with a distance of approximately 1 cm between the boards. The bath utilisation can be done up to 3 dm^2 board surface per liter bath. The resulting deposition rate will be around 5 $\mu m/h$. After the electroless coppering, the boards are water rinsed for 5 min and dried.

Further processing: Pattern transfer is done in the usual manner either by screen-printing or by photoprinting. The electrolytic copper plating is a similar process as applied in ordinary plated-through hole processes. The only difference is the higher deposition thickness required since the full conductor thickness comes through additive processing.

Where there is a need for a metal resist on the pattern, it can be added in the semi-additive processes. Fully additive processes do not offer this option.

After stripping, the differential etching finally removes copper from the complete board in the thickness of the initial electroless copper plating (typically 5 μm). A prerequisite for the successful application of differential etching is the uniformity of the electrolytically deposited copper on both the sides of the board. It can be improved by avoiding high concentrations of copper pattern areas which would have a reduced plating thickness if compared with narrow conductors in low concentrated pattern areas. This point has to be cared for during the design stage of the PCB.

18.5.3 Why Semi-Additive Processes

If semi-additive processing is compared with the conventional plated-through hole processing using metal etch resist, a few simplifications in the semi-additive processing become obvious:
— no need for a metal etch resist
— no deburring required
— multifold etchant economy and simple etchant system (ferric chloride or cupric chloride)
— no underetching or overhang, permitting finer pattern details
— less costly base materials because no copper foil needed
— applicable also to paper based laminates with punched holes (cheap mass consumer electronics).

Because of such advantages, semi-additive processes have become an economically attractive alternative where PCBs with plated-through holes are required in larger quantities.

18.6 ADDITIVE PROCESSES

18.6.1 Scope

The development of additive processing started already in the late '50s and the first production trials on a large scale were carried out in 1963. But only in the early '70s, additive processing was established and fully recognised as an alternative method in PCB processing. The main problems which had to be solved at the beginning were the adhesion to the base laminate and the development of satisfactorily and stably working electroless copper plating baths with a sufficiently fast deposition rate.

To come to a better understanding of the advantages of additive PCBs some limitations of the still common subtractive processes are given below:

Underetching: It reduces the conductor width, resulting in less conductor adhesion on the laminate and making it very difficult to have an exact control on fine details in the pattern.

Limited board-thickness to hole-diameter ratio: In electroyltic plating processes, the metal deposition is closely related to the electric field distribution. In holes with a diameter of considerably less than the board thickness, the deposition in the middle of the holes is therefore less than in other portions.

Irregular plating thickness on the conductors: The differences might not be only among tracks on the same side but also between front and back sides of the board. Basically, conductors near the edges of the boards and in areas with a low conductor density will get more plating thickness than conductors in the board centre or areas with a high pattern density.

Metal resist breaking down: Because of the mushroom-like overhang of the platings on the underetched conductor tracks, the danger of its breaking down arises and this can cause short-circuits. A remedy often practiced is solder fusing. This has, however, the disadvantage of subjecting the laminate to heat which reaches the thermal limits of the laminate.

Etching: The etching of large quantities of copper requires an investment in etching equipment, consumes considerable quantities of chemicals and regenerants and creates severe disposal and pollution problems.

Advantages and Possibilities with Additive Processes

Base materials: A wide range of laminate systems is suited for additive processing, e.g., paper phenolic, paper epoxy, glass epoxy, glass polyester and many others. Since the laminate is supplied without copper foil, an economic advantage is already obtained here.

PCB feasibility: The non-existence of overhang and undercut permits finer conductors and patterns, hence a higher circuit density. With the adhesives available today, it is possible to achieve peel-off strengths which are even better than the ones with the copper-clad base material for subtractive processes. The exact control of plating thickness over the entire pattern including inside holes and the thickness uniformity also opens up new possibilities in high-frequency applications.

Simpler processing: Since no copper surface must be cleaned and no holes deburred prior to chemical processing, the preparation of the base material is much simplified. The acceptance of punched holes makes the additive PCBs further attractive for high-volume processing of dense, paper based boards. The whole additive processing requires less process steps (Fig. 18-10), an important cost consideration. The solderability of the highly pure, electroless deposited copper is good and is usually preserved with a solderable lacquer. For higher demands in corrosion resistance, solder dip with hot-air levelling is often applied. Additive processing is well suited for medium and large scale PCB production and the reported cost savings compared to subtractive processing are in the range of 10-40%. Last but not least, from the environmental point of view, the complete elimination of copper etching means one big step forward.

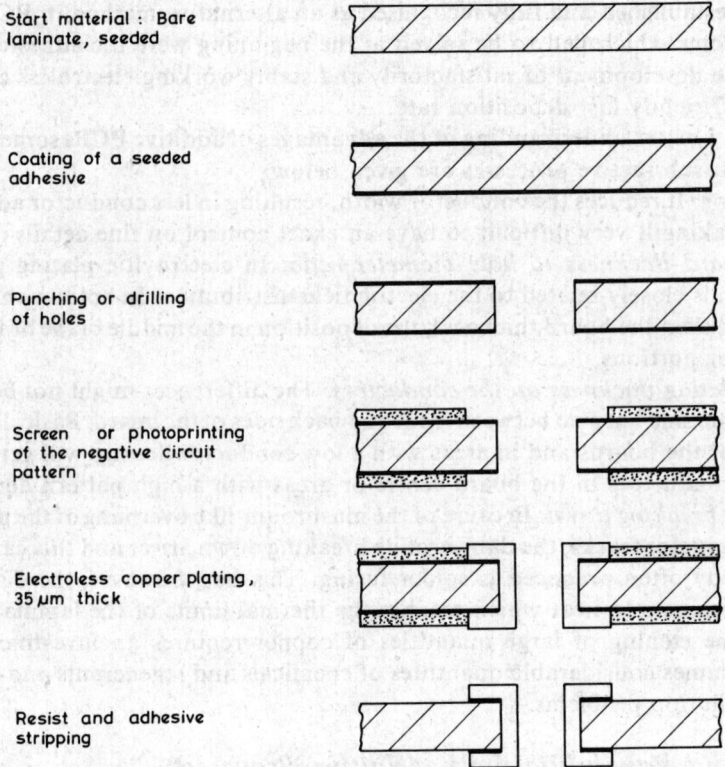

Fig. 18-10 Principle steps of additive process

18.6.2 Base Materials and their Preparation

Filler/resin systems: The laminates for additive processing must be carefully selected. Although a wide variety of filler/resin systems are basically suited, a critical material parameter is the dimensional stability of the thickness (z-axis) which is particularly critical during the soldering operation. Plated-through holes in laminates with grade specification FR-3 (paper epoxy), FR-4 (glass epoxy) and G-10 (glass epoxy) should not give much problems. If grade FR-2 (paper phenolic) has to be used, a careful choice of the material and its supplier should be made.

Catalyst: To start the electroless copper deposition process on the insulation material surface, a catalyst is necessary. In the case of seeded base materials, the catalyst is already available with the resin in the laminate. The adhesive subsequently coated on a seeded laminate must therefore also contain a catalyst. The drilling or punching of holes follows thereafter.

With unseeded laminates, an activation in a solution containing catalyst is done after drilling or punching of the holes.

Laminate surface preparation: In Section 18.5.1 we have already dealt with the laminate surfaces typically supplied for additive processing (swell-and-etch; sacrificial foil; adhesively coated).

Adhesive coating has become relatively popular and is often executed directly prior to processing through the board manufacturer. Coating methods are dipping, roller coating, laminating and 'curtain-flow' coating. The adhesive surface is thereafter dried (prehardening) in a suitable oven and further chemical treatment follows to provide surface texture and surface qualities for maximum copper adhesion.

18.6.3 Additive Processing

As an example, a typical process for additively made PCBs shall be given (Table 18-2). (There are also variations possible but the major steps will be found in any type of additive processing.)

Table 18-2 Additive PCB process.

Step	Operation	Details
1	Laminate cutting to board processing size	Nonseeded base material
2	Solvent rinse, drying	Trichloroethylene
3	Curtain flow coating of adhesive and predrying	Coating of 2nd side after predrying of 1st side
4	Prehardening of adhesive	Final thickness of adhesive approximately 25 μm
5	Punching of registration holes	
6	NC drilling or punching of component holes	
7	Mechanical roughening and cleaning of board surface and drying with warm air	Abrasive brushing machine with abrasive slurry (40 l H_2O; 10 l pumice power)
8	Negative-pattern printing	Screen-printing or photoprinting
9	Surface activation	Hydrofluoboric acid, sodium-bichromate
10	Chromium neutralisation	Sodium-bisulfite
11	Cascade rinsing	
12	Electroless coppering	20–24 hrs for 35 μm Cu, rocking and air bubbling
13	Cascade rinsing	
14	Pickling	Sulphuric acid (5% concentration)
15	Cascade rinsing	
16	Resist stripping	
17	Final hardening of the adhesive	60 min at 160°C
18	Printing solder mask	
19	Deoxidation of pattern	Solution of sodium persulfate (22%) and sulphuric acid (2%)
20	Solderable lacquer coating	
21	Contour cutting	

The particulars of the electroless copper bath are [5]:

Copper sulfate, $CuSO_4 \cdot 5H_2O$, highly pure (< 7 ppm Fe)	7g/l
Ethylene diamine tetra sodium acetate	25–30g/l
Sodium cyanide, NaCN (ductility promoter)	60mg/l
Formaldehyde, HCHO	6.5ml/l
pH (adjusted with NaOH)	12.6
Baumé gravity (adjusted with H_2O)	9–10° Bé
Bath temperature	68 ± 1°C
Bath circulation	> 1x/hr

The details given on the above process are just enough to understand the major production aspects. The most important criteria in any additive process is the copper adhesion (adhesive) and the quality and bath control of the electroless-deposited copper. When deciding to go in for an additive process, it is a usual practice to get the complete process know-how from one source. The know-how transfer includes also necessary assistance in the introductory phase and support in trouble-shooting later on. The fees for such a know-how transfer, however, can assume substantial magnitudes; this in particular with imported know-how.

18.7 FLEXIBLE PRINTED CIRCUIT BOARDS

18.7.1 Scope

Flexible PCBs have become an established interconnection concept in many applications where their physical flexibility and their weight, as well as their advantages in saving space and

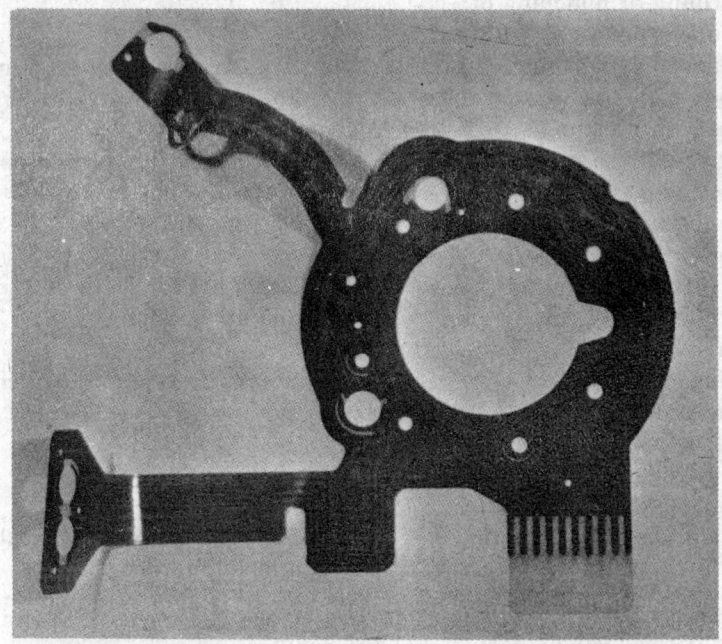

Fig. 18-11 Flexible PCB (Courtesy: Hans Kolbe & Co., Gittelde, Germany)

assembly-time lead to an economic advantage over traditional solutions. Flexible PCBs may have the function of a harness or similar to a rigid PCB or a combination of both. If designed for, they can repeatedly be flexed or folded. This is of special interest to the designers of systems with plug-ins and other moving devices. Flexible PCBs were earlier mainly used in military and aerospace applications but now they are incorporated in many other fields like computers, industrial, consumer and automotive applications. In the US market, the flexible PCBs form a respectable 10% share of all the PCBs produced [15].

An optimum use of flexible PCBs is made if they replace complete interconnection systems rather than only a part thereof. This of course needs rethinking during the early equipment design stage. The advantage will not only be in a considerable weight and space reduction but also in a higher reliability, complete elimination of wiring errors and less testing and assembly costs. Even connectors can be eliminated if pins, wires or plated-through holes are provided at the termination points.

A disadvantage of the flexible PCBs is their limited suitability for high-frequency applications. The characteristic impedance of the transmission lines formed by the laminated system is very difficult to control because of variations in thickness of the conductors, adhesives and films. The early flexible circuits had still further drawbacks like insulating films which were unstable in processing. This gave a poor yield with respect to dimensions and solderability. But with the availability of polyimide laminates, the material problems were at least partly solved.

Processing chemistry is basically not different from the methods used in rigid PCB production. Only the handling of the laminate has to be adapted to the flexible material and a final, insulating top layer must be provided. Flexible PCBs in single, double or multilayer mode are therefore produced both by subtractive as well as additive processes.

18.7.2 Flexible Base Materials

The requirements asked for in flexible base materials cover a very wide range and no material has been found till now which incorporates all these qualities:
—Low cost
—Dimensionally stable under mechanical and thermal stress
—Good tear strength
—Good bond strength between resin layer and reinforcing material
—Good bond with copper
—Chemical/physical stability under temperature and humidity cycling
—Good resistance to thermal shock (soldering)
—Good electrical properties
—Flame retardancy
—Punchability
—Low water absorption
—Compatibility with covercoat materials
—High flexibility.

There is a limited choice of laminates which can be considered for flexible PCB applications. The laminate chosen will in any case form a compromise of the various desirable qualities. A few of these materials are shown in Table 18–3 along with some of their properties.

Epoxy resin/glass mat: It offers the highest adhesion for the copper foil and shows little tensile expansion. However, it is not suitable for a continued bending-unbending stress.

Polyester foil: It combines a low price with a good overall electrical/mechanical behaviour. Its use is much restricted by the sensitivity to soldering.

Polyimide foil: An excellent material for flexible PCB's with flame retardation and good mechanical, electrical and thermal qualities.

Teflon foil: It shows excellent electrical qualities and is flame retardant. The material is mechanically mediocre and tends to flow in cold condition.

Table 18-3 Some typical specifications of base materials currently used for flexible PCBs [1].

Parameter	Epoxy resin/ glass mat	Polyester foil	Polyimide foil	Teflon foil
Soldering temp. [°C]	260	230	260	260
Soldering time [sec]	10	1	10	10
Tensile strength [kP_a/cm^2]	1750	1500	1700	200
Tensile expansion [%]	3	130	70	
Cu peel-off strength [kP_a]	4.5	1.8	1.3	3.4
Water absorption [%]	0.5	0.8	2.5	0.01
Thermal expansion [ppm/°C]	11	15	20	
Dielectric constant, 60 Hz	3.4	3.25	3.5	2.1
Loss factor, 1 kHz	0.037	0.006	0.003	0.0002
Specific resistivity [Ω/cm]	16×10^{13}	10^{17}	4×10^{16}	10^{15}
Relative price	1.4–2	1	2–3	

18.7.3 Design Constraints

For the design of flexible PCBs, most of the design rules for rigid PCBs have to be applied. There are however a few exceptions plus some new considerations to be taken into account. A few of them will be given here.

Current-carrying capacity of conductors: Because of less cooling capability by the board itself (when compared to rigid PCBs), sufficient conductor width has to be provided. A guideline for currents of more than one ampere is given in Fig. 18-12. Where several conductors with a high current are placed opposite or neighbouring, the heat concentration has to be taken care of by giving additional conductor width or extra spacing.

Contours: Wherever possible, rectangular shapes are preferred because of the better base material economy achieved. There should be sufficient free border space near edges due to the possible dimensional changes with the base materials. Inward looking corners in the contour should be rounded; sharp inward corners could initiate tearing of the board.

Bending: As a general rule, the bending radius should be designed as wide as possible. The possibility to undergo many cycles is also further improved with thinner laminates (e.g., 50 μm foil instead of 125 μm foil) and larger conductor widths. If subjected to a high number of bending cycles, single-sided flexible PCBs show in general a better performance.

Solder pads: Around the soldered solder pad, there will be a transition from flexible to rigid material. This zone is highly prone to conductor breakage. Solder pads are therefore avoided in active bending zones.

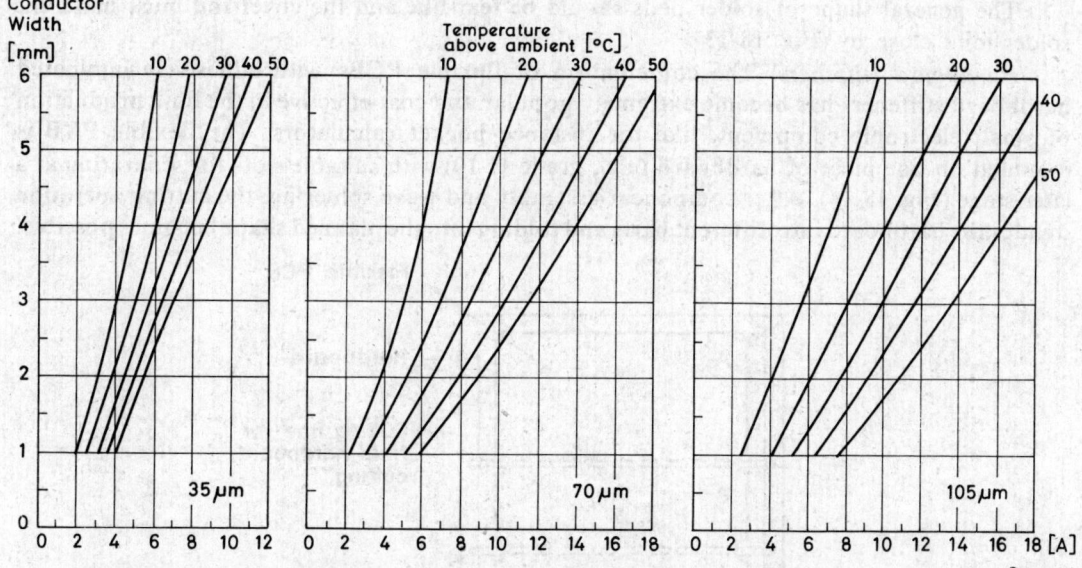

Fig. 18-12 Conductor widths for safe operating temperatures with flexible PCBs [1]

A) Shape of solder pads

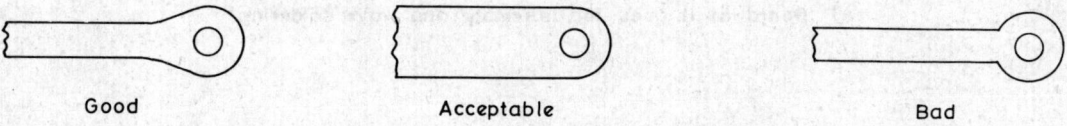

Good   Acceptable   Bad

B) Solder joint masking with cover film

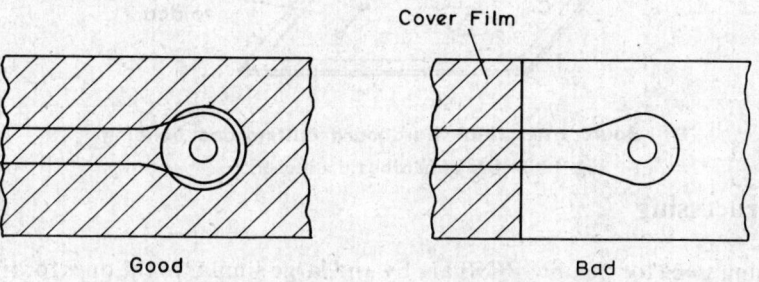

Good   Bad

Fig. 18-13 Shape and masking of solder joints

The general shape of solder pads should be tear-like and the cover foil must mask the solder joint close by (Fig. 18-13).

Hardboard stiffeners: The combination of flexible PCBs with adhesively laminated hardboard stiffeners has become extremely popular and cost-effective in the bulk production of small electronic equipment, like for instance pocket calculators. The flexible PCB is mounted on one piece of hardboard (e.g., grade G-10) with suitable slots for separating at a later stage (Fig. 18-14). After component assembly and wave soldering, the cutting operation divides the hardboard into different parts and folding into the planned shape becomes possible.

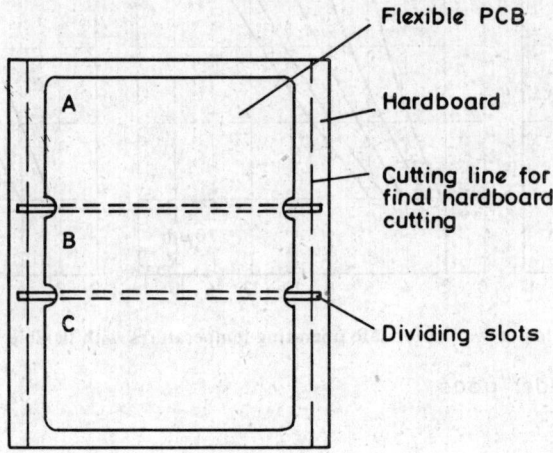

A) Board as it goes for assembly and wave soldering

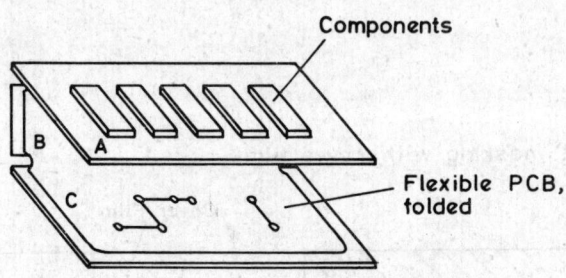

B) Board after final hardboard cutting and bending

Fig. 18-14 Use of hardboard stiffeners

18.7.4 Processing

Processing steps for flexible PCBs are by and large similar to the ones for rigid PCBs. The majority of flexibles is subtractively processed but also additive processes are employed. One main difference is in the handling of the base material. The flexible material is supplied in rolls of different widths. Hence, the processing equipment is sometimes geared for a reel-to-reel production for efficient bulk fabrication. The reel-to-reel processing, however, requires a large investment into special equipment and is therefore economically attractive only for extremely large production quantities. For the processing of flexible boards with equipment meant for the processing of rigid PCBs, the flexible boards have, if necessary, to be given sufficient mechanical strength, e.g. by spanning them in a suitable frame.

Holes in flexible boards are mostly punched; this leads to high tooling costs. However, drilling is also possible. Soldering is in general more critical due to the limited heat-resistance of the base materials. Where possible, wave soldering is employed but hand soldering can be done as well. An absolutely clean and easy solderable surface is still more important with flexible than with rigid PCBs. The hand soldering of flexible PCBs especially on a polyester film base needs sufficient experience. The lamination of a cover film with suitable cut-outs for solder joints is recommended. It does not only act as a solder mask but it also improves the board characteristics under bending stress.

18.8 METAL CORE CIRCUIT BOARDS

In metal core boards, the typical resin/filler laminate of ordinary PCBs is replaced by sheet metal covered with insulation material. The metal core is mostly made of steel or aluminium sheet of 0.7–1.2 mm thickness, into which holes are punched according to the component layout of the circuit. Another possibility is to use standardised core boards into which full arrays of holes are punched, forming a grid pattern. Insulation between pattern and metal core is provided by an electro-static- or fluidised bed coated insulation layer. The further generation of the conductor patterns is done by subtractive-additive or additive techniques.

18.8.1 Scope

In multilayer boards, full metallic layers are utilised for ground and voltage supplies or for electrical screening. However, these thin layers do not act as a mechanical support to the board. In metal core techniques, the metal core gives primarily the full mechanical strength to the board. In addition to giving physical strength, the metal core acts as a heat dissipator with a low thermal resistance. This permits the densest realisation of heat producing circuitry, for which otherwise considerably more board area and heatsinking would have been required. Also, current carrying capacity of the conductors can be kept 2–3 times higher than the usual values for ordinary PCBs.

Metal cores are comparatively simple to produce with shearing/punching techniques. For smaller series, the holes may be drilled. The excellent dimensional stability of the core with respect to temperature is a feature which is of special interest in the Z-axis (thickness direction) for highly reliable plated-through holes.

The costs for the metal cores in mild steel are a fraction of the costs otherwise caused with epoxy laminate. Still only few companies have facilities for metal core PCB production. The reason is probably that processing steps are completely changed as compared with the established concepts.

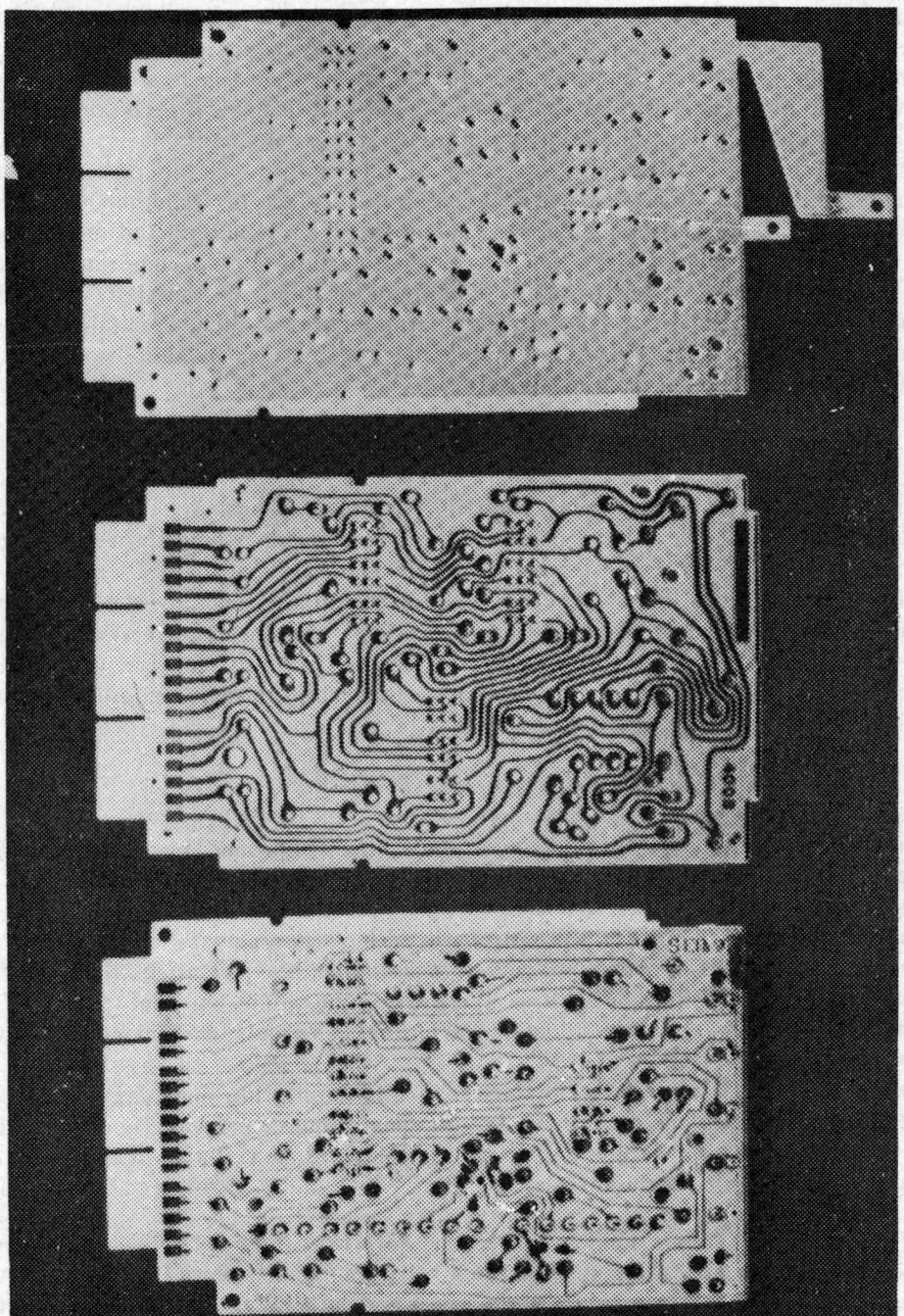

Fig. 18-15 Metal core circuit board at three different stages: coated with epoxy layer; with finished conductive pattern; after soldering (this photo was originally published in the book 'Leiterplatten-Herstellung and Verarbeitung', edited by Eugen G. Leuze Verlag, D-7968 Saulgau/Wurtt., Germany)

Advantages
- Lower costs for the base material (metal core plus insulation coating)
- High mechanical strength: PCBs become thinner
- High dimensional stability (for highly reliable plated-through holes)
- Easy machining of the metal core (sheet-metal techniques)
- Built-in heat sinking capability especially for power electronics applications
- Constant characteristic wave impedance
- Excellent electrical decoupling of the conductive patterns.

Drawbacks
- Higher initial investment: Besides most of the facilities for conventional processing, additional equipment for sheet metal work and insulation coating is required
- Low control possibilities and variation of insulation layer thickness
- Variations of hole diameters
- Unprotected metal cross-sectional surface after final cutting, in particular with mild steel
- Imperfect insulation coating can cause short-circuits. This problem is however solved if core is made of anodised aluminium.
- Higher weight of the finished board.

Metal core boards will therefore mainly bring an economic advantage with higher production volumes where the lower base material costs will offset the higher equipment investment and tooling costs.

18.8.2 Metal Core Board Production

Core Materials

Mild steel: It is a cheap material and sufficient experience with its handling is available.

Aluminium: Aluminium is also an unproblematic material in such applications. After the holes have been punched, it is usually anodised and gives therefore an extra insulation in addition to the subsequently coated insulation layer.

Sandwich technique: To reduce the weight and also to obtain the advantages of a steel core board, a hard board insulation sheet is laminated on both sides with a thin sheet of mild steel (e.g., 0.2 mm thick).

Fig. 18-16 shows three possible ways how the metal cores can be realised for further processing. The grid-type punched standard core enables also an economic production of smaller series of metal core PCBs since only minor additional tooling costs are involved. In framing-and strip technique, the unwanted material is cut off as a last step thus leaving only small side portions unprotected.

Insulation Layer

The powders used to coat the metal cores are usually of epoxy type. The cured layer has typically a breakthrough voltage of 40 kV/mm, a dielectric constant of 4 and a water absorption of 0.5–1%. Deposition thickness should be atleast 0.25 mm on the plate surface.

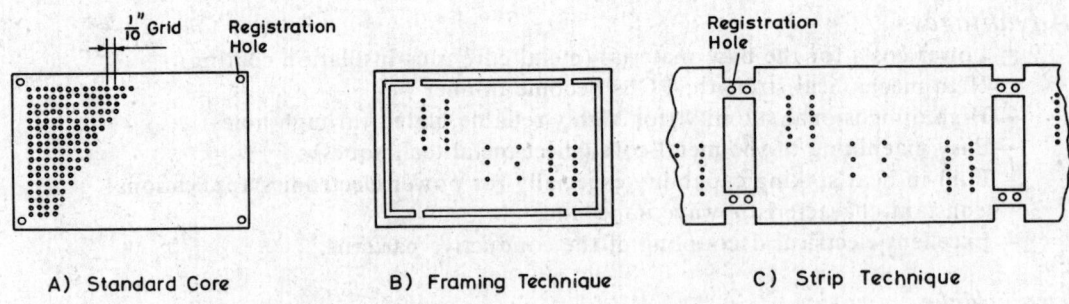

A) Standard Core   B) Framing Technique   C) Strip Technique

Fig. 18-16 Metal core preparation

Two coating methods which have gained importance in metal core coating are: electro-static and fluidised-bed coating.

In *electro-static coating*, the powder gets attracted on the board surface by the voltage potential difference of approximately 50 kV. The epoxy-powder coated boards are thereafter cured in an oven. In *fluidised-bed coating*, the boards are first heated up to 250°C and then exposed to an air stream containing the epoxy powder. Curing is again done in an oven thereafter.

In both these methods, the coating layer inside the holes gives limitations to the application of metal core PCBs: The built-up powder layer is not even. While the electro-static coating gives minimum thickness in the middle of the holes, fluidised-bed coating does it at the beginning and end of the holes (Fig. 18-17). The coating thickness difference itself depends on the ratio of core thickness to hole diameter and the type of epoxy powder used. Holes with a very small diameter will get closed with powder and therefore cannot be realised. This will limit the density of circuit patterns which can be realised on metal core PCBs. Even on the flat surfaces of the metal core, the coating thickness may vary in a range of 0.05-0.1mm.

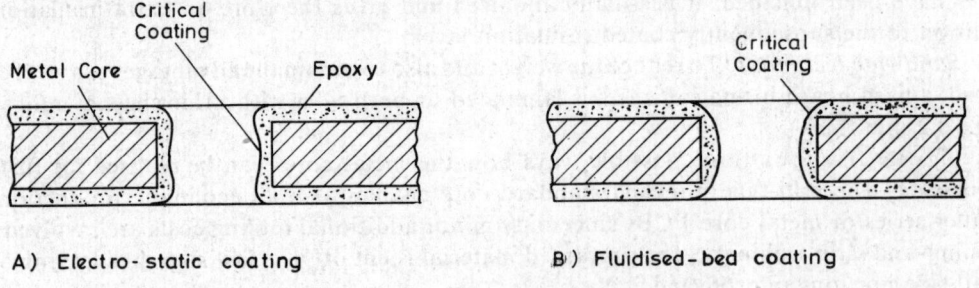

A) Electro-static coating   B) Fluidised-bed coating

Fig. 18-17 Insulation layer within holes of metal core boards

The hole diameters in the metal should be about 0.6 mm more than the component lead to compensate for the built-up insulation layer and the plating.

The adhesion properties of epoxy on metal surfaces are fairly well known from other applications and do not cause problems with metal core boards. Before powder coating, the burrs around the holes are removed by sand blasting. Thereafter, degreasing with a solvent (e.g., trichloroethylene) or with a soak cleaner is carried out.

18.8.3 Generation of the Circuit Pattern

Subtractive-Additive Processing

The further processing of the insulated metal core boards can be carried out in subtractive-additive processing and will therefore be quite similar as already dealt with in Section 18.4. Where asked for, electrolytic solder plating on the pattern can also be included.

Table 18-4 Subtractive-additive process for metal core PCBs.

Step	Operation	Details
1	Precutting of metal core	Mild steel or aluminium
2	Punching of holes and contour	
3	Sand blasting and degreasing	Aluminium cores are additionally anodised
4	Coating with epoxy powder and curing	
5	Surface swelling	Solvent
6	Surface etching	Chromic sulphuric acid: 25% CrO_3 and 13% (weight) sulphuric acid
7	Activation	Palladium
8	Electroless copper	Panel-plating, 2–3 μm
9	Pattern printing, negative	Screen or photoprinting
10	Electrolytic copper	Pattern-plating, 35 μm
11	Stripping	
12	Flash etch	

Additive Processing with Catalytic Adhesive

The insulated metal core board is treated by a dip in a liquid adhesive which also contains a catalyst. After drying and curing, the negative circuit pattern is printed. The pattern is then further treated in chromic-sulphuric acid to prepare its surface for the electroless copper plating. The additive processing is otherwise similar to the additive PCB fabrication as described in Section 18.6.3.

18.9 MECHANICAL MILLING OF PCB'S

In mechanical PCB milling, the copper surface of the copper-clad material is divided into conductor paths by means of copy-milling technique. The separating channels and component holes are machined according to a 1:1 production drawing (with manual equipment) or according to an NC- or CNC programme (with automatic equipment).

18.9.1 Equipment

The most relevant part is the milling head which holds the tool for milling and drilling. The tungsten-carbide tool rotates with approximately 10,000 rpm. The small laboratory-type

362   PRINTED CIRCUIT BOARDS: DESIGN AND TECHNOLOGY

Fig. 18-18 Mechanically milled PCB, double-sided (Photo CEDT)

machines use an optical system with a light point which is manually moved along the channel lines on the sketch while the milling head executes simultaneously the channel milling. Fig. 18-19 shows such an equipment where two milling heads are employed. The light-point scanner and consequently the milling heads, are running in straight lines in X- or Y-direction, because the movement in either direction can be electromagnetically blocked. For an automatic series production, NC- or CNC equipment with 6 or more milling spindles is available.

18.9.2 Circuit Patterns which can be Realised

The process as such is suitable for single-sided and, with a suitable registration method, also for double-sided PCBs without plated-through holes. Through-contacting can be done with special rivets. Medium-dense patterns with conductor widths of less than 1mm can be realised, but require a certain amount of skills especially with the manual equipment.

Various channel widths are obtained according to the milling tool used (typically 0.2–1mm). Where more spacing is required, as an example for the connector pattern, several channels are milled closely neighbouring each other. Holes are drilled with the same tool after the conductor pattern has been milled.

Since most of the copper layer will remain on the base laminate, mechanical milling is well suited for circuitry with higher currents. The author has had good experience with such PCBs for power supply and control circuitry.

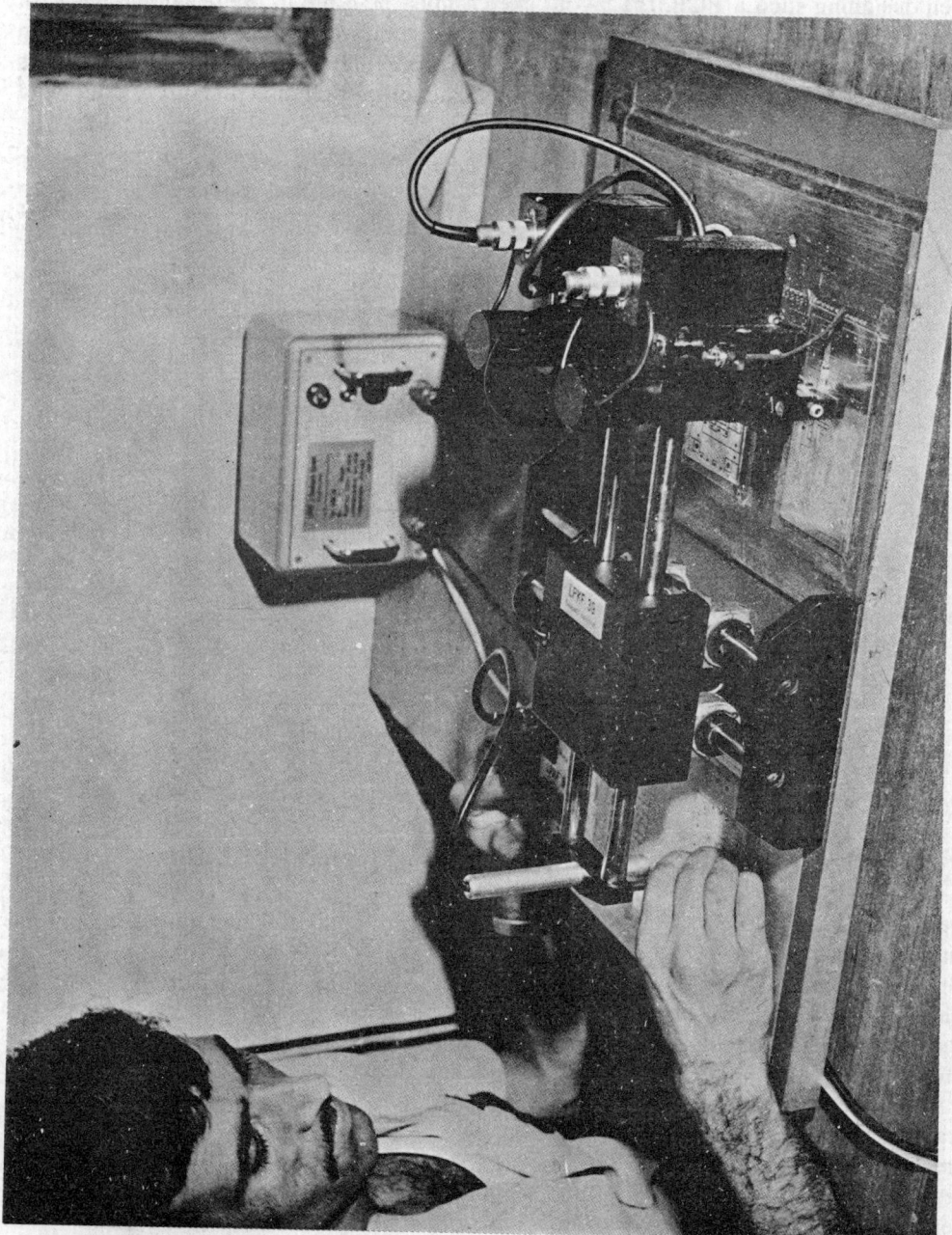

Fig. 18–19 PCB milling with a 2-spindle machine (Photo CEDT)

When designing such a PCB, the layout sketch must be done in the usual manner with pencil lines corresponding to the conductors. Only now a separate production sketch is made in which just the shape of the conductors is changed to rectangular shapes while the pencil lines represent the insulating channels (Fig. 18-20). The production sketch is preferably done on a transparent graph sheet by using pencil or Indian ink.

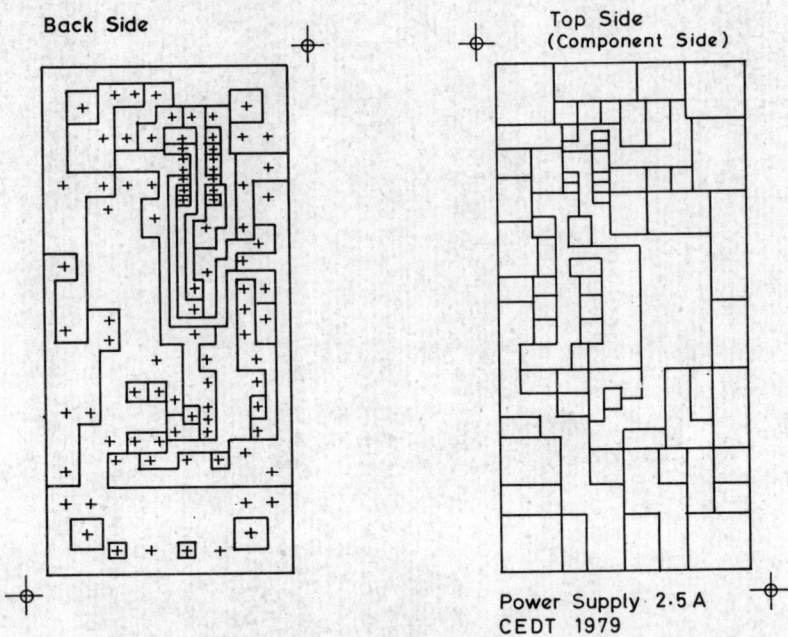

Fig. 18-20 Production drawing of pattern for mechanically milled PCB

18.9.3 Where to Employ PCB Production by Mechanical Milling

The mechanical milling of PCBs has a few distinct advantages which makes it a unique solution for small-quantity PCB consumers:
—Small and compact: Only a bench, is required for the complete PCB set-up.
—No chemicals involved and no pollution.
—Production can start immediately after the production sketch is available: No artwork and no film master.
—Drilling of the component holes is done immediately after finishing the milling of the separating channels with the same fixing and the same tungsten-carbide tool.
—Where later a bigger quantity of the same PCB should be produced with conventional processing, the production sketch in Indian ink can still be used as an artwork for the film master production.
—Electrolytic surface plating can be carried out prior to milling. Gold plating of connectors after milling is possible by providing a conducting bar outside the pattern area as in conventional processing.

Limitations

The milling/drilling operation takes approximately 10-40 min time for a board of 100 × 150 mm size. The actual production time for one board is obtained by dividing the operation time by the number of milling spindles utilised for board production.

Digital circuits with plenty of ICs and smaller conductor widths are better produced in series by standard chemical processing. The limitation is given here also by the non-availability of plated-through holes.

BIBLIOGRAPHY

1. 'Flexible Gedruckte Schaltungen (Datenblatt 10108)', Hans Kolbe & Co., 3371 Gittelde (Germany), 1974.
2. FRIBERG G.G.: 'Leiterplatten mit ultradunner Elektrolytkupferschicht', *INDUSTRIE, ELEKTRIK + ELEKTRONIK*, Heidelberg, Nr. 6-1975, pg. EP 53/54.
3. HAMILTON J.R. et al.: 'Semi additive and ultra-thin copper clad technologies for fabrication of printed wiring boards', *PROCEEDINGS OF THE WORLD PRINTED CIRCUIT CONVENTION*, Volume I, London, 1978.
4. HENKEL E.: 'Multiwire—Schaltungen', *EIPC PROCEEDINGS*, Zurich, Nov. 1976, pg. 1-15 (III).
5. HERRMANN G.: *LEITERPLATTEN*, Leuze Verlag, Saulgau/Wurtt. (Germany), 1978.
6. LYMAN J.: 'Fine-line printed circuits catch on', *ELECTRONICS*, McGraw-Hill, New York, Jan. 19, 1978, pg. 84-86.
7. LYMAN J.: 'Flexible circuits bend to designers' will', *ELECTRONICS*, McGraw-Hill, New York, Sept. 15, 1977, pg. 97-105.
8. LYMAN J.: 'New design doubles wiring density of multiwire process', *ELECTRONICS*, McGraw-Hill, New York, April 12, 1979, pg. 39/40.
9. MANSVELD J.F.: 'The PD-R additive process and its use in the manufacture of rigid printed circuit boards' *PROCEEDINGS OF THE WORLD PRINTED CIRCUIT CONVENTION*, Volume I, London, 1978.
10. 'Mehrlagenschaltungen (Datenblatt 10107)', Hans Kolbe & Co., 3371 Gittelde (Germany), 1974.
11. MILLER W.B.: 'The use of microthin coppers in through-plated boards', *PROCEEDINGS OF THE WORLD PRINTED CIRCUIT CONVENTION*, Volume I, London, 1978.
12. 'Multiwire (Datenblatt 10112)', Hans Kolbe & Co., 3371 Gittelde (Germany), 1974.
13. NITSCH H. + ACKERMANN D.: 'Optimum design of flexible printed circuit boards with respect to flexural strength', *PROCEEDINGS OF THE WORLD PRINTED CIRCUIT CONVENTION*, Volume II, London, 1978.
14. 'System LPKF 39 (Data Sheet I)', Jurgen Seebach GmbH, Scheffelstr. 17, 3000 Hannover (Germany).

15. 'US Markets Forecast 1979', *ELECTRONICS,* McGraw-Hill, New York, Jan. 4, 1979, pg. 114–120.
16. WEAVER H.A.: 'High volume production of flexible PCBs', *PROCEEDINGS OF THE WORLD PRINTED CIRCUIT CONVENTION,* Volume I, London, 1978.
17. WEINHOLD M.: 'Trockenresist fur das Additiv-Verfahren', *EIPC PROCEEDINGS*, Zurich, Dec. 1974, pg. 118–132.

19
MULTILAYER BOARDS

19.1 INTRODUCTION

19.1.1 Background

With the advent of the plated-through hole technology, conductor cross-overs on different planes and consequent reduction in space requirements and increase in packaging density of electronic components has been rendered possible. When the MOS/LSI, VLSI and multi-pin configurations made their appearance, the packaging density increased to such an extent that a high concentration of interconnecting lines was the result. This gave rise to unpredictable design problems like noise, cross-talk, stray capacitances and drops due to parallel signal lines without adequate decoupling. The design aimed at minimising the length of signal lines, avoiding parallel routing and efficient decoupling by bringing ground close to them. These could not be realised satisfactorily in a single-sided or even a double-sided board on account of the limited cross-overs which could be realised. The idea of the cross-overs was extended beyond the two-plane approach and the result is the multilayer circuit board. The multilayer board has more than two planes of interconnections and the electrical circuit is most commonly completed by means of interconnecting the different layers by plated-through holes, transverse to the board, at appropriate places. The manufacturing technology is highly sophisticated in view of the fact that much of the reliability depends on the plated-through holes and the bonding techniques employed.

19.1.2 Where to Apply Multilayers

If the printed wiring boards do not have at least four to six layers, one cannot begin to capitalise on the high component density that is possible with integrated circuits, flat packs or TO-5 packages. Single-sided or double-sided designs are inadequate. And once the restraints of the V_{CC}- and ground connections have been removed, it is possible to package at least 50% more components than in a simple double-sided board.

a) Whenever weight and volume are the overriding considerations, multilayers are the likely answer. The trade-off is simply cost for space and weight versus board costs.

b) For very complex interconnection problems, multilayer boards are less costly than other approaches on a production scale.
c) The addition of a third or more circuit planes to accommodate 'cross-circuit runs' frequently eases design problems and saves time and money.
d) When coupling or shielding of a large number of connections is necessary, multilayers are ideal. The high capacitance distributed between the different layers gives a good decoupling of power supply and this permits high speed circuits.
e) Multilayers provide the answer where frequency requirements call for careful control and uniformity of conductor wave impedances with minimum distortion in signal propagation, and where the uniformity of these characteristics from board to board is of high importance.
f) Several conventional PC modules can be incorporated into one multilayer module. Reliability of the total equipment can be improved.
g) The layout and artwork designs are greatly simplified on account of the absence of the V_{cc} and ground lines on the signal planes.
h) Multilayers provide low impedance of supply lines and good distribution of the supply currents.

19.2 DESIGN AND TEST CONSIDERATIONS

19.2.1 Design

The listed advantages of the multilayers can be realised once the designer knows the pitfalls and avoids them.

The most common problem with a poorly fabricated multilayer is the interlayer connection discontinuity—generally an opening between an interlayer conductor and the plated-through hole. Delamination and/or measling which might happen during soldering shows inadequate care in bonding.

The other likely problems can be:
— inadequate plating in holes which causes a high resistance
— poor drilling quality
— unimaginative design with cramming of too much circuitry into too little space thus giving unrealistic tolerances, high interlayer capacitances and possibly a compromised quality.

To dispense with such problems, it is important that the manufacturer is experienced, quality-dedicated and that he has stringent process controls. The boards stored before soldering for instance must be dried for two hours at 120°C to eliminate absorbed moisture. Performance specifications should allow complete evaluation of thermal shock, insulation resistance, solder resistance and microresistance of interlayer connections.

To understand the conception of a multilayer board, it will be most convenient to attempt the design of a four-layer board.

The circuit taken as an example consists of 4 layers: The two external layers comprise the conductor patterns and the two internal layers represent the supply and ground areas.

In Fig. 19-1, the layers 1 and 4 are the conductor planes which interconnect the component leads. The layers 2 and 3 are supply and ground planes comprising copper and absence of copper in the form of round lacunae. It is therefore possible to connect each 0.25 mm conductor from layer 1 and 4 to the layers 2 and 3. Fig. 19-2 represents the geometry of layers 2 and 3.

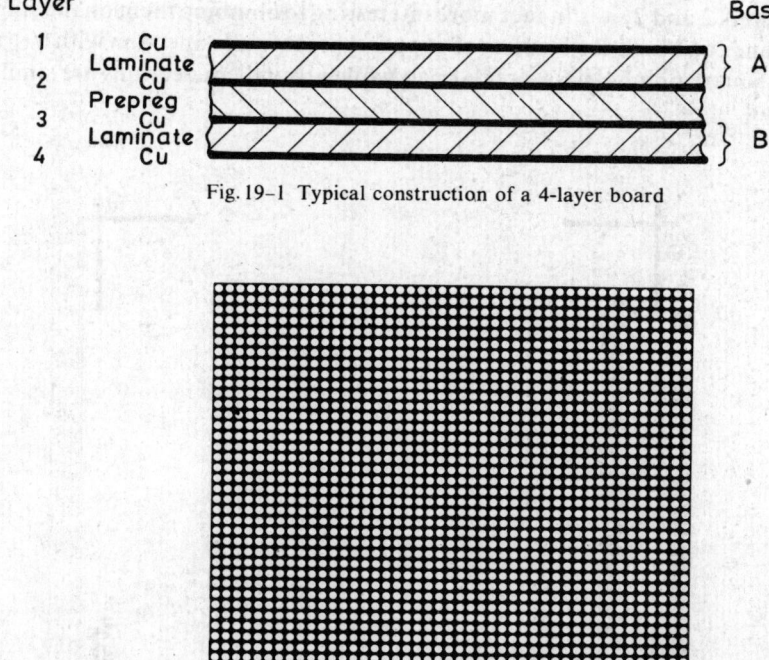

Fig. 19-1 Typical construction of a 4-layer board

Fig. 19-2 Geometry of layers 2 and 3 representing V_{cc} and GND

19.2.2 Artwork Methods

The design determines the dimensions of the circuit. The draftsman makes a drawing on dimensionally highly stable polyester foil at an appropriate scale.

This drawing must comprise trimming lines, fixing holes, reference symbols, edge-connectors, the logo of the company, the space reserved for the logo of the circuit fabricator, side identification for solder side and component side (S1–S2).

Fig. 19-3 illustrates an artwork of this kind. From this artwork, it is necessary to take a paper copy (contact print) which would serve to indicate all the particulars of the layers 2 and 3 with grid position, placement of targets for fixing holes, etc. The artwork and the paper copy will then be sent to the circuit manufacturer who through his photographic department makes the step-and-repeat negatives and adds the coupons necessary for electrical and metallographic tests. A set of film master positives will then be delivered to the user. Having received the positives, the draftsman has two possibilities to execute the remaining artwork. He can employ the red/blue taping technique on 2 different polyester foils by using black pads, red and blue tapes. The other possibility is to use totally 4 polyester foils: The artwork for the layers 1 and 4 is done with the 3-layer artwork method using only black pads and tapes. The artwork for layers 2 and 3 is then carried out with the red/blue taping technique on one polyester foil.

370   PRINTED CIRCUIT BOARDS: DESIGN AND TECHNOLOGY

For the layers 2 and 3, it is in fact more interesting to combine them on the single polyester foil on which one can block the lacunae of copper wherever a connection with V_{CC} is desired by sticking a red square or a holeless circular pad while ground connections are similarly marked with a blue pad.

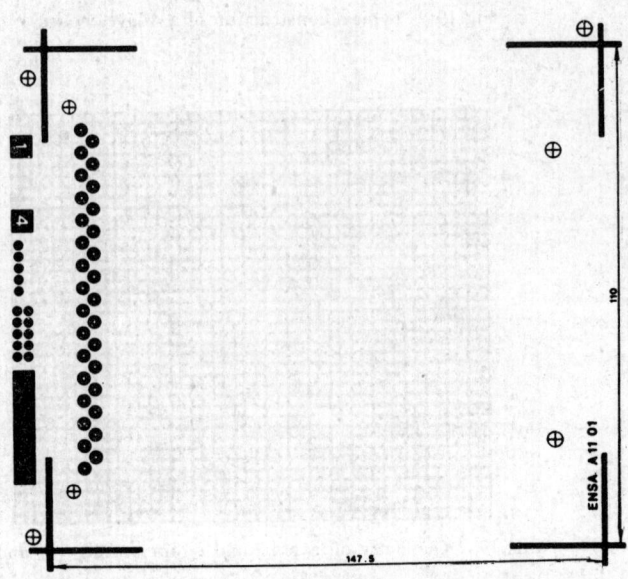

Fig 19-3 Typical basic artwork for outer layers

In order to avoid a registration error, the draughtsman must position terminal pads on planes 1 and 4 by using planes 2 and 3 as the reference grid.

For board dimensions beyond 380 mm, a few tenths of a millimetre shift is possible and this must be kept in view to avoid any short-circuit.

Fig. 19-4 represents the artwork (red and blue) of planes 1 and 4 of a typical circuit.

Fig. 19-5 represents the artwork (red and blue) for planes 2 and 3 of the same multilayer board.

19.2.3 Tolerances

Fig. 19-6 along with Table 19-1 gives certain tolerances for an optimum design of multilayers for maximum reliability.

19.2.4 Board Thickness

For boards with plug-in connectors, the thickness is important. In general, plug-in connectors allow a thickness tolerated between 1.4 and 1.8 mm.

MULTILAYER BOARDS

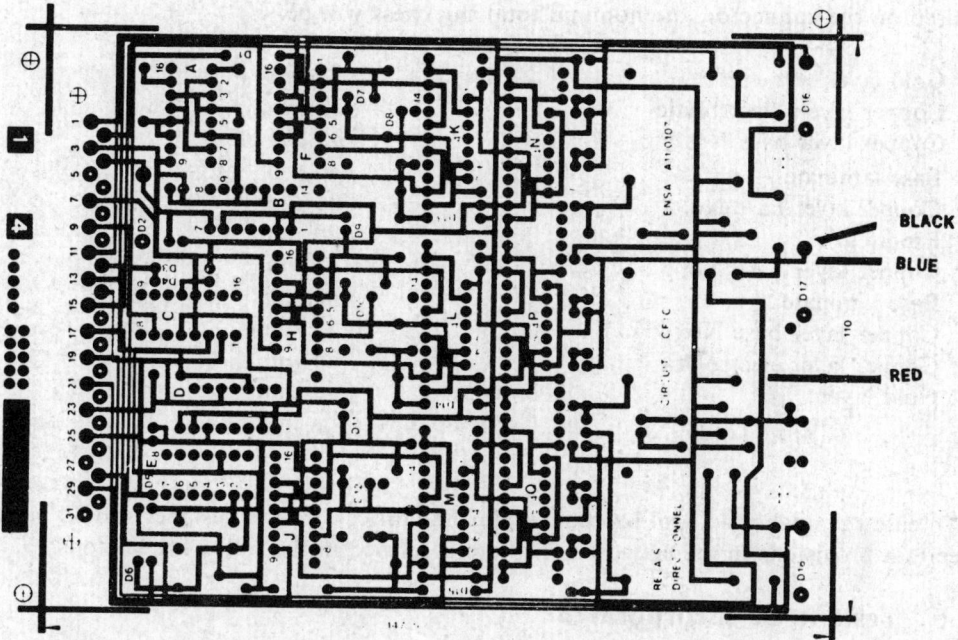

Fig. 19-4 Example of a 2-layer artwork for outer layers

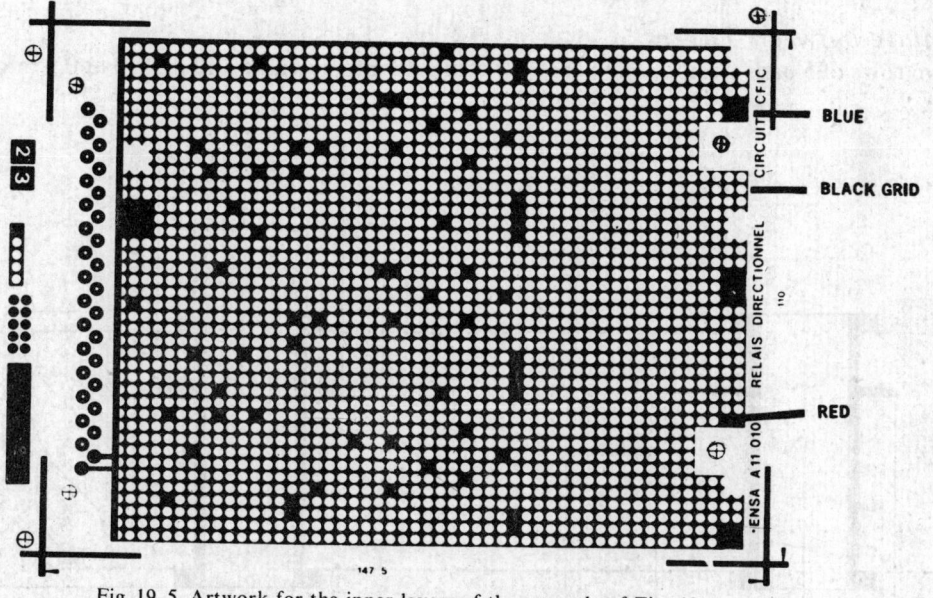

Fig. 19-5 Artwork for the inner layers of the example of Fig. 19-4

Measured on the connector, the nominal total thickness will be:

Gold layer	3 μm
Copper layer, electrolytic	25 μm
Copper layer base No. 1	35 μm
Base laminate	640 μm
Copper layer base No. 2	35 μm
Pre-preg	120 μm
Copper layer base No. 3	35 μm
Base laminate	640 μm
Copper layer base No. 4	35 μm
Copper layer electrolytic	25 μm
Gold layer	3 μm
Total	1596 μm

To ease the quality control procedures during fabrication, the designer should incorporate certain provisions in the artwork. They are described in the following sections.

19.2.5 Tests to be Incorporated

Electrical and Metallographic Tests

For purposes of these tests, the designer adds on the outer sides of the board a special geometry (Fig. 19-7).

Continuity between Layers

Two rows of 5 pads positioned in the 2.54 mm grid intersections and drilled at 0.9 mm are

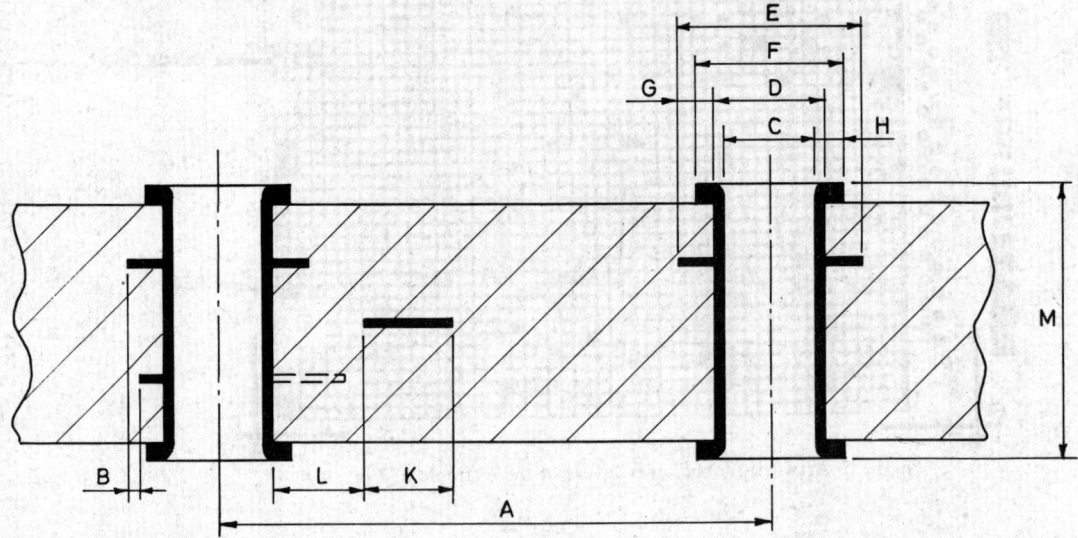

Fig. 19-6 Dimensions and tolerances for multilayer boards. Specifications in Table 19-1.

Table 19-1 Dimensions and tolerances for multilayer boards.

Dimension	Tolerances and minimums for hole spacing in mm	
	< 2.5	≥ 2.5
A[1] < 150	0.1 R of TP (± 0.075)	0.175 R of TP (± 0.125)
A[1] ≥ 150	0.175 R of TP (± 0.125)	0.250 R of TP (± 0.175)
B	0.075 of TP	0.125 of TP
C[2],[3]	± 0.075 for 90% of holes ± 0.125 for remainder	± 0.125 for all holes
D	± 0.025	±0.050
E	0.35 + nominal drilled hole diameter	0.76 + nominal drilled hole diameter
F	—	0.76 + nominal drilled hole diameter
G	0.025 min.	0.025 min.
H	0.050 min.	0.125 min.
K[4],[5]	± 0.050	± 0.075 internal ± 0.100 external
L	0.075 min.	0.250 min.
M[6],[7] < 1.80	± 0.18	± 0.18
> 1.80	± 10% of nominal composite thickness	± 10% of nominal composite thickness

Remarks:

1) Recommended hole space centre-to-centre 0.125 and 0.250 mm
2) For 0.125 mm boards, minimum diameter is 0.46 mm or 1/3 of the composite board thickness or maximum lead diameter + 0.125 mm, whichever is more
3) For 0.250 mm boards, minimum diameter is 1/3 of the composite board thickness or maximum lead diameter + 0.250 mm, whichever is more
4) For 0.125 mm boards, more than 0.250 is preferred
5) For 0.250 mm boards, more than 0.50 is preferred
6) Maximum thickness is 3× minimum hole diameter after plating
7) Note: Compensation may be required for the particular process used.

TP = true position R = radius

utilised for this test. Fig. 19-8 represents the electrical schematic. On passing a current of 2A and measuring the voltage drop in millivolts, a good indication of the metallisation of the holes and the continuity between layers is obtained.

Insulation Resistance between Layers

A row of 5 pads positioned in the 2.54 mm grid intersections and drilled at 0.9 mm is used for this test. Fig. 19-9 represents the electrical schematic.

Fig. 19-7 Test coupon for electrical and metallographic tests

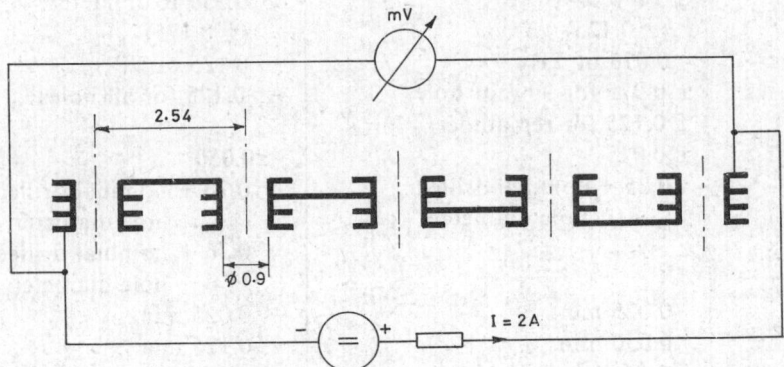

Fig. 19-8 Test schematic for continuity between layers

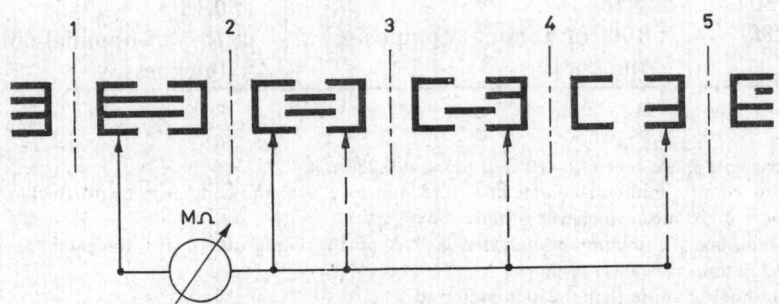

Fig. 19-9 Test schematic for insulation resistance

The principle of the test is to measure the insulation resistance in megohms between pad 1 which has a plated-through hole short-circuiting all layers, and the pads 2 and 3 which have the positive and negative layers isolated; pad 4 is connected to the positive terminal, pad 5 is connected to the negative terminal. It is herewith possible to locate an insulation defect, for example due to a bad bonding and migration of chemical copper between the layers.

Voltage Breakdown Tests

A similar row of pads is used for the breakdown voltage test. For example, IPC A-600A advises a continuous voltage of 250-500V. In principle, this must be agreed with the user at the time of order.

Test for Delamination and Solderability

A copper surface of 10 × 50 mm is reserved for the solderability test. A check for delamination is also effected.

After the electrical tests, the coupons are used for metallographic study.

It is evident that the electrical tests have to be performed on the actual multilayer board itself.

19.3 MULTILAYER CONSTRUCTION

There are many types of multilayer constructions. The different types are distinguished by the method used to make connections from one layer to another. For example, there are clearance holes, plated-through holes, fused tubelets or eyelets, plated risers, plated tabs, chemical milled expanded eyelets and silver-brazed terminals.

To select from this variety, the designer must consider the fabrication facilities, funds and manpower available. He must also consider if the equipment is to be designed for maximum performance and reliability or is cost a significant factor. Fig. 19-10 compares layer interconnection methods for military applications.

After selecting the type of layer interconnection, the designer must select suitable materials. In rigid multilayer circuits, layers of cured boards are separated by sheets of 'B-stage' (semi-cured) resin, commonly known as prepregs. The purpose of the B-stage layer is to bind the rigid layers together and seal the printed circuit between the layers.

19.3.1 Specifications

Before starting the design of a multilayer board, the applicable military specification should be reviewed. *MIL-STD-275* and *MIL-P-55110* which apply to single-layer printed circuits can be used as a guide for laying out the circuits on each layer.

However, this information must be supplemented because layouts for multilayer circuit boards must also provide holes and pads for layer interconnections. The American Army Electronics Command has issued a specification, *SCL7503 : Printed Circuit Wiring*, which provides this additional guidance. The Institute for Interconnecting and Packaging Electronic Circuits has prepared a specification on multilayer printed wiring: *IPC-ML-950 : Performance specification, multilayer printed circuits*. Heat dissipation has also to be considered. A multilayer has lower heat dissipation capability than the equivalent number of single-sided boards. Nevertheless, voltage and current ratings can be upgraded because the circuits are encapsulated.

19.3.2 Interconnection Techniques

The designer who is a beginner would do well to start with the clearance-hole interconnection method as this is simpler than others, and the usual PCB skills, equipment and design experience are directly applicable. The important limitation with the clearance-hole method is that it is less efficient in its use of space because the layers are interconnected by external jumpers. After some experience, nevertheless, a layer can be added that contains interconnecting strips terminated at the desired layers by plated-through hole connections. Of

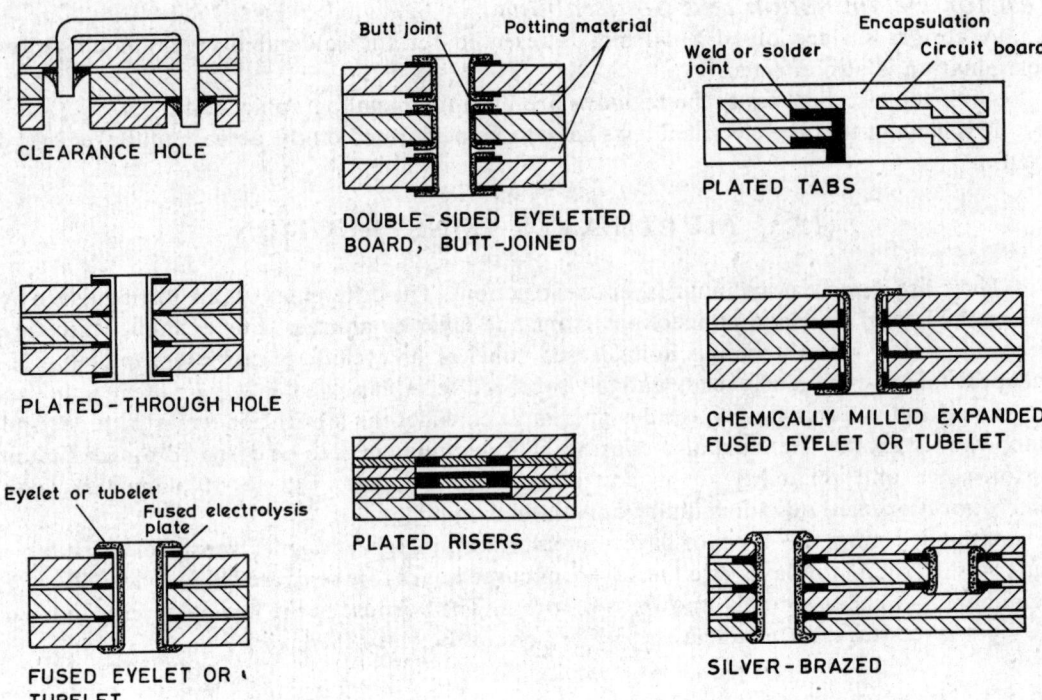

Fig. 19-10 Interconnection methods for multilayer boards. Characteristics in Table 19-2

all the methods, the plated-through hole connection has come to stay as it provides the maximum packaging density and its reliability can be achieved with assiduous quality control. Further treatment in this chapter therefore restricts itself to this method.

19.4 EQUIPMENT

Fabrication equipment and procedures are essentially the same for all glass reinforced, thermosetting-resin base laminates. As the majority of multilayers fall into the glass-epoxy category, we shall discuss the equipment associated with it.

The pressure needed to make laminates with void-free intermediate epoxy layers is in the range of 200 to 500 psi. The pressure is applied to drive the air bubbles out from the B-stage resin towards the edges of the board. As some air bubbles get trapped in the edges by the time curing starts, a border of perhaps 25 mm on each side should be left for trimming. If, for example, the size of the multilayer board is 250×300 mm (including the air-bubble margins gives 275×325 mm) a total force of 42900 lbs at 300 psi is required to form the laminate. A 25 ton press is adequate for this purpose. A typical multilayer press for such purposes is shown in Fig. 19-11.

The temperature needed for laminating of epoxy is typically 180° C. The press must be capable of maintaining an even temperature over the entire laminate area. It has therefore massive platens with thick press plates. A clearance of about 150 mm between the platens is necessary to enable easy insertion and removal of the circuit boards.

Table 19-2 Characteristics of the intercorrection methods in Fig. 19-10.

Characteristic	Clearance hole	Plated through hole	Fused eyelet or tubelet	Double-sided eyeletted boards, butt-joined	Plated risers	Plated tabs	Chemically milled expanded fused eyelet	Silver-brazed
Reliability of joint	Excellent	Fair	Good	Poor	Excellent	Excellent	Good	Good
Single- or double-sided circuits	Single	Both	Both	Both	NA	NA	Both	Both
Soldered circuits	Easy	Easy	Easy	Easy	NA	Easy	Easy	NA
Welded circuits	Easy	Hard	Hard	Hard	NA	Easy	Difficult	NA
Number of layers (max)	8	20	10	No limit	8	10	10	No limit
Registration	Not critical	Not critical	Not critical	Not critical	Critical	Not critical	Not critical	Critical
Hole drilling	Not critical	Not critical	Not critical	Not critical	NA	NA	NA	Critical
Mechanical strength	Good	Good	Good	Poor	Fair	Fair	Good	Good
Environmental resistance	Poor	Fair; good if conformal coated	Fair; good if conformal coated	Poor	Excellent	Good	Fair; good if conformal coated	Fair to excellent
Number of process steps	Few	Many	Few	Few	Many	Few	Many	Few
Packaging density	Poor	Good	Good	Poor	Excellent	Good	Good	Fair to excellent
Weight	Heavy	Light	Medium	Heavy	Light	Light	Medium	Light to medium
Thickness	Thick	Thin	Medium	Thick	Thin	Thin	Medium	Thin to medium
Sockets or added terminals	Easy	Easy	Hard	Hard	Hard	Not possible	Hard	Easy
Flexible circuits	Easy	Easy	Easy	Hard	Hard	Hard	Hard	Easy
Design changes	Easy	Hard	Hard	Easy	Hard	Hard	Hard	Hard
Cost	Low	Low	Low	Low	High	High	High	Low
Production time	Low	Low	Low	Low	High	High	High	Low
Automation possible	Yes	Yes	Yes	Yes	Yes	No	Yes	Yes
Equipment readily available	Yes	No	Yes	Yes	No	Yes	No	Yes
Visual inspection	Easy	Not possible	Hard	Easy	Easy	Easy	Hard	Hard
Repair	Easy	Hard	Moderately hard	Easy	Hard	Easy	Moderately hard	Hard
Proprietary process	No	Yes	No	No	Yes	Yes	Yes	Yes

Note: NA means not applicable.

The press plates are usually of steel or aluminium and about 6.4 mm thick steel plates are preferred which provide better dimensional stability of the boards. The surfaces of the plates should be free of irregularities such as pits, dents and high spots. For alignment of top and bottom plates, pins of 4.8 mm diameter are press-fitted into each corner of the lower plate. The pins are slightly tapered above the plate (about 2°). The upper plate will have holes with a clearance of 0.025 mm to 0.125 mm. To align and maintain registration of layers during lamination, additional pins of about 1.6 mm diameter and spaced 150 mm apart are installed.

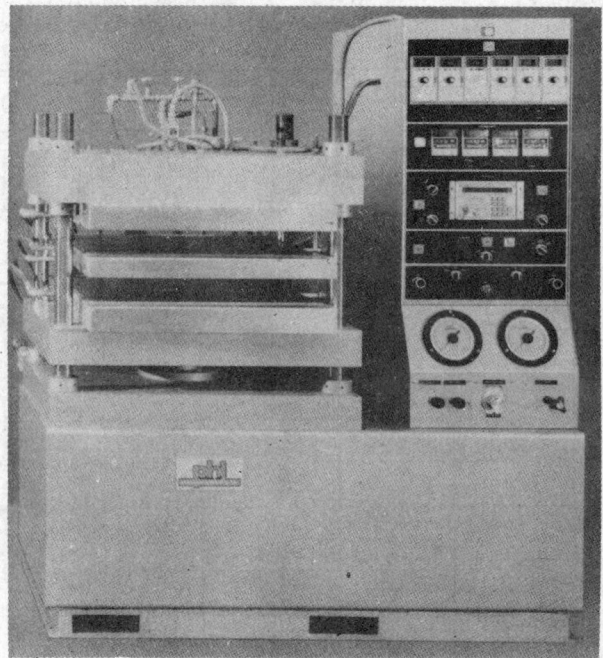

Fig. 19-11 Typical multilayer press with controls (Courtesy: Pasadena Hydraulics Inc., City of Industry, Ca. 91744, USA)

19.5 LAMINATING PROCESS

19.5.1 Preparations

The circuit is prepared and etched on the inner layers by standard printed circuit techniques. Since each layer of B-stage and board substrate requires a different hole arrangement, it is important to do a very careful layout for each one of the layers in order to prevent masking of needed holes. The tooling and alignment holes on the prepregs must be 1.25 mm larger in diameter than the conductor pads and the holes in the laminates must be 1.25 mm smaller than pads over which they are to be placed. This will prevent flow of resin into aligning pins thus rendering an easy disassembly after lamination.

Although high-flow of the B-stage materials is desirable to provide clear, void-free laminates, the flow of the prepregs used for clearance hole boards should be held between 4% and 10% to prevent contamination of conductor pads. B-stage flow with other types of multilayer boards can be as high as 30%.

19.5.2 Preheating

The platens and the press plates are preheated in the press to a temperature of 180° C. The lower plate is then removed and placed on an asbestos pad or a hot plate. The aluminium release

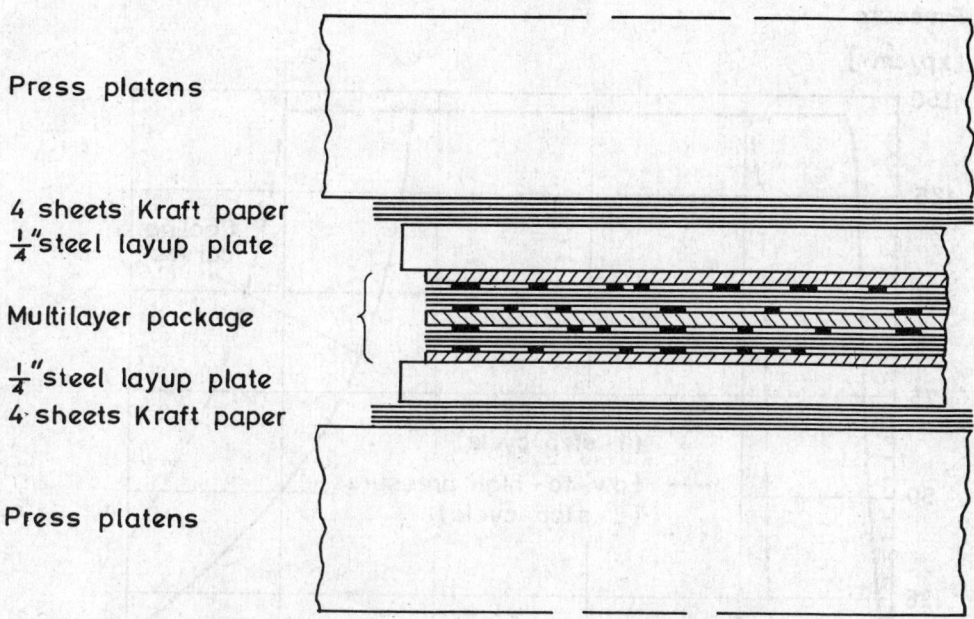

Fig. 19-12 Cross-section of a multilayer package ready to go into the preheated press

foil and the successive layers of laminate and prepregs are aligned and pressed carefully into position. Another aluminium foil is kept on top of the package and the top press plate is placed over the entire sandwich. Kraft paper which is required to equalise pressure as well as control the temperature rise as per desired profile, is interposed both below the bottom plate and above the top plate and the entire assembly is placed between the platens, as shown in Fig. 19-12.

If low-flow B-stage material is used for prepregs, press can be closed and full pressure applied for not less than 45 minutes. To prevent warping of the finished boards after curing, it is a good practice to allow the board to cool in the press under full pressure down to about 60° C or even lower. When high-flow B-stage material is used, knowing the flow and gel time, the timing of the application of full pressure has to be studied to prevent resin-starved condition. The time and temperature cycle is shown in Fig. 19-13.

19.5.3 Post-Laminating Inspection

After removal from the mould, the laminate is inspected for insulation resistance as design requirements. The alignment of layers can be studied to advantage by radiography. The board is then trimmed of excess material and drilled. It is important to use a sharp drill bit and to adjust feed and speed to minimise epoxy smear.

19.6 FURTHER PROCESSING

19.6.1 Inspection of the Holes

The boards then need a thorough inspection to verify the absence of epoxy smear in the

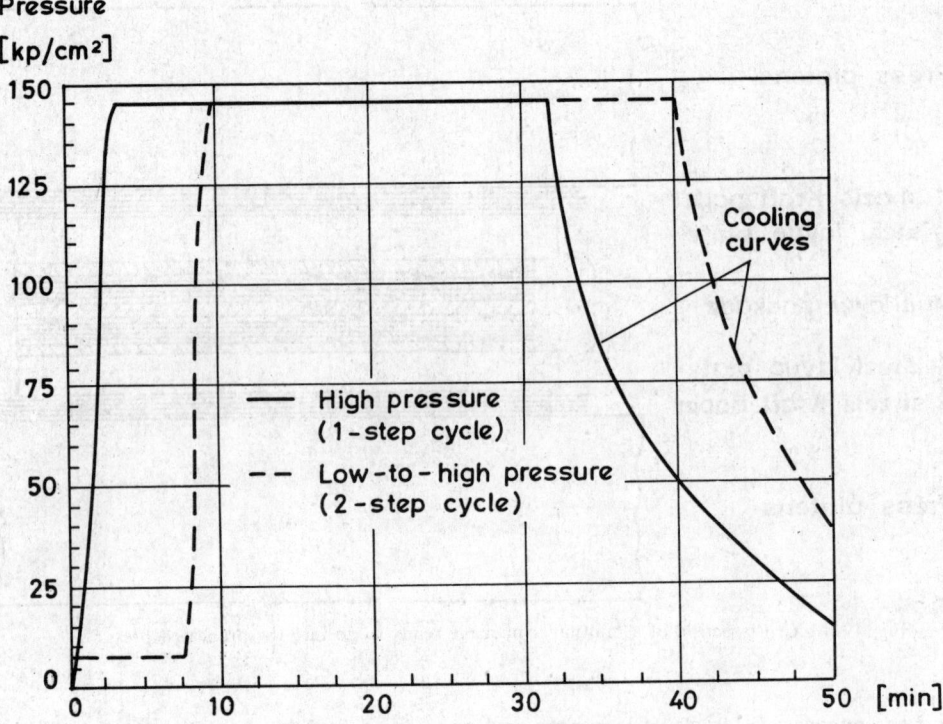

Fig. 19-13 Time-temperature profile in multilayer laminating

holes. After the inspection they are given for further treatment to build up the external layers and the plated-through holes as it is similarly done with ordinary double-sided boards.

19.6.2 Etch Back

The common technique by which the epoxy smear is removed and the exposed glass fibres prepared for copper deposition is known as etch back.

Etchants

It has been found that sulphuric acid of a concentration exceeding 85% is efficient in eroding away epoxy smear. As it removes epoxy smear not only from the hole walls of the inner copper layers but also from the interleaving epoxy laminate, glass fibres also get exposed. A following hydrofluoric acid dip removes these exposed fibres. It is a standard practice to immerse the boards for 10 seconds in 85% H_2SO_4 and for 10 seconds in hydrofluoric acid, alternately two or three times with high-pressure deionised water rinsing between each step. This is a hazardous operation and hence the operator has to be adequately protected.

19.6.3 Copper Plating

The copper plating in the through holes of multilayer boards is of a specially ductile quality to withstand thermal shock due to soldering. It is mostly plated from copper pyrophosphate baths or acid sulfate baths with proprietary addition agents for grain refinement and stress relief.

Fig. 19-14 illustrates a cross-section of a multilayer board plated-through hole, showing how the conductors on internal layers are connected to the wall of the hole by electroplated copper.

Multilayer ciruit boards, because of their inherent advantages are today produced in thousands by using automated equipment in the technologically advanced countries.

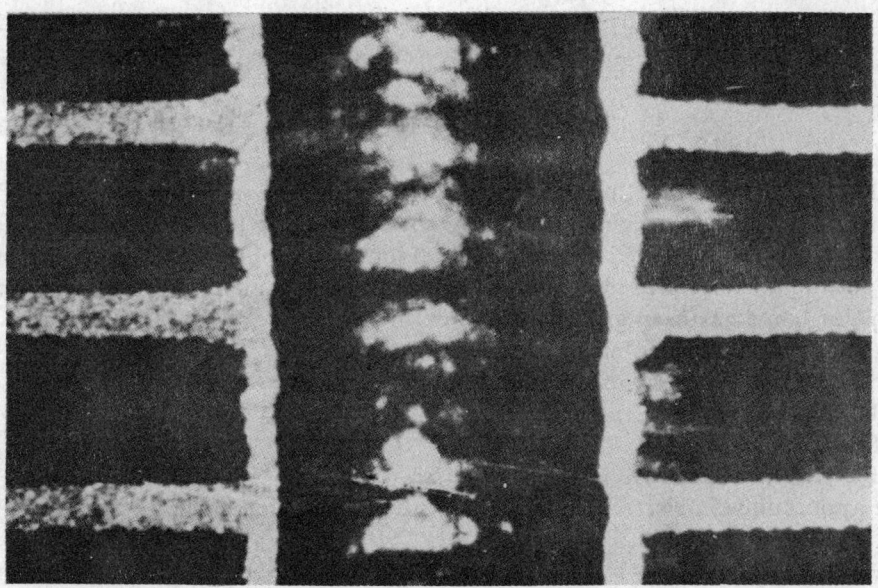

Fig. 19-14 Microsection of a multilayer plated-through hole

BIBLIOGRAPHY

1. COOMBS C. F.: *PRINTED CIRCUITS HANDBOOK*, McGraw-Hill Book Co., New York, 1967.
2. MITTER V.: 'Conception of a multilayer circuit board', *ELECTRONICS FOR YOU*, Vol. VI No. 10, New Delhi, 1976.
3. RIGLING W. S.: 'Designing and making multilayer printed circuits', *ELECTRO-TECHNOLOGY*, USA, May 1976.

20
SOLDERS AND SOLDERING TECHNIQUES

20.1 INTRODUCTION

Solders have been in use since time immemorial, the main use being for household and domestic applications. In Pharaoh's regime, solder was used in making jewellery. Centuries later, the Romans used this technique to join the lead pipes for their public baths. During all these times, the composition of tin and lead in the various solders has remained almost the same.

A major boost has been given to the soldering techniques with the ushering of canning industry followed by automobile and refrigeration markets.

Today, we can put most of the solder applications into three major groups:
— Mechanical joining
— Sealing
— Electrically conductive joining.

For applications in electronics, it is the third group which has to be explored in greater detail.

The production of highly reliable solder joints entails a good understanding of all the basic requirements involved in the process. In spite of widespread use of soldering, large unknown areas exist both in theory and practice. The reason for this is that originally soldering was a very simple technique which stimulated neither the scientist nor the technologist. It was only with the adoption of mass-scale soldering techniques that these gaps became evident, resulting in systematic investigations of the ancient soldering technology.

In this chapter, an attempt is made to provide such essential information on soldering in a compact form. It is primarily meant for engineers and students in electronics to facilitate a proper designing and production of solder connections.

20.2 PRINCIPLES OF SOLDER CONNECTIONS

'Soldering is a process for the joining of metal parts with the aid of a molten metal (solder), where the melting temperature is situated below that of the material joined, and whereby the surface of the parts are wetted, without them becoming molten'. (The definition is from the International Institute of Welding, 1955.)

Soft soldering or *soldering* generally implies that the joining process occurs at temperatures below 450° C: The filler metal (solder) wets and alloys with the base metals and gets drawn, by capillary action, into the gap between them. This process forms a metallurgical bond between the parts of the joint. Therefore, solder acts by

—*wetting* of the base metal surfaces forming the joint
—*flowing* between these surfaces which result in a completely filled space between them
—metallurgical *bonding* to these surfaces when solidified.

If the basic constituents in making a soldered joint are represented in a diagram, it will look as follows:

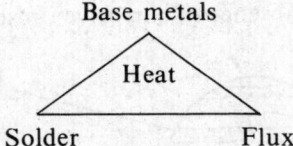

On the face of this diagram, soldering is a simple operation. It consists of the relative positioning of the surfaces to be joined, wetting of these surfaces with molten solder and allowing the solder to cool down until it has solidified. It is essential that the surfaces to be joined are sufficiently wetted in order to ensure adequate adhesion and a sound electrical contact (Fig. 20-1). Wetting, however, is possible only if the solder can come into direct contact with the metal surfaces to be soldered: The atoms of the solder must be able to come within atomic distances from the component parts to be soldered to ensure sufficient attraction.

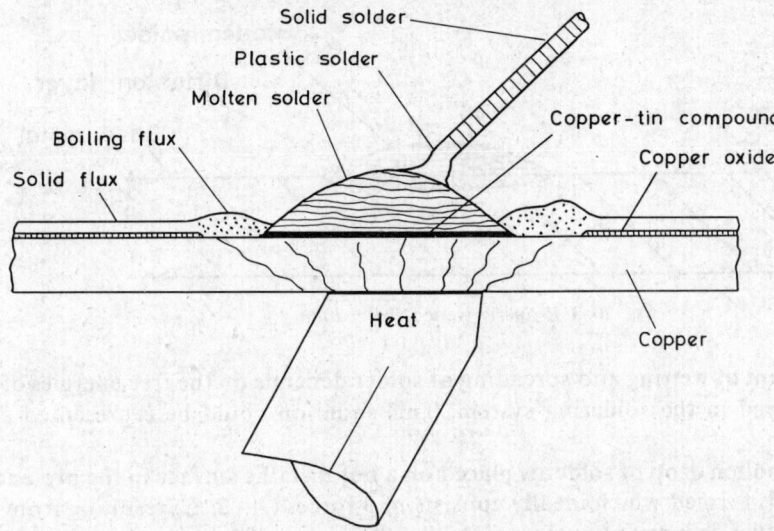

Fig. 20-1 Events during wetting of a solder joint

Any *contamination* firmly adhering to the surfaces to be soldered (e.g., oxides) will act as a barrier to metallic continuity and thus prevent wetting. A drop of molten solder on a contaminated surface will behave like an isolated drop of mercury (Fig. 20-2). After solidifying, this drop can easily loosen when subjected to mechanical shocks. The quality of the electrical contact in such a case will be very poor: The current has to pass through more or less insulating oxide film.

If the surfaces are clean (with their metal atoms thus lying immediately at the interface), wetting takes place and the solder will flow across the surface (Fig. 20-3). The solder atoms are now able to come very close to the atoms of the base metal; hence, they are attracted. The solder thus gradually diffuses into the solid metal and provides a good adhesion. The direct metallic contact with the atoms of the base metal, furthermore, ensures a good electrical contact. But strength and reliability of solder joints are not only dependent on the wetting: Design/geometry of the joint as well as presence or absence of corrosive substances are the influencing factors.

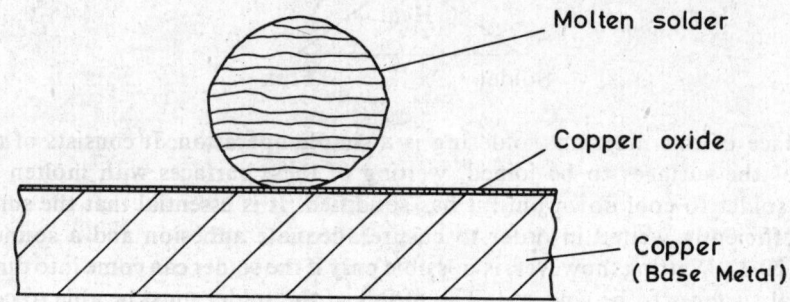

Fig. 20-2 A contaminated surface will prevent wetting

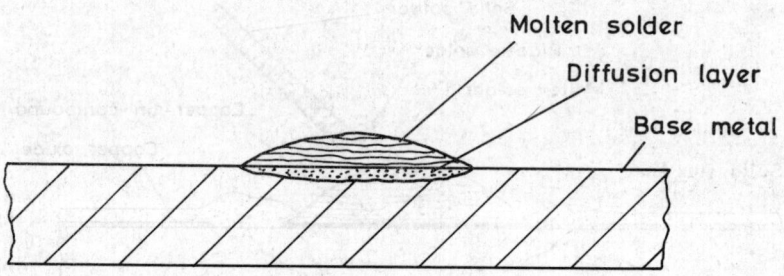

Fig. 20-3 Properly wetted solder joint

The amount of wetting and spreading of solder depends on the free energies of the various surfaces involved in the soldering system. This situation could be represented as shown in Fig. 20-4.

When a molten drop of solder is placed on a hot metallic surface in the presence of flux, a configuration is formed which ideally consists of 3 forces (A, B, C) resulting from the surface energies or rather interfacial surface energies. Every interface (metal-air, liquid solder-air, metal-liquid solder) has a quantity of energy which is proportional to the size of the surface. As every system tends towards a minimum of free energy, these interfaces tend to become as small as possible. While doing so, however, they counteract one another: Reduction of one interface leads to the enlargement of another.

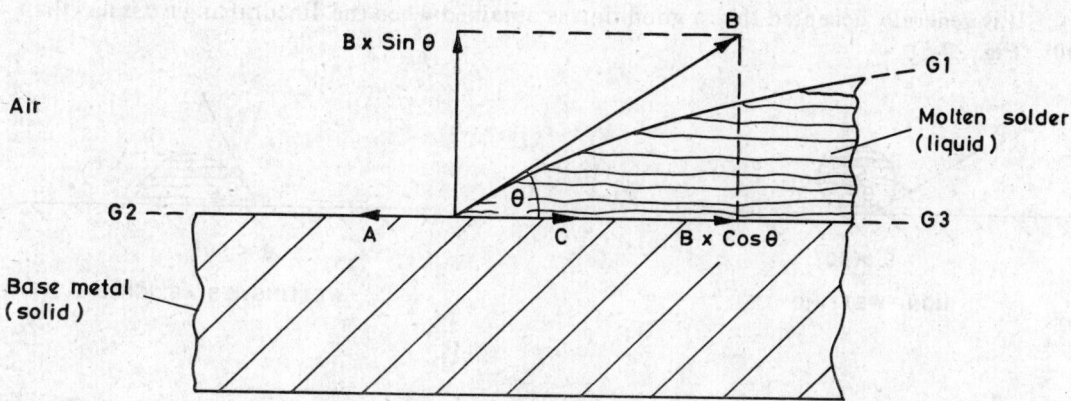

Fig. 20-4 Geometric analysis of a wetted solder joint

The boundary lines between the 3 phases (G1, G2, G3) adjust themselves equally and in such a way that the whole configuration has a minimum of interfacial surface energy. The three imaginary forces on the boundary lines are as follows:

Force A — for the interface between the part to be soldered and the air (also referred to as the *surface tension of the metal*)

Force B — for the interface between the solder and the air (also referred to as the *surface tension of the solder*)

Force C— for the interface between the solder and the component to be soldered (also referred to as the *solder metal interfacial tension*).

Starting from the dihedral angle θ, force B can be resolved into
— a force $B \times \cos \theta$, acting along the surface, and
— a force $B \times \sin \theta$, perpendicular to the surface. This component force is compensated for by the rigidity of the base metal.

If the forces acting on the border of the solder drop and along the surface of the component to be soldered *keep each other in equilibrium*, neither further wetting nor dewetting will occur. In this case,

$$A = B \times \cos \theta + C \text{ and thus } \cos \theta = \frac{A - C}{B}$$

Cos θ is a measure of the wetting. If cos θ is large, then θ is small and wetting is good. Conversely, if cos θ is small, then θ is large and wetting is poor. Consequently, wetting is influenced by changes in the forces A, B, C. Such changes could occur
—as a result of chemical changes on the surface of either the solder or the component to be soldered. Such chemical changes can, for instance, be caused by the diffusion of tin from the solder into the base metal, thereby forming alloys. These alloys are solid solutions, occasionally with chemical compound of the base metal (e.g. Cu_3Sn on copper or Fe_2Sn on iron base metal).
—due to the replacement of the air by another gas or by vapour (e.g., flux).

It is generally accepted that a good flow is obtained when the dihedral angle θ is less than 10° (Fig. 20-5).

Fig. 20-5 Wetting and dihedral angle

20.3 SOLDER ALLOYS

All solder alloys have a low melting point with the liquidus temperature below the melting point of pure lead (327°C). The bulk of the solders even melt below 250°C. In any solder system, lead (Pb) is mainly used as a dilutant only to lower the costs. The wetting phenomenon is dependent on tin only : The higher the tin content, the better would be the wetting. There are various systems of solders in practical use; the most important ones are mentioned below.

20.3.1 Tin-Lead

The binary mixture of tin and lead constitutes a simple and classical eutectic system with the eutectic point at 61.9% tin and the eutectic temperature at 183°C.

Point C in Fig. 20-6 is the eutectic point. Compositions which vary from the eutectic one have rather a melting range than a melting or solidification point: The solder which corresponds to the point D, as an example, has a melting range of 183–273°C and a tin content of 30%. If a solder has a melting range, it is only fully molten at the highest temperature of this range. Similarly, cooling off must come down to the lowest temperature of the range for a complete solidification of the alloy; above this temperature, the mechanical strength is zero. If a soldered joint is disturbed during solidification, it becomes unreliable. Therefore, one generally chooses a solder with a narrow melting range for the soldering of electronic components since wide-range solders increase the soldering time.

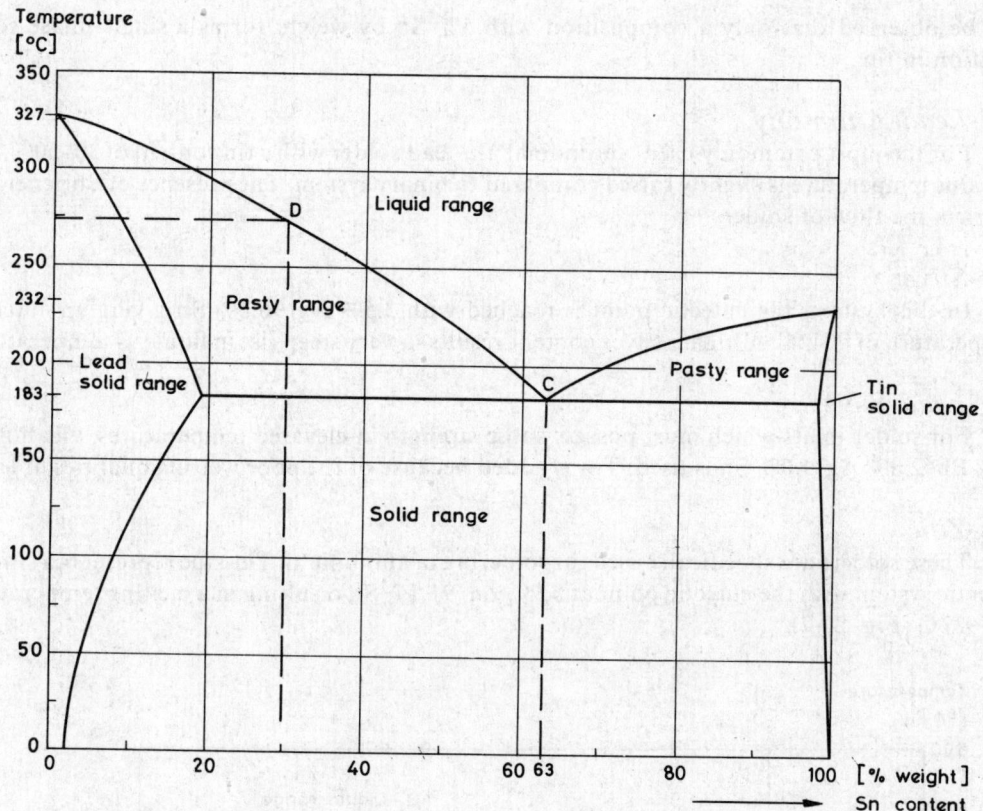

Fig. 20-6 Metallurgical equilibrium diagram for tin-lead system

		100% tin	63% tin 37% lead	37% tin 63% lead	100% lead
Specific gravity	[kg/dm^3]	7.3	8.4	9.5	11.4
Latent heat of melting	[cal/g]	14	13	10.5	5.5
Specific heat	[cal/°C]	0.06	0.05	0.04	0.03
Final melting point	[°C]	232	183	257	327
Required energy from 20°C	[cal/g]	30	24	21	17
Required energy from 20°C	[cal/cm^3]	219	292	200	191

20.3.2 Other Solder Systems

Tin-Antimony

From the metallurgical equilibrium diagram for the binary tin-antimony alloy system, it can be observed that only a composition with 5% Sb by weight forms a single-phase solid solution in tin.

Tin-Lead-Antimony
For the most commonly used 'antimonial' tin-lead solder with a tin content of 30–50%, the liquidus temperature is slightly raised compared to binary system. The presence of Sb generally restricts the flow of solder.

Tin-Silver
In this system, the eutectic point is reached with 3.5% Ag/96.5% Sn giving an eutectic temperature of 221°C. A higher silver content results in very steep rise in liquidus temperature.

Tin-Lead-Silver
For solder joints which must possess some strength at elevated temperatures, an alloy of 97% Pb/2.0% Ag/1.0% Sn is used. Tin is added because of the poor wetting qualities of lead.

Tin-Zinc
These solders are specifically used for soldering of aluminium. Tin-zinc represents a simple eutectic system with the eutectic point at 8.9% Zn/91.1% Sn resulting in a melting temperature of 198°C (Fig. 20-7).

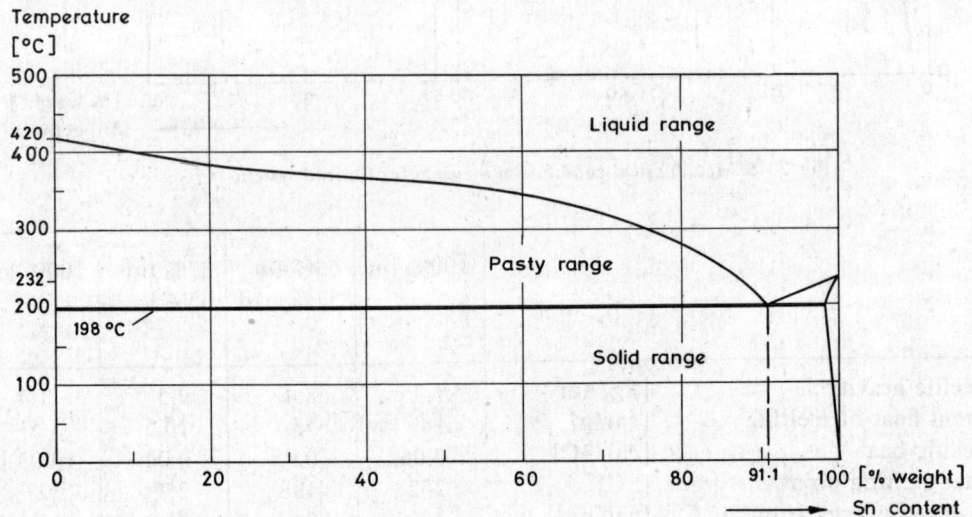

Fig. 20-7 Metallurgical equilibrium diagram for tin-zinc system

Low-Melting-Point (Fusible) Alloys
The most commonly used alloy of this category, featuring in several specifications, is the ternary eutectic composition 50% Sn/32% Pb/18% Cd which melts at a temperature of 145°C. The use of cadmium in solders, however, poses certain health hazards and will be dealt with separately.

Another useful but quarternary system, melting at only 70°C, is 12.5% Sn/25% Pb/50% Bi/12.5% Cd. This alloy is also called *wood metal*.

Special Solders

Near-eutectic Sn-Pb solder dissolves such metals like copper and silver, especially if these are present in thin films. Therefore, in order to reduce the solubility of these metals, the near-eutectic Sn-Pb system is loaded with a small percentage of these metals, i.e., 1.5% Cu or 2% Ag (% weight).

A gold-tin eutectic system, containing 20% Sn and melting at 280°C, has a special application in soldering of semiconductors. Pure tin, tin-cadmium, cadmium-silver or cadmium-zinc-silver systems are some of the other systems used mainly for obtaining a higher strength of soldered joints.

20.3.3 Influence of Impurities on Tin–Lead Solder

Copper: The final melting point of solder is very much raised even by only a very small addition of copper. The molten solder becomes easily dirty and sticks to all nonmetallic materials. Consequently, such solder has much less capillary action which is a disadvantage. The appearance of copper-containing solder is also less fine because of the tin-copper crystals. The copper solubility of solder with 60% Sn/40% Pb is around 0.4% at the temperature of 250°C. If the copper content is above 0.29%, the solder will show grittiness. It is therefore suggested to maintain the copper content in eutectic Sn-Pb solder at less than 0.25%.

Iron: The iron content should be kept below 0.1%.

Antimony: Above 2.5% (% weight), flow properties are reduced but wetting does not decrease significantly.

Aluminium: It is an oxide promoting element and its presence above 0.001% results in surface deterioration of molten solder (grittiness).

Zinc: Zinc also reduces the flow of solder and its content should be kept below 0.02%.

20.3.4 Mechanical Properties of Solder Alloys

The mechanical strength of solder is comparatively small, amounting to only a few percents of that of normal construction materials. The mechanical strength is particularly small at elevated temperatures. In this connection, room temperature must be considered as already a relatively high temperature for these alloys.

Shear strength

Composition	Shear strength [N/mm^2]		
	130°C	20°C	150°C
100% Sn	100	20	7
62% Sn/38% Pb	90	30	8
40% Sn/60% Pb	70	30	9
10% Sn/90% Pb	50	20	10

Tensile strength: Results of tensile strength are dependent on the rate of loading the test specimen. As an example, the increase of the loading rate by a factor 400 at 20°C, results in an increase of the tensile strength by a factor 4. Moreover, the permissible stress in soldered joints depends to a large degree on the geometry of the gap between the parts joined by the solder and on the quantity of solder in the joint. *Note that a large quantity of solder does not necessarily result in stronger joints.*

Creep strength: At long-term loading, the mechanical strength decreases further. This is best illustrated by an example, measured at a temperature of 20°C:

10 hrs loading	10 N/mm^2
1,000 hrs loading	4 N/mm^2
100,000 hrs loading	2.5 N/mm^2

By alloying the solder with antimony, the strength can be increased to some extent but it does not change essentially.

Note: Tin-lead solders are generally not suitable for mechanical joining if the load at 20°C lies permanently above 1 N/mm^2. In case of higher temperature, the mechanical strength of solder gets reduced to a fraction of the strength at room temperature (e.g., a strength of 1 N/mm^2 at 20°C will become only 0.1 N/mm^2 at 110°C). Where mechanically stronger connections are required, some other mechanical fixation devices have to be used along with soldering.

20.4 SOLDERING FLUXES

During the soldering operation, an auxiliary medium is mostly used to increase the flow properties of molten solder or to improve the degree of wetting. Such a medium is called *flux*.

Following characteristics are required in a flux:
— It should provide a liquid cover over the materials and exclude air up to the soldering temperature.
— It should dissolve any oxide on the metal surface or on the solder and carry such unwanted elements away.
— It should be readily displaced from the metal by the molten soldering operation.
— Residues should be removable after completion of the solder.

Fluxes are usually divided into 3 groups according to the nature of their residues: corrosive, intermediate and non-corrosive fluxes.

20.4.1 Corrosive Fluxes

The corrosive or acid fluxes consist mainly of inorganic acids and salts. They are generally used to solder 'difficult' materials, e.g., where rapid wetting by molten solder is required or where the surface conditions are critical. Corrosive fluxes may also be needed when using lead based solders.

Note: Because of the corrosive nature of such fluxes, they should only be used on components and in conditions where the residues can be completely removed after soldering.

The most basic ingredient in corrosive fluxes is zinc chloride since it melts at soldering temperatures and forms a convenient source of HCl to perform fluxing action. Ammonium chloride, sodium chloride and tin chloride may also be present to increase the activity or lower the fusion point of the mixture.

Example:

Composition (% weight)	Melting temp.
87% $ZnCl_2$ / 13% NH_4Cl	232°C
73% $ZnCl_2$ / 27% NH_4Cl	180°C
82% $ZnCl_2$ / 18% NaCl	262°C
23% $ZnCl_2$ / 77% $SnCl_2$	171°C

As solvent, mostly water is used but also bromides or fluorides are applied. Phosphoric acid is particularly useful for soldering of stainless steel.

Two typical general-purpose, corrosive flux compositions shall be given:

1) $ZnCl_2$ 100 g/l   2) $ZnCl_2$ 33 g/l
   NH_4Cl 10 g/l   NaCl   8 g/l
   HCl   ca.25 ml/l NH_4Cl 4 g/l
                                HCl   ca.10 ml/l

The rapid heating of aqueous fluxes produce spattering due to violent evolution of steam. This can be reduced by using polyethylene glycol as a solvent, perhaps mixed with iso-propanol. A suitable mixture is 70% polyethylene glycol and 30% iso-propanol (% volume). Such solvents boil gently over a wide range of temperature. Residues of these halogen base corrosive fluxes, however, have often caused serious corrosion problems: These residues, being hygroscopic in nature, absorb water and then get strongly dissociated thus liberating halogen radicals which corrode the base metal.

Since the danger of corrosion of the soldered joint with corrosive fluxes is not immediately noticeable, corrosive fluxes should be used only where adequate facilities for water washing are available. It is also possible to reduce the corrosivity to some extent by using petroleum jelly, lanolin, tallon, etc., along with zinc chloride. The proportion used may be around one-third zinc chloride and two-thirds carrier medium (by weight).

Note: Because of their corrosive nature, corrosive fluxes are hardly applied in electronics industry.

20.4.2 Intermediate Fluxes

They are weaker in nature if compared to inorganic fluxes. They consist of organic acids and certain of their derivatives such as hydrohalides. These fluxes are active at soldering temperature but, with some of them, the activity period is short because they volatile or oxidise.

Typical intermediate fluxes are the following:
— Lactic, oleic, glutamic, stearic and succinic acids. Some of them are active also below soldering temperature.
— Hydrohalides or amines or amino acids from which halogen acids are liberated during heating, e.g., diethyl ammonium chloride, cetyl pyridinium bromide, anilin hydrochloride or glutamic acid hydrochloride.
— Amines or amides such as triethanolamine, ethylene diamine and urea, etc.

20.4.3 Non-Corrosive Fluxes

For electrical components, electronic assemblies, delicate instruments and all other parts where it is difficult to wash off flux residues after soldering, a non-corrosive flux is required. Pure water-white rosin, 20–25% by weight respectively volume, dissolved in a solvent such as iso-propanol, is the closest approach to a non-corrosive flux solution. Colophony or natural rosin (obtained from a particular type of pine wood) contains chiefly abeitic acid and its isomers which become mildly active at soldering temperatures, being a weak acid. It is then in a form which will attack not too strong impurities such as the natural oxide layers of rolled copper. Impurities which will not be removed are the oxides of iron, chromium, aluminium, etc.

The principal difference between rosin (colophony) and most other fluxes is due to the following properties:
—Rosin is not dissociated at room temperature.
—Rosin is practically insoluble in water; thus rosin cannot form an electrolyte.

Non-corrosive fluxes would be the ideal solution if they would show a better activity and a better degree of flow. To overcome these limitations, activators are added to the rosin flux to improve both the degree of flow as well as the activity without significantly increasing the corrosivity or lessening the electrical insulation resistance. Suitable activators are glycol acid, urea and its derivatives, ammonium chloride and halogen containing amines. Such liquid fluxes containing organic halides should not contain more than 0.5% halide, as chloride, based on the solid contents of the flux. Solderability is generally increased in proportion to the halide content in the flux.

The most important criteria for the selection of a flux are:
—flow promoting properties of the flux
—solder composition
—composition of materials to be soldered
—surrounding in which soldering has to take place
—soldering technique
—form in which the flux has to be applied
—soldering temperature
—properties of the flux residues
—possible health hazards.

20.4.4 Testing of Fluxes

For measuring flux as well as corrosion characteristics of a flux, a number of methods have been specified. Generally, a known weight of solder and flux are placed on a freshly (chemically) cleaned copper sheet and left floating on molten solder for a specified time. The area of solder spread determines the flow-property of the flux. Resistance to corrosion is then assessed by placing the specimen into the humidity chamber at 40°C and 98% RH for a duration of 100 resp., 500 hrs. The presence of green-blue copper corrosion products indicates the corrosion behaviour of the flux.

20.5 SOLDERING TECHNIQUES

To achieve a soldered joint, the solder and the base metal must be heated above the melting

point of the solder used. The method by which the necessary heat is applied, depends among other things on
— nature and type of the joint,
— melting temperature of the solder,
— flux.

Generally applied soldering methods are iron soldering, torch soldering, mass soldering, electrical soldering (high-frequency soldering, resistance soldering), furnace soldering, and other methods. Here, we shall have a closer look at the iron soldering and the mass soldering.

20.5.1 Iron Soldering

The most commonplace method for general soldering is the use of the soldering iron. This technique has reached a reasonably high state of perfection as far as the design of the iron is concerned. Basically, a soldering iron consists of an insulated handle, connected via a metal shank to the bit. The face of the bit actually makes contact with the component parts of the joint and the solder, and heats them up. The electrical heating element is usually located in the hollow shank or in the handle and may be thermostatically controlled to give a pre-set temperature in the bit. Soldering irons are produced in an enormous variety of designs and sizes, and the selection of an appropriate type depends on the intended application. But in all soldering irons, the key to the function of the iron is in the bit itself.

Function of the Bit

The bit of the soldering iron has to perform the following functions:
— It stores heat and conveys it from the heat sources to the work.
— It may be required to store and deliver molten solder, and often flux, to the work. However, it is usually recommended that solder is applied to the workpieces, as they are heated by the bit.
— It may be used to remove surplus solder from the joint.

It is essential that the bit surface is perfectly wetted (tinned): This encourages flow of solder into the joint. When the surface of the work also becomes wetted by solder, a continuous film of liquid metal bridges the gap between the soldering iron bit and the work which provides a path of high thermal conductivity along which the heat can flow into the workpiece.

Soldering bits are usually made of copper; this metal combines good wetting properties with the optimum heat capacity and thermal conductivity. There is, however, an erosion problem in long-term use: Tin-lead solders will attack the copper and dissolve it during the soldering operation. Time has thus to be spent on re-sharpening of the bit. Various approaches have been tried to overcome the erosion problem; the most successful solution is to protect the copper bit with a thick iron coating followed by nickel/tin plating. The life of the bits can this way be increased by a factor 10 to 15.

Soldering Iron Design

The size and to a certain extent also the shape of the bit are largely determined by the amount of heat that has to be supplied during each joining operation. The bit must be sufficiently large to enable soldering of the required number of joints, before the heat is drained away, thereby reducing its temperature below that necessary for a sound joint. The heat input of an electric soldering iron (wattage) is thus mostly related to the intended rate of working. The proper choice of a soldering iron, therefore, depends on whether occasionally a small number of individual joints have to be made or whether continuous production line soldering of many joints has to be undertaken.

Soldering with an Iron

After the parts to be joined have been cleaned, fluxed where necessary and assembled for soldering, the soldering iron at soldering temperature is held on the workpieces to heat them. The solder, usually in the form of wire, stick or preform, is applied to the work close to the bit where it should melt immediately and become bright and fluid. If enough solder has been applied, it should completely penetrate and fill the gap of the joint. Surfaces to be joined should, after cleaning, be first treated with liquid flux unless flux-cored solder wire is used. When the joint appears to be sufficiently filled, the soldering iron is removed and the joint allowed to cool down undisturbed. The time required to keep the soldering iron on the work is entirely dependent on the nature of the joint and the characteristics of the soldering iron. It is also well known that gentle movements of the bit during soldering can assist in rapid penetration of the solder into the joint spaces, especially if seams of some length have to be soldered.

20.5.2 Mass Soldering

Mass soldering incorporates those techniques by which large numbers of joints are made simultaneously using a solder bath, both as the source of heat as well as of filler metal. The most important mass soldering techniques employ some form of immersion or contact with a molten solder bath. The major advantage of mass soldering techniques, apart from the high productivity, is that a more rigorous control is possible over all the individual stages of soldering. This contributes to the reliability of the final assemblies.

Mass soldering techniques are much used in electronics industries, particularly for assembling of PCBs: Components are mounted on one side of the boards so that their connecting leads pass through the component holes. These leads are then soldered to the conductive tracks on the other side of the boards. For high production rates, making all these contacts manually would be a very slow and costly task but mass soldering can provide an economic solution.

Initially, flux must be applied to the PCBs. This flux is usually of rosin-type, dissolved in an organic solvent. Methods used to apply flux include
—dipping the board onto the surface of a bath with flux
—brushing
—spraying, using a special spraying cabinet
—rolling, in contact with a plastic foam rubber roller impregnated with flux
—wave fluxing, i.e. by passage over a standing wave of liquid flux
—foam fluxing, where the standing wave is of foamed flux.

For flux coating, the PCBs are usually heated to remove the bulk of the solvent prior to soldering. If this is not done, gases may form air locks during soldering which may be trapped in the solidifying solder or which may result in areas of the board not being wetted by molten solder. In the case of in-line soldering machines, drying is carried out by passage over controlled infrared heaters or hot-air blowers.

Dip Soldering

In simple dip soldering, the prefluxed assembly is lowered vertically onto the clean solder surface until it makes contact and is then immersed in the solder bath to the required depth (Fig. 20-8). The surfaces become wetted by the solder, and, wherever the interstices are sufficiently narrow, solder penetrates and is retained between them by capillary forces. It is essential that the surface of the solder is freshly cleaned. This can be done by manual or automatic skimming to sweep the oxides and flux residues to one side of the bath immediately before dipping the assembly.

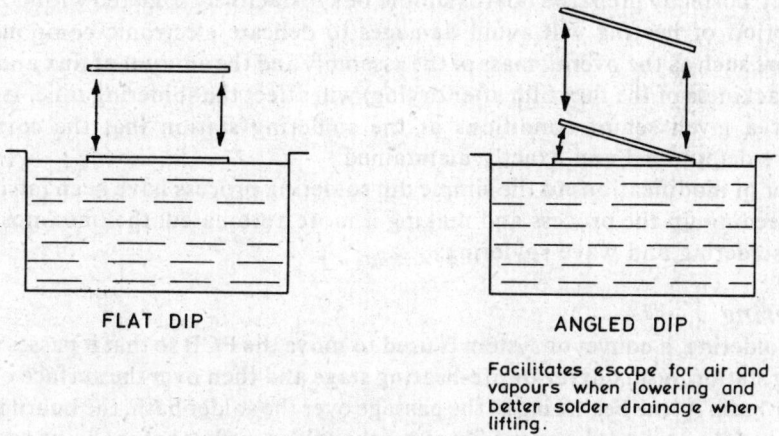

FLAT DIP

ANGLED DIP

Facilitates escape for air and vapours when lowering and better solder drainage when lifting.

Fig. 20-8 Dip soldering principle

The solder pot is usually made of cast iron or steel and is electrically heated. It should be large enough to take the intended size of boards and additionally to provide a large enough mass of molten solder to counteract local cooling effects when the board is immersed. The bath temperature should be suited to the nature of the assembly being processed and other factors such as mass, thermal conductivity and specific heat. The solder bath temperature is normally within the range of 220–260°C for binary tin-lead alloys close to eutectic composition. The temperature may be increased to 350–400°C for lead-rich alloys with a high liquidus temperature or where it is required to burn off the insulation and to pre-tin polyurethane-enamelled copper wires. The time of contact with the solder should be the minimum necessary for complete wetting of all surfaces by the solder and for filling all the joint spaces by capillary action. Both the required time of contact and the optimum temperature of the solder bath are best determined by trials before the beginning of production runs. Once the optimum parameters are determined, they have to be strictly maintained; an automatic control for time and temperature will be an advantage.

The solder composition for dipping baths in electrical and electronics applications is normally 60% Sn/ 40% Pb, or the eutectic alloy. An alloy of 65% Sn / 35% Pb may be used for adjusting the tin content of the bath should it fall in the course of operation. An alloy of 50% Sn /50% Pb is used for less demanding applications and tin contents as low as 20% may be employed for low-cost production work. It is inevitable that some impurities will be picked up by the solder bath after it has been used for some time. However, the better the purity of the initial solder, the longer it will last until the impurity level becomes significant.

For soldering of PCBs by the simple dipping method, it is advisable that one edge of the board is lowered into contact with the solder first, to allow escape of flux and remaining solvent vapours (angled dip). Complete contact can then be made after another 2–3 seconds. Upon withdrawing the board, an angled path should again be used to assist solder drainage and to minimise icicles and solder bridges between adjacent conductors.

In general, the temperature for mass soldering of PCBs will be around 240–250°C. It should not be necessary, with average electronic assemblies, to exceed a contact period of 5 seconds; in fact, correctly prepared boards should be satisfactorily soldered within 2–3 seconds. This short period of heating will avoid damages to delicate electronic components on the boards. Factors such as the overall mass of the assembly and the amount of flux and its residual solvent (i.e., tackiness of the flux film after drying) will affect the soldering time. It is therefore important for a given set of conditions at the soldering station that the correct fluxing parameters are determined and exactly maintained.

A number of modifications to the simple dip soldering process have been introduced with the aim of speeding up the process and making it more automated; the ones mostly used are termed drag soldering and wave soldering.

Drag Soldering

In drag soldering, a conveyor system is used to move the PCB so that it passes successively over a fluxing station, a flux dryer or pre-heating stage and then over the surface of a long and narrow solder bath. At the beginning of the passage over the solder bath, the board is lowered at a small angle and then travels horizontally along the solder surface before being withdrawn at a small angle to assist solder drainage (Fig. 20-9).

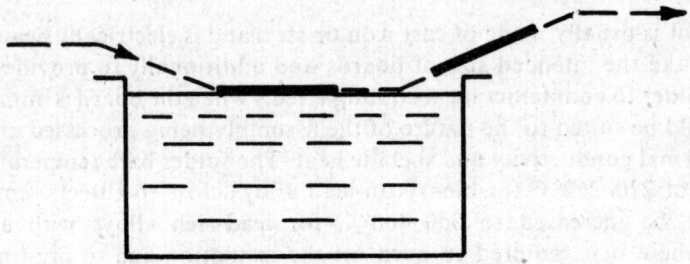

Fig. 20–9 Drag soldering principle

Control systems can vary the speed of travel over the fluxing station and solder bath, and there are often facilities for increasing the time of contact with the solder by a predetermined dwell period. Before the PCB makes contact with the solder surface, a horizontal scraper bar travels along the bath to remove oxide films and any flux residues, leaving a clean solder surface to facilitate wetting.

Wave Soldering

In wave soldering, instead of lowering the boards onto a solder bath, solder is pumped out of a narrow slot to create a standing wave in the solder bath (Fig. 20-10). The boards, after passing over the usual fluxing and drying sections, are conveyed across the crest of the solder wave by a conveyor system which follows a straight-line path. This path may be inclined upwards at a small angle to the horizontal (e.g., 5–15°) in order to assist solder drainage after the boards have passed the wave. The drainage and distribution of solder may also be improved by choosing a particular wave form; double-crested, flat-topped or uni-directional solder flow against the direction of board movement may be the preferred wave types for certain board geometries.

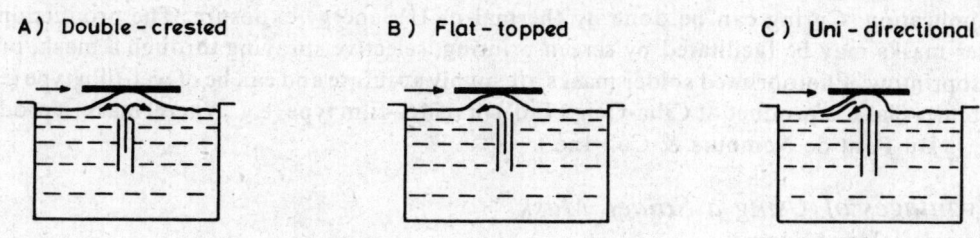

Fig. 20–10 Wave soldering principles

The use of a solder wave brings the advantage of a virtually oxide-free surface being continuously generated on the solder while air, flux and flux vapours are dislodged by the rapid movement of the solder.

In automatic mass soldering methods, there are usually a number of successive stations at which sequentially the operations of board loading, flux application, flux drying, soldering, cooling, flux removal and finally off-loading are carried out. To avoid costly re-work to remedy poor solder joints, surfaces having good solderability are essentially required; capillary rise of solder through eyelets or plated-through holes should then readily occur. Penetration may be helped by carefully adjusting the depth to which the boards are immersed when in contact with the solder.

20.5.3 Flux Removal after Soldering

If a corrosive flux has been used, it is very important that the flux residues are removed after cooling of the assembly to avoid corrosion during storage. Since corrosive fluxes are generally water-soluble (e.g., zinc chloride), special care must be taken to avoid hydrolising while using mild neutralising rinse. Corrosive flux must be removed by first washing the assembly with 0.5–1% HCl or 5–10% citric or acetic acid, followed by neutralisation in dilute alkali. Addition of detergents will also help as long as they are not harmful to the finished assembly.

Non-corrosive or rosin-based fluxes, generally used for electronic assemblies, can be removed with hot organic solvents. However, certain flux activators (particularly halides being water-soluble) can only be removed with aqueous cleaning. Iso-propanol could be used for such a purpose.

Ultrasonic transducers with organic/aqueous solvents are extensively used to assist dislodging flux residues during washing. However, they may cause certain risks to some type of electronic components such as electrolytic capacitors, etc. More details on flux removal are given in Chapter 21.

20.6 SOLDER MASK

The solder mask, also called solder resist or stop-offs, are organic coatings which are applied selectively to those areas where no solder wetting is desired. Thus, if a **PCB** is selectively coated with a solder mask, only such areas to which components are to be soldered are left exposed.

The solder masks are usually one- or two- component systems which have to be cured prior to application. Curing can be done by thermal or UV energy exposure. The production of solder masks may be facilitated by screen-printing, selective spraying through a mask, or by photoprinting. Photoprinted solder masks are highly accurate and can be of wet-film type (e.g., Probimer mask, a product of Ciba-Geigy Ltd.) or of dry-film type (e.g., Vacrel mask, a product of E.I. Du Pont de Nemours & Co., Inc.).

Advantages of Using a Solder Mask
— Avoids solder bridging
— Less contamination of solder alloy
— Better solder fillets resulting in more reliable joints
— Lower costs for solder per unit assembly
— Improved mechanical properties of the boards, i.e. flexural strength, etc.
— Acts as a barrier between the atmosphere and the board thus eliminating danger of corrosion and leakage currents.

Disadvantages
— It is an extra operation.
— Repair work on the PCB with solder mask is very difficult. The solder mask has to be mechanically removed (scrapping) before soldering can be performed.
— Its compatibility with flux must be verified before its use.

Surfaces under the solder mask have to be absolutely clean; otherwise, the mask will have the tendency to peel off or bubble up in the area of dirt. This in turn might cause leaks and other defects.

During thermal curing of solder masks, the bare copper surface sometimes shows a poor solderability inspite of having a lacquer coat. In such cases, the PCBs have to be activated after application of the solder mask. In special cases, it is possible to apply solder masks over the protective coating.

Solder mask coatings are normally deeply coloured but are still sufficiently translucent to show the substrate: This is essential for observing any disconnection or other defects in the circuit pattern below.

20.7 REFLOW SOLDERING PRACTICE

By reflow soldering or resoldering, it is understood to make a soldered joint by means of remelting a previously applied solder deposit without the addition of any more solder during the soldering process. This also applies to similar processes whereby the solder is previously added in the form of preforms (rings, etc.).

Only if certain conditions are complied with, is it possible to obtain acceptable soldering results without a flux. The conditions for this are that at least one surface is practically free of oxide. This means, in practice, that one of the surfaces to be joined has been provided with a coating of noble metal. But generally, it is better to apply a flux.

For the various reflow soldering processes, a definite minimum quantity of solder is required (layer thickness of the solder or tin plating), depending on the design of the components to be soldered.

Soldering Methods

The required quantity of heat for the melting of the solder can be supplied in different ways: Iron soldering, hot gas soldering, radiant heat soldering, induction heat soldering and resistance heat soldering. The first 2 ways (iron soldering, hot gas soldering) are well known, hence only the following 3 ways will be discussed here briefly.

Radiant heat soldering: White or infrared light is a good heat source for reflow soldering. In many cases, white light causes a cumbersome glare and impedes the evaluation of the adjustment or progress of the soldering process. A major advantage of radiant heating is, just as with heating by hot gas, that there is no mechanical contact with the soldering point. By using reflectors, the light beam can be focussed onto the workpiece. In most cases, halogen lamps and projection lamps with built-in mirrors are used. In principle, the control is simple: The energy flow and the time can be varied.

Induction heat soldering: In soldering with inductive heating, a high-frequency alternating current is used to generate an electric field in and around the coil: Metal objects placed inside this field are heated. By a directed design of the coil, frequency and field location, it is possible to accomplish different kinds of heating effects. Characteristic is the rapid transfer of great quantities of energy; the result is a quick heating, also of larger objects. With special designs of such heat sources, it is also possible to heat only the surface of the objects. As with radiant heat soldering, the control must also be automated.

Advantages: —Rapid transfer of great quantities of energy, if necessary.
 —Local heating of surface zones is in some cases also possible.
Disadvantages: —Fixation of the object with respect to the coil is required for each specific application. As a rule, a different design of coil is required and for certain applications, the coil shape is very complicated.
 —Overheating must be prevented.
 —Sometimes too expensive for small series.
 —Complicated technique requiring the advice of specialists.
 —In certain cases less suitable for copper.

Resistance heat soldering: In resistance heat soldering, an electric current passes through the parts to be soldered; due to the contact resistance and the resistance of the parts themselves, heat is developed. The most important resistance heat reflow soldering methods are the parallel-gap soldering, the single-point soldering and the multiple-lead soldering.

Parallel-gap soldering facilitates the soldering of polyurethane insulated wires to tin plated soldering pads of PCBs. Single-point soldering or multiple-lead soldering methods are only variations of the parallel-gap soldering; they have been specially developed for mass-scale soldering of miniature packs onto PCBs.

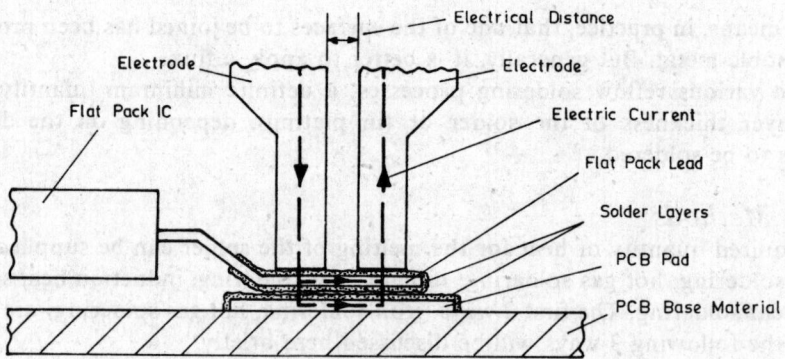

Fig. 20-11 Parallel-gap reflow soldering

In conclusion, it may be stated that reflow soldering methods, especially that of resistance heat soldering, are being more and more used in electronic assemblies. The soldering time has been reduced from a few seconds to milliseconds, thus providing a valuable method for high-speed mechanised assembly methods. It is expected to grow further in dimension in the near future.

20.8 TESTING AND QUALITY CONTROL

For achieving a high quality and reliability in soldered assemblies, it is essential that a properly planned quality assurance scheme is imposed. All constituents and factors taking part in any soldering operation or associated with it play a specific part in making a good solder joint:
—Solder
—Flux
—Solderability of surfaces to be joined
—Soldering apparatus
—Design.

20.8.1 Solderability Test Methods

Almost in all electronic assemblies, the soldering operation needs to be completed in 2–3 seconds only. Solderability testing of components is therefore based on measurement of the time in contact with molten solder to achieve good wetting under given conditions. Most of such tests monitor the forces acting upon a specimen to be soldered as it becomes wetted.

Among the many tests for solderability, the most commonly used ones are:
—Solder bath test
—Solder globule test
—Surface tension balance
—Meniscus rise test
—Soldering iron test.

Solder Bath Test

The *rotary dip test*, developed at the Tin Research Institute, U.K., is used as a standard test method in the specifications IEC 68-2-20 and BS 4025. This test method is suitable for assessing the solderability of PCBs, sheet metal, tags, etc., used by the electronics industry.

The test procedure consists essentially of immersing individually a number of pre-fluxed square test pieces of 30 mm length. The immersion time is successively increased until a distinct 'end point' is reached with a smooth solder coating which is free from pinholes and from non-wetting or de-wetting. The solder bath temperature, for example, can be maintained at 235°C. The testing apparatus comprises a horizontally rotating spindle with a radial arm to carry the test pieces horizontally and face down across the surface of the solder bath. The speed of rotation can be varied to enable different contact times with the solder. The portion of the non-wetted surface will become less with increasing time in the solder bath, until a time of contact is found at which complete wetting occurs.

Certain types of surface contaminations may cause the solder to dewet, resulting in globules and ridges of solder resting on a solder-coloured surface. Test specifications usually require that this defect is not noted below a certain soldering time (e.g., 5 sec) and that this test may be made sometimes at higher-than-usual solder bath temperatures (e.g., 260°C). With very severe oxidation or contamination, simultaneous non-wetting and de-wetting may occur which does not give an end point for the procedure: This indicates a very poor surface preparation of the material.

The rotary dip test, although requiring some training for the operator in recognising the visual features of wetting and de-wetting, is still the simplest and most versatile technique for a quick and cheap assessment of the solderability of PCBs and plated-through holes.

Solder Globule Test

The solder globule test is specifically used for component terminations, i.e., round wires, and is covered by the specifications IEC 68-2-20, BS 2011 and DIN 40016.

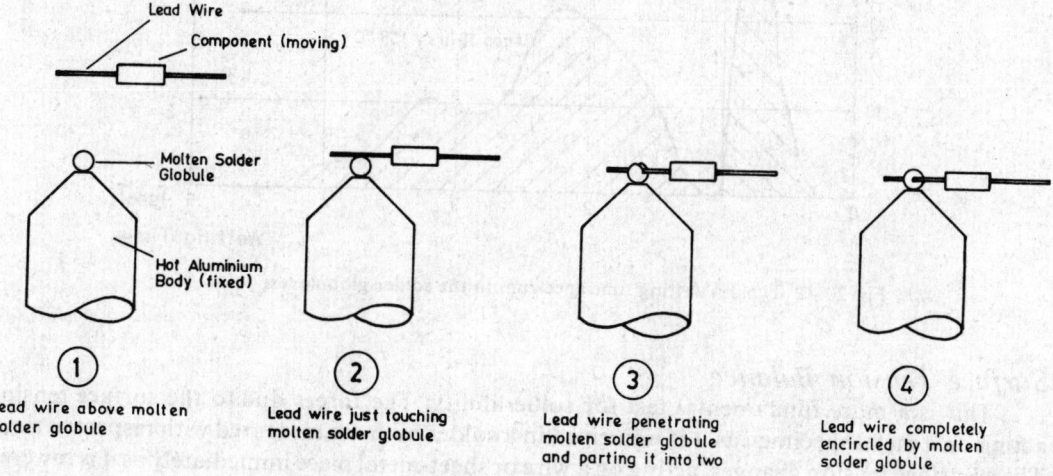

Fig. 20-12 Principle of the solder globule test

A globule of molten solder, its volume related to the wire diameter, rests on the flat top of an iron pin set in an electrically heated aluminium block of test temperature (e.g., 235 ± 2°C for

IEC test). The test wire, after prefluxing, is quickly lowered into the solder globule until it rests horizontally on the top of the heated block so that the solder globule is initially divided by the wire. The solderability of the wire is indicated by the time it takes for the molten solder to wet the surface and flow over the top of the wire.

The procedure, however, does not readily permit to detect a tendency for de-wetting. Wetting times exceeding about 2 seconds indicate a solderability which may not be satisfactory for mass soldering. To simulate a prolonged natural storage for the test pieces, accelerated ageing of the wires for 16 hrs at 155°C in air (IEC 68-2-20, BS 2011) or suitable steam ageing (MIL-STD 202D, BS 1972) may be used before solderability testing. Such ageing procedures will not significantly change the wetting time of good-quality wires.

In Fig. 20-13, a typical wetting time spectrum is shown for solder plated wires when fresh and after an artificial ageing treatment: The wetting times increase considerably after a 'dry' treatment. A small number of wires have extremely long wetting times which would represent the small proportion in a given production batch which is 'unsolderable'.

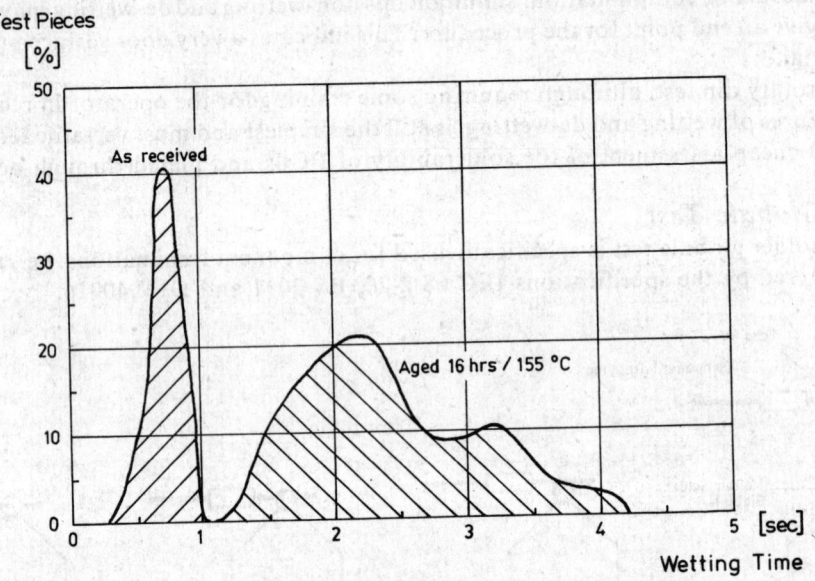

Fig. 20-13 Typical wetting time spectrum in the solder globule test

Surface Tension Balance

This is a more fundamental test for solderability: The forces due to the surface tension acting on a metal specimen as it is immersed in a solder bath are measured with respect to time. These surface tension changes, acting on a wire or sheet-metal piece immediately as it is lowered into molten solder are electronically amplified and recorded on a strip-chart recorder. Initially, before the specimen becomes wetted, an Archimedean upthrust acts on it; as wetting occurs, this upthrust is counteracted by an increasing downward pull due to surface tension forces until equilibrium is achieved with a stable solder meniscus formed on the specimen.

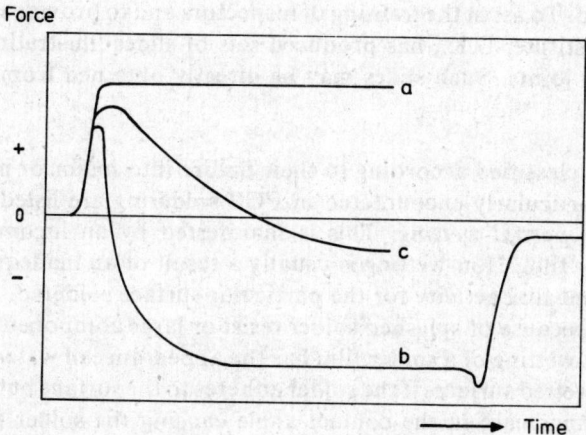

Fig. 20-14 Typical force-time curves obtained with a surface tension balance

Fig. 20-14 shows different force-time curves which may typically result in the surface tension balance test:

Curve *a*: Positive force persists (upthrust); no intercept by curve of zero-force line means no wetting.
Curve *b*: A very short time before force becomes negative means good solderability.
Curve *c*: If a fairly long time before wetting occurs, mediocre solderability is indicated.

Meniscus Rise Test

When a wire or sheet-metal specimen is immersed in a bath of molten solder, the formation of a meniscus of solder on the sample is observed through a microscope. The microscope has a graduated eyepiece to measure the height of rise of the meniscus at equilibrium or after a given period of time: This height can then be used as a measure of the solderability.

Soldering Iron Test

Where assembly methods require the use of manual iron soldering, it is sometimes decided to conduct soldering tests with an iron of specified wattage, bit size and bit temperature. This can give a useful guide to the likely performance on the production line, provided that such tests are carried out by experienced personnel.

20.8.2 Inspection

Visual examination may be made with the aid of a magnifier. A high magnification is not essential and may, in fact, be undesirable in order to minimise the effect of surface markings that may become visible and counted as defects in the soldered joints. The magnification should be the minimum that is suitable. Magnification of 2× to 5× are adequate for normal purposes but 10× to 15× may be required for micro-circuits.

Visual assessment of soldering quality demands inspectors trained in the standards of quality to be attained. To assist the training of inspectors and to provide standards for industry, the Tin Research Institute, U.K., has produced sets of slides illustrating good and defective soldering and solder joints. Such slides may be directly obtained from there.

Types of Defects

Defects may be classified according to their nature into major or minor defects. Some of the major defects particularly encountered in PCB soldering are listed here:

Non-wetting or partial wetting: This is manifested by an incomplete coverage of the surface by the solder film. Non-wetting is usually a result of an inadequate time-temperature cycle or of insufficient flux activity for the particular surface soldered. Non-wetting may also arise in small areas because of splashed solder resist or large components acting as a heatsink.

De-wetting: De-wetting of a solder film has the appearance of water on a greasy surface. It arises with initially wetted surfaces if the solder adheres to the surface but retracts after a certain time due to a rapid increase in the contact angle causing the solder to collect into discrete globules and ridges. Although the remaining base metal surface retains the colour of solder, this solder layer is thin and the surface has poor solderability.

De-wetting arises from a contaminated base metal surface, for example because of embedded cleaning abrasive particles. Sometimes, also an excessive thickness of the intermetallic-compound layer formed between the solder and the base metal may cause de-wetting of the overlying solder.

In extreme cases of surface contamination, non-wetting and de-wetting occur simultaneously; it is then most unlikely that any time-temperature cycle to produce good soldering can be discovered.

Poorly filled joints: An incomplete fillet formation or insufficient solder filler metal in the joint are usually manifestations of other defects such as poor wetting of the surfaces or excessive clearances between the joint members thus inhibiting capillary flow. In the case of soldering components onto PCBs, the incorrect spacing could be the result of excessive clearance between the lead and the hole (incorrect hole diameter) or off-centre positioning of the lead through the hole. Poorly filled joints may also result from incorrect fluxing or soldering conditions.

Other defects which may occur at various parts of the solder joints are
—Excess solder
—Bridging
—Icicles
—Porosity
—Overheated joints.

Most of these defects can easily be recognised and corrective action can be taken as these originate due to defects in the soldering technique.

Solderability of PCBs

In mass soldering of PCBs, a large number of joints have to be soldered simultaneously in one operation. Conflicts arise from the fact that many of the joints to be made have specific differences, yet they receive the same fluxing, heating, solder contact time, etc. Such differences may be in
—the metals to be joined
—the cleanliness of surfaces and speed of wetting

—the size and shape of the gaps to be penetrated and filled
—the facilities for a rapid dispersal of gases
—the thermal capacity of the joint and the rate of heat conduction from the joint
—the spacing between adjacent joints
—the physical shape of the joint.

If not kept under observation and control, the difference may readily vary and produce those catastrophic periods which every factory experiences when soldering quality suddenly deteriorates. To achieve good solder joints on PCBs, it is therefore necessary that the parameters below are carefully studied and controlled.

Solderability of the PCB surface: During the manufacturing process, PCBs are likely to be contaminated with corrosive substances unless the PCBs have been perfectly washed. Impurity traces of halides such as chloride ions particularly have to be faced often: They cause discoloration and lower the solderability. This phenomena is independent of the type of protective lacquer present on the PCB, as the corrosion starts below this protective layer.

Therefore, to improve and to retain the solderability of non-plated PCB surfaces, the following 2 factors must be carefully controlled to enable a storage life of 3–6 months for normal PCBs:

—Adequate rinsing of the PCBs during manufacture. The last rinse should be in de-ionised water.
—The protective coating must have an adequate thickness and should not corrode the copper surface itself.

Solderability of terminals: The excentricity of the tin or tin-lead coating on the lead wires of the components may cause major soldering problems: The copper from the substrate forms a copper-tin compound underneath the tin coating. This copper-tin compound gets exposed through the pores of thin top tin coatings thereby reducing the surface solderability to zero. For brass substrates, it is the migration of zinc through tin coating which causes the same problems. To overcome such drawbacks, it is necessary to use electroplated wires with a barrier layer.

Points of Concern in Mass Soldering of PCBs
—Correct thickness and chemical activity of the flux
—Proper soldering technique with respect to solder composition, bath temperature, angle of PCB dipping, etc.
—Proper pre-heating of the boards
—Adequate heat storage capacity of the soldering equipment
—No damaging of copper around the holes, before and during assembly
—Limited lead wire length under the PCB
—Correct lead to hole diameter (hole diameter 0.1–0.3 mm larger than lead diameter)
—Correct tag to hole ratio
—No movement of the components with respect to the PCB during the solidification period of the molten solder on the PCB
—Minimum warpage of the PCB.

With an average hole density of approximately 200 holes per dm^2 on the PCB, it is necessary to solder about 100 components at any one time within 2–3 seconds. Touch-up jobs are very expensive and must be avoided as far as possible. Awareness of the various soldering parameters as stated above, design characteristics, proper PCB manufacturing and soldering equipment, etc., will substantially lower the necessity of rework.

20.9 SAFETY, HEALTH AND MEDICAL ASPECTS IN SOLDERING PRACTICE

20.9.1 Safety

The temperatures employed in soldering techniques are high enough to cause significant burning of human skin. This fact must be recognised by all soldering operators and supervisors. Accidental contact with hot solder wire, molten solder or soldered assemblies will immediately require medical attention.

In mass soldering, splashing of hot solder due to accidental dropping of ingots or careless handling gets frequently reported. In order to protect the eyes from such damages, it should be compulsory to wear suitable protective eye glasses or face shields. Traces of moisture may also produce an instant steam formation causing severe splashing of solder. The contact of water, in any form, with liquid solder must therefore be strictly avoided. The use of heat-resistant gloves and armlets is very much advisable. To avoid accidents, sign boards with the letters 'DANGER HOT' must be put at suitable places.

20.9.2 Health and Medical Aspects

In soft soldering, the base metal, solder and flux are heated to 200–400°C. Under this condition, some of the constituents are vaporised or dissociated; the rising warm air becomes impure as a vapour or more specifically as an aerosol, consisting of particles of 0.01–1 μm in size. Whether health is affected by inhalation of these particles depends completely on the nature and concentration of the matter. As a standard for such effects, the experts resort to the *MAC-value* (Maximum Admissible Concentration): This is by definition the concentration which a healthy person can inhale for 8 hrs a day without incurring permanent adverse effects. Unfortunately, MAC-values are not known for all such matters and certainly not for complicated mixtures; the situation must therefore be judged upon day-to-day experience.

Base Metal

The composition of the base metal is of little importance since it is only the surface layer which is brought to the highest temperature and which can vaporise. The vapour pressure of most solderable metals such as copper, nickel and zinc remains far below the MAC-value at 250°C. The vapour pressure of cadmium, however, exceeds the MAC-value 0.1 mg/m^3 at about 150°C.

At soldering temperatures of approximately 250°C, the cadmium concentration becomes dangerously high. For this reason, cadmium plated parts or use of cadmium in solders have been limited in most of the countries by law.

The organic coatings sometimes used on the parts to be soldered (e.g. enamel, oil, etc.) should also be looked into carefully: The dissociation or decomposition of these coatings could give rise to objectionable fumes.

Solder

Apart from cadmium, lead contamination is the major source of danger with a MAC-value of 0.2 mg/m^3. In mass soldering techniques, there is practically no danger of reaching the

MAC-value because of a solder temperature of only about 250°C. However, higher soldering temperatures are usually applied in manual soldering and problems may arise. As a basic recommendation, vigorous flow of fresh air must be provided.

The addition of thallium to solder should be restricted as this metal is very toxic.

Fluxes

Corrosive fluxes containing strong acids can cause irritation to eyes and nose. The contact with the skin may also result in deramatities. Therefore, contact of the skin with corrosive fluxes must be avoided.

Non-corrosive (rosin based) fluxes are non-toxic. Most of the solvents used for such fluxes have also a very high MAC-value with the exception of methyl alcohol (MAC-value of 260 mg/m^3); methyl alcohol should therefore not be used as a solvent for rosin based fluxes.

The decomposition of rosin based fluxes could give rise to a certain characteristic smell which will cause nausea or vomiting to some persons. This danger, however, is likely to affect very few people and as such could be well ignored.

Note: Although the vapours from flux decomposition are not very harmful, it is better not to inhale them. Therefore, the vapours must be channeled outside. Where this is not possible, the vapours should be blown away so that less impure air is inhaled.

BIBLIOGRAPHY

1. AINSWARTH P.A.: 'Solderable finishes for electronic assembly', *METAL FINISHING JOURNAL*, Vol. 19, 1973, pg. 114–117.
2. BAILEY G.L.J. AND WATKINS J.C.: 'The flow of liquid metals on solid metal surfaces and its relation to soldering, brazing, and hot-dip coating', *JOURNAL OF THE INSTITUTE OF METALS*. Vol. 80/1951, pg. 57–81.
3. BUD P.J.: 'Testing of solderability—an international review', *EIPC CONFERENCE PROCEEDINGS*, Zurich, Nov. 1970.
4. COOMBS C.F.: *PRINTED CIRCUITS HANDBOOK*, McGraw-Hill Book Co., New York, 1967.
5. DISQUE F.C.: 'The use of rosin and activated rosin fluxes', Symposium on Solder, *A.S.T.M.—S.T.P.*, No. 189/1956, pg. 71–79.
6. DUNN B.D.: 'Producing highly reliable joints for spacecraft electronics', *ELECTRONIC PRODUCTION METHODS AND EQUIPMENT*, March/April 1978.
7. JOHNS A.A. AND MILLER E.S.: 'Dip soldered printed circuit joint characteristics', Symposium on Solder, *A.S.T.M.-S.T.P.*, No. 189/1956, pg 115–126.
8. LEWIS W.R.: 'The action of fluxes that assist tinning and soldering', *TIN AND ITS USE*, No. 72, 1966, pg. 3–6.
9. MANKO H.H.: 'How to choose the right solder flux', *PRODUCT ENGINEERING*, June 1960, pg. 43.
10. MILNER: 'A survey of the scientific principles related to wetting and spreading', *BRITISH WELDING JOURNAL*, March 1958.
11. NEKERVIS R.J.: 'Tin and its alloys', '*INDUSTRIAL AND ENGINEERING CHEMISTRY*, Oct. 1953, pg. 2253–2260.

12. *PHOTOGRAPHIC GUIDE TO SOLDERING QUALITY*, I.T.R.I. Publication No. 555, International Tin Research Institute, London.
13. THWAITES C.J.: *SOFT SOLDERING HANDBOOK,* I.T.R.I. Publication No. 533, International Tin Research Institute, London 1977.
14. THWAITES C.J.: 'Some effects of abrasive cleaning on the solderability of printed circuits', *METAL FINISHING JOURNAL,* Sept. 1968.
15. THWAITES C.J.: 'The philosophy of quality and soldering operation', *CIRCUIT WORLD*, April 1975, Wela Publication Ltd., U.K., pg. 3.
16. THWAITES C.J.: 'The solderability of some tin, tin alloy and other metallic coatings', *TRANSACTIONS OF THE INSTITUTE OF METAL FINISHING.* Vol. 36, 1959, pg. 203.

21
COMPONENT ASSEMBLY TECHNIQUES

The careful assembly of the PCB is as relevant for the final equipment reliability as the circuit design or PCB design and fabrication. When going through equipment failure statistics, this fact can easily be verified.

Assembly techniques can vary widely from case to case. While small-scale production is fully based on manual performance, automation plays an important role in high-volume assembly, especially in countries where manual labour is a main cost factor. Assembly techniques also undergo changes in the course of time with the accumulation of experience and with the introduction of new component types. A review on assembly techniques can therefore give an account of the present state of the art.

To fit the scope of this book, the emphasis will be laid more on smaller production volumes with manual assembly techniques.

21.1 PREPARATION AND MOUNTING OF COMPONENTS

The guidelines given here have mainly the purpose to minimise the cracking of solder joints due to mechanical stress on the joint. Mechanical stress on a solder joint is mostly caused by mechanical inputs but also thermal effects do lead to the same stress situation. Adequate precautions taken during the PCB design stage can be an effective remedy to minimise stress on solder joints. The optimum component preparation as well as the mounting has to be planned and considered hand in hand with the PCB design. In any case, hole diameters should at the maximum be 0.5 mm more than the lead diameter for non-plated holes or 0.7 mm for plated-through holes to give still reliable solder joints.

21..1.1 Component Lead Preparation

The bending of the axial component leads is done in a manner to guarantee an optimum retention of the component on the PCB while a minimum of stress is introduced on the solder joint. In no case should any damage to the component or its leads be caused while bending the leads; any stress on the component leads where they directly come out from the component body (component-lead junction) has to be avoided. The lead bending-radius should be approximately two times the lead diameter. The bent leads should fit into the holes perpendicular to the board so that any stress on the component-lead junction is minimised.

Fig. 21-1 shows a few of the very common shortcomings which provide stress on the component as well as on the solder joint. Therefore suitable bending tools must be made available for easy but perfect component preparation. A simple hand-held device for such purposes is suggested in Fig. 21-2. In this device, a change from one bending slot to the next will automatically give 2.54 mm or 0.1" increased spacing between the bent component leads to take care of hole patterns in a grid system on the same dimensional base.

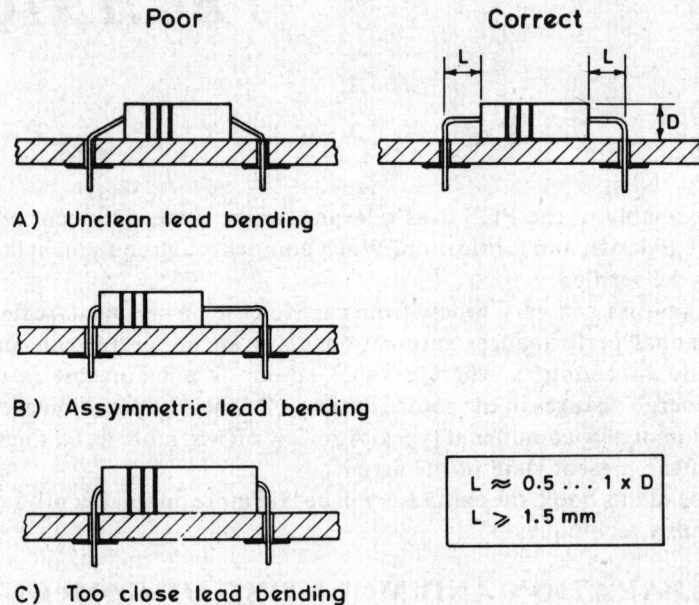

Fig. 21-1 Lead bending for axial 2-lead components

Polarised 2-lead components (e.g., electrolytic capacitors) have their leads bent in a manner to show the polarity symbol on the top, after mounting, for easy readability.

Where large quantities of boards have to be assembled, automatic lead bending machines are employed with a throughput of thousands of components per hour. They are available in different degrees of sophistication and automation. For small equipment, the components have usually to be supplied in belt form with both the lead-ends sticking on strips of adhesive tape.

21.1.2 Component Mounting

Components are basically mounted on only one side of the board. In double-sided PCBs, the component side is usually opposite to the major conductor pattern side, unless otherwise dictated by special design requirements.

COMPONENT ASSEMBLY TECHNIQUES

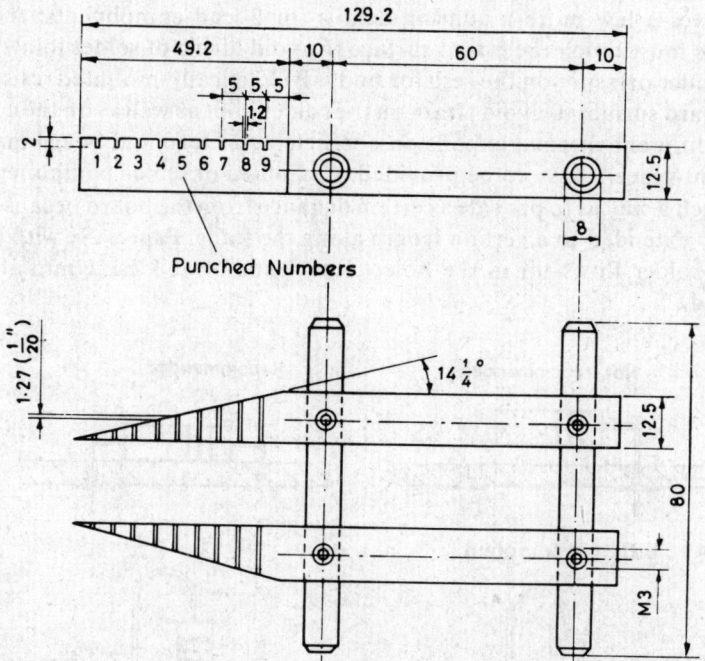

Fig. 21-2 Simple axial component lead bending device

Nonpolarised 2-lead components are mounted to give the marking or colour code the *same orientations* throughout the board (Fig. 21-3). The component orientation can be both horizontal as well as vertical but uniformity in reading directions must be maintained.

The uniformity in orientation of polarised components (diodes, capacitors, transistors, ICs, etc.) is determined during the design of the PCB. Its importance is not only for assembly convenience but also later during testing or servicing.

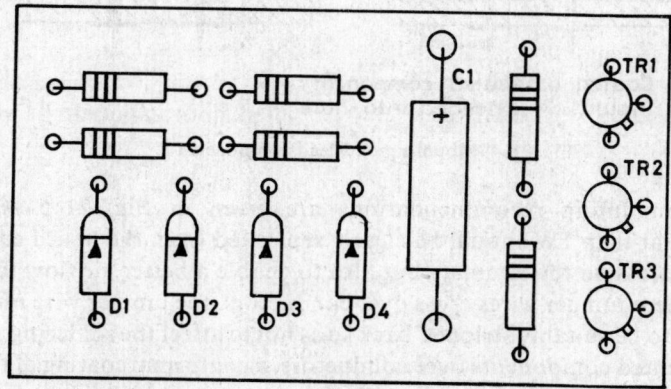

Fig. 21-3 Component orientation to be kept uniform

Fig. 21-4 gives a few more mounting details for 2-lead components: A) Horizontally-mounted resistors must touch the board surface to avoid lifting of solder joints along with the copper pattern under pressure on the resistor body. B) Vertically-mounted resistors should not be flush to the board surface to avoid strain on the solder joint as well as on the component-lead junction due to different thermal expansion coefficients of lead and board materials. Where necessary, resilient spacers have to be provided. C) Coated or sealed components too, have to be mounted in such a way as to provide a certain distance from the board because the insulation coating is usually extended to a certain length along the leads. Especially with plated-through holes where the solder flows up in the hole, clean leads of at least 1 mm above the board are recommended.

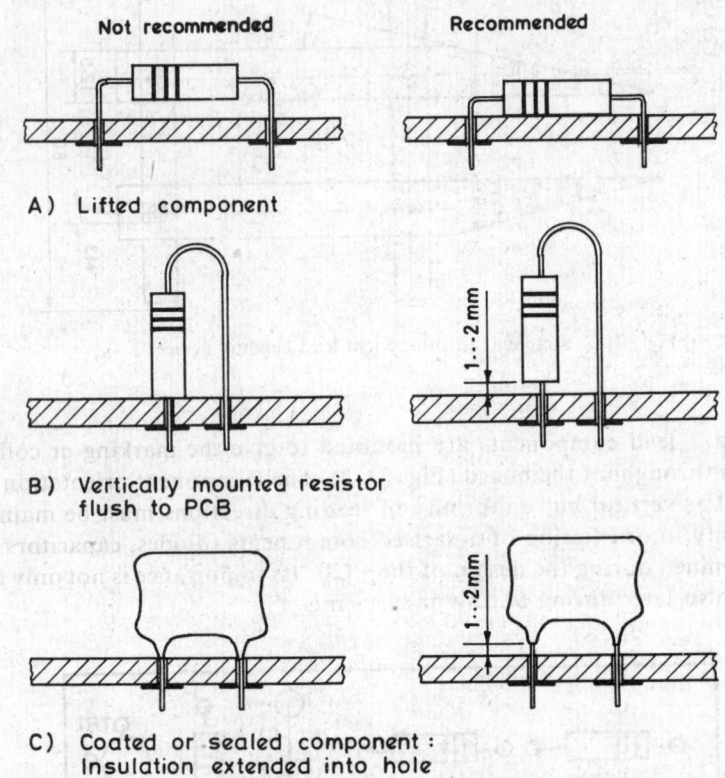

Fig. 21-4 Mounting of 2-lead components

Some more mounting recommendations are given in Fig. 21-5 A:. Components dissipating more heat than 1W should be clearly separated from the board surface. This is to minimise thermal stress on the laminate but also to enable a better air flow for cooling of the component. B) Where jumper wires cross over conductors, the jumper wire must be insulated. The insulation has to be suitably stripped back so as not to affect the soldering. C) In the case of flush to board mounted components over conductors, a conformal coating of the finished PCB is recommended to prevent the formation of moisture traps which could be harmful to the electrical functioning of the circuit.

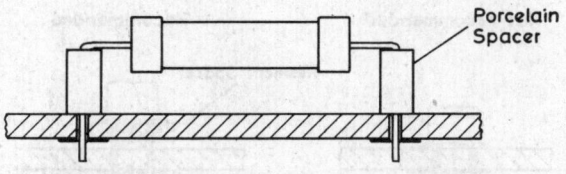

A) Heat producing component separated from board surface

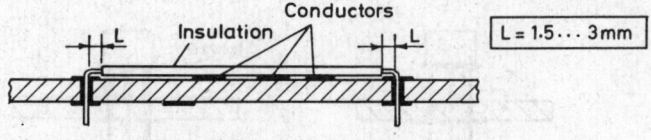

B) Jumper wire insulation stripped back

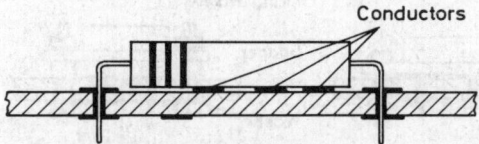

C) Conformal coating to prevent conductor bridging by moisture traps below components

Fig. 21-5 Various component mounting recommendations

Transistor mounting should never be done flush to the board: This could give considerable stress on the solder joints and to the lead junctions besides the possible overheating during the soldering operation (Fig. 21-6). There is a wide range of resilient spacers and spreaders available. With single-sided PCBs, the use of sleeves is also a common practice. If large transistors (e.g., TO-3 case) have to be mechanically secured to the board, spacers, washer and spring washers are preferably used as indicated in Fig. 21-6 C.

Mounting of integrated circuits: To enable an easy insertion, the leads of integrated circuits in TO-can or dual-inline-package (DIP), the leads have usually to be readjusted. Simple jigs are available for this purpose, for each particular package type.

Mechanical securing of components must be considered as soon as the component weight is more than 10 g. This is not only to withstand the shocks and vibrations in the ordinary use of the electronic equipment later, but also to survive the transport and shipment handling prior to the final use. Special clips, clamps and brackets are available in a wide variety for such purposes.

Straight-through mounting or lead clinching: The simple straight-through mounting of components (Fig. 21-7) is compatible with the common mass soldering techniques. The replacing of straight-through mounted components is simple since they can be readily removed after heating up the solder joints. Straight-through mounting is preferred for single-sided PCBs because it provides an optimum 360° symmetrical solder joint. A drawback is the necessity to hold the component firmly in its position while soldering it.

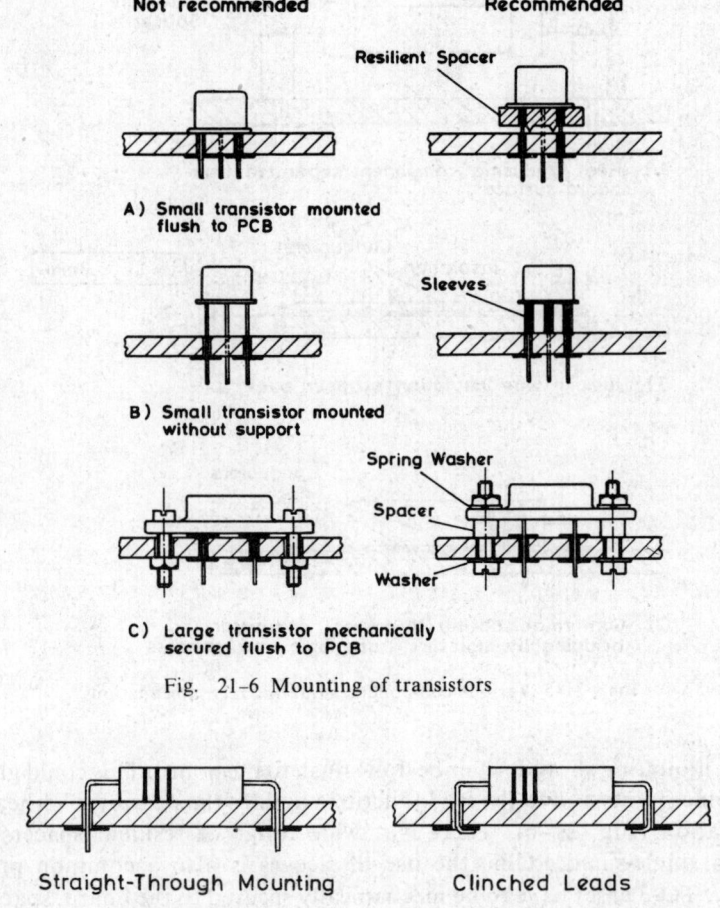

Fig. 21-6 Mounting of transistors

Fig. 21-7 Straight-through mounted and clinched component leads

The clinching of leads means an additional operation after the leads are in proper position for soldering. But clinching avoids the need for special pressure soldering frames. Also, hole-to-lead clearance becomes less critical for soldering and the more the clearance, the easier will be the insertion of the component. In addition, plated-through hole solder connection reliability is improved due to the larger wetted area and better mechanical contact. As a rule, the clinched lead portion should not be extended beyond the annular ring but it is an accepted practice to extend it further along conductor tracks.

Lead preforming: A compromise between straight-through mounting and clinching is given with appropriate component lead preforming. The dimple offset- and dimple lead preformations are shown in Fig. 21-8. Other shapes are used as well. A good component retention on the board prior to soldering is achieved while the advantages of straight-through mounting are still maintained. In addition, the leads are cut to final length in the same operation. Therefore, lead preforming with semi-automatic equipment has become very popular with bulk electronic equipment manufacturers.

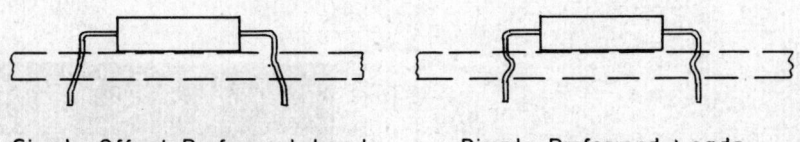

Fig. 21-8 Component lead preforming

21.2 ORGANISATION OF NON-AUTOMATIC PCB ASSEMBLY

To organise an efficient assembly means to care about innumerable factors on an individual base. There are no two identical assembly organisations. Not only must technical factors be considered but also human factors play their role and directly affect the reliability of the operations. In spite of such irregularities, there are a few commonly accepted practices which should be mentioned here.

Workplace convenience: The board to be assembled is held in a suitable frame right in front of the operator. Slight tilting of the board of approximately 15° is often preferred. The sorted components are in easy reach as well as the insertion tools, if required. The components are kept in bins or trays in the order of assembly sequence. Special rotary bins or rotary trays have been designed considering the ergonomical aspects.

Assembly sequence: A preferred methodology is the assembly sequence from the right top to the left bottom corner of the PCB. But in any sequence followed, it is a must to keep it the same at every assembly workplace and not to change it after some time. This is to minimise the learning time required by the operators. In a first run, the components that are not preformed are assembled, followed by the preformed components. Only thereafter, the more voluminous, heavier and mechanically secured components are assembled.

Assembly splitting: For PCBs with a large number of components, the assembly is usually divided among different operators. This does not apply for prototype assembly where usually the same operator finishes the assembly on an individual base. The maximum number of different components for one operator should not be more than 20. Above this number, the fault probability will go up hand-in-hand with the increasing component number.

Visual assembly aids: When switching over from one board type to another, the operator's learning time should be as short as possible. The assembly information to the operator, therefore, has to be provided in a suitable manner. It can be simply in the form of an assembled sample board but assembly drawings or photographs are preferred in which the related components are predominantly marked or coloured along with sequence numbers. More sophisticated aids include film or slide projections illuminating the component position on the board while simultaneously only the related component is made accessible in the component storage device.

Inspection: If an assembly error remains undetected, it will come to light only during the electrical testing of the board. To rectify it at this late stage will be a fairly troublesome and costly procedure. Because of such reasons, it is a common practice to have the assembled boards checked up to 100 % prior to soldering. This will also take care of the experienced fact, that once a fault occurs, a repetition is very likely to occur again. The assembly inspector has therefore to be located directly at the end of the assembly line for immediate inspection of the assembled boards. Any fault has to be set right directly by the operator concerned.

Fig. 21-9 Hand-assembly frame (Photo CEDT)

A suitable aid for an efficient inspection is a transparent plastic foil on which the component outlines are shown. Besides checking the presence of all the components, the inspection must also include verifying of component polarity, orientation, value and physical mounting.

21.3 LEAD CUTTING AND SOLDERING

The performance and reliability of solder joints give best results if lead cutting is carried out before soldering so that the lead ends get protected and also covered with solder and herewith contributing to the actual solder connection. However, lead cutting after soldering is still common in particular in smaller industries where hand soldering is used.

The specific aspects of soldering have already been covered in Chapter 20 of this book. Therefore only procedural steps in connection with lead cutting will be considered here.

In small-scale assembly, manual techniques with simple hand tools are common for all the assembly operations. The choice of a particular procedure is usually determined by the requirements of the PCB end user.

Typically followed manual procedures are:
A) INSERT-CUT-SOLDER
B) INSERT-CUT-CLINCH-SOLDER
C) INSERT-SOLDER-CUT.

If the production volumes are high (e.g., as typically for cheap mass-consumer electronics), the use of automatic wave soldering machines is economically justified. But the above procedures basically will still be applicable.

A further rise in production calls for concentration on the cutting operation as far as lead clinching is not a necessity. Depending on the range of components to be used, cutting can be eliminated by prebending/precutting of the leads prior to component mounting. On the other side, high-speed cutters may be employed. High-speed cutters are available, working on different principles such as circular saw, band saw, rotating or oscillating knife arrays. Among the various principles, circular sawing with tungsten-carbide inserts brazed into a steel disc seems to work most satisfactorily. Since automatic lead cutting exerts a certain pressure on the leads, an additional step to fix the components before cutting then becomes a must. This can be done first by dip soldering but also component fixation with wax or other stabilising systems are applied.

Procedures with mechanised cutting for high-volume PCB assembly:
A) INSERT-DIP SOLDER-CUT-WAVE SOLDER
B) INSERT-WAVE SOLDER-CUT
C) INSERT-STABILISE-CUT-DESTABILISE-WAVE SOLDER.

The equipment used for cutting has to be well maintained in order to provide the desired clean cutting without bending the leads. This includes a frequent resharpening of the tungsten-carbide cutting edges. The lead length after cutting is set to be 1–3 mm above board.

The latest development in high-volume PCB assembly is to use automatic insertion equipment which simultaneously performs lead preforming and, if desired, lead cutting after insertion and also clinching. However, a more detailed description of assembly automation is beyond the scope of this book. Besides, assembly automation is economically justified only with very large production volumes and, in the author's view, should never be attempted in countries with considerable unemployment problems.

21.4 PCB CLEANING AFTER SOLDERING

With the soldered PCB, many contaminants can be found which may produce difficulties with the functioning of the circuit. The problems usually arise at a much later date rather than during the final functional testing of the board in the factory. Among the contaminants, we can typically find flux, chips of plastics, metals and other constructional materials, plating salts, oils, greases, environmental soils and other processing materials.

The following performances are expected from the cleaning procedure with the appropriate cleaning medium:
— Dissolution or dissolving of organic liquids and solids, e.g., oils, greases, resin flux
— removal of plating salts and silicone oils
— displacing of particulate and other insoluble matters, e.g., chips, dust, lint
— no severe attacks on boards and components to be cleaned, no alteration of ink or paint notations and last but not least, compatibility with healthy environmental working conditions.

A wide range of cleaning media are commercially available; they usually form a certain compromise on the desirable attributes. They are preferably used with rinsing equipment or ultrasonic tanks. When using these chemicals, properly designed air-suction facilities are a must, besides the provision of a vigorous flow of fresh air.

Chemicals mostly used in the cleaning media:
— acetone
— alcohols
— aliphatic hydrocarbons
— aromatic hydrocarbons
— fluorinated hydrocarbons
— trichlorotrifluorethane
— de-ionised water plus detergents.

Each one of these chemicals offers advantages as well as disadvantages. Only a careful testing in relation with the final requirements will reveal the economically optimum choice.

A few guidelines will nevertheless be given here: As balanced-performance solvents we can name methyl alcohol, ethyl alcohol, acetone and some of the chlorinated solvents. The most stable organic cleaning media are trichlorotrifluorethane and fluorcarbon blends; they are also compatible with all the materials normally used for boards, components and notations Where a strong solvency is required, aromatic hydrocarbons are employed. For simple resin flux removal, alcohols are preferably used. Grease and wax removal requires the strong solvent power of chlorocarbons but damages can be caused to plastics, elastomers and notations.

BIBLIOGRAPHY

1. BOYNTON K.G.: 'Recent developments in automatic component lead trimming methods and materials and GBS—a new concept in wave soldering', *PROCEEDINGS OF THE WORLD PRINTED CIRCUIT CONVENTION*, Volume II, London, 1978
2. COOMBS C.F.: *PRINTED CIRCUITS HANDBOOK*, McGraw-Hill Book Co., New York, 1967.
3. HERRMANN G.: *LEITERPLATTEN*, Leuze Verlag, Saulgau/Wurtt. (Germany), 1978.
4. JOWETT C.E.: *RELIABLE ELECTRONIC ASSEMBLY PRODUCTION*, Business Books Ltd., London, 1970.
5. *PRINTED BOARD COMPONENT MOUNTING* (Revision A), Publication No. IPC-CM-770A. The Institute for Interconnecting and Packaging Electronic Circuits, 1717 Howard Street, Evanston, IL. 60202, March 1976.

22
GUIDELINES FOR STARTING PCB FACILITIES

22.1 INTRODUCTION

In this chapter, particular attention is given to the planning and starting of small-scale PCB facilities with limited funds available. With the growing electronics market in India, there is still a good scope for small and flexible PCB manufacturers which can turn out highly precise PCBs for professional equipment manufacturers. Also in the consumer type of PCB market, new requirements are created in states where consumer electronics production is introduced. There are also dozens of engineering colleges which could greatly benefit their electronics engineering students by operating simple in-house PCB facilities. We shall therefore concentrate in this chapter on the production of single- and double-sided PCBs without plated-through holes. However, facilities for the production of plated-through holes may still be added at a later stage, after sufficient basic experience has been acquired and a market position has been established.

The planner of new PCB facilities is confronted with a wide range of questions which have to be dealt with on an individual base: There are no two PCB set-ups alike although the final PCBs of one manufacturer might appear almost undistinguishable from the ones of the other manufacturer. There are many questions of a technical nature but they must be solved hand-in-hand with economic considerations. With the use of harmful chemicals control of pollution is a requirement which directly influences both the technical as well as the economic sides. To fight pollution is therefore included in the planning right at the beginning and not as one of the less urgent or secondary questions.

22.2 MARKET CONSTRAINTS

It is a basic advantage to locate a PCB industry geographically near the customers to be served. With respect to the conditions in India, where electronics industries are more or less centralised in a few places, a close proximity or at least a short and direct transport link to the particular centre to be served is an essential requirement. Also, the planning of the *State Electronics Development Corporations* must be taken into account, since their efforts may create new PCB markets in a very welcome *decentralised* fashion.

A direct link to the customer does not only quicken the turn-around time for PCBs but it also facilitates an easy way to communicate with the customer. This need should not be underestimated; it is of particular importance in dealing with small-scale customers.

When the location for a PCB industry is under planning, it should be borne in mind that the materials required and final products to be supplied to customers do not require any railway link. PCBs are a product which can easily be transported by car or minibus. The material content in volume and weight is comparatively low while the content in labour and processing is considerable. This makes PCB industries well suited for a decentralised industrial growth, under the condition that electronics companies themselves are also growing in a decentralised manner.

A look shall now be given to the typical structure of PCB users. They can be roughly put into six categories:
1) Electronics company without any internal PCB facilities.
2) Electronics company with in-house PCB facilities for their prototype requirements. The PCBs required for equipment series production are ordered from outside.
3) Electronics company with in-house PCB facilities both for equipment series production as well as prototype requirement.
4) R & D establishment without any PCB facilities.
5) R & D establishment with in-house PCB facilities covering their entire needs.
6) Open market: General purpose boards, PCBs for hobby kits, etc.

Categories 3 and 5 are not directly of further interest here since they take care fully of their own needs. All the other categories depend on PCB suppliers.

It is now important for the planner of PCB facilities to know exactly in which categories there is a market gap. Is there mainly a need of PCBs for equipment series production or is also a prototype PCB service needed? It might also be that the series production needs are well taken care of but nobody can offer a 24 h-turn-around prototype board service. In Western countries, many of the now well established PCB manufacturers have started literally as a one-man-show for fast-turn-around PCBs with additional charges inversely proportional to the turn-around time.

Another concern, after the market gap has been identified, is to see whether consumer-type or professional-type of PCBs are predominantly required. *Consumer-type PCBs* are mostly single-sided boards for entertainment and mass consumer electronics. They are produced in large quantities with a minimum profit margin. Paper phenolic laminate is used because of price considerations. Consumer PCBs do not need the highest precision in production; screen-printing is therefore usually applied. *Professional-type PCBs* are mostly used in professional electronic equipment. Such boards have to be very precisely made, photoprinted, and have a double-sided pattern on glass epoxy laminate. The choice for the type of boards to go for or whether both should be included, depends further on the typical quantities required. When the series quantities are considered, check also the total annual requirements of the customers. Only once we know the typical spectrum of customers to be served with their present and future needs, can we proceed to the technical planning of the PCB set-up.

22.3 EQUIPMENT

22.3.1 Local versus Imported Make

After it has been decided whether to go for consumer or professional type of boards, suitable equipment has to be selected. One has to be well informed about the equipment generally available along with the relevant prices. The addresses given in the Appendix can support such a survey.

The pieces of equipment required can be put into three categories, depending on the origin:
1) Equipment readily available from local manufacturer.
2) Equipment not readily available locally but which could be locally manufactured if details and drawings are supplied.
3) Equipment which must be imported.

About the equipment in category 1, we can say that it is usually somewhat cheaper in price than the imported equivalent, mostly because of the taxes and duties on an import. Wherever possible, the local make should be given preference. This also serves as an incentive to the local manufacturer to improve the quality standard. Equipment items which are locally available, are only ordered from abroad if the local make cannot meet the specifications for the purpose envisaged.

An often considered compromise, to avoid the import of locally not available equipment, is to get it custom-made in a local company. In this case, the PCB industrialist has to know, in detail, about the functioning, materials and design of the equipment to be built. It will hardly give the desired end result if just a colourful leaflet with performance specifications of a foreign equipment is given to the local equipment manufacturer. A close cooperation between equipment user and equipment manufacturer is essential besides the professional expertise which should exist on both sides. Another point is that the equipment manufacturer should not be too far away from where the equipment is later used: Especially at the beginning when the equipment is put into operation, frequent trouble-shooting is very likely to be required.

The import of an equipment can only be considered if it is a must and if at all the necessary amount of foreign exchange is available. When an order for an import is prepared, spare parts in sufficient quantity for several years must be included! Do not believe all the promises of the local representative about the assistance given in case of a breakdown. You can rely only on your own provisions and ingenuity. A later import of spare parts takes often a very long time and considerable time of yours will be wasted for endless administrative procedures in addition to the nonavailability of that particular equipment.

In the selection of any piece of equipment, we have to keep in mind the abundant availability of labour at comparatively low salaries in India. Where the funds for the equipment are minimum, the question of automation cannot arise at all. But in cases with sufficient funds available, the planner of PCB facilities shares a high degree of responsibility of the long-term development of the nation. The awareness of such a responsibility gets reflected in deciding for the employment of manual labour and the exclusion of unduly automated processing equipment.

22.3.2 Simple Professional PCB Facilities

Where such a set-up is planned, the following points should first become clear:
— Quantity per series
— Total output annually
— Maximum size of boards
— Minimum conductor width and spacing required
— Surface finish typically used

Will plated-through hole facilities be added later?

These are all variables that change from case to case. It is not possible here to go into all the variations possible. Nevertheless, an idea of the basic equipment needed shall be given.

For professional PCBs, the use of photoresist is recommended because of the high accuracy of patterns to be produced. A screen preparation is also fairly expensive if only a small number of boards are required. The artwork is usually supplied by the customer and is prepared at a 2 : 1 or 4 : 1 scale. The film masters have first to be made for which a *reprographic camera* in a darkroom is a prerequisite. There are various types and sizes of reprographic cameras manufactured in India. Where the relatively high costs of such a ready-made camera have to be avoided, a less comfortable but still workable solution is to build such a camera with suitable means. Fig. 22-1 can give some ideas.

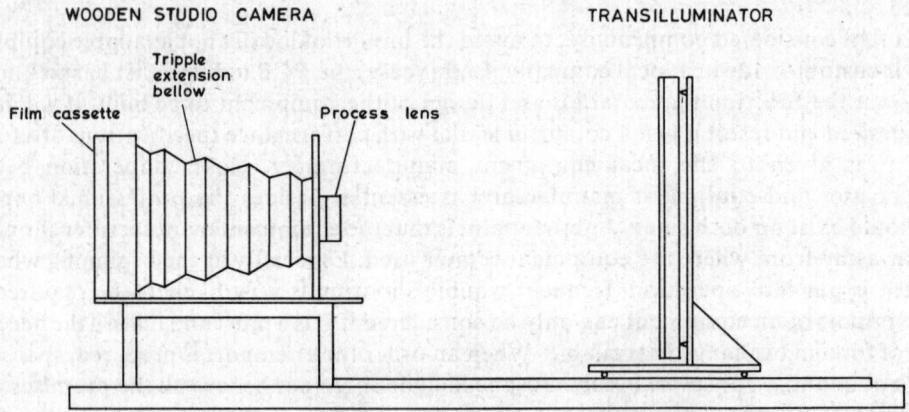

Fig. 22-1 Simple camera set-up for film master fabrication

In a smaller PCB set-up, only *manual board cleaning* processes should be considered. This involves only minor investment (tanks, electrically powered hand tools, compressed air, etc.). The cutting of the boards is carried out on a *hand-operated plate shear*. Only where larger quantities of laminate have to be cut, a motor-driven plate shear or a circular saw are further required.

Photoresist coating (wet-film) is preferably done by *dip-coating*. An alternative is spray coating since an *air compressor* has in any case to be made available. Dip coaters are commercially available; a simple slit tank with a speed-controlled withdrawal device can also be built. The use of dry-film resist is recommended where plated-through hole facilities are added soon. For the lamination with dry-film resist, a laminator has to be custom-made or imported.

For the exposure, a *UV exposure unit* with vacuum frame is recommended. A low-cost solution here could be to purchase a vacuum frame and two mercury arc lamps and to design and fabricate a suitable mounting oneself. To develop the resist image with wet-film resists, a simple *cascaded tank developing* will meet the requirements. For dry-film resists, a spray developing machine has probably to be imported.

In the etching process, both the equipment and the etchant are a point of concern. *Spray etching* is highly desirable; a substantial investment into the etching machine cannot be avoided here. Splash-type of etching machines are found to be limited in their practical use by insufficient etching regularity. Where smaller quantities of boards are etched, the laboratory type of spray etching machine can meet the purpose up to an annual requirement of approximately 10,000 PCBs. For larger quantities, conveyorised spray etching machines have to be employed. There are indigenous ones available in India, costing almost Rs. 100,000. Imported ones having smaller size and costing less, must also be evaluated. As an etchant, *cupric chloride* is the preferred choice. Only for small quantities of a few boards per day, ferric chloride can be envisaged.

Drilling of professional PCBs, mostly made of glass epoxy laminates, has to be done at speeds of typically 40,000 rpm with *tungsten-carbide drill bits*. Only imported drilling machines can meet such speed requirements. The drill bits too have to be imported. Machines for drilling by direct sight are cheaper than those for drilling by optical sight. The drilling of component holes is very time-consuming and very often, the overall PCB production capacity is limited here. Depending on the output, several drilling machines might be necessary. For paper-phenolic laminate boards and for small quantities of glass-epoxy boards, high-speed steel drill bits can be used on drilling machines with a speed range up to at least 15,000 rpm.

The *stripping* of the photoresist is again done by manual means. Suitable solvents, chemicals and abrasive aids are used.

Where gold plating of connectors is required, all the prerequisites are indigenously available in India. This is also valid for *nickel plating* which is recommended prior to gold plating to reduce the gold plating thickness required and hereby to save gold.

The simplest way of surface protection is spraying of a *solderable lacquer*. This, however is only a short-term solution. Other and better solutions are *electroless tin* or *roll-solder coating*. These requirements are usually specified by the customer.

Other items required, but not yet listed are the *refrigerator* for the storage of photoresist, films and chemicals; the *hot air oven* for baking of the photoresist after developing; *exhaust facilities* where solvents or other harmful chemicals are applied and finally also the *water deioniser*.

The essential equipment listed here in a minimum solution with partly self-built equipment will cost at least Rs. 100,000 but where all the equipment is bought ready-made up to Rs. 400,000 or even more is needed.

22.3.3 Simple Consumer PCB Facilities

In consumer type of boards, conductor widths are not less than 1 mm and mostly single-sided paper phenolic laminates are used. Here screen-printing will be the most economic method of pattern transfer, especially for the large quantities usually involved in consumer electronics boards.

Once the screen is made available, PCBs can be produced with very simple means involving only limited capital investment. Screen-printing is done with a simple *hand screen-printing frame*. The boards have to be cut priorly on a *hand-operated plate shear* or on a circular saw followed by *manual cleaning*. The screen-printed boards are allowed to dry and are kept for this purpose in specially designed *drying shelves*. The drying process can be accelerated by using a fan, warm air blower or drying oven into which the drying shelves are put. After the initial cleaning of the boards up to this stage, any dirt and dust could harm the results; clean rooms are therefore highly desirable.

Since etching of consumer type of boards is much less critical than the etching of professional type boards, processes like *bubble-tank etching*, *splash etching* as well as *spray etching* can serve the purpose. For a large throughput, conveyorised spray etching is usually employed: In such a case, a major share of the overall investment will be absorbed by the spray etching equipment.

Etchant: Here also the author basically recommends *cupric chloride*. It combines good general etching performance with easy disposal of the saturated etchant to chemical industries. Drilling of the paper-phenolic boards is done with *HSS drill bits* on indigenously available *drilling machines*. Such drilling machines are offered with speeds ranging up to 22,000 rpm. Where many thousands of boards of the same design have to be fabricated, *hole punching* will probably be more economic. A *punching press* must in this case be available and the *punching tool* has to be supplied from outside.

Many customers of consumer PCBs require a *solder-stop mask* on the PCB surface for economic wave-soldering of a large series of PCBs. The solder-stop mask is, again, produced with the screen-printing process. *Solderable lacquer coating* is a usual requirement for consumer boards. It is applied by spraying.

It is possible to start the production of consumer PCBs with an investment into equipment and tools of hardly Rs. 10,000 under the condition that the *film master* and the *screen* are prepared somewhere else. For film master and screen preparation facilities, a full-fledged darkroom with reprographic camera and a high-energy light source for screen exposure are the usual assets. But screen exposure can also be done in a suitable exposure frame using sunlight.

22.4 ROOM AND BUILDING REQUIREMENTS

A few guidelines shall be given about rooms and buildings in which the PCB equipment has to be operated. The desirable features for each room are mentioned; however, it is known that for very small facilities, less rooms or even only one room may have to serve all the operations.

22.4.1 Darkroom

The room size depends quite a bit on the type of camera chosen; Vertical-type cameras need little floor area but there are limitations in artwork size while horizontal camera set-ups are suited for large artworks but considerable floor area has to be provided. Besides the camera, also space for film processing, artwork storage, sinks etc., must be included (see also Fig. 10-14). Two sinks with running water are highly desirable: One is needed for film washing and the other one for hand washing. More details on how to implement a darkroom are given in Chapter 10.

22.4.2 Room for Photoresist Work

All the handling of photoresists is done in this room under subdued illumination (yellow-orange spectrum): This includes photoresist coating, PCB exposure, developing, dyeing and baking. In this room, running water has to be provided for rinsing of the boards after dyeing.

Special exhaust facilities are required for the place of wet-film resist coating and developing because of the solvent vapours produced. A sufficient flow of fresh air can otherwise usually not be achieved because of dust and illumination restrictions. A slight atmospheric overpressure to keep out dust has to be obtained by an inlet fan fitted with a suitable dust filtering medium.

Fig. 22-2 shows an exhaust hood with photoresist dip coater and developing tanks: The exhaust chamber can be completely closed with doors to avoid solvent vapours in the room when the exhaust fan is not switched on.

Fig. 22-2 Exhaust hood for a clean photoresist working environment (Photo CEDT)

22.4.3 Screen-Printing Room

In this room, too, dust and dirt must be minimised although it is less critical than in photoresist applications. At the same time, a rigorous fresh air flow is required to remove the vapours from drying screening inks and solvents. A frequent wet cleaning of the room is suggested and windows on opposite sides of the room should be kept open.

22.4.4 Etching/Plating Room

In this room, considerable quantities of strong and harmful chemicals are used. Special provisions have to be implemented in order to avoid, even under worst-case conditions, any of these chemicals reaching the sewer system. A first recommendation is to use acid-proof tiles for the floor and, as far as possible, also for the walls. In Fig. 22-3, two methods are shown to avoid any pollution of the sewer system in case of a tank breaking: A safety tank can be built around e.g., the etching machine with bricks and acid-proof tiles. Or a whole room can be made safe with a safety drain which is not directly connected to the sewer system. If a tank should break, the chemicals will get collected in this drain from where they can be removed. The drain, however, has also an outlet to the sewer system which is normally closed. Only while cleaning the room with water, will this valve be opened.

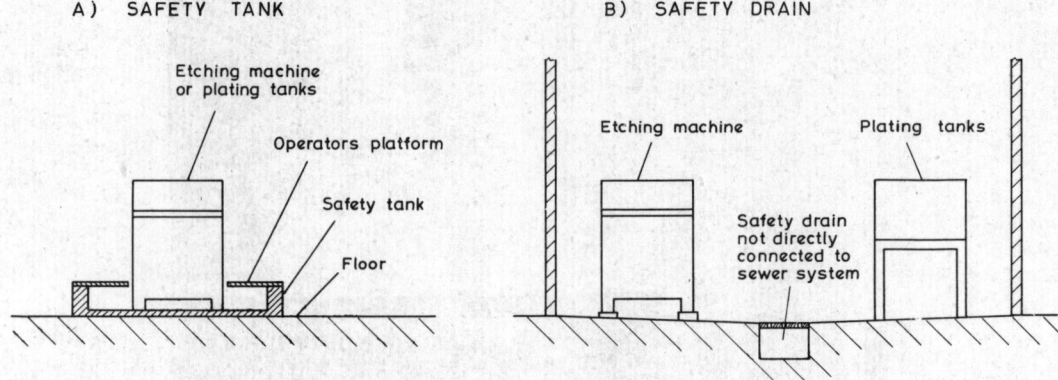

Fig. 22-3 Safety measures to prevent chemicals from directly reaching the sewer system

Exhaust facilities have to be provided for the places where harmful chemical fumes are produced. In any case, sufficient air circulation must be guaranteed under all circumstances: This is achieved with opened windows and with ceiling fans

22.4.5 Chemical Store Room

Where large quantities of chemicals have to be stored, an extra room must be made available. The store room should be located closely to the etching/plating room where most of these chemicals are used.

This room is again fitted with acid-proof tiles and provisions are made that no chemicals can reach the sewer system. In this room, we also store the spent etchant until further disposal. Continuous air circulation must be provided and no inflammable materials are kept in this room.

22.4.6 Mechanical Fabrication Room

In this room, operations like drilling, punching, laminate cutting and final board trimming are executed. The laminates, as supplied from the manufacturer, are also stored here.

For all these operations, very good room illumination must be provided. Dust collectors are recommended for the drilling machines and for the circular saw. Special care must be taken if solderable lacquer spraying is also done here: The drying should not be affected by dust. For cleaning of boards prior to processing, a separate and dust-free section is very important. Furthermore, exhaust facilities are required to remove solvent vapours.

We include here the plans of a small professional set-up for prototype PCBs as it has been implemented in an almost similar fashion at the Centre for Electronics Design Technology (CEDT).

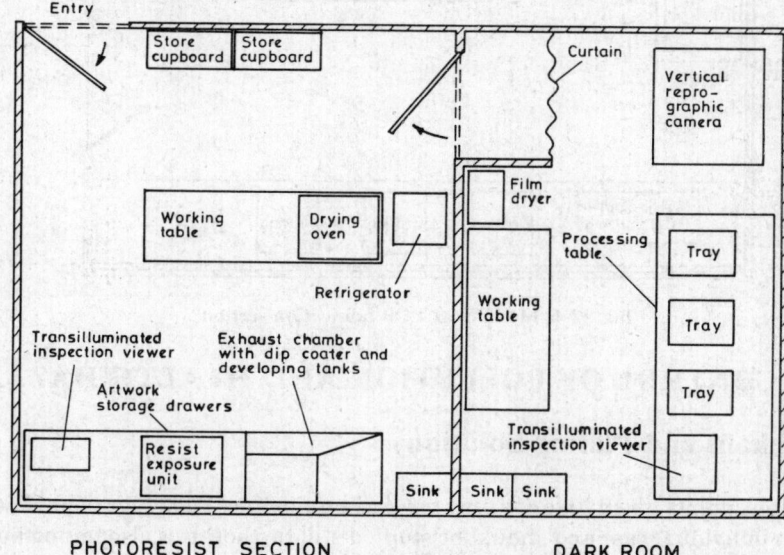

Fig. 22-4 Photoresist section and darkroom at CEDT

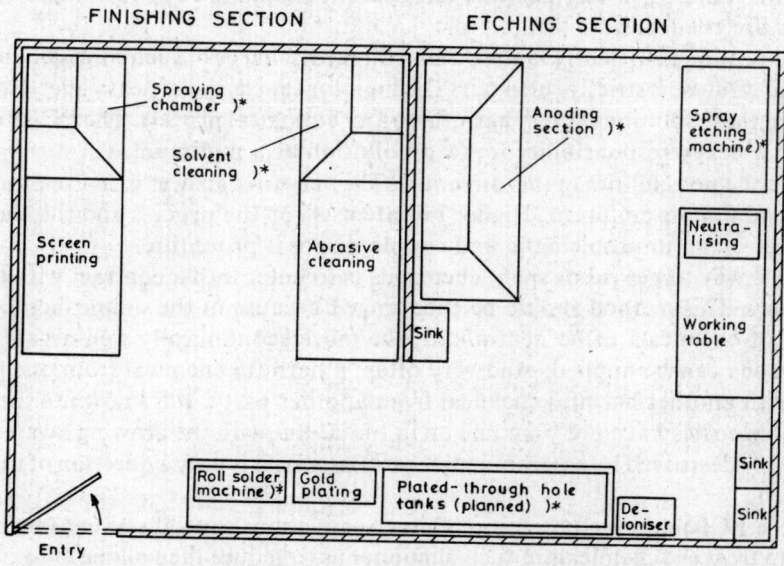

Fig. 22-5 Etching and finishing section at CEDT

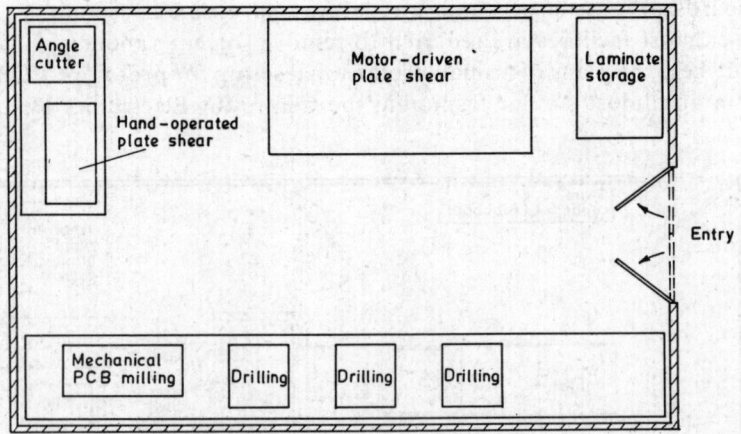

Fig. 22-6 Mechanical PCB fabrication section

22.5 CONTROL OF POLLUTION AND HEALTH HAZARDS

22.5.1 Etchant and Plating Solutions

The various points given here can only serve as a guideline: Each type of chemical to be disposed of is an individual case and should be studied still further. It is also important to observe that many chemicals cannot be mixed together without hazardous reactions taking place, such as a violent heat generation or the evolution of toxic gas. The prior advice of someone with a profound understanding of any possible implication should always be obtained before some new methods are tried out.

It should now be sufficiently known that a direct discharge of such solutions into the sewer system, even if diluted, is strictly ruled out: The high concentration of heavy-metal ions (e.g., Cu, Fe, Cr, etc.) in these solutions is very harmful to any biological process. The PCB manufacturer has therefore a heavy responsibility to keep pollution at a minimum.

What are the possibilities in this direction? The very first answer is to use processes which cause only a minimum pollution. Higher initial costs for the process and the chemicals may easily be offset by an unproblematic and simple disposal procedure.

A possible way to get rid of spent chemicals is to enter into a contract with a recognised disposal agency. This method should be encouraged because of the simple fact that the larger the quantity of chemicals to be neutralised, the more economically well-organized and less harmful methods can be applied. And very often, a harmful chemical from one party can be neutralised with another harmful chemical from another party. It is known to the author that such disposal agencies have not yet come up in India. But with the growing awareness of what gets irrepairably destroyed by environmental carelessness, it is only a question of time until they get established.

Where the PCB manufacturer himself has to care about neutralisation of spent chemicals, it is his task to treat the chemicals in such a manner as to reduce the volume of toxic material to the minimum. The aim is to get a minimum quantity of highly concentrated and water-insoluble sludge. This sludge is then buried in specially selected dumping sites where no ground water can be affected. This procedure must be considered as a worst-case treatment. Luckily, in many of the saturated chemicals, there are some byproducts which can be chemically separated and used again after suitable processing. Such techniques are gaining popularity also because of the constantly increasing prices for metals and other materials which can, in this way, be reclaimed.

The basic process of heavy-metal precipitation is shown in Fig. 22-7. Where most of the chemicals to be disposed of are of acidic type, the adding of sulfuric acid becomes superfluous. The principle of heavy-metal precipitation is to raise the pH value to a certain value where all the heavy metals like Cu, Ni, Pb, Fe or Cr get precipitated as water-insoluble hydroxides. This will work under the condition that the metals are not complexed in the solution with ammonia, EDTA or other complexing agents. The sludge from the settling tank is then still further concentrated by filtration or centrifugal action.

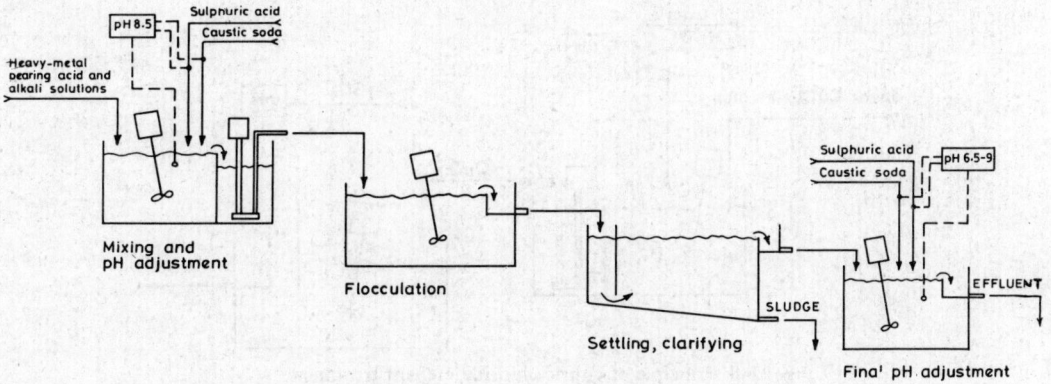

Fig. 22-7 Principle arrangement for heavy-metal precipitation

The pH range in which precipitation occurs is not the same for different metals. The pH ranges for some metals are given below [3]. Note that certain metals go again in solution at a higher pH value.

Metal ion	Precipitation starts at pH	Metal goes again in solution at pH	Suggested pH for precipitation	Metal content still in solution at suggested pH
Fe (III)	2.8	—	3.5	2 mg/l
Sn (II)	3.9	10.6	—	
Cr (III)	5.5	9.2	6.3–6.5	2 mg/l
Cu (II)	5.8	—	7.5	1 mg/l
Ni (II)	7.8	—	9.3	3 mg/l
Pb (II)	7.0	—	9.5	1 mg/l

The actual neutralisation (hydroxide formation) occurs almost immediately after the critical pH value has been reached. However, the flocculation and settling of the hydroxides may require several hours. This must be considered when planning the size of the tanks required.

Cupric Chloride

Cupric chloride is probably the least problematic etchant since valuable copper sulfate can be obtained by a comparatively simple procedure (Fig. 22-8). Cupric chloride is first neutralised with NaOH to form CuO which is thereafter dissolved in H_2SO_4. The resulting copper sulfate can be used in the preparation of electroless and electrolytic copper plating baths and as the copper supplying material for additive PCB processes or many other chemical processes.

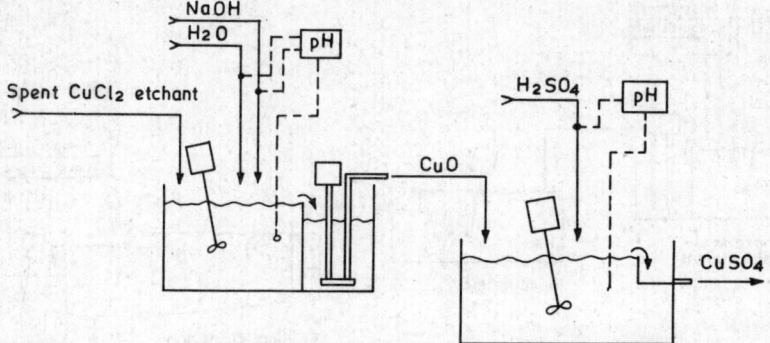

Fig. 22-8 Principle of cupric chloride etchant treatment

Chromic Acid

The treatment of chromium-bearing solutions (e.g. chromic-sulfuric etchant) is done in two stages. In the first stage, the reduction stage, the hexavalent chromium is reduced to the less harmful trivalent chromium (Fig. 22-9). Since the reduction occurs much faster at a low pH value, the pH adjustment to a value below pH 2.5 is done with H_2SO_4. The reduction chemical added can be sulfur dioxide, sodium bisulfite or hydrazine. The actual reduction is controlled via the redox potential. The second stage of the treatment is the heavy-metal precipitation process as already shown in Fig. 22-7 in which chromium gets precipitated as chromium hydroxide.

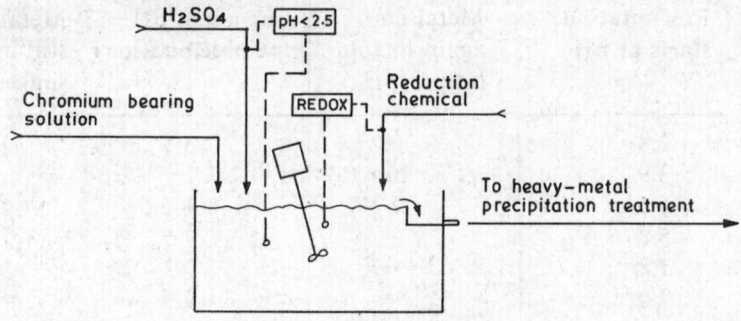

Fig. 22-9 Chromium treatment, reduction stage

Ammonia Etchant

Where small quantities of alkaline etchant with ammonia have to be disposed, the best solution is to return it to the supplier who can treat it for reuse.

For large quantities, distillation is recommended (Fig. 22-10): The NH_3 gas produced in the distillation is welcome in the alkaline ammonia etching process. The remaining CuO can be dissolved in H_2SO_4 which finally gives the valuable copper sulfate ($CuSO_4$).

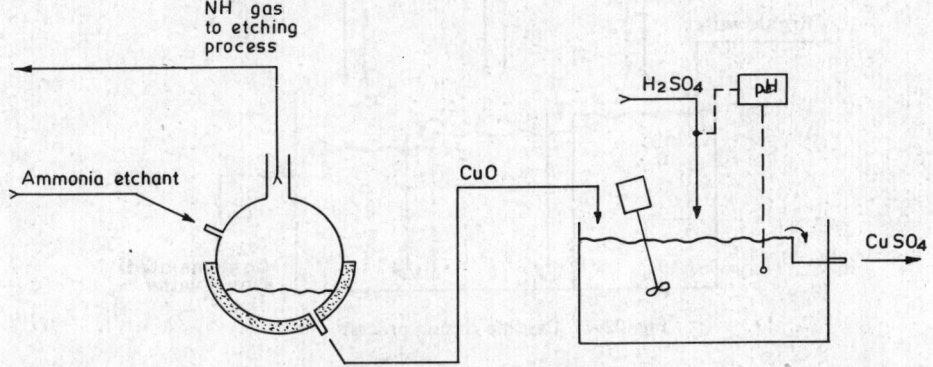

Fig. 22-10 Distillation of ammonia etchant

Sludge Disposal

One possible solution to get rid of the highly concentrated sludge is to bury it in a dumping site which is meant for such purposes. The selection of the dumping site must be done in consultation with the official authorities. The dumping site should not be near to ground waters and the effects of possible floods or earthquakes must be studied as far as possible and be considered in the selection of the site.

The heavy metal-bearing hydroxides in the sludge form a water-insoluble compound. Although this compound is considered to have little potential for environmental harm or dissolution, only a deposition site which is really and under all circumstances safe can be tolerated.

Another alternative to burial is incineration. This depends usually on whether there is a firm with its own incinerator which is ready to accept the sludge. A disadvantage is the HCl formation where chlorides are burnt as well as the volatilisation of metals like lead. Also the problem of disposal of the water-soluble ash must be solved.

22.5.2 Rinsing Water

In any PCB processing, larger quantities of rinsing water are required which, after the rinsing process, are contaminated. Such rinsing water cannot be directly fed into the sewer system because it also bears heavy-metal containing solutions. A prior treatment is required to reduce the contamination to a level acceptable for sewer waters. For the pollution limiting values, check also under Section 16.4.2.

The very first rule is always to reduce the rinse water consumption to its minimum. This can be implemented by cascade rinsing as well as reusing of rinsing water in another rinsing step with less stringent purity requirements. Cascade rinsing, sometimes also called 'counter-current rinsing', is shown in Fig. 22–11. The water flow is opposite to the direction of board movement. The last rinsing dip for the boards is herewith in almost fresh water. For the reuse of rinsing water, it must be verified that the contamination of the previous use does not harm the requirements of the following rinsing step.

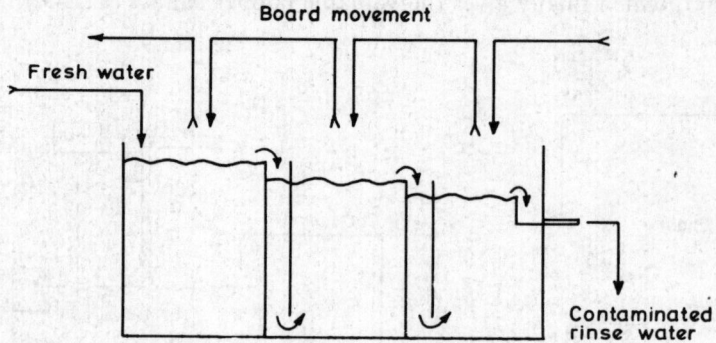

Fig. 22–11 Cascade rinsing process

The purifying treatment for rinsing water can be done by several methods. In smaller PCB set-ups, a batch treatment is done after collecting of all rinsing water for a certain period. For large rinsing water consumption, a continuous treating plant must be considered.

For heavily contaminated rinsing water, the heavy-metal precipitation process (Fig. 22–7) must be employed. If the rinsing water contains cyanides and heavy-metal complexed cyanides, the neutralisation by reduction is done with hypochlorite solution at a pH value of 11.

A highly recommended solution for rinsing water treatment is *de-ionising* with a two-bed ion-exchange system. This is also an excellent solution because 95–98% of the water treated will again be available as highly purified water for reuse.

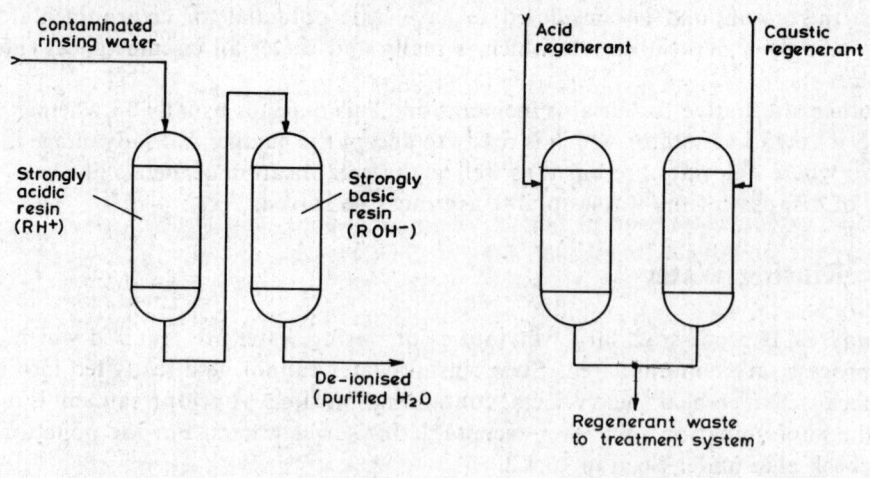

Fig. 22–12 Ion-exchange system (2-bed)

The ion exchange process makes use of the fact that most of the water-soluble chemicals used in PCB production get ionised in water thereby forming anions and cations:

$$CuSO_4 \rightarrow \underset{\text{(cation)}}{Cu^{+2}} + \underset{\text{(anion)}}{SO_4^{-2}}$$

When such rinsing water passes through the ion exchange system, cations replace hydrogen ions in the cation resin. Similarly, anions replace OH ions in the anion resin (R = resin):

$$RH^+ + Cu^{+2} \rightarrow 2H^+ + RCu^{+2}$$
$$ROH^- + SO_4^{-2} \rightarrow 2OH^- + RSO_4^{-2}$$

Thereafter, hydrogen ions (H^+) and hydroxyl ions (OH^-) combine together, thereby forming water

$$H^+ + OH^- \rightarrow H_2O$$

After a certain time of use, the resins will get saturated. This can be detected by a rising conductivity of the de-ionised water. The regeneration of the resins is carried out by adding a strong acid to the cation resin and a strong base to the anion resin: A reverse reaction now takes place in the system, thus returning it to the original state. In the regeneration process, the metals and ions get washed out in a concentrated form. The outcoming solution has thereafter to go for heavy-metal precipitation.

Another rinsing water treatment is the *reverse osmosis*, although not yet so well known. In such systems, most of the salts dissolved in water are removed under high pressure in a separate stream. The *Lancy process* is one more promising alternative: Instead of water, a special solution is used for rinsing. The solution has to match exactly the type of contamination existing on the board surface after processing. After rinsing in this special solution, another rinse with de-ionised water has to follow. The heavy metals in the Lancy solution can be comparatively easily removed as heavy-metal hydroxides.

22.5.3 Air

Dust

In the various mechanical operations which a PCB has to undergo (e.g., drilling, sawing with circular saw), dust will be produced. There are comparatively simple means to collect such dust which otherwise would harm personnel and equipment. Industrial dust collectors work mostly on the cyclon separation principle: The incoming air stream is brought into a double vortex where the stream starts rotating. The dust particles get separated by inertial forces and settle at the bottom of the equipment from where they are collected. The air stream leaves the double vortex at the top.

Another process makes use of a baghouse through which the air stream is pressed (vacuum cleaner principle).

In practice, it is good to collect the dusty air as near as possible to the point where it is produced. If necessary, individual dust collectors are provided for each dust producing equipment to avoid lengthy pipes (which will always reduce the suction) to a central dust collector.

Etching and Plating Fumes

Where *large* quantities of such fumes are produced, the exhaust facilities usually provided that do not allow for treatment of such air, are not sufficient: The air has to be cleaned before it can be released into the open air.

The method for air cleaning to be applied is usually *wet-scrubbing*: The air contaminated with acid fumes, chlorides or ammonia is brought into close contact with water or an aqueous solution. In this process, the contaminants get absorbed in the cleaning fluid before the air stream gets released. There are various techniques in use to perform the wet-scrubbing: The air stream can either be pushed or pulled through the wetted cleaning column which contains surface multiplying balls or rings; the air stream may also be passed through a high-velocity liquid spray or fog.

The fluid recommended for acidic type of air is a caustic soda solution or water while ammonia fumes are always treated with water. After the washing solution is saturated, it has to be replaced and neutralised.

Chlorinated Solvent Fumes

The most often used chlorinated solvents are the chlorinated hydrocarbons like trichloroethylene, trichloroethane and methylene chloride.

With little concentration of such solvent fumes in the air, an effective exhaust is sufficient. The exhaust outlet, however, should be at least 1.5 m above the roof, 10 m above ground and 5 m above any building within a 50 m radius [2].

Where large quantities of such fumes are produced, there will be a need for an active carbon absorption plant which has also the advantage of reclaiming the chlorinated hydrocarbons for reuse.

22.5.4 Health of Personnel

Basic Precautions

The use of different kinds and of highly concentrated chemicals in PCB production involves high risks and health hazards if not handled properly. It is therefore a must to have an experienced person thoroughly familiar with all aspects in handling of chemicals to supervise all such handling operations.

All protective devices for personal safety and accident prevention should be made available exactly at the place where they are practically used. Their use is no luxury and they are not meant only for the workers afraid of chemicals. Remember that most of the accidents could have been avoided by observing and acting upon the known safety rules.

The devices mostly required for personal safety in handling chemicals include goggles, acid-proof rubber gloves and aprons, face shields and filter masks fitted with the proper filter medium for the particular chemicals to be handled.

In all the places where chemicals are handled and stored, smoking, eating as well as the presence of eatables is strictly forbidden. The hands are washed immediately after the handling of chemicals is completed. A steady stream of fresh air through the rooms must be guaranteed. Chemical bottles and containers are always kept tightly closed.

Health affecting Chemicals

An account of the chemicals typically used in PCB processing is given now. Each one of the chemicals is mentioned with the formula, name of the chemical, specific danger and handling instructions.

$CCl_2:CCl_2$, *perchloroethylene*

Health-affecting vapours.
Store only in tightly closed bottles. Do not breathe in the vapours.
Care for sufficient air circulation or wear filter mask.

$CHCl:CCl_2$, *trichloroethylene*

Health-affecting vapours.
Store only in tightly closed bottles. Do not breathe in the vapours. Care for sufficient air circulation or wear filter mask.

C_6H_6, *benzene*

Easily inflammable, vapour-air mixture is explosive, highly toxic vapours.
Store only in tightly closed bottles in a well ventilated place.
When using, don't eat or smoke. Do not pour it into sewer water.
Prevent from electrostatic charges. Keep away from open fires, heat sources and sparks. Avoid contact with skin, eyes or clothes. Remove infected clothes immediately. Provide sufficient air circulation or wear filter mask. In case of fire, use only CO_2-type extinguishers or sand but never water. Call for the doctor if not feeling well.

CH_3COOH, *acetic acid (> 25% concentration)*

Causes burns and etchings.
Keep bottles dry, tightly closed and away from any eatables. Handle bottles carefully. Avoid contact with skin, eyes and clothes. Clean floor and dirty equipment with suitable means.

CH_3OH, *methyl alcohol*

Easily inflammable, vapour-air mixture is explosive, serious toxic danger if breathing in or swallowing occurs.
Store bottles tightly closed in a cool and well-ventilated place. When using, don't eat or smoke. Keep away from open fires, heat sources and sparks. Don't breath in vapours. Remove infected clothes immediately. In case of fire, use only CO_2-type of fire extinguishers or sand but never water. Call for the doctor if not feeling well.

C_2H_5OH, *ethyl alcohol*

Easily inflammable, vapour-air mixture is explosive. Store bottles tightly closed in a cool and well-ventilated place. Do not smoke when working. Keep away from open fires, heat sources and sparks. Remove infected clothes immediately. In case of fire, use only CO_2-type fire extinguishers or sand but never water.

Cl_2, *chlorine*

Highly toxic gas; attacks skin, eyes and breathing path. Store bottles tightly closed in a cool and well-ventilated place. Keep bottles in upright position and open carefully. Avoid contact with skin and eyes. Care for sufficient air circulation and wear always filter mask. Call for the doctor if not feeling well.

CrO_3, *chromic anhydride*

Danger of fire if in contact with inflammable materials; causes serious burns and etchings.

Keep bottles well closed but not gas-proof. Don't smoke when working. Keep away from inflammable materials. Avoid contact with skin, eyes or clothes. Immediate hand washing after working is necessary. Thorough cleaning of floor and dirty equipment with water.

HCHO, *formaldehyde*

Serious toxic danger if breathing in or swallowing occurs. Causes burns and etchings and attacks skin, eyes and breathing path.

Store tightly closed bottles in a cool place. Don't eat or smoke when working. Keep away from eatables. Don't breath in gases. Avoid contact with skin, eyes or clothes and wear goggles. Immediate hand washing after work is necessary. Call for the doctor if not feeling well.

HCl, *hydrochloric acid (> 25% concentration)*

Causes burns and etchings.

Store bottles tightly closed in a cool place and handle them carefully. Do not breathe in vapours. Avoid contact with skin, eyes or clothes. Wash out splashes on skin and eyes with plenty of water. Clean floor and dirty equipment with suitable means.

$HClO_4$, *perchloric acid (10–50 % concentration)*

Causes burns and etchings.

Store in tightly closed bottles and handle them carefully. Avoid contact with skin, eyes or clothes. Clean floor and dirty equipment with plenty of water.

$HClO_4$, *perchloric acid (> 50% concentration)*

Danger of fire if in contact with inflammable materials; causes burns and etchings.

Store in tightly closed bottles. Do not smoke when working. Handle the bottle carefully. Keep away from inflammable materials. Avoid contact with skin, eyes or clothes. Remove infected clothes immediately. Wash hands immediately after the work is finished. Clean floor and dirty equipment with plenty of water.

NCN, *hydrocyanic acid*

Easily inflammable; vapour-air mixture is explosive; highly toxic vapours.

Store in tightly closed bottles in a cool and well-ventilated place. Do not eat or smoke when working. Keep away from eatables. Handle bottles carefully. Keep away from open fires, heat sources and sparks. Avoid contact with skin, eyes and clothes. Use filter mask when working. In case of a fire, use only CO_2-type fire extinguishers or sand but never water. Call for the doctor if not feeling well.

HCOOH, *formic acid (> 25 % concentration)*

Causes burns and etchings.

Store in tightly closed bottles and handle them carefully. Do not breathe in vapours. Avoid contact with skin, eyes or clothes. Rinse splashes on skin or eyes with plenty of water. Clean floor and dirty equipment with suitable means. In case of an accident, call for the doctor.

HF, *hydrofloric acid*

Serious toxic danger if breathing in or swallowing or skin contact occurs; causes heavy burns and etchings.

Store tightly closed bottles in a cool and well ventilated place. Keep away from eatables. Avoid contact with skin, eyes or clothes. Do not eat or smoke when working. Rinse splashes on skin or eyes with plenty of water. Wear impermeable safety suit and gloves when working. Care for sufficient air circulation and wear filter mask. Immediate hand washing after work necessary. Cleaning of floor and dirty equipment with suitable means. Call for the doctor if not feeling well.

HNO_3, *nitric acid (20–70 % concentration)*

Causes heavy burns and etchings.

Store tightly closed bottles in a well ventilated place. Careful handling of the bottles is important. Do not breathe in vapours. Avoid contact with skin, eyes or clothes. Remove infected clothes immediately. In case of an accident, call for the doctor.

HNO_3, *nitric acid (>70 % concentration)*

Danger of fire if in contact with inflammable materials; causes heavy burns and etchings.

Store tightly closed bottles in a cool and well ventilated place. Do not smoke when working. Handle bottles carefully. Keep away from inflammable materials. Avoid contact with skin, eyes and clothes. Remove infected clothes immediately. Care for sufficient air circulation and wear filter mask. Clean floor and dirty equipment with suitable means. In case of an accident, call for the doctor.

H_3PO_4, *phosphoric acid (> 25% concentration)*

Causes burns and etchings.

Store in tightly closed bottles. Avoid contact with skin, eyes or clothes. Clean floor and dirty equipment with suitable means. In case of an accident, call for the doctor.

H_2SO_4, *sulfuric acid (>20% concentration)*

Causes heavy burns and etchings.

Store in tightly closed bottles. Should never be put into sewer water. Never add water to it. Handle bottles carefully. Avoid contact with skin, eyes or clothes. Remove infected clothes immediately. When working, use impermeable safety suit, gloves and goggles. Clean floor and dirty equipment with suitable means. In case of an accident, call for the doctor.

KOH, *potassium hydroxide (anhydrous and solutions with >10% KOH)*

Causes heavy burns and etchings.

Keep bottles dry and tightly closed. Do not mix with acids. Avoid contact with skin or eyes. Rinse splashes on skin or eyes with plenty of water. Remove infected clothes immediately. When working, use impermeable safety suit, gloves and goggles. In case of an accident, call for the doctor.

$Me^I CN$, $Me^{II}(CN)_2$, *cyanides*

Serious toxic danger if breathing in or swallowing occurs or if in contact with skin; highly toxic gas is developed if in contact with an acid.

Keep bottles tightly closed and dry. Do not eat or smoke when working. Do not mix with acids. Do not breathe in vapours or dust. Avoid contact with skin or eyes. Immediate hand washing after work necessary. Clean floor and dirty equipment with plenty of water. If not feeling well, call for the doctor.

$Me^I F$, $Me^{II} F_2$, *fluorides (soluble)*

Serious toxic danger if swallowed.

Store in tightly closed bottles. Do not eat or smoke when working. Keep away from eatables. Do not breath in dust. Avoid contact with skin or eyes. Rinse splashes on skin or eyes with plenty of water. Wear safety suit when working. Call for the doctor if not feeling well.

NaOH, *sodium hydroxide (caustic soda)*

Causes heavy burns and etchings.

Store in dry and tightly closed bottles. Do not mix with acids. Avoid contact with skin or eyes. Rinse splashes on skin or eyes with plenty of water. Remove infected clothes immediately. When working, wear impermeable safety suit, gloves and goggles. In case of an accident, call for the doctor.

NH_3, *ammonia (gas)*

Toxic gas; attacks skin, eyes and breathing path.

Store in tightly closed bottles in a cool and well ventilated place. Handle bottles carefully. Do not open valve with force. Keep bottles upright and open carefully. Avoid contact with skin or eyes. Care for sufficient air circulation and wear filter mask. Call for the doctor if not feeling well.

NH_4OH, *ammonia (10-35% NH_3 concentration)*

Attacks skin, eyes and breathing path.

Store in tightly closed bottles. Do not mix with acids. Do not breathe in vapours. Avoid contact with skin and eyes. Rinse splashes on skin or eyes with plenty of water.

NH_4OH, *ammonia (>35% NH_3 concentration)*

Causes burns and etchings; attacks skin, eyes and breathing path.

Store in tightly closed bottles. Do not mix with acids. Avoid contact with skin, eyes or clothes. Remove splashes on skin or eyes with plenty of water.

NO_2, N_2O_4, *nitric oxide*

Highly toxic gas; attacks skin, eyes and breathing path. Store in tightly closed bottles in a cool and well ventilated place. Keep bottles upright and open them carefully. Do not breathe in the gas. Avoid contact with skin or eyes. Care for sufficient air circulation and wear filter mask. Call for the doctor if not feeling well.

BIBLIOGRAPHY

1. COOMBS C.F.: *PRINTED CIRCUITS HANDBOOK*, 2nd edition, McGraw-Hill Book Co., New York, 1979.
2. HERRMANN G.: *LEITERPLATTEN*, Leuze Verlag, Saulgau/Wurtt. (Germany), 1978.
3. *PRAKTISCHE GALVANOTECHNIK*, 3rd edition, Leuze Verlag, Saulgau/Wurtt. (Germany), 1975.

APPENDIX
LIST OF ADDRESSES

A.1 PROFESSIONAL ORGANISATIONS

IPC
The Institute for Interconnecting and Packaging Electronic Circuits, 1717 Howard Street, Evanston, Illinois 60202, USA

ICT
The Institute of Circuit Technology
Secretary: Bruce Routledge
5 Windrush Close, Ickenham, Middlesex UB10 8EJ, England

EIPC
European Institute of Printed Circuits
Via Maraini 17, CH-6900 Lugano, Switzerland

A.2 STANDARD ORGANISATIONS

ASTM
American Society for Testing and Materials
1916 Race Street, Philadelphia, PA. 19103, USA

EIA
Electronic Industries Association
2001 Eye Street N.W., Washington, DC. 20006, USA

IEEE
Institute of Electrical and Electronics Engineers, Inc.
345 East 47th Street, New York, NY. 10017, USA

IPC
The Institute for Interconnecting and Packaging Electronic Circuits, 1717 Howard Street, Evanston, Illinois 60202, USA

NEMA
National Electrical Manufacturers Association
155 East 44th Street, New York, NY. 10017, USA

UL
Underwriters' Laboratory, Inc.
207 East Ohio Street, Chicago, Illinois 60611, USA

ANSI
American National Standards Institute
1430 Broadway, New York, NY. 10018, USA

IEC
International Electrotechnical Commission
1 rue de Varembe, Geneva, Switzerland

ISI
Indian Standards Institution
Manek Bhavan, 9 Bahadur Shah Zafar Marg, New Delhi-110 001, India

Metal Finishers' Association of India
Parijat Block 24, 95 Marine Drive, Bombay-400 002, India

A.3 PCB EQUIPMENT AND MATERIAL MANUFACTURERS

A.3.1 Introduction

With the publication of this collection of addresses, an attempt is made to assist all those who are planning to start new activities in PCB technology or those who would like to review their own position in this field.

The inclusion of such a wide range of addresses in a book like this might appear rather unusual. For the dedicated reader, however, to get otherwise access to these addresses is considerably more difficult in a country like India as compared to Western countries. There are many enquiry letters sent to the author which can easily prove this fact.

The list of addresses given here is in no way complete. It just represents a collection done by the Process Technology Group at the Centre for Electronics Design Technology (CEDT). There are definitely many more addresses which are not included here. But to work out a really complete catalogue could be the task of a professional organisation with the necessary support.

The order in which the addresses are given does not say anything, neither about the quality nor about the price or the performance of the product. Where there are Indian manufacturers for a specific product, their addresses are listed first. Many of the foreign companies mentioned might have a representative in India: As a rule and as far as known, the addresses of the actual manufacturers have been given.

A.3.2. Alphabetical Key-Word Index

Adhesive Artwork Materials, Sticking Patterns and Tapes
Adhesive Artwork Materials, Transfer Foil Type
Artwork Base Foil, Polyester
Artwork Scriber
Artwork Tables, Illuminated
Assembly Workplaces

Base Materials for Multilayer Boards
Base Materials for PCBs, Flexible
Base Materials for PCBs, Rigid
Brushing Machines, Laminate Cleaning

Circular Sawing Blades, Tungsten Carbide
Circular Sawing Machines
Cleaning of assembled PCBs, Flux Removal
Component Insertion Equipment
Component Lead Cutting Equipment
Component Preparation Equipment, Insertion
Contour Milling Equipment

Darkroom Furniture and Equipment
Desoldering Equipment
Developing Equipment, Photoresist
Diamond Sawing Tools
Dip Coating Equipment, Wet-Film Resist
Diazo Film
Drill Bit Grinding Machines
Drill Bits, Tungsten Carbide
Drilling Machines, PCBs
Drilling Spindles
Dry-Film Photoresist
Drying Equipment, Transfer-Type
Dust Collectors

Etchant Regeneration Systems
Etchants
Etching Machines, Spray-Type
Exposure Equipment, Actinic Light

Film Materials, Polyester Base
Filter Pumps, Plating
Fluxes, Soldering

Gold Recovery Equipment

Hand Soldering Equipment
Hot Air Leveling

Insulation Stripping Devices

Mechanical PCB Milling Equipment
Multilayer Presses

Plated-Through Hole Process Chemicals
Plate Shearing Equipment
Plating Equipment
Plating Thickness Measuring Equipment
Precious Metal Plating Process Chemicals
Protective Lacquer Coatings

Reprographic Cameras
Roller Coating Equipment, Wet-Film Resist
Roller Tinning Equipment

Safety Equipment, Personnel
Screen Fabrics
Screen Preparation Chemicals
Screen Printing Inks
Screen Printing Machines and Equipment
Silver Recovery Equipment
Solder Reflow Equipment
Stripping Equipment
Stylus Plating Equipment, Selective Plating

Through-Contact Revets, Double-Sided PCBs

Waste Water Purification Equipment
Water Deionising Plants
Wave Soldering Equipment
Wet-Film Photoresists
Whirl Coating Equipment, Wet-Film Resist

A.3.3 Addresses

Adhesive Artwork Materials, Sticking Patterns and Tapes
Crotech Systems, Cottage 27, West Patel Nagar, New Delhi-110 008, India
Jain Electronics, F31 Nandham Industrial Estate, Marol, Bombay-400 059, India
Stikon Systems, 355 West of Chord Road, Rajajinagar, Bangalore-560 010, India
S.V. Products, Hridgat IInd floor, Woollen Mill Lane, Dadar, Bombay-400 028, India

Bishop Graphics Inc., 20450 Plummer Street, Chatsworth, California 91311, USA
Chartpak Europe, Didcot, Berkshire, England
Mecanorma GmbH, Elisabethstr. 3, D-4000 Dusseldorf, Germany
Quest Automation Ltd., Wimborne BH22 9HQ, England
Webtek Corporation, 4326 West Pico Blvd., Los Angeles, California 90019, USA
Wolfson Microelectronics, King's Building, Edinburgh, England

Adhesive Artwork Materials, Transfer Foil Type
Invento Graphics, 1-1719 Chittaranjan Park, New Delhi-110 019, India
Mecanorma GmbH, Elisabethstr. 3, D-4000 Dusseldorf, Germany
Zentak AG, Solothurnstr. 72, CH-4002 Basel, Switzerland

Artwork Base Foil, Polyester
Kanva Industries, 2 Metagalli Industrial Estate, Mysore-570 002, India
Bishop Graphics Inc., 20450 Plummer Street, Chatsworth, California 91311, USA

Artwork Scriber
Linton Laboratories Ltd., 4 Bartlow Rd., Linton, Cambridge CB1 6LY, England

Artwork Tables, Illuminated
Balaji Fabricators & Consultants, 23 Rammohanpura, Srirampura, Bangalore-560 021, India
Doschen India Pvt. Ltd., Bangli Naka, Sandor, Post Vasai-401 201, Dist. Thane, India
Monotype India Ltd., 27th Cross Banashankari 2nd Stage, Bangalore-560 070, India
The Standard Printing Machinery Co., D-11 Industrial Estate, Ambattur, Madras-600 058, India

Assembly Workplaces
AAT Aston GmbH, Rothenburger Str. 33, D-8500 Nurnberg, Germany
Dr. K. Schleuniger & Co., Schongrunstr. 27, CH-4500 Solothurn, Switzerland
Electrautom Ltd., Etom House, Queens Rd., Maidstone, England
EMP AG, Klosbachstr. 83, CH-8030 Zurich, Switzerland
ERSA, Leonhard-Karl-Str. 24, D-6980 Wertheim 2, Germany
FARCO, Girardet 29, CH-2400 Le Locle, Switzerland
Ferco S.p.A., Via Rovani 6, I-20052 Monza, Italy
Kontakt-Systeme AG, Gewerbestr. 16, CH-8132 Egg/Zurich, Switzerland
PAF Systems Ltd., Paterson Rd., Wellingborough, England
Polytronik GmbH, Quagliostr. 6, D-8000 Munchen 90, Germany
Rotawinder Ltd., Forest Rd., Hainault Essex, England
Trescomp-Brauer, Burchardstr. 8, D-2000 Hamburg 1, Germany
Universal Instruments GmbH, Alte Str. 37, D-6368 Bad Vilbel, Germany

Base Materials for Multilayer Boards
Atlantic Laminate, Oak Materials Group, 174 No. Main St., Franklin, NH. 03235, USA
E. I. DuPont de Nemours & Co. Inc., Fabrics & Finishes Dept., Industrial Products Div., Wilmington, DE. 19898, USA
Formica Limited, Coast Rd., North Shields, Tyne & Wear, England

Fortin Laminating Corp., 1323 Truman Street, San Fernando, USA
Matsushita Electric Works Ltd., Kakoshin Division, 1048 Kadoma Ohaza Kadoma, Osaka, Japan
Mitsubishi Gas Chemical Co. Inc., Mitsubishi Main Building, 5-2 Marunouchi 2-chome, Chiyoda-ku, Tokyo, Japan
Nikkan Industries Co. Ltd., 3-44-4 Maenocho, Itabashi-ku, Tokyo, Japan

Base Materials for PCBs, Flexible

Brittains-Riegel Ltd., Wrexham Industrial Estate, Wrexham, England
Chase-Foster Div/Keene Corp., 199 Amaral St., P. O. Box 4305, East Providence, RI. 02914, USA
Circuit Materials Company, Research Park, 1101 State Rd. Bldg. J., Princeton, NJ. 08540, USA
Detakta, D-2000 Hamburg 60, Germany
E. I. DuPont de Nemours & Co. Inc., Fabrics & Finishes Dept., Industrial Products Div., Wilmington, DE. 19898, USA
Felten & Guilleaume Dielektra AG, Kaiserstr. 127, D-5050 Porz, Germany
Formica Limited, Coast Rd., North Shields, Tyne & Wear, England
Fortin Laminating Corp., 1323 Truman Street, San Fernando, USA
Matsushita Electric Works Ltd., Kakoshin Division, 1048 Kadoma Ohaza Kadoma, Osaka, Japan
Mitsubishi Gas Chemical Co. Inc., Mitsubishi Main Building, 5-2 Marunouchi 2-chome, Chiyoda-ku, Tokyo, Japan
Nikkan Industries Co. Ltd., 3-44-4 Maenocho, Itabashi-ku, Tokyo, Japan
Sheldahl Ltd., Eastern Rd., Bracknell, Bershire, England

Base Materials for PCBs, Rigid

Bakelite Hylam Ltd., Tiecicon, 18 Dr. Moses Rd., Bombay-400 011, India
Caprihans India Ltd., Block D, Shivsagar Estate, Dr. Annie Besant Rd., Worli, Bombay-400018, India
Formica India Ltd., P. B. 64, Poona-411 001, India
Contiflex AG, Florastr. 17, CH-8700 Kusnacht-Zurich, Switzerland
Detakta, D-2000 Hamburg 60, Germany
Dynamit Nobel AG, D-5210 Troisdorf, Germany
E.I. DuPont de Nemours & Co. Inc., Fabrics & Finishes Dept., Industrial Products Div., Wilmington, DE. 19898, USA
Ferrozell-Ges. Sachs & Co. mbH, Postfach 101569, D-8900 Augsburg 1, Germany
Fortin Laminating Corp., 1323 Truman Street, San Fernando, USA
General Electric Company, Laminated and Insulating Materials Business Dept., Coshocton, Ohio 43812, USA
Kawecki-Billiton Metaalindustrie N.V., Westervoortsedijk 67d, Arnhem, Holland
Lamination Technology Inc., 2720 South Main Street, Santa Ana, CA. 92707, USA
Nikkan Industries Co. Ltd., 3-44-4 Maenocho, Itabashi-ku, Tokyo, Japan
Norplex Europe, Abteilung der UOP GmbH, Postfach 1109, D-5272 Wipperfurth, Germany
N.V. Cincinnati Milacron S.A., 3650 Dilsen Fabriekspark, Rotem, Belgium

Matsushita Electric Works Ltd., Kakoshin Division, 1048 Kadoma Ohaza Kadoma, Osaka, Japan
Mitsubishi Gas Chemical Co. Inc., Mitsubishi Main Building, 5-2 Marunouchi 2-chome, Chiyoda-ku, Tokyo, Japan
Perstorp AB, Fack, S-28400 Perstorp, Sweden
Quartz & Silice S.A., 8 rue d'Anjou, F-75008 Paris, France
Schjeldahl Ltd., Eastern Rd., Bracknell, Berks., England
Schweizerische Isola-Werke, CH-4226 Breitenbach, Switzerland
Sumitomo Bakelite Co. Ltd., 2-2, 1-chome, Uchisaiwaicho, Chiyoda-ku, Tokyo 100, Japan
Synthane Taylor, Valley Forge PA 19482, La Verne, CA. 91750, USA
The Mica Corporation, 10900 Washington Blvd., Culver City, CA. 90230, USA

Brushing Machines, Laminate Cleaning
Bernhard Schmiedeskamp, Blucherstr. 18, D-4970 Bad Oeynhausen 1, Germany
Chemcut Corporation, 500 Science Park, State College, Pennsylvania 16801, USA
FAG S.A., 7 rue de Geneve, CH-1001 Lausanne, Switzerland
Gebr. Schmid, Robert-Bosch-Str.; D-7290 Freudenstadt, Germany
Hans Hollmuller GmbH & Co., Kappstr. 69, D-7033 Herrenberg, Germany
International Supplies Co., Via Zanardi 7, I-43100 Parma, Italy
Nubal Electronics Ltd., 10 Tribune Drive, Trinity Trading Estate, Sittingbourne, Kent ME10 2PG, England
Peter Jordan, Mainstr. 5-7, D-6050 Offenbach, Germany
Resco, Via Massena 2/A, I-20145 Milano, Italy
Wesero GmbH & Co. KG, D-4322 Sprockhovel, Germany

Circular Sawing Blades, Tungsten Carbide
Andreas Maier KG, D-7959 Schwendi-Horenhausen, Germany
FAG S.A., 7 rue de Geneve, CH-1001 Lausanne, Switzerland
Roeck AG, CH-4600 Olten, Switzerland
Tempress Microelectronics B.V., Marconistraat 14, Hoogeveen, Holland

Circular Sawing Machines
Photo Goods Service, 6465 Katra Baryan, Delhi-110 006, India
FAG S.A., 7 rue de Geneve, CH-1001 Lausanne, Switzerland
Georg Ott, Postfach 3240, D-7900 Ulm/Do., Germany
ITS GmbH, D-8000 Munchen 60, Germany
Oswald Boll GmbH, Rothenbaumchaussee 140, D-2000 Hamburg 13, Germany
Smid SA, 25 rue des Roses, F-68051 Mulhouse Cedex, France
Roeck AG, CH-4600 Olten, Switzerland

Cleaning of Assembled PCBs, Flux Removal
International Meters & Electronics Corporation, 19-C Shalimar Industrial Estate, Matunga, Bombay-400 019, India
Ralsonics, 3 Khetani Textile Industrial Compound, 106/7 Bazarward, Kurla, Bombay-400 009, India

Branson Ultraschall GmbH, Bavariaring 8, D-8000 Munchen 2, Germany
Dawe Instruments Ltd., Concord Rd., Western Avenue, London W3 OSD, England
Dr. K. Schleuniger & Co., Schongrunstr. 27, CH-4500 Solothurn, Switzerland
Farbwerke Hoechst AG, Frigen Informationsdienst, D-6230 Frankfurt (Main) 80, Germany
GLT Gesellschaft fur Lottechnik mbH, Kreuzstr. 150, D-7534 Birkenfeld, Germany
Greiner Electronic AG. Gaswerkstr. 33-35, CH-4900 Langenthal, Switzerland
ICI Mond Division, P.O. Box 13, The Heath, Runcorn, Cheshire WA7 4QF, England
Kerry Ultrasonics Ltd., Hunting Gate, Wilbury Way Hitchin, Hertfordshire, England
KLN-Ultraschall GmbH, Siegfriedstr. 124, D-6148 Heppenheim, Germany
Laytron S.A., 21 rue du Mont-Blanc, CH-1201 Geneve, Switzerland
Linton Laboratories Ltd., 4 Bartlow Rd., Linton, Cambridge CB1 6LY, England
Resco, Via Massena 2/A, I-20145 Milano, Italy
Shipley Europe Ltd., Humber Avenue, Coventry CV3 1JL, England

Component Insertion Equipment
AAT Aston GmbH, Rothenburger Str. 33, D-8500 Nurnberg, Germany
Bullnheimer & Co., Schaezlerstr. 6, D-8900 Augsburg, Germany
Electrautom Ltd., Etom House, Queens Rd., Maidstone, England
ERSA, Leonhard-Karl-Str. 24, D-6980 Wertheim 2, Germany
FARCO, Girardet 29, CH-2400 Le Locle, Switzerland
Herbert Streckfuss KG, Kruppstr. 1, D-7514 Eggenstein-Leop. 1, Germany
Kontakt-Systeme AG, Gewerbestr. 16, CH-8132 Egg/Zurich, Switzerland
Peter Jordan, Mainstr. 5-7, D-6050 Offenbach, Germany
Rotawinder Ltd., Forest Rd., Hainault Essex, England
Trescomp-Brauer, Burchardstr. 8, D-2000 Hamburg 1, Germany
Universal Instruments GmbH, Alte Str. 37, D-6368 Bad Vilbel, Germany

Component Lead Cutting Equipment
AAT Aston GmbH, Rothenburger Str. 33, D-8500 Nurnberg, Germany
Anglade S.A., 7 rue A. Briand, F-92300 Levallois Perret, France
Dr. K. Schleuniger & Co., Schongrunstr. 27, CH-4500 Solothurn, Switzerland
Evomec Oy, SF-08700 Virkkala, Finland
FARCO, Girardet 29, CH-2400 Le Locle, Switzerland
Herbert Streckfuss KG, Kruppstr. 10, D-7514 Eggenstein-Leop. 1, Germany
Polytronic GmbH, Quagliostr. 6, D-8000 Munchen 90, Germany
Resco, Via Massena 2/A, I-20145 Milano, Italy
Smid S.A., 25 rue des Roses, F-68051 Mulhouse Cedex, France
Surpro, Sportlaan 76, Hertogenbosch, Holland
Terima Steiner and Co., Kriesbachstr. 5, CH-8304 Wallisellen, Switzerland

Component Preparation Equipment, Insertion
AAT Aston GmbH, Rothenburger Str. 33, D-8500 Nurnberg, Germany
Fico Fischer and Co. GmbH, Untere Augasse 41, D-7530 Pforzheim, Germany
GKN Saneky Limited, Greeway Rd., Bilston, West Midlands, England
Polytronik GmbH, Quagliostr. 6, D-8000 Munchen 90, Germany

Herbert Streckfuss KG, Kruppstr. 10, D-7514 Eggenstein-Leop. 1, Germany
Kontakt-Systeme AG, Gewerbestr. 16, CH-8132 Egg/Zurich, Switzerland
Resco, Via Massena 2/A, I-20145 Milano, Italy
Rotawinder Ltd., Forest Rd., Hainault Essex, England
S.E.S. Elektromaterial GmbH, Alte Str. 22, D-7858 Weil/Rhein, Germany
Tekma Kinomat S.p.A., Via E. Fermi 635, I-21042 Caronno P., Italy
Zevatron GmbH, D-3548 Arolsen, Germany

Contour Milling Equipment
Excellon Europe GmbH, Lise-Meitner-Str. 7, D-6072 Dreieich, Germany
FAG S.A., 7 rue de Geneve, CH-1001 Lausanne, Switzerland
Peter Jordan, Mainstr. 5-7, D-6050 Offenbach, Germany
Retab AB, P. O. Box 153, S-18121 Lidingo, Sweden
Sarcem Nlle. Soc. S.A., Centre Commercial, B.P. 371, CH-1217 Meyrin-Geneve, Switzerland
W F. Klingelnberg, Berghauser Str. 54/63, D-5630 Remscheid, Germany

Darkroom Furniture and Equipment
Doschen India Pvt Ltd., Bangli Naka, Sandor, Post Vasai-401 201 Dist. Thane, India
Monotype India Ltd., 27th Cross Banashankari 2nd Stage, Bangalore-560 070, India
Lefra-Frank, Bonnigheimer Str. 53, D-7000 Stuttgart 40-Zuffenhausen, Germany
Metallum Pratteln AG, CH-4133 Pratteln, Switzerland

Desoldering Equipment
AAT Aston GmbH, Rothenburger Str. 33, D-8500 Nurnberg, Germany
Dr. K. Schleuniger & Co., Schongrunstr. 27, CH-4500 Solothurn, Switzerland
ERSA, Leonhard-Karl-Str. 24, D-6980 Wertheim 2, Germany
Hans Knurr KG, Ampfingstr. 27, D-8000 Munchen 80, Germany
PACE Inc., 9329 Fraser Street, Silver Spring, MD. 20910, USA

Developing Equipment, Photoresist
Adam Pill, Industriestr. 7, D-7151 Auenwald bei Stuttgart, Germany
Bernhard Schmiedeskamp, Blucherstr. 18, D-4970 Bad Oeynhausen, Germany
Chemcut Corporation, 500 Science Park, State College, Pennsylvania 16801, USA
Gebr. Schmid, Robert-Bosch-Str., D-7290 Freudenstadt, Germany
Hans Hollmuller GmbH & Co., Kappstr. 69, D-7033 Herrenberg, Germany
Herbert Streckfuss KG, Kruppstr. 10, D-7514 Eggenstein-Leop. 1, Germany
Hermann Wolf GmbH & Co. KG, Kieler Str. 33/35, D-5600 Wuppertal 1, Germany
International Supplies Co., Via Zanardi 7, I-43100 Parma, Italy
Laif Electronic GmbH & Co. KG, D-5202 Hennef-Happerschoss, Germany
Resco, Via Massena 2/A, I-20145 Milano, Italy
Tempress Microelectronics B.V., Marconistraat 14, Hoogeveen, Holland

Diamond Sawing Tools
Crouzet S.A., rue Jules Vedrines, F-26010 Valence Cedex, France
FAG S.A., 7 rue de Geneve, CH-1001 Lausanne, Switzerland
Gunter Effgen, Postfach 80, D-6581 Herrstein, Germany

Hager & Meisinger GmbH, Kronprinzenstr. 5-11, D-4000 Dusseldorf, Germany
Joisten & Kettenbaum GmbH, D-5060 Bergisch Gladbach 4, Germany
R. Becker, Senserstr. 8, D-8031 Neu-Esting/Munchen, Germany
Roeck AG, CH-4600 Olten, Switzerland
Tempress Microelectronics B.V., Marconistraat 14, Hoogeveen, Holland

Dip Coating Equipment, Wet-Film Resist
Doschen India Pvt. Ltd., Bangli Naka, Sandor, Post Vasai-401 201, Dist. Thane, India
Linton Laboratories Ltd., 4 Bartlow Rd., Linton, Cambridge CB1 6LY, England
Procirc Co. Ltd., Station Rd. West Haddon, Warwickshire, England

Diazo Film
Kanva Industries, 2 Metagalli Industrial Estate, Mysore-570 002, India

Drill Bit Grinding Machines
Christen, & Co. AG, Quellenweg 15, CH-3084 Wabern-Bern, Switzerland
Hawera Probst GmbH & Co., Schutzenstr. 3, D-7987 Weingarten, Germany
Meteor Engineering Works Ltd., Moosstr., CH-8803 Ruschlikon, Switzerland
Paul Kemmer GmbH & Co. KG, Postfach 452, D-7070 Schwabisch Gmund, Germany

Drill Bits, Tungsten Carbide
Andreas Maier KG, D-7959 Schwendi-Horenhausen, Germany
Diametal AG, Solothurnstr. 136, CH-2500 Biel, Switzerland
Dymet Alloys Ltd., Frimley Rd. Camberley, Surrey, England
Excellon Europe GmbH, Lise Meitner Str. 7, D-6072 Dreieich, Germany
FARCO, Girardet 29, CH-2400 Le Locle, Switzerland
Hager & Meisinger GmbH, Kronprinzenstr. 5-11, D-4000 Dusseldorf, Germany
Hawera Probst GmbH & Co. Schutzenstr. 3, D-7987 Weingarten Germany
Paul Kemmer GmbH & Co. KG, Postfach 452, D-7070 Schwabisch Gmund, Germany
Peter Jordan, Mainstr. 5-7, D-6050 Offenbach, Germany
Sphinxwerke Muller & Co. AG, CH-4500 Solothurn, Switzerland
Tulon Inc., 15209 South Broadway, Gardena, CA. 90248, USA

Drilling Machines, PCBs
Indian Mini Drills Pvt. Ltd., 8 J.C. Road, Vyayamshala Building, Bangalore-560 002, India
Punjab Recorders Ltd., B-17 Industrial Focal Point Phase II, S.A.S. Nagar, Mohali-160 051, India
Anton Kirner, Postfach 1340, D-7828 Neustadt/Schwarzwald, Germany
Bracker AG, CH-8330 Pfaffikon-Zurich, Switzerland
Elmar Wessel KG, D-4930 Detmold, Germany
Ernst Lenz KG, Jahnstr. 6, D-6349 Sinn, Germany
Excellon Europe GmbH, Lise Meitner Str. 7, D-6072 Dreieich, Germany
FAG S.A., 7 rue de Geneve, CH-1001 Lausanne, Switzerland
FARCO, Girardet 29, CH-2400 Le Locle, Switzerland
Ferco S.p.A., Via Rovani 6, I-20052 Monza, Italy

Hans Knurr KG, Ampfingstr. 27, D-8000 Munchen 80, Germany
HAPRO-Hanggı, Pfaffikerstr. 78, CH-8623 Wetzikon, Switzerland
Heinz Schmoll Ing., Westerbachstr. 4, D-6242 Dronberg, Germany
H.P. Heeb, D-7141 Oberriexingen, Germany
Joisten & Kettenbaum GmbH, D-5060 Bergisch Gladbach 4, Germany
Kager, Konstanzer Str. 73, Postfach 610324, D-6000 Frankfurt/M. 61, Germany
KEMA Elektronik, Lindenstr. 8, CH-8304 Wallisellen, Switzerland
Kontakt-Systeme AG, Gewerbestr. 16, CH-8132 Egg/Zurich, Switzerland
Laif Electronic GmbH & Co. KG, D-5202 Hennef-Happerschoss, Germany
Lasag AG, Schweizerweg 8, CH-3600 Thun, Switzerland
Oswald Boll, Rothenbaumchaussee 140, D-2000 Hamburg 13, Germany
Pluritec Italia S.p.A., Via Candossina 2, I-10010 Burolo d'Ivrea, Italy
Posalux AG, Opplingerstr. 18, CH-2500 Biel, Switzerland
Retab AB, P.O. Box 153, S-18121 Lidingo, Sweden
Rhone-Poulenc, 21 rue Jean-Goujon, F-75360 Paris Cedex 08, France
Surpro, Sportlaan 76, Hertogenbosch, Holland
Terima Steiner & Co., Kriesbachstr. 5, CH-8304 Wallisellen, Switzerland
W.F. Klingelnberg, Berghauser Str. 54/63, D-5630 Remscheid, Germany

Drilling Spindles
Punjab Recorders Ltd., B-17 Industrial Focal Point Phase II, S.A.S Nagar, Mohali-160 051, India
Federal-Mogul Westwind Air Bearings Ltd., Dalling Rd. Branksome, Poole, Dorset BH12 1LG, England
KAVO Elektrotechnisches Werk GmbH, Postfach 1206, D-797 Leutkirch/Allgau, Germany

Dry-Film Photoresists
Du Pont Far East Inc., 410 III Block Jayanagar, Bangalore-560 011, India
Dynachem International Ltd., Longley Lane, Sharston Industrial Estate, Wythenshawe, Manchester M22 4SY, England
Shipley Chemicals Ltd., Humber Avenue, Conventry, England

Drying Equipment, Transfer-Type
Adam Pill, Industriestr. 7, D-7151 Auenwald bei Stuttgart, Germany
Chemcut Corporation, 500 Science Park, State College, Pennsylvania 16801, USA
Ekra-Siebdrucktechnik GmbH & Co., Wilhelmstr. 45, D-7125 Kirchheim/N., Germany
E.T. Marler Ltd., Deer Park Rd., Wimbledon SW19 3UE, England
Hans Hollmuller GmbH & Co., Kappstr. 69, D-7033 Herrenberg, Germany
Holec Furnaces-Smit Ovens Nijmegen B.V., Post Box 68, NL-6500 AB Nijmegen, Holland
International Supplies Co., Via Zanardi 7, I-43100 Parma, Italy
Laif Electronic GmbH & Co. KG, D-5202 Hennef-Happerschoss, Germany
Resco, Via Massena 2/A, I-20145 Milano, Italy
Svecia GmbH, Happurger Str. 88, D-8500 Nurnberg, Germany
Tempress Microelectronics B.V., Marconistraat 14, Hoogeveen, Holland
W.C. Heraeus GmbH, D-6450 Hanau, Germany
Wesero GmbH & Co. KG, D-4322 Sprockhovel, Germany

Dust Collectors
Sur Electrical Company Pvt. Ltd., 21 Seal Lane, Calcutta-700 015, India

Etchant Regeneration Systems
Chemcut Corporation, 500 Science Park, State College, Pennsylvania 16801, USA
DEA Products Inc. 945 W. 23rd Street, Tempe, Arizona, USA
Dr. E. Durrwachter, Doduco KG, Westliche Karl-Friedrich-Str. 61, D-7530 Pforzheim, Germany
Hans Hollmuller GmbH & Co., Kappstr. 69, D-7033 Herrenberg, Germany
Hunt Chemicals GmbH, Rheinufer 70, D-6500 Mainz, Germany
I.T.C International GmbH, Scheibmeirstr. 32a, D-8000 Munchen 82, Germany
Oswald Boll GmbH, Rothenbaumchaussee 140, D-2000 Hamburg 13, Germany
Resco, Via Massena 2/A, I-20145 Milano, Italy

Etchants
Roxy Metal Finishers Pvt. Ltd., A-158 (A) Peenya Industrial Estate, Bangalore-562 140, India
Allied Chemical International S.A. N.V., Research Park, B-3044 Haasrode, Belgium
Chemcut Corporation, 500 Science Park, State College, Pennsylvania 16801, USA
Dr. E. Durrwachter, Doduco KG, Westl. Karl-Friedrich-Str. 61, D-7530 Pforzheim, Germany
E. Merck, Postfach 4119, D-6100 Darmstadt 1, Germany
Hans Hollmuller GmbH & Co., Kappstr. 69, D-7033 Herrenberg, Germany
Hunt Chemical N.V., Europark Noord 21-22, B-2700 Sint-Niklass, Belgium
Shipley Chemicals Ltd., Humber Avenue, Conventry CV3 1JL, England

Etching Machines, Spray-Type
Balaji Fabricators & Consultants, 23 Rammohanpura, Srirampura, Bangalore-560 021, India
Doschen India Pvt. Ltd., Bangli Naka, Sandor, Post Vasai-401 201, Dist. Thane, India
Jyoti Plastic Works, 94 B.T. Compound, Malad (West), Bombay-400 064, India
Adam Pill, Industriestr. 7, D-7151 Auenwald bei Stuttgart, Germany
Chemcut Corporation, 500 Science Park, State College, Pennsylvania 16801, USA
Colight, 820 Decatur Avenue North, Minneapolis, Minnesota 55427, USA
Compagnie Europeenne de Machines pour Circuits Imprimes et Microelectronique, 19 Boulevard du Lycee, F-92170 Vanves, France
FAG S.A., 7 rue de Geneve, CH-1001 Lausanne, Switzerland
Gebr. Schmid, Robert-Bosch-Str., D-7290 Freudenstadt, Germany
Hans Hollmuller GmbH & Co., Kappstr. 69, D-7033 Herrenberg, Germany
Hunt Chemical N.V., Europark Noord 21-22, B-2700 Sint-Niklaas, Belgium
I.T.C. Intercircuit GmbH, Scheibmeirstr. 32a, D-8000 Munchen 82, Germany
Kontakt-Systeme AG, Gewerbestr. 16, CH-8132 Egg/Zurich, Switzerland
Laif Electronic GmbH, Happerschoss, D-5202 Hennef/Sieg, Germany
Resco, Via Massena 2/A, I-20145 Milano, Italy
Tempress Microelectronics B.V., Marconistraat 14, Hoogeveen, Holland

Exposure Equipment, Actinic Light
Balaji Fabricators & Consultants, 23 Rammohanpura, Srirampura, Bangalore-560 021, India
Doschen India Pvt Ltd., Bangli Naka, Sandor, Post Vasai-401 201, Dist. Thane, India

Hansa Enterprises, 102 Williams Town Extn., P.B. 4607, Bangalore-560 046, India
Macc Engineering Co., 759/117 Deccan Gymkhana, Poona-411 004, India
Monotype India, 27th Cross Banashankari 2nd Stage, Bangalore-560 070, India
The Standard Printing Machinery Co., D-11 Industrial Estate, Ambattur, Madras-600 058, India
Colight, 820 Decatur Avenue North, Minneapolis, Minnesota 55427, USA
DEK Printing Machines Ltd., 1 Euston Centre, London NW1, England
Dynachem International Ltd., Longley Lane, Sharston Industrial Estate, Wythenshawe, Manchester M22 4SY, England
Fotoclark, Friedrich Grun KG, Industriepark Kottenforst, D-5309 Meckenheim bei Bonn, Germany
Gyrex Corporation, 436 East Gutierrez Street, Santa Barbara, CA. 93101, USA
Hibass Photomec Ltd., Denington Industrial Estate, Wellingborough, Northants., England
I.T.C. Intercircuit GmbH, Scheibmeirstr. 32a, D-8000 Munchen 82, Germany
Klimsch & Co., Schmidtstr. 12, Postfach 3434, D-6000 Frankfurt/M. 1, Germany
Kontakt-Systeme AG, Gewerbestr. 16, CH-8132 Egg/Zurich, Switzerland
Laif Electronic GmbH & Co., D-5202 Hennef Happerschoss, Germany
Littlejohn Circuit Equipment Ltd., Gt. Park Street, Wellingborough, Northants. NN8 4DH. England
Procirc Co. Ltd. Station Rd., West Haddon, Warwickshire, England
Resco, Via Massena 2/A, I-20145 Milano, Italy

Film Materials, Polyester Base
Agfa Gevaert India Ltd., Merchant Chambers, 41 New Marine Lines, Bombay-400 020, India
Hindustan Photo Films Mfg. Co. Ltd., Indu Nagar, Ootacamund-643 005, India
Kodak Ltd., Kodak House, Dr. D.N. Rd., Bombay-400 001, India

Filter Pumps, Plating
Candorchemie GmbH, Prinz-Regent-Str. 50/60, Postfach 100144, D-4630 Bochum, Germany
Dr. E. Durrwachter, Doduco KG, Westliche Karl-Friedrich-Str. 61, D-7530 Pforzheim, Germany
Flow Laboratories GmbH, Diezstr. 10, D-5300 Bonn 3, Germany
M & T Chemicals GmbH, Hechinger Str. 68, D-7000 Stuttgart 80, Germany
Serfilco U.K. Ltd., P.O. Box 11, Eccles, Manchester M30 9LA, England
W. Wirth AG, Maulbeerstr. 6, CH-4021 Basel, Switzerland

Fluxes, Soldering
Billiton Metals and Ores Deutschland GmbH, D-4300 Essen 1, Germany
Detakta, D-2000 Hamburg 60, Germany
ERSA, Leonhard-Karl-Str. 24, D-6980 Wertheim 2, Germany
Litton Precision Products International Inc., Oberfohringer Str. 8, D-8000 Munchen 80, Germany
Multicore Solders Ltd., Hemel Hempstead, Herts, England
Zevatron GmbH, D-3548 Arolsen, Germany

Gold Recovery Equipment
Dr. E. Durrwachter, Doduco KG, Westliche Karl-Friedrich-Str. 61, D-7530 Pforzheim, Germany
Laytron S.A., 21 rue du Mont-Blanc, CH-1201 Geneve, Switzerland
Schering AG, Mullerstr. 170-178, D-1000 Berlin 65, Germany

Hand Soldering Equipment
Reliance Electronics, 767 Sadashiv Peth, Poona-411 030, India
Suberb Products, 866 Sadashiv Peth, Poona-411 030, India
AAT Aston GmbH, Rothenburger Str. 33, D-8500 Nurnberg, Germany
ADOLA AG, Etzelstr. 3, CH-8800 Thalwil, Switzerland
AMTEX (Electronics) Ltd., Armada Way, Plymouth, Devon, England
Cooper Group Deutschland GmbH, Postfach 140, D-7122 Besigheim/Neckar, Germany
Engel GmbH, Rheingaustr. 34-36, D-6200 Wiesbaden-Schierstein, Germany
ERSA, Leonhard-Karl-Str. 24, D-6980 Wertheim 2, Germany
GLT Gesellschaft fur Lottechnik mbH, Kreuzstr. 150, D-7534 Birkenfeld, Germany
Hans Knurr KG, Ampfingstr. 27, D-8000 Munchen 80, Germany
Kager, Konstanzer Str. 73, Postfach 610324, D-6000 Frankfurt/M. 61, Germany
Lotring, Kantstr. 115, D-1000 Berlin 12, Germany
PACE Inc., 9329 Fraser Street, Silver Spring, MD. 20910, USA
Polytronik GmbH, Quagliostr. 6, D-8000 Munchen 90, Germany
Weld Equip Deutschland, Alte Allee 52, D-8000 Munchen 60, Germany
Weller Elektro-Werkzeuge GmbH, D-7122 Besigheim/Wurtt., Germany

Hot Air Leveling Equipment
Electrovert Inc., International Division, 3285 Cavendish Blvd., Montreal, Quebec H4B 2L9, Canada
Gyrex Corporation, 436 East Gutierrez Street, Santa Barbara, CA. 93101, USA
Gebr-Schmid, Robert-Bosch Str., D-7290 Freudenstadt, Germany

Insulation Stripping Devices
AAT Aston GmbH, Rothenburger Strasse 33, D-8500 Nurnberg, Germany
Bullnheimer & Co., Schaetlerstrasse 6, D-8900 Augsburg, Germany
Dr. K. Schleuniger & Co., Schongrunstrasse 27, CH-4500 Solothurn, Switzerland
Hans Knurr KG, Amfingstrasse 27, D-8000 Munchen 80, Germany
Polytronik GmbH, Quagliostrasse 6, D-8000 Munchen 90, Germany
Terima, Steiner & Co., Kriesbachstrasse 5, CH-8304 Wallisellen, Switzerland

Mechanical PCB Milling Equipment
LPFK Jurgen Seebach GmbH, Scheffelstr. 17, D-3000 Hannover 1, Germany

Multilayer Presses
Europrim, 19 Boulevard du Lycee, F-92170 Vanves, France
FAG S.A., 7 rue de Geneve, CH-1001 Lausanne, Switzerland
Given-Phi, 200 N. Berry Street, Brea, CA. 92621, USA
Hull International Ltd., Grangestone Industrial Estate, Girvan, England

Lauffer & Butscher, D-7241 Muhlen-Horb, Germany
MAS Deutschland GmbH, Industriestr. 3, D-6233 Kelkheim/Ts., Germany
Oswald Boll GmbH, Rothenbaumchaussee 140, D-2000 Hamburg 13, Germany
Robert Burkle & Co., D-7290 Freudenstadt, Germany

Plated-Through Hole Process Chemicals
Grauer & Weil (India) Ltd., Sukh Sagar, S. Patkar Marg, Bombay-400 007, India
Metal Chem, 173A 1st N Block, Rajajinagar, Bangalore-560 010, India
Roman Industries, Kakad Chambers, Dr. Annie Besant Rd., Worli, Bombay-400 018, India
Roxy Metal Finishers Pvt. Ltd., A-158 (A) Peenya Industrial Estate, Bangalore-562 140, India

Plate Shearing Equipment
Machinery Impex Corporation, 271-A Nagdevi Street, Bombay-400 003, India
International Machine Tools Corporation, 5 Bank Street, Fort, Bombay-400 023, India
IMETEX, Havenkade 9, P.O. Box 78, Venio, Holland
UNICUM, Paul Julien, Brandrijsweg 6-8, Beekbergen, Holland
von Arx AG, CH-4450 Sissach, Switzerland

Plating Equipment
ARA Pvt. Ltd., 41 Hamam Street, Bombay-400 001, India
Delta Chemicals, Vijay Kailash Estate, Sonawala X Road, Goregaon (East), Bombay-400 063, India
Grauer & Weil (India) Ltd., Sukh Sagar, S. Patkar Marg, Bombay-400 007, India
Jyoti Plastic Works, 94 B.T. Compound, Malad (West), Bombay-400 064, India

Plating Thickness Measuring Equipment
Helmut Fischer GmbH & Co. KG, Industriestr. 21, D-7032 Sindelfingen, Germany
UNIT Process Assemblies Inc., 60 Oak Drive, Syosset, New York 11791, USA

Precious Metal Plating Process Chemicals
Arora Matthey Ltd., 166 Netaji Subhas Chandra Bose Rd., Tollygunge, Calcutta-700 040, India
Canning Mitra Phoenix Ltd., Eucharistic Congress Building III, 5 Convent Street, Bombay-400 063, India
Delta Chemicals, Vijay Kailash Estate, Sonawala X Rd., Goregaon (East), Bombay-400 063, India
Grauer & Weil (India) Ltd., Sukh Sagar, S. Patkar Marg, Bombay-400 007, India
Metal Chem, 173A 1st N Block, Rajajinagar, Bangalore-560 010, India
Platewel Processes & Chemicals Ltd., Padra Rd., Atladra, P.B. 70, Baroda-390 001, India
Roman Industries, Kakad Chambers, Dr. Annie Besant Rd., Worli, Bombay-400 018, India
Roxy Metal Finishers Pvt. Ltd., A-158 (A) Peenya Industrial Estate, Bangalore-562 140, India

Protective Lacquer Coatings
Electronaids, 43 6th Main Vth Block, Jayanagar, Bangalore-560 041, India
HAUNSA Graphic Arts Chemicals, Radiant Industrial & Commercial Co., Bombay-400 059, India

KEONICS Ltd., Emlyn Haven, 30 Race Course Road, Bangalore-560 001 India
Laktham Enterprises, 53/3 I & III Block East, Byrasandra Layout, Bangalore-560 011, India
Seshasayee Brother's Pvt. Ltd., 10 Rutland Gate, Fourth Street, Madras-600 006, India
Sukant Electronics Pvt. Ltd., Mahajan Silk Mill Compound, L.B. Shastri Marg, Vikhroli, Bombay-400 079, India
Airco Temescal B.V., Stadionweg 93, Rotterdam, Holland
Detakta, D-2000 Hamburg 60, Germany
Dexter GmbH, Ingolstadter Str. 62, D-8000 Munchen 46, Germany
Engelhard Industries Limited, Valley Rd., Cinderford, England
ERSA, Leonhard-Karl-Str. 24, D-6980 Wertheim 2, Germany
Hans Hollmuller GmbH & Co., Kappstr. 69, D-7033 Herrenberg, Germany
Niederrheinische Lackfabrik, Werner Peters KG, Untergath 24, D-415 Krefeld, Germany

Reprographic Cameras
Doschen India Pvt. Ltd., Bangli Naka, Sandor, Post Vasai-401 201, Dist. Thane, India
Monotype India Ltd., 27th Cross Banashankari 2nd Stage, Bangalore-560 070, India
Photo Goods Service, 6465 Katra Baryan, New Delhi-110 006, India
The Shevade's Camera Works Ltd., Tilakwadi, Belgaum-590 006, India
The Standard Printing Machinery Co., D-11 Industrial Estate, Ambattur, Madras-600 058, India
Agfa-Gevaert AG, D-5000 Leverkusen 1, Germany
Hans Sixt KG, D-6909 Walldorf bei Heidelberg, Germany
Klimsch & Co., Schmidtstr. 12, Postfach 3434, D-6000 Frankfurt/M. 1, Germany
Mollidor & Muller, Weisser Str. 161, D-5038 Rodenkirchen, Germany

Roller Coating Equipment, Wet-Film Resist
Gyrex Corporation, 436 East Gutierrez Street, Santa Barbara, CA. 93101, USA
Hans Hollmuller GmbH & Co., Kappstr. 69, D-7033 Herrenberg, Germany
Resco, Via Massena 2/A, I-20145 Milano, Italy

Roller Tinning Equipment
Balaji Fabricators & Consultants, Rammohanpura, Srirampura, Bangalore-560 021, India
Hansa Enterprises, 102 Williams Town Extn., P.B. 4607, Bangalore-560 046, India
Electrovert Inc. International Division, 3285 Cavendish Blvd., Montreal, Quebec H4B 2L9, Canada
ERSA KG, Leonhard-Karl-Str. 24, D-6980 Wertheim 2, Germany
Linton Laboratories Ltd., 4 Bartlow Rd., Linton, Cambridge CB1 6LY, England

Safety Equipment, Personnel
Francies Leslie & Co., 105 Apollo Street, Fort, Bombay-400 001, India

Screen Fabrics
Woodpeck Industries, 81 Sheriff Devji Street, Bombay-400 003, India
Albert-Frankenthal Aktiengesellschaft, Johann-Klein-Str. 1, D-6710 Frankenthal, Germany
Argon Service Ltd., Via Malpighi 4, I-20129 Milano, Italy
Buckbee-Mears Europe GmbH. Renkenrunstr., D-7840 Mullheim/Baden, Germany

DEK Printing Machines Ltd., 1 Euston Centre, London NW1, England
G. Bopp & Co. AG, Bachmannweg 20, CH-8046 Zurich, Switzerland
Schweiz. Seidengazefabrik AG, CH-9425 Thal/SG, Switzerland
Sporl & Co., Staudenweg 757, D-7485 Sigmaringendorf, Germany
ZBF Zuricher Beuteltuchfabrik AG, CH-8803 Ruschlikon, Switzerland

Screen Preparation Chemicals
Lunar Caustic Pvt. Ltd., P.B. 806, Poona-411 004, India
Radiant Industrial & Commercial Co., Jafferbhoy Industrial Estate, Kurla-Andheri Road, Bombay-400 059, India
Roman Industries, Kakad Chambers, Dr. Annie Besant Rd., Worli, Bombay-400 018, India

Screen Printing Inks
Radiant Industrial & Commercial Co., Jafferbhoy Industrial Estate, Kurla-Andheri Road, Bombay-400 059, India
Rainbow Ink & Varnish Mfg. Co. Pvt. Ltd., 133C Vakola, Santacruz (East), Bombay-400 055, India
Argon Service Ltd., Via Malpighi 4, I-20129 Milano, Italy
Cermalloy, Division Bala Electronics Corp., 14 Fayette Street, Conshohocken, PA. 19428, USA
DEK Printing Machines Ltd., 1 Euston Centre, London NW1, England
Detakta, D-2000 Hamburg 60, Germany
Engelhard Industries Limited, Valley Rd., Cinderford, England
E.T. Marler Ltd., Deer Park Road, Wimbledon SW19 3UE, England
Hans Hollmuller GmbH & Co., Kappstr. 69, D-7033 Herrenberg, Germany
H. Wiederhold, Am Stadtpark 69, D-8500 Nurnberg, Germany

Screen Printing Machines and Equipment
Magmo Textile Equipments Pvt. Ltd., 2M Steel Centre, Sant Tukaram Road, Iron Market, Bombay-400 009, India
Albert-Frankenthal Aktiengesellschaft, Johann-Klein-Str. 1, D-6710 Frankenthal, Germany
Argon Service Ltd., Via Malpighi 4, I-20129 Milano, Italy
Chemcut Corporation, 500 Science Park, State College, Pennsylvania 16801, USA
DEK Printing Machines Ltd., 1 Euston Centre, London NW1, England
Ekra-Siebdrucktechnik GmbH & Co., Wilhelmstr. 45, D-7125 Kirchheim/N., Germany
E.T. Marler Ltd., Deer Park Rd., Wimbledon SW19 3UE, England
Hans Hollmuller GmbH & Co., Kappstr. 69, D-7033 Herrenberg, Germany
I.T.C. Intercircuit GmbH, Scheibmeirstr. 32a, D-8000 Munchen 82, Germany
mpm n.v., Terbekehofdreef 55-59, B-2610 Wilrijk/Antwerpen, Belgium
Peter Jordan, Mainstr. 5-7, D-6050 Offenbach, Germany
Resco, Via Massena 2/A, I-20145 Milano, Italy
Svecia GmbH, Happurger Str. 88, D-8500 Nurnberg, Germany
Technitron GmbH, Martin-Luther-Str. 20, D-8000 Munchen 90, Germany

Silver Recovery Equipment
Engelhard Industries Limited, Valley Rd., Cinderford, England

Dr. E. Durrwachter, Doduco KG, Westliche Karl-Friedrich-Str. 61, D-7530 Pforzheim Germany
Laytron S.A., 21 rue du Mont-Blanc, CH-1201, Geneve, Switzerland

Solder Reflow Equipment
Argus Engineering Co., P.B. 38, Hopewell, New Jersey 08525, USA
Electrovert Inc., International Division, 3285 Cavendish Blvd., Montreal, Quebec H4B 2L9, Canada
Glo-Quartz Ovens Inc., 7074-7100 Maple Street, Mentor, Ohio 44060, USA
I.T.C. Intercircuit GmbH, Scheibmeirstr. 32a, D-8000 Munchen 82, Germany
Peter Jordan, Mainstr. 5-7, D-6050 Offenbach, Germany
Research Inc., P.B. 24064, Minneapolis, Minnesota 55424, USA

Stripping Equipment
Bernhard Schmiedeskamp, Blucherstr. 18, D-4970 Bad Oeynhausen, Germany
Chemcut Corporation, 500 Science Park, State College, Pennsylvania 16801, USA
Gebr. Schmid, Robert-Bosch-Str., D-7290 Freudenstadt, Germany
Hans Hollmuller GmbH & Co., Kappstr. 69, D-7033 Herrenberg, Germany
International Supplies Co., Via Zanardi 7, I-43100 Parma, Italy
Laif Electronic GmbH & Co. KG, D-5202 Hennef-Happerschoss, Germany
Resco, Via Massena 2/A, I-20145 Milano, Italy

Stylus Plating Equipment, Selective Plating
Selectrons Ltd., 38 Walkers Rd., Moons Moat North, Redditch, Worcestershire B98 9HD, England

Through-Contact Revets, Double-Sided PCBs
Elektromin, 276/277 Narayan Peth, N.C. Kelkar Rd., Poona-411 030 India
General Tube & Hardware Mart, 138 Linghi Chetty Street, Madras-600 001, India

Waste Water Purification Equipment
Friedr. Blasberg GmbH & Co. KG, Merscheider Str. 165, D-5650 Solingen, Germany
Gebr. Schmid, Robert-Bosch-Str., D-7290 Freudenstadt, Germany
Laytron S.A., 21 rue de Mont-Blanc, CH-1201 Geneve, Switzerland
Resco, Via Massena 2/A, I-20145 Milano, Italy
Schurmann GmbH & Co., Ringstr. 17-19, D-5620 Velbert 15, Germany
Wesero GmbH & Co. KG, D-4322 Sprockhovel, Germany

Water Deionising Plants
Ion Exchange (India) Ltd., Tiecicon House, Dr. E. Moses Rd., P.B. 6273, Bombay-400 011, India
The Starit Engineering Co. Pvt. Ltd., 18 Park Street, Calcutta-700 071, India

Wave Soldering Equipment
Balaji Fabricators & Consultants, 23 Rammohanpura, Srirampura, Bangalore-560 021, India
Cooper Group Deutschland GmbH, Postfach 140, D-7122 Besigheim/Neckar, Germany

DEE Electric Company Inc., 2501 North Wayne Ave., Chicago, Illinois 60614, USA
Dr. K. Schleuniger & Co., Schongrunstr. 27, CH-4500 Solothurn, Switzerland
Electrovert Inc., 3285 Cavendish Blvd, Montreal, Quebec H4B 2L9, Canada
EPM AG, Klosbachstr 83, CH-8030 Zurich, Switzerland
ERSA, Leonhard-Karl-Str. 24, D-6980 Wertheim 2, Germany
GLT Gesellschaft fur Lottechnik mbH, Kreuzstr. 150, D-7534 Birkenfeld, Germany
Holec Furnaces-Smit Ovens Njimegem B.V., P.O. Box 68, NL-6500 AB Njimegen, Holland
Kontakt-Systeme AG, Gewerbestr. 16, CH-8132 Egg/Zurich, Switzerland
Laytron S.A., 21 rue du Mont-Blanc, CH-1201 Geneve, Switzerland
Peter Jordan, Mainstr. 5-7, D-6050 Offenbach, Germany
Polytronik GmbH, Quagliostr. 6, D-8000 Munchen 90, Germany
Rotawinder Ltd., Forest Rd., Hainault, Essex, England
Sarcem Nlle. Soc. S.A., Central Commercial, B.P. 371, CH-1217 Meyrin-Geneve, Switzerland
SEHO GmbH, D-6983 Kreuzwertheim, Germany
Tekma Kinomat S.p.A., Via E. Fermi 635, I-21042 Caronno P., Italy
Tempress Microelectronics B.V., Marconistraat 14, Hoogeveen, Holland
Zevatron GmbH, D-3548 Arolsen, Germany

Wet-Film Photoresists
Agfa-Gevaert India Ltd., Merchant Chambers, 41 New Marine Lines, Bombay-400 020, India
Kodak Ltd., Kodak House, Dr. D.N. Rd., Bombay-400 001, India
Lunar Caustic Pvt. Ltd., P.B. 806, Poona-411 004, India
Prima Industrial & Commercial Corp., 29 Ansa Industrial Estate, Saki Vihar Rd., Sakinaka, Bombay-400 072, India
Roman Industries, Kakad Chambers, Dr. Annie Besant Rd., Worli, Bombay-400 018, India
Unilab, 45/6 VI Block, Rajajinagar, Bangalore-560 010, India
Vista Graphics, 110 Sarojini Devi Rd., Secundarabad-3 (A.P.), India
Detakta, D-2000 Hamburg 60, Germany
Hunt Chemical N.V., Europark Noord 21-22, B-2700 Sint Niklaas, Belgium
Kalle Niederlassung der Hoechst AG, Rheingaustr. 190, D-6200 Wiesbaden 1, Germany
Shipley Chemicals Ltd., Humber Avenue, Coventry CV3 1JL, England

Whirl Coating Equipment, Wet-Film Resist
Doschen India Pvt. Ltd., Bangli Naka, Sandor, Post Vasai-401 201, Dist. Thane, India
Monotype India Ltd., 27th Cross Banashankari 2nd Stage, Bangalore-560 070, India
The Shevade's Camera Works Ltd., Tilakwadi, Belgaum-590 006, India
The Standard Printing Machinery Co., D-11 Industrial Estate, Ambattur, Madras-600 058, India

INDEX

Abrasive cleaning, 228
Abrasive slurry, 229
Acetate-based film, 171
Acetic acid, 435
Acid copper electrolyte, 276
Acid dip, 227
Acid, care, 435
Additive process, 348
Addresses, 439
Adhesively coated laminate, 346
Ageing, film, 180
Air, 433
Air conditioner, 185
Air dryer temperature, 180
Air effect, microstrip, 124
Air flow, 95
Alkali cleaning, plating, 275
Alkaline ammonia, 297, 305, 431, 438
Alloys, solder, 386
Alternative finishes, plating, 283
Aluminium frame, screen-printing, 258
Aluminium shielding, 52
Ammonia, alkaline: *see* "Alkaline ammonia"
Amplifier,
   broadband, 66, 80
   differential, 102
   feedback, 99
   high-frequency, 96
   high-gain, dc, 99
   low-level, 100
   selective, 65
Analog circuits, design rules, 95
ANSI, 440
Aqueous-developing dry-film resist, 253
Aperture, 187
Approach, artwork, 130
Approximation, inductive and capacitive line, 67, 73

Artwork, 129
Artwork, draughting, automated, 152
Artwork, multilayer, 369
Assembly drawing, 16
Assembly techniques, 409
Assembly, power electronics, 112
ASTM, 439
Automated artwork draughting, 152
Automation PCB design, 151

B-stage, 378
Back etching, 380
Backing plate, 329
Base foil, artwork, 135
Base material: *see* "Laminate"
Base, film, 170
Basic PCB processes, 232
Batch-production etching, 308
Bath test, solder, 400
Baume, gravity, 304
Bending, leads, 410
Benzene, 435
Betaray backscatter, 237, 288
Betascope: *see* "Betaray backscatter"
Bit, drilling, 328
   iron soldering, 393
Black taping, artwork, 133, 139
Blades, circular saw, 316
   shearing, 315
Blanking, 320
Blistering, 220
Board cleaning, 225
Boric acid, analysis, 280
Broadband amplifier, 66, 80
Bromide, 179
Brushes, abrasive, 229
Bubble etching, 293

Building requirements, 424
Bushes, drilling, 327

C-MOS, 37
C_{equ} fast pulse, 70
CAD, 160
Camera, 181, 482
Capacitance, 20
Capacitive approximation, line, 67, 73
Capacitive conductor, 22
Capacitive coupling, 22, 101
Capacitors, printed, 75
Carbon arc lamp, 245
Carousel, light trap, 186
Cascade etching, 310
Cascade rinsing, 432
Cascode stage, 79
Catalyst, 350
Catalytic surface, 272
Caustic soda, 438
Centrifugal tinning, 283
Characteristic curve, film, 172
Characteristic impedance, 114, 116, 121, 124
Check,
  artwork, 148
  layout, 31
Chemical reduction, 273
Chemical store, room, 426
Chloride, analysis, 277, 280
Chlorinated solvent fumes, 434
Chlorine, 435
Chromic acid, 297, 303, 430
Chromic anhydride, 435
Chromium treatment, 430
Circle of confusion, 188
Circuit diagram, 14
Circular saw, 316
Clamp-down device, drilling, 327
Classification, thermal, laminate, 203
Clean cutting, 315
Cleaning,
  assembly, 417
  board, 225
  screen, 268
Clearance hole, 376
Clearing angle, 332
Clinching, 413
Closed-loop regeneration, 311
CNC-drilling, 325
Coating,
  screen, 261
  wet-film resist, 239
Coaxial cable, 85

Code,
  component, 15
  layout sketch, 15
Colometric comparison, 300
Component list, 14
Component,
  placing, mounting, 29, 95, 158, 163, 409
  polarity identification, 148
Computer-aided design, 156
Computer, PCB design, 151, 154
Conductor,
  capacitance, 22
  configuration, 27
  ground, supply: see "Ground/Supply conductors"
  inductance, 23, 25
  orientation, 145
  routing, 145, 159, 163
  spacing, 24, 110, 145, 147
  wave impedance, 60
  width, 25, 109, 355
Confusion circle, 188
Connection, terminal, 107
Construction, multilayer, 375
Consumer PCB, 421, 423
Contamination, soldering, 383
Continuity between layers, test, 372
Continuous-feed etching, 309
Contour sharpness, screen, 267
Contrast, film, 174
Control, pollution, 428
Conveyorised etching, 294
Cooling requirements, 30, 95
Coordinatograph, 141
Copperclad laminate: see "Laminate"
Copper-to-base laminate bond, 208
Copper analysis, 277
Copper bath, electroless, 273, 344, 347, 357
Copper electrolyte, acid, 276
Copper foil, 200
Copper foil thickness, 107
Copper plating, 276, 380
Copper surface standards, 210
Copy board, 183
Core materials, 359
Corner, artwork, 144
Corrosive flux, 390
Corrosiveness, etchant, 297
Costs,
  etching, 297, 308
  labour, 161
Coupled transmission line, 125
Coupling,
  capacitive, 22, 101
  magnetic, 103

INDEX

Coupling, resistive, 110
Creep strength, solder, 390
Crosstalk, 44, 101
Cupric chloride, 297, 300, 430
Current spike, 48
Cut-and-strip artwork, 141
Cut corners, artwork, 144
Cutting speed, drilling, 329
Cutting, laminate, 315
Cyanide, 437

Darkroom, 181, 184, 424
Data input, CAD, 156
De-ionising, 432
Deburring, 228
Decoupling capacitor, 49
Decreasing line width, film, 193
Defects, soldering, 404
Degreasing,
    boards, 227
    screen, 260
Delamination, test, 375
Density, film, 171, 175
Depth of field, 188
Design rules,
    analog, 95
    digital circuit, 33
    high-frequency and fast pulse, 59
    microwave, 114
    power electronics, 105
Design,
    automation and computers, 151, 161
    multilayer, 368
Development,
    film, 191
    resist, 247
Diallyl pthalate laminate, 214
Diameter,
    hole: *see* "Hole diameter"
    solder pad, 148
Diazo film, 139
Dielectric constant, 20, 205
Dielectric losses, 89, 127, 205
Dielectric materials, microwave, 127
Dielectric strength, 204
Digital circuit, design rules, 33
Digitiser, 152
Dihedral angle, 385
Dimensional hysteresis, film, 178
Dimensional stability, film, 176, 179, 181
DIN-beaker, 237
DIN standard, size, 9, 12
Dip coating, 241
Dip soldering, 394
Direct-sight drilling, 322

Direct method, screen, 261
Disposal,
    etchant: *see* "Neutralisation"
    sludge, 431
Dissipation factor, 205
Distillation, ammonia etchant, 431
Documentation, 14
Double-sided, boards, 6, 232
Drag soldering, 396
Draughting artwork, automated, 152
Drill bits, 328
Drilling, 320
Drilling spindles, 327
Dry-film resist, 250
Dry heat, test, 220
Drying, film, 180, 193
Dust, 433
Dust, darkroom, 185
Dyeing, 248

ECL, 40
Economy,
    CAD, 161
    etchant, 308
EIA, 439
EIPC, 439
Electrical properties, laminate, 204
Electrical test, multilayer, 372
Electro deposition, laminate, 200
Electrode arrangement, laminate test, 204, 207
Electrographic porosity test, 287
Electroless copper plating, 233, 272, 344, 347
Electroless plating, 272
Electrolytic copper plating, 233
Electromagnetic (E.M.) interference, 50, 53, 54, 112
Electroplating, 274
Electrostatic coating, 360
Emitter follower stage, 97
Emulsion parameters, 171
Emulsion,
    film, 170, 175
    screen, 262
Equipment import, 420
Epoxy laminate, 213
Etch back, 380
Etch factor, 292
Etch resist, 234, 297
Etching, 292
Ethyl alcohol, 435
Eutectic system, solder, 386
Example,
    artwork, double-sided, 137
    artwork, multilayer, 370
    room layout, 427

Expansion coefficient, film, 177
Exposure control,
  resist, 246
  screen, 262
Exposure latitude, film, 174
Exposure time, film, 190
Exposure,
  resist, 244, 253
  screen film, 263

Fabrication, microwave, 127
Fabrics, screen-printing, 256
Faraday's law, 274
Fast-pulse, design rules, 59
Feedback capacitance, 80
Ferric chloride, 298
Filler, 199
Film emulsion, 175
Film master, 170
  example, 138
Film processing, 187
Film registration, 195
Film, screen, 263
Filter, 50, 125
Fine-line conductors, 333
Finishes, alternative to plating, 283
Fixed light-source, resist exposure, 244
Fixing bath, 192
Fixing, screen to frame, 259
Flashover, 111
Flame resistance, 209
Flammability, 220
Flash etching, 335
Flat-field lens, 183
Flexible base materials, 353
Flexible PCB, 352
Flexural strength, 208, 221
Flow coating, 239
Fluidised-bed coating, 360
Fluoboric acid, analysis, 282
Fluorides, 437
Flux, 390
  removal, 397
Focal length, 183
Focussing, camera, 188
Formaldehyde, 436
Formic acid, 436
Frame preparation, screen-printing, 258
Frame, screen-printing, 258
Free fluoboric acid, analysis, 282
Free mixed acids, analysis, 279
Front-to-back registration, 195
Fumes, 434
Fusible alloys, 388
Fusing, solder, 293

Gamma, film, 172, 175
Globule test, solder, 401
Gold,
  analysis, 281
  electroplating, 281
  immersion plating, 272
Grain, film, 175
Gravity, baumé, 304
Grid sheet, artwork, 142
Grid system, 5
Ground conductor, 79, 104
  noise, 48
  width, 25
Ground line: see "Ground conductor"
Guard line, 79, 101, 103
Guidelines,
  artwork taping, 142
  starting PCB facilities, 419

Hand assembly, 416
Hand screen-printing, 266
Hardboard stiffener, 356
Hardness, water, 193
Hazards, health: See "Health care"
Health-affecting chemicals, 434
Health care, 228, 265, 304, 406
Heat shock, 220
Heatsink, 30, 112
Heavy-metal precipitation, 429
High-density interconnections, 342
High-frequency amplifier/oscillator, 96
High-frequency losses, 81
High-frequency PCB design, 59
High-power output stage, 98
High-power part, 106
Hole diameter, 148, 318, 321, 409
Hole inspection, multilayer, 379
Horizontal-type camera, 181
Horizontal burning test, 222
Hot-air leveling, 286
Hydrochloric acid, 436
Hydrocyanic acid, 436
Hydrofloric acid, 436
Hysteresis, dimensional, film, 178

ICT, 439
IEC, 440
IEEE, 439
Illuminance density, film, 171
Illumination. darkroom, 185
Immersion plating, 271
Impedance: see "Wave impedance" or "Characteristic impedance"
Imperfection, copper surface, 210
Import, equipment, 420

INDEX

Impregnation, 201
Impurities, solder, 389
Increasing line width, film, 193
Indexing pins, 194
Indirect method, film, 263
Inductance, conductor, 23
Induction heat soldering, 399
Inductive approximation, line, 67, 73
Inductors, printed, 75
Infection effect, film, 175
Ink drawn artwork, 131
Ink, screen-printing, 265
Input data, CAD, 156
Inspection,
   artwork, 150
   assembly, 415
   holes, multilayer, 379
Insulation layer, metal core, 359
Insulation resistance, 206
Insulation resistance between layers, test, 373
Interconnection techniques, 375
Interfacial surface energy, 385
Interference,
   E.M., 50, 53, 54, 112
   magnetic, 52
Intermediate flux, 391
Ion-exchange, 432
IPC, 439
Iron soldering, 393
ISI, 440

Jig drilling, 322
Jumper wire, 413

Knife, artwork, 142

Laboratory-type etchers, 294
Labour cost, 161
Laminate, 199
   dielectric constant, 20
   microwave, 127
   semi- and additive processes, 345
   thickness, 106, 209
Laminating, multilayer, 378
Lamination, dry-film, 251
Lancy process, 432
Latitude, film exposure, 174
Layout, 17
Layout check, 31
Layout materials, 12
Layout procedures, 13
Layout sketch, 12, 15
Layout,
   approach, 12
   scale, 4

Lead cutting, 416
Lead preparation, component, 409, 414
Lead, analysis, 282
L_{equ}, fast-pulse, 70
Light sources, resist exposure, 245
Light trap carousel, 186
Limitations,
   design automation, 164
   manual designing, 151
Line defects, inductive/capacitive approximation, 73
Line emulsion, 176
Line losses, 91
Line width changes, film, 193
Liquid-type resist, 236
List of addresses, 439
Lith emulsion, 175
Load reflection factor, 65
Loss angle, 205
Loss factor, 206
Losses,
   dielectric, 90, 127
   line, 91
   radiation, 86
   skin effect, 83
Low-level signal amplifier, 100
Low-melting-point alloys, 388
Low-power part, 106

Machine screen-printing, 267
Machine,
   cleaning, 228
   drilling, 321
   etching, 293
Magnetic coupling, 103
Magnetic interference, 52
Magnification, camera, 188
Mains filter, 50
Manual board cleaning, 226
Manual designing, limitations, 151
Manual plating installation, 276
Manual regeneration, cupric chloride, 302
Manufacture, laminate, 199
Market constraints, 419
Mask, soldering, 398
Mass soldering, 394
Matching,
   counductors, 59
   high-frequency case, 61
   wave impedance, 40, 43
Material flow, darkroom, 184
Materials, microwave, 127
Mechanical drawing, 15

Mechanical machining operations, 314
Mechanical milling, 361
Mechanical properties, solder, 389
Medical aspects: see "Health care"
Melamine laminate, 215
Meniscus rise test, 403
Mercury-vapour lamp, 245
Mesh per cm, screen, 257
Metal core PCB, 357
Metal precipitation, 429
Metal resist, 297
Metallic soil, 225
Metallographic test, 372
Methyl alcohol, 435
Microsection, 286
Microstrip, 62, 85, 88, 89, 122
Microwave, design rules, 114
Miller capacitance, 81
Miller effect, 98
Milling, 320
Mismatch,
    effect on rise-time, 72
    fast-pulse case, 61
    high-frequency case, 61
    pulse, 64
    pulse circuits, 67
Mixed acids, analysis, 279
Modes of propagation, 117
Modular machine concept, 231, 295
Moisture impact, surface resistivity, 213
Moisture trap, 412
Monofile polyester fabrics, 257
Motorboating effect, 98
Mounting, components, 29, 410
Moving light-source, resist exposure, 245
Multilayer, 337, 367
Multiple reflections, pulse circuits, 67
Multistage amplifier, 98
Multiwire, 337

NC drilling, 324
Negative-working resist, 237
NEMA, 440
Neutralisation, etchant, 297, 300, 303, 306
Nickel bath, 279
Nickel plating, 279
Nickel analysis, 280
Nitric acid, 437
Nitric oxide, 438
Noise, ground/supply line, 48
Non-automatic PCB assembly, 415
Non-corrosive flux, 392

Off-contact distance, 267
Open-loop regeneration, 309

Optical-sight drilling, 322
Organic soil, 225
Organisation, non-automatic assembly, 415
Organisations, professional, 439
Orientation, conductors, 145
Orthochromatic film emulsion, 185
Oscillator, high-frequency, 96
Osmosis, reverse, 432
Overhang, 251, 283

Package density, 30
Packs, laminate, 201
Pads, artwork, 133
Panchromatic film emulsion, 185
Panel-plating process, 233, 334
Parallel-gap soldering, 399
Particulate matter, board cleaning, 225
Pattern plating, 234
Pattern transfer, screen, 260
PCB processes, basic, 232
Perchloric acid, 436
Perchloroethylene, 435
Permittivity, 205
Personnel, artwork, 129
pH, 273
Phenolic laminate, 211
Phosphoric acid, 437
Photoplotter, 154
Photoprinting, 232
Photoresist, 235
Photospectrum, 236
Physical characteristics, laminate, 208
Piercing, 318
Placing, components, 29, 95, 158, 163
Plate resist, 234
Plated-through holes, 6, 230
Plating, 270
    installation, 276
    quality control, 286
Plotter, photo, 154
Point angle, 332
Polarisation slots, 317
Polarity identification, components, 148
Pollution minimising, 307, 428
Polyester-based film, 171
Polyester fabric, monofile, 257
Polyester laminate, 214
Polyimide laminate 215
Porosity test, 286
Positive-working resist, 238
Post-laminating inspection, 379
Postbaking, 248
Potassium hydroxide, 437
Power electronics, design rules, 105
Power factor, 206

ppm, 177
Precipitation, heavy-metal, 429
Preheating, multilayer, 378
Preplate treatment, 274
Press, multilayer, 376
Primary clearing angle, 332
Print-and-etch process, 232
Printed capacitors/inductors, 75
Process film, screen, 263
Processes for PCBs, 232
Processing, film, 178, 187
Professional organisations, 439
Professional PCBs, 420
Propagation,
  constant, 117
  modes, 117
PTEF laminate, 215
Pull-off strength, 219
Pulse circuits, 67
Pulse interference, 50, 53, 54
Pulsed xenon lamp, 245
Pumice slurry, 227
Punchability, 209
Punching, 318
Puppets, 12
Purifying, water, 432
Purity, water, 275
PVC, 295

Q-factor, 77
Quality control, soldering, 400

Radiant heat soldering, 399
Radiation losses, 86, 89
Raw foil, 200
Reactance, 205
Reclamation, screen fabric, 264
Recommendations, PCB design, 56, 92
Red/blue taping, 136
Reducing agent, 272
Reduction ratio, camera, 183
Reduction,
  chemical, 273
  chromium treatment, 430
Reflection coefficient, 117
Reflections, 33, 67, 119
Reflow soldering, 398
Regeneration,
  alkaline ammonia, 306
  closed-loop, 311
  cupric chloride, 302
Regeneration,
  de-ioniser, 432
  open-loop, 309
Registration accuracy, 193, 257, 267

Relative aperture, camera, 189
Relative dielectric constant, 20
Relative humidity, film, 177, 179, 188
Remote sensing lines, 99
Reprographic cameras, 181, 422
Resharpening, drill bit, 330
Resin, 200
Resist exposure control, 246
Resist thickness, screen, 267
Resist,
  photo, 235
  screen ink, 265
Resistance heat soldering, 399
Resistance,
  conductors, 17, 19
  temperature, 19
Resistive coupling, 110
Resistivity, 18
Resolving power,
  film, 173
  lens, 189
Reverse osmosis, 433
Ring-and-disc pattern, 217
Rinsing water, 431
Rise-time, 60
Rise-time,
  limitations, 81, 85, 87, 89
  mismatch effect, 72
Roller coating, 239
Roller tinning, 283
Room requirements, 129, 181, 424
Room, camera, 181
Rosin, 392
Rotary abrasive brushes, 229
Rotary dip test, 400
Round-bent corners, artwork, 144
Routing,
  clean cutting, 320
  conductors, 145, 159, 163

Sacrificial foil, laminate, 346
Safe light,
  darkroom, 187
  resist, 246
Safety measures, sewer, 426
Safety: *see also* "Health care"
Safety, chromic acid, 304
Sanding, 229
Sandwich, film, 196
Sawing, 316
Scale, artwork, 130
Scratch resistance, film, 175
Screen-printing, 255
Screen fabrics, 256, 267
Screen process film, 263

INDEX

Scrubbing, 229
Seeded laminate, 346
Selective amplifier, 65
Self-spanning frames, 258
Semi-additive processes, 345
Sensing lines, remote, 99
Separation, analog/digital ground, 29
Separator, dry-film, 250
Sharpening, drill bit, 330
Shear strength, solder, 389
Shearing, 315
Shielding, aluminium, 52
Signal conductors, 96
Signal line: *see* "Signal conductors"
Silicon laminate, 215
Silver halides, 178
Simple PCB facilities, 421
Single-sided boards, 5
Size, PCB, 7, 9
Sketching, layout, 12, 152
Skin-depth, 52
Skin-effect, 83, 85, 87
Slit-tank developing, resist, 247
Sludge disposal, 431
Slurry, pumice, 227
Soak cleaning, 228
Sodium hydroxide, 438
Soil, board cleaning, 225
Solder float test, 221
Solder fusing, 293
Solder pad diameter, 148
Solder techniques, 382
Solderability chart, 285
Solderability test, 288, 375
Solders, 382
Solvent-developing dry-film resist, 251
Squeegee, 259, 267
Spacing, conductors, 24, 110, 145, 147
Specifications, multilayer, 375
Spikes, supply/ground line, 48
Spindle speed, 330
Spindle, drilling, 327
Splash etching, 293
Spray developing, resist, 248
Spray etching, 293
Spraying, resist, 242
Stability, dimensional, film, 176, 179, 181
Stack drilling, 329
Stainless steel fabrics, 257
Standards,
    copper surface, 210
    laminate, 216
    multilayer, 375
    PCB, 7, 9
    spacing, 147

Starting PCB facilities, guidelines, 419
Steel fabrics, stainless, 257
Step tablet, resist exposure, 247
Stiffener, hardboard, 356
Stop bath, 192
Straight-through mounting, 413
Stress, component lead, 410
Strip line, 62, 119
Strip, exposure test, film, 190
Stripping, 248
Strong-acid dip, 227
Subtractive-additive process, 342
Sulfuric acid, 437
Sulfuric acid, analysis, 277
Supply conductor,
    noise, 48
    width, 25, 79, 104
Supply lines: *see* "Supply conductor"
Surface energy, solder, 384
Surface resistivity, 206, 218
Surface tension balance, 402
Surface, laminate, 201
Swash plate, 294
Swell-and-etch laminate, 345

Table, transilluminated, 142
Tank developing, resist, 247
Tank etching, 293
Tape handling, artwork, 144
Taping, artwork, 133
Technique, film development, 192
Temperature variation, film, 176
Temperature,
    air dryer, film, 180
    film developer, 191
    resistance, 18, 19
Template, artwork, 132
Tensile strength, solder, 390
Tension, screen, 260
Tenting, 235
Terminal connections, 107
Test method,
    laminate, 216
    solderability, 400
Test pattern, resolving power, 173
Test strip, film exposure, 190
Test, multilayer, 372
Testing, flux, 392
Thermal classification, laminate, 203
Thermal effects, high-gain dc amplifier, 99
Thickness test, plating, 288
Thickness,
    multilayer, 370
    resist, 237
Three-conductor system, crosstalk, 44, 45

INDEX

Throughilluminator, 183
Tin-antimony, solder, 387
Tin, lead-antimony, solder, 388
Tin-lead-silver, solder, 388
Tin-lead plating, 281
Tin-lead,
  electrolyte, 282
  solder, 386
Tin-silver, solder, 388
Tin-zinc, solder, 388
Tin bath, 271, 278
Tin electroplating, 278
Tin immersion plating, 271
Tin, analysis, 278, 282
Tolerances, multilayer, 370
Touch-up, 248
Toxicity, etchant, 297
Transilluminated table, 142
Transilluminator, 183
Transition,
  C-MOS, 38, 39
  ECL, 40
  TTL, 35, 36
Transmission density, 171
Transmission line, 59, 114, 116, 125
Treater, 201
Trichloroethylene, 435
TTL, 34
Twist and warp, 209
Twisted pairs, wires, 61

UL, 221, 440
Ultra-thin copper foil, 333
Underetching, 292
Underwriters laboratories, tests, 221
Uniformity, resist coating, 237
UV fluorescent tube, 245
UV light emission spectrum, 246

Vapour degrease, 275
Vertical burning test, 223
Vertical camera, 181
Via-hole, 6

Viscosity,
  screen resist, 267
  wet-film resist, 237
Voltage breakdown test, 374
Voltage considerations, spacing, 147
Voltage standing wave ratio, 117
Volume loss tangent, 218
Volume permittivity, 218
Volume resistivity, 207, 218
VSWR, 117

Warp and twist, 209
Washing,
  film, 192
  screen, 262, 264
  screen film, 264
Waste water, 311, 431
Water absorption, 208, 221
Water pollution: see "Waste water"
Water resistance, epoxy laminate, 214
Water,
  hardness, 193
  purity, 275
Wave impedance, 34, 56, 60
Wave impedance,
  microstrip, 62
  strip line, 62
  wires, 63
Wave soldering, 396
Wet-film resist, 236
Wetting, solder, 383
Whirl coating, 243
Width,
  conductors, 19
  ground lines, 25
  signal lines, 26
  supply lines, 25
Wirewrap, 342
Wood metal, 389

Xenon, arc lamp, 245

Z_{even}, 46
Z_{odd}, 46
Z_w: see "wave impedance"